Quantitative approaches to phytogeography

Tasks for vegetation science 24

HELMUT LIETH
University of Osnabrück, F.R.G.

HAROLD A. MOONEY
Stanford University, Stanford Calif., U.S.A.

Quantitative approaches to phytogeography

Edited by

P. L. NIMIS AND T. J. CROVELLO

KLUWER ACADEMIC PUBLISHERS

DORDRECHT / BOSTON / LONDON

Library of Congress Cataloging in Publication Data

Quantitative approaches to phytogeography / edited by P.L. Nimis and T.J. Crovello.
p. cm. -- (Tasks for vegetation science ; 24)
Proceedings of a symposium, organized during the 1987 Botanical Congress in Berlin.
ISBN 0-7923-0795-X (printed on acid-free paper)
1. Phytogeography--Congresses. 2. Phytogeography--Mathematics--Congresses. I. Nimis, P. L. II. Crovello, Theodore J.
III. Series.
QK101.Q36 1990
581.9--dc20 90-4621

ISBN 0-7923-0795-X

Published by Kluwer Academic Publishers,
P.O. Box 17, 3300 AA Dordrecht, The Netherlands.

Kluwer Academic Publishers incorporates the publishing programmes
of D. Reidel, Martinus Nijhoff, Dr W. Junk and MTP Press.

Sold and distributed in the U.S.A. and Canada
by Kluwer Academic Publishers,
101 Philip Drive, Norwell, MA 02061, U.S.A.

In all other countries, sold and distributed
by Kluwer Academic Publishers Group,
P.O. Box 322, 3300 AH Dordrecht, The Netherlands.

Printed in The Netherlands

CONTENTS

PREFACE

Many aspects of phytogeography have gained greatly from the recent development of analytical and numerical methods. The new methods have opened up new avenues of research, leading to a better understanding of the distribution and evolutionary patterns of species and communities.

During the 1987 Botanical Congress in Berlin, Drs Nimis and Haeupler organized a symposium in which examples of present-day phytogeographic work were discussed. After the symposium it was agreed that a proceedings volume should be edited by Drs Nimis and Crovello. From the lectures presented, those dealing primarily with numerical methods were selected for the book. This is the second volume of the T:VS series that deals with new aspects and methods of phytogeography. The first volume, Box, T:VS Volume 1, 1981, dealt with environmental correlations to types of physiognomic vegetation on a global scale. The present volume shows examples of phytogeographic applications that are oriented towards the compositions of species. The approaches collected rely heavily on newly developed numerical techniques which allow the combination of quantitative floristic and vegetational analyses with mapping and causal or evolutionary deductions. The papers selected for the book show approaches for higher and lower plant forms. Several papers dealing with relevant information on vegetation for the respective areas appear for the first time. The combination of new approaches successfully applied to new problems should be very stimulating to young scientists as many papers demonstrate how to make efficient use of the new developments in information science for species-oriented phytogeography.

While the book does not intend to serve as a textbook, it can be viewed as a guide to the future application of phytogeographic work and the series editors hope that this volume will give our colleagues in the field new stimulus in their tasks.

Osnabrück,
February 1990

H. Lieth
Series editor

1. ARTIFICIAL INTELLIGENCE AND EXPERT SYSTEMS IN PHYTOGEOGRAPHY

THEODORE J. CROVELLO

Dean Graduate Studies and Research, 714 Administration Building, California State University, Los Angeles, CA 90032, USA

1. INTRODUCTION

As a separate discipline Artificial Intelligence (AI) is only about thirty years old. Since its growth has been exponential, many of its concepts and results are quite recent and still undergoing revision. AI is an overlap of the fields of computer science and cognitive science (Gardner, 1986). AI workers have adopted as their major paradigm the information-processing model. This model dominates cognitive psychology, one of the supporting disciplines of cognitive science. As long as other aspects and factors of intelligence are excluded from the AI model, it may be difficult for AI to duplicate higher aspects of intelligence (e.g. creativity), but useful knowledge bases are still possible to construct in fields like phytogeography today.

AI can be viewed as a set of problem-solving procedures. A more widely used definition is that it allows computers to exhibit behaviors that if shown by a human being would be considered intelligent. This begs the question of what intelligent behavior is. Some feel that there is a clear, unambiguous separation between intelligent and nonintelligent behavior. But at least operationally it seems best to view intelligence quantitatively and multidimensionally. Perhaps the best general definition of AI is one that I have used for the computer itself; AI is simply an extension of the user's mind. Such a definition is positive, open-ended and not threatening to our sense of uniqueness.

AI may be as important an advance to phytogeography as the use of computers themselves. Neither the computer nor AI is a panacea, but the wise use of computers has helped many biogeographers achieve their goals. The same can be expected from AI. The essential modifier in the above sentence is "wise"; wise use by thinking phytogeographers.

This paper has several purposes: to describe basic AI concepts; to suggest why AI in phytogeography is particularly timely today; to provide some examples of how AI can be used in phytogeography; to indicate the potential of AI for phytogeography; and to discuss the different ways in which knowledge bases can be created using today's program packages. Finally, for this paper phytogeography is defined broadly to include floristics, vegetation classification and mapping. It includes both its descriptive and explanatory aspects (e.g. description of a region's flora and also hypotheses to explain its origins).

Effective use of AI in phytogeography requires a clear understanding of what the goals, procedures and knowledge of phytogeography are. This last statement may seem unnecessary, since we know our own field and what we are doing. While we presumably know about phytogeography, it is essential that such knowledge be made *explicit*, i.e. able to be understood clearly by others and now also by a computer. In addition, the ultimate goal is to have such explicit knowledge at a level of detail sufficient to permit completion of aspects of phytogeographic analyses without further human consultation.

Every practicing phytogeographer is an expert, someone very knowledgable about that field. Such knowledge has been acquired through study

P. L. Nimis and T. J. Crovello (eds): Quantitative Approaches to Phytogeography, pp. 1–13.

and practical experience. An expert's "knowledge base" can be viewed as a model of how that particular researcher organizes and understands phytogeography. Just as it is important to test other types of hypotheses and models, so too is it important to test an expert's knowledge base. The process of making the knowledge base explicit is also the way to evaluate, correct and expand such expertise. Once such a knowledge base has been evaluated and corrected to an acceptable level, other workers can use it with confidence. In contrast, use of a knowledge base with serious errors in it will confuse more than contribute to phytogeography.

To summarize, the goal of AI in phytogeography is to make the knowledge of expert phytogeographers explicit enough so that it can be stored and used by a computer in ways that might be described as intelligent.

2. THE RELEVANCE OF AI TO PHYTOGEOGRAPHY

As scientists we must investigate any methodology that may deepen understanding of phytogeography, may increase efficiency, or may allow not only phytogeographic data but also the expert knowledge of experts to remain available and useful longer. AI has proven effective in other disciplines with characteristics analogous to phytogeography, so its investigation is appropriate.

More specifically, this is the time to consider AI in phytogeography for several reasons. First, for some phytogeographic purposes there never seems to be enough relevant data to achieve the goals of specific studies. Any approach that would increase the use and value of the available data would be welcome. Paradoxically, at the same time phytogeography often suffers from too many data. For example, published sources such as floras, monographs and articles contain important data and knowledge. Museums and computer data bases contain valuable data. But they still are not effectively available to many biogeographers and to others who could benefit from intelligent analyses of their contents. The problem is to find the time and other resources to locate, search and analyze such data and knowledge sources when they exist.

A major bottleneck are biogeographers ourselves. Each active phytogeographer is unique, and until recently, most all analyses required the active participation of an intelligent human at each small step of the study. One definition of AI is that it can create nonliving "intelligent agents" in the form of computer knowledge bases. By electronically cloning some of the expertise of biogeographers, in some cases it then can be applied without direct human participation. The important perspective to keep in mind is that knowledge bases can help biogeographers, not replace them. By electronically cloning some lower levels of expertise, AI can free biogeographers to concentrate on what such nonliving, intelligent agents cannot do.

Another reason to consider AI in phytogeography now is that the number of biogeographers being trained today is decreasing. For example, Takhtajan (1986) divided the world into detailed floristic regions. As a second example, Kuchler has produced summaries of the world's vegetation or parts of it at various levels of detail. A well-known example is the classification of the conterminous United States (Kuchler 1964). Such works are based on a full lifetime of fieldwork and other scholarship in that area. But the demand and external rewards of many areas of phytogeographic expertise are decreasing rapidly. In addition, with the extinction of large segments of the world's biota the opportunities and experiences of earlier workers may never be possible again. Researchers thus will have less opportunity to derive phytogeographic knowledge from direct field observations in nature. As a result of these trends, fewer and fewer people will be confronted with larger and larger phytogeographic tasks and challenges, both academic and applied.

The need to acquire phytogeographic data and knowledge about the living world is not the only pressing one. An urgent need also exists to acquire and preserve the expertise of biogeographers themselves. Such expertise includes not only factual or "declarative" knowledge but also "procedural" knowledge required to plan and to carry out phytogeographic research.

Finally, this is an opportune time to explore AI in phytogeography because special AI applications programs are available even for microcomputers. Such "expert system shells" can effectively serve at least the initial needs of phytogeography since a sound knowledge base is the most important component of an expert system. And the most useful and reliable knowledge bases are possible only when biogeographers themselves play critical roles in their development.

3. DATA VERSUS KNOWLEDGE

AI distinguishes between data and knowledge so the uses of each term must be understood and not used interchangeably. "Data" are defined as basic facts which most people or sensing equipment can observe or otherwise note. Examples include the presence or absence of certain species of fish in a sample, the altitude at a given geographic location, and replacement lists of species in a study of community succession.

In contrast, knowledge requires higher levels of cognition. It often involves analysis or synthesis of data to determine trends, generalizations, rules, laws, etc. For example, the "generalized tracks" of vicariance biogeography as well as estimates of floristic or faunistic similarities between OGUs each are "knowledge chunks" of varying size. An OGU, or Operational Geographic Unit, is a unit of geography chosen for study (Crovello 1981). It is analogous to Operational Taxonomic Unit They and other knowledge chunks usually result from the data processing in combination with an expert's previous knowledge and methods of "knowledge processing". The latter includes evaluation, comparison and integration with what one's brain already knows.

Several types of knowledge are valuable in biology.

1. Declarative, domain-dependent knowledge summarizes rules or laws relevant to a given domain, such as biology. For example, IF food-web diversity increases THEN species survival should increase. (Throughout this paper pieces of knowledge presented as IF... THEN rules will be capitalized, as indicated, for emphasis.)

2. Procedural, domain-dependent knowledge describes part of a process that is specific to a given field. For example, IF species diversity is low but the area of a region is very high THEN it still can be a major floristic classification unit.

3. Procedural, domain-independent knowledge involves knowledge useful during activities in many fields. For example, IF data sources differ in reliability THEN use the most reliable one available.

4. KNOWLEDGE-BASED SYSTEMS

The area of AI with the most immediate value to phytogeography focuses on so-called expert or knowledge-based systems (these terms will be used synonymously in this paper). Waterman (1986) provides a thorough and readable introduction to this topic, including many examples.

Buchanan and Duda (1983) define a knowledge-based or expert system and also describe three of its important properties. "An expert system is a computer program that provides expert-level solutions to important problems and is:

"1. Heuristic, i.e. it reasons with judgmental knowledge as well as with formal knowledge of established theories

"2. Transparent, i.e. it provides explanations of its line of reasoning, and answers to queries about its knowledge

"3. Flexible, i.e. it integrates new knowledge incrementally into its existing store of knowledge."

Let us consider each of these three properties in more detail. Individual pieces of knowledge are often "heuristics". Each is a rule of thumb, a unit of "judgmental knowledge". That is, its correctness or relevance is not known to be absolutely correct or to be most effective in all situations, but for many situations it seems to be. To say that an expert system is heuristic means that while it may contain some exact data and knowledge, it also contains knowledge that may not be correct in every instance. But it is still useful to incorporate such knowledge chunks into an expert system and use them in phytogeographic problem solving. The following is an example of such a knowledge chunk represented as a "rule": IF species of the same genus are found in eastern North America and in eastern Asia, THEN their distribution in the past probably was more complete in OGUs located between them.

Being transparent, an expert system can explain its reasoning procedures to a user. This usually means that it can answer two questions beyond those that a conventional program can: why and how. During use of an expert system a user can ask the program why it wants to know a certain piece of information (e.g. does the site in question have deciduous trees as a dominant component of the vegetation?). The computer will respond by explaining the chain of reasoning that makes the requested information important to achieve the goal of the use of the expert system.

After an expert system has produced a final result the user may ask how the program arrived at this conclusion. It then would support it by displaying its reasoning processes in a stepwise fashion. Again it should be emphasized that no mysterious process is involved in an expert system. Rather, the reasoning process displayed for a given use will be one instance of the general reasoning heuristics and domain knowledge base of the particular human expert from whom they were obtained.

Being flexible, an expert system can be modified easily. Many of today's phytogeographic data bases can be modified, but a knowledge base goes a step further since its knowledge as well as its data can be modified.

An expert system has two essential components: the domain knowledge base and the inference engine. The latter will be described later. Using the definitions of the previous section, a domain knowledge base contains knowledge of a specific domain, e.g. biology. In actuality no current knowledge base completely includes such a wide domain or even a major subdiscipline like phytogeography. Knowledge bases exist for very narrow segments of a domain. Examples possible today include procedures to choose taxa for use in a floristic study; choice of the best procedure for vegetation or floristic classification; and identification of the phytogeographic classification unit to which a site belongs.

The domain knowledge base has several types of entities. It contains data, much like that found in any phytogeographic data base. For example, it may have data on the distribution of taxa in a series of OGUs, or about climatic or geological data for each OGU, etc. But in addition to data it also contains knowledge.

To understand the contents of a knowledge base and to underscore the difference between data and knowledge, consider a species by OGU phytogeographic data matrix. The item in each cell is one piece of data, i.e. the presence or absence of a particular species in a particular OGU. But phytogeographic data matrices are not sets of random data. Rather, analysis of such matrices usually reveals patterns or structures among the data that are important to phytogeography. For example, certain phylogenetically related taxa may share similar geographic distribution patterns. Or, geographically distant OGUs may be found via multivariate analysis to be very close in terms of the taxa found in each (e.g. Crovello 1982). The resulting phytogeographic knowledge base would contain statements ("knowledge chunks") about which specific OGUs are closest to each other, etc. Such derived knowledge is based on data and data analysis, including interpretation of the results in the context of previous knowledge.

As another example, a phytogeographer may discover that taxa known to have a certain leaf shape also have a similar distribution pattern. This last situation is an example of an expert combining prior knowledge (what taxa possess a certain leaf shape) with new data to create new knowledge. This example also emphasizes that not everyone presented with the same sets of data and knowledge can use it effectively or can create new knowledge or insights. So we would not expect a biochemist, or anatomist, or even another phytogeographer viewing the same series of phytogeographic data based to always derive the same kinds and amounts of phytogeographic knowledge from them.

Another major and more challenging goal of AI and more specifically of expert systems research is to develop procedures to allow computers to "process knowledge" just as a domain expert would. The concept of knowledge processing includes all stages, from the planning of knowledge processing to its actual accumulation, representation, verification, storage, retrieval, use, modification and replacement. To continue an earlier example, Takhtajan (1986) describes his floristic classification of the world. How was the classification produced? Most likely he repeatedly, consistently and cleverly applied a number of pieces of domain-procedural knowledge to the huge amount of floristic data accumulated over his productive career. One such rule might have been the following: IF family endemism is high THEN the OGU should be considered as a possible floristic kingdom.

4.1. Some uses of knowledge-based systems

In other fields knowledge bases already have been created to carry out one or more tasks. The same uses are relevant to phytogeography although some of the names may seem peculiar. The following list of uses is based primarily on that found in Waterman (1986):

- Design (configure objects under constraints, e.g. resource allocation studies)
- Implementation (put a design into action, e.g. carry out steps in a phytogeographic study)
- Interpretation (knowledge from sensory data, e.g. from remote sensing or laboratory instruments)
- Classification (creation of classes of objects, e.g. hierarchies in systematics or floristics)
- Identification or Diagnosis (determination of membership in a class or group; determination of system malfunctions, etc., e.g. to which plant community type a site belongs, or why a biological preserve is not continuing to serve as a critical habitat)
- Debugging (choice of remedies for malfunctions, e.g. creating a new management plan)
- Repair (execution of plans to administer the remedy, e.g. carrying out the new management plan)
- Monitoring (comparison of observations to expectations, e.g. monitoring changes in environmental characters at a series of environmental sites)
- Control (government of overall system behavior, e.g. making decisions during a phytogeographic study)
- Prediction (infer likely consequences, e.g. the likely results of a proposed phytogeographic experiment)
- Machine learning (the computer acquisition of more knowledge, including knowledge of relationships, tasks, etc., within biogeography, e.g. building a computer-based semantic network among phytogeographic concepts)
- Education (teaching people, e.g. learning via computer about new phytogeographic knowledge produced from others' research)

Not all expert systems act autonomously. Human interaction during the decision-making process can be frequent, e.g. to allow additional data input, to decide which hypothesis to explore next; or to provide additional knowledge that the expert system does not yet possess.

4.2. Some important concepts of expert systems

As mentioned earlier, the inference engine is the second essential part of an expert system. It contains general problem-solving procedures that are domain independent. That is, certain procedures exist that are useful to solve problems of the same type regardless of the domain of each specific instance. This approach stems from basic research findings in several fields, but most notably from cognitive psychology.

A relevant example is identification. The claim

is that a person with the problem of determining to which class an object belongs would solve it more or less in the same way, regardless of the domain of the object and its relevant characters. Thus the procedural knowledge used in identification would remain the same while the domain dependent knowledge would change in each of the following examples: identification of an OGU with respect to soil type or to floristic or faunistic classification; identification of a twig as to the species to which it belongs: determination of which ailment a patient has; determination of the best financial investment alternative for a given person at a given time; determination of the best clustering procedure to create a phytogeographic classification of a large region.

Domain independent knowledge can be classified according to its level of task applicability. Some knowledge is useful only for a specific task (e.g. identification) while other knowledge has wider usage. An example of the latter is, IF several knowledge sources exist THEN use the most reliable source.

Historically, AI workers felt that the inference engine was the expert system component that had to be complex, because it was assumed that experts must always use involved reasoning procedures. But over the last decade the belief has changed to the point that now most researchers and practitioners see that what allows an expert system to emulate a human expert more closely is a sophisticated knowledge base. This is extremely important for phytogeographers, since it implies that they need not study cognitive science in great depth to create a valuable expert system. Rather, since the inference engine usually is part of any expert system shell, the only real component that phytogeographers must supply to create a meaningful phytogeographic expert system is the domain knowledge base. This is not a trivial task, but the procedure can be mastered with reasonable effort.

Given the comments in the previous paragraph, another advantage of AI in phytogeography now can be stated: the process of creating a phytogeographic knowledge base can deepen one's understanding of phytogeography, and probably even more so than the resulting product (i.e. the knowledge base). This is analogous to findings of systems analysts, in and outside of biology. They often learn more about the system of interest from creating and refining the model of some biological system than the people who might use the finished systems model itself. To paraphrase, the actual process of creating an expert system will be more valuable to the expert and to new knowledge in the discipline, while the product will be more valuable to other researchers. Finally, the above comparison of a knowledge base with systems analysis again emphasizes that a knowledge base can profitably be used as a particular expert's model of a field like phytogeography and of how the knowledge about that field is organized.

4.3. Knowledge representation in an expert system

Decisions about how to represent knowledge is an important step in creation of a knowledge base, and thus of an expert system. As the term suggests, it concerns ways in which knowledge can be represented in a computer, and begs the question of how it might best be represented to achieve a given function. Analogies exist with data representation and organization in the design of maximally effective data bases.

Knowledge can be represented in many ways: prose statements; an equation such as a regression relating number of species and island area; a graph or picture; etc. While people can process knowledge that is represented in many diverse ways, computers are not yet so versatile. For the most part, current knowledge bases cannot accommodate ambiguity, lack common sense and have difficulty integrating knowledge from sources that represent it in different ways. Consequently, knowledge representation in AI most frequently takes the form of rules presented in an "IF... THEN" format, as in the following example:

IF species i has not been reported from OGU q AND

IF species i has been reported from 3/4ths or more of OGUs adjacent to OGU q AND
IF OGU q contains habitat suitable for species i
THEN species i is in OGU q (CF=.85)

Each rule represents one knowledge chunk. Rules consist of two parts: one or more premises (or conditions); and one or more conclusions (or actions). When an expert system is being used in a given situation, IF all of the premises of the rule being evaluated are true THEN the conclusion is true. CF=.85 describes the value of a Certainty Factor, an indication of the expert's estimate of the reliability of the conclusion based on those premises.

Several rules can be linked together to form a rule chain, defined as a set of rules wherein the conclusion of one rule is a premise of the next rule. When the conclusions of several rules form the premises of another rule, etc., a rule network is created. Synonyms for these structures are inference chains and inference networks.

This concept of an inference network emphasizes how the inference engine in an expert system can use a knowledge base that represents its knowledge as rules. For example, in the rule just given above, a user may ask the expert system if species *i* is in OGU *q*. This is referred to as the "goal rule". The inference engine would "chain backward", through a "chain" or network of rules to determine if all the premises of the goal rule are true. When all the premises have been determined as being true then the goal rule is considered true (or false if one or more of the premises are false). The rule format is natural and common in our normal thinking, as emphasized by the many IF... THEN statements in this paper.

To paraphrase the process of backward chaining, the inference engine first would search for rules that have the user's question as a conclusion. Why? Because IF its premises can be shown to be true THEN its conclusion also is true. IF it finds a rule with the user's question as a conclusion THEN the next task is to determine the correctness of each premise of that rule. This requires a search of the rule base to find a rule that has one of the goal rule's premises as a conclusion. That rule's premises then must be verified, with the program chaining backward up through as many levels of the rule network as needed.

IF a premise of some rule in a chain relevant to the user's current question cannot be verified from the knowledge base THEN the program may pause and ask the user to supply the information. IF the user can not supply the information THEN the program will continue to reason, but may indicate that the resulting final answer has less certainty than if such information were known.

IF in its search of the knowledge base the inference engine discovers more than one rule with the same conclusion THEN it must determine which rule to evaluate first. Procedures for such conflict resolution also often take the form of sets of rules themselves. So rules can be found both in the domain knowledge base and in the inference engine.

This introduction to basic concepts of expert systems of necessity omits many aspects, such as rule generalization, rule hierarchies, introspection, and other types of knowledge representation and reasoning. Interested readers should consult some of the many recent books on expert systems.

4.4. Effective use of AI in phytogeography

As mentioned earlier, the goals and characteristics of phytogeography are similar to those in which expert systems already have been used effectively. These fields share common characteristics: knowledge or data are incomplete or unreliable; some knowledge is qualitative; theories in the topic are incomplete; the task requires decision making; the task is neither too easy nor too hard for today's expert system programs; and knowledge is or should be treated as a dynamic entity.

While AI can contribute to phytogeography, it still is inappropriate for many phytogeographic tasks. For example, today's knowledge based systems have little common sense. They can reason deeply about a narrow topic, but have great difficulty when a broad perspective is required.

5. WHAT BIOGEOGRAPHERS EXPLICITLY DO

Before AI can contribute to phytogeography, the goals, procedures and knowledge of phytogeography must be stated explicitly; that is, unambiguously and in sufficient detail such that someone (or some computer!) not trained in phytogeography could carry out the required phytogeographic tasks successfully without further human assistance.

The general goals of phytogeography include the following: to describe the patterns of variation in nature; to explain how these patterns arose and are maintaining themselves; to predict future patterns; and at times to suggest management strategies of biota and of OGUs. These goals can apply both to natural patterns as well as OGUs, communities, etc., that have been strongly influenced by society (e.g. farming or forestry activities).

To accomplish these general goals we engage in many diverse activities, each of which contains subgoals. Such activities include the following: phytogeographic theory formation; classification of OGUs; identification of OGUs; historical phytogeography; and OGU character analysis. Crovello (1970, 1981) suggested that every phytogeography study involves a multistage decision process. That is, it involves many stages and substages, and each step requires choices or decisions among options. These are based on knowledge about the subject area (here phytogeographic principles, the OGUs and taxa under study, etc.) and on general procedural knowledge. This is an ideal context for expert systems.

Young biogeographers often have a relatively small set of options to consider. Interestingly, experienced biogeographers may have thought about many options over their professional lives but have forgotten some of them. The computer need never forget (this also has disadvantages but they can be overcome).

5.1. AI and phytogeographic identification

As an example of AI in phytogeography, consider the phytogeographic goal of creating an expert system to identify to which floristic region a specific geographic site might belong using the detailed floristic classification of the world in Takhtajan (1986). Alternatively, some readers may prefer to think about creation of an expert system to identify to which biome or formation an OGU belongs. Although identification is one of the easier tasks a phytogeographer encounters, it still is challenging and can illustrate basic concepts of expert systems.

Several papers already have appeared in the biological literature that focus on taxonomic identification using expert systems. Atkinson and Gammerman (1987) describe several uses of EXPERT KEY, an expert system program that accommodates reasoning with uncertainty. The authors demonstrate the increased efficiency of the expert system over conventional identification programs. Mascherpa and Pellegrini (1987) introduce concepts of AI relevant to biological identification and provide examples of possible uses, such as with the flora of Switzerland. Woolley and Stone (1987) discuss a prototype knowledge base consisting of 114 rules. It was created for the identification of individuals belonging to one species group of the insect genus, *Signophora*.

A standard, published, dichotomous key contains knowledge derived from analysis of large amounts of data and thus can be considered a knowledge base. The procedure to use that knowledge for the goal of identification is straightforward; begin at the start of the key and work through it until the goal is reached (e.g. determination of the floristic province to which an OGU belongs). A knowledge base represented in the form of a published key can be directly transformed into a rule-based representation since each part of a couplet can be stated as an AI rule. For example,

"Trees present . . . 2,"
can be restated as, "IF trees are present THEN go to rule 2".

Computer based identification in taxonomy introduced several alternatives to the dichotomous key (e.g. Pankhurst 1975, 1978). Representation of identification knowledge inside the computer can vary. For example, the program might simply have the equivalent of a printed key available as data. The sequence of the questions asked by the computer would be fixed, just as occurs in published keys.

Alternatively, the computer could "dynamically" choose the next character about which to request information. This usually requires the presence of a full data matrix within the computer. It also requires a procedure to determine which question the computer is to ask next. For example, choose that character among those not yet asked whose states will divide the remaining taxa into two groups, each with more or less the same number of taxa.

If a complete, reliable data matrix is available within the computer, then AI is not needed. But if data for certain characters are missing for some formation, or if the variation in its character states is not fully understood, then the AI rule-based approach is useful to determine the best identification. In other words, if complete, absolute knowledge is not available, then the judgmental knowledge that an expert system can accommodate may contribute to the correct identification.

Given representation of identification knowledge in a rule format, it is clear how a set of rules derivable from a key can become a rule chain or network. In fact an alternative, succinct definition of a phytogeographic or other key is the following: knowledge for the purpose of identification that is represented as rules and forms an inference network.

It is important not to dismiss AI in phytogeography simply as a different way to represent knowledge of dichotomous keys. This would be like equating all use of AI in medical diagnosis only with the ability to distinguish among nine easily identifiable childhood diseases. Such uses are "low-end" AI. As the biogeographer's identification goal becomes more challenging due to incomplete information, the need for expertise, for judgemental knowledge, increases. Unlike computer based keys that rely only on data and some decision rules to choose what data to request next, AI can accommodate a biogeographer's heuristics, i.e. the rules of thumb that the phytogeographer has found useful in previous identifications, even if totally similar situations have not been encountered before. In such cases an expert's domain independent, procedural knowledge also can be incorporated into the knowledge base.

Simple examples useful in many domains when the goal is identification include the following heuristics, represented here as rules:

IF several samples from the same object (e.g. an OGU) are available THEN evaluate all samples.

IF some characters are more reliable than others THEN first use those with higher reliability.

Such "metalevel" rules are obvious once they are read, and seem to be simply common sense. They are obvious because they are so ingrained in our problem solving procedures that we take them for granted. Yet such presumed knowledge must be explicitly stated before it can be entered into a computer knowledge base.

In summary, like other computer approaches to identification, an expert system has several advantages over a book of expertise. An expert system is interactive, allowing the user to more quickly obtain an answer, and allowing the computer to dynamically optimize each interaction during a session. Such programs can be easily duplicated to allow distribution to other research and service laboratories. In addition, unlike most other computer identification programs, the rule format of knowledge representation in an expert system is easily understood by the user. In addition, it can be readily understood by other computer programs and permits direct access and use of data bases. The knowledge base can be electronically updated quickly, without editing the program itself (because the knowledge is not integrated into the program code). An expert system is testable for inconsistencies in its procedural knowledge, and allows easy revision of it. An expert system can explain its reasoning procedures, and only to the level of detail desired by each user.

5.2. How an expert system is created and maintained

Biogeographers can create and maintain an expert system in various ways, depending on the level of sophistication of the program product used. At one extreme no expert system program product would be involved. Based on the purposes of the expert system, the designer would devote considerable time to the organization and contents of the inference engine as well as of the domain knowledge base. Decisions would include what data and knowledge to include in the knowledge base and how to represent it. Design of the inference engine would involve the ability to recognize rule formats, to determine when all the premises of a rule are correct, to determine what rules are relevant at different moments in a given session, and how to resolve rule conflicts. In addition, the program should be able to explain its reasoning to the user and allow easy modification of the rule base. Finally, additional programming is required to permit efficient, high level communication between the computer and the user.

While an expert system can be created using any well-known computer language (e.g. BASIC or Pascal), the preferred languages of AI workers have been LISP in the United States and PROLOG in Europe and Japan, although this is changing. These are preferred for several reasons, including the fact that data (knowledge) and program statements are interchangeable. More precisely, program statements can be considered either as instructions or as data. This allows more efficient programming. LISP is a powerful language because it allows programmers to create hundreds and thousands of primitive instructions, which in turn allows greater program flexibility and power. In this sense LISP is closer to assembly language than it is to compiled languages like Pascal. PROLOG is more like a compiler language but has special builtin abilities useful in expert system development. For example, it has a builtin procedure to carry out backward chaining, thus eliminating the requirement for the knowledge-based system developer to create an inference engine.

Few biogeographers will choose to build an expert system from scratch using the above languages. Most will opt to use an expert system "shell". Shells are utility programs allowing users to enter only knowledge and data relevant to their particular use. That is, they do not have to create an inference engine, a user interface, etc. Familiar examples of utility programs are data base, word processing and statistical programs, since the user only enters words or numbers in an appropriate format and the program carries out the required analysis. Expert system shells available for microcomputers range in price from $100 to over $10,000.

The cheapest expert system shells allow an expert system developer (e.g. a knowledgeable phytogeographer) to enter perhaps a maximum of 50–300 rules, and allow only backward chaining. Mode of rule input may be via a word-processing program, somewhat analogous to batch processing on mainframe computers. That is, after a set of rules has been typed into the word processor they are read as a batch into the expert system shell program. Depending on the sophistication (and thus price) of the shell, inconsistencies among the rules may be determined and the designer notified.

More expensive expert system shells permit some amount of graphic input to speed the design of the knowledge bases. Just as people store knowledge in chunks and groups in their brains, advanced shell programs allow the designer to represent knowledge in several ways. This not only increases efficiency during creation and use of the expert system but also can increase the probability of it providing the best response to a given task. Expert system "development tools" provide additional features but cost considerably more.

The trend in expert system shells is to allow compatibility with major commercially available data base programs are well as with such special programs as HYPERCARD. Used on Apple Computer Company's Macintosh computer, this linkage with HYPERCARD allows easy access to graphics and pictorial screens, including maps, photographic images of OGUs, etc.

5.3. Transforming data bases into knowledge bases

Many computerized data bases useful to phytogeography already exist. One example is, "GEOECOLOGY", a county level environmental data base for the conterminous United States (Olson et al. 1980). It contains information not only on distribution of many taxa, but also on climatic, topographic, geological and other characteristics.

A major thrust in many disciplines in the next decade will be to convert data bases to knowledge bases. This will occur first via direct interaction of domain experts as well as AI experts (called knowledge engineers). But work is under way to create intelligent programs that will automatically create knowledge bases from data bases. Naturally, any such programs that are successful will have been provided with a considerable amount of a human expert's knowledge about how to accomplish such a challenging task.

To understand how data bases can be transformed into knowledge bases, consider again the difference between data and knowledge. Most data bases about the real world are not random. That is, their contents reveal repeated patterns of variation. This should be expected since patterns of variation are the rule in nature. For centuries biogeographers have been identifying such patterns. The question is, how do they do it. If the data base consists of concrete observations then some procedures are obvious. For example, one procedure might be as follows:

1. Choose certain pairs of characters that are of phytogeographic interest (e.g. number of species and latitude; number of species and island area).
2. Subject such series of paired observations to regression analyses.
3. IF some regressions are significant THEN formulate rules summarizing the relationship. For example, IF island area increases THEN number of species increases as twice the square of island area.
4. Have such rules reviewed by a phytogeographic expert.
5. Compare rules accepted in step 4 with those already in the knowledge base to determine rule conflicts.
6. IF rule conflicts occur THEN resolve them EITHER with a conflict resolution program AND/OR with human expert intervention.

The above procedure obviously is only one of many possible, and is incomplete. For example, it does not allow consideration of the relative reliability of each data base or of its parts. But as an example it illustrates that while today's phytogeographic data bases are a valuable resource, they can become even more valuable as the major source of phytogeographic knowledge bases. Interestingly, this was the same rationale used to convert data stored in museums, data about ecological releves, etc., into electronic data bases. Now the value of such data can be enhanced further by converting data bases into knowledge bases.

5.4. Different levels of AI in phytogeography

Another way to summarize how AI might contribute to phytogeography is to consider it in the following three roles:

1. As an *intelligent assistant*, a knowledge-

based system can perform routine tasks. For example, determination of site type based on frequency and other data. This frees researchers to do what the computer cannot yet do well.

2. As an *intelligent associate*, a knowledge-based system can perform challenging tasks. For example, such a system might be requested to perform floristic analyses on a given data set. Based on the procedural knowledge in the AI program, it will determine how to carry out the specifics of the request. So unlike conventional programs, with AI one need only give a high level request to the computer. Increasingly, the phytogeographer can tell the computer what is to be done without telling it how to do it.

3. As an *intelligent colleague*, a knowledge-based system can perform creative tasks. For example, a researcher might have a new idea about historic relationships among the biota of the world or for a new floristic classification. It would be valuable to have the opinions not only of one's human colleagues but also of another "intelligent agent"; i.e. a relevant phytogeographic knowledge-based system.

The concept of an intelligent colleague may seem like science fiction to some. But it is scientific reality today in at least one area of biology. For example, at Stanford University, knowledge engineers collaborated with molecular geneticists to create a prototype knowledge base that not only can evaluate the results of previous experiments but also suggest the next best experiment to do. This was accomplished by entering into the computer some of the relevant knowledge of Professor Charles Yanovsky, a well-known molecular biologist. In this case a knowledge engineer was electronically trying to clone a genetic engineer who was genetically trying to clone a gene!

6. CONCLUSIONS

The wise use of AI in phytogeography promises to be as valuable as the use of computers themselves. Expert systems is the area of AI that will be most useful. Effective development and use of expert systems will be possible only after phytogeographers reflect on how we do what we do, and are able to make both our procedural and declarative knowledge sufficiently explicit to be processed by computer. AI thus offers an opportunity for phytogeographic knowledge and expertise both to increase and to live longer.

Neither computers nor expert systems are phytogeographic panaceas. But as a logical extension of use of computers in phytogeography, AI can be a valuable addition. Just as important, the design, development and use of expert systems will help biogeographers gain deeper insights into both phytogeographic concepts and procedures. For some this may be the most valuable result of AI in phytogeography.

There is nothing mysterious about AI. It will be useful in phytogeography as it has proved useful in other disciplines simply because relevant knowledge (not just data) of one or more phytogeographic experts will have been entered into the computer. The advantage is that every subsequent use of such expertise will not require the presence of the expert phytogeographer.

More specifically, since AI can clone electronically more phytogeographic expertise than previously possible, the resulting knowledge bases can have the following results:

- An expanded lifetime for phytogeographic knowledge. The expertise of phytogeographers need not be lost when they retire.
- Increased simultaneous use. The expertise easily can be duplicated electronically and distributed to those who need it. This is similar to the effects of the printing press, but exceeds it since it has the advantages of being in an interactive mode, of dealing with knowledge instead of just data, and of being in a form that can quickly focus on that small amount of knowledge that is relevant to the task at hand.
- Increased performance levels by both phytogeographic experts and paraprofessionals.
- Expanded and better use of phytogeographic knowledge by more people, be they academic or applied biologists or others.
- Emphasis on the centrality and ubiquity of problem solving in phytogeography. The AI

paradigm clearly shows that even deciding to what floristic category a site belongs is problem solving. It involves repeated use of the "generate a hypothesis and test it" cycle until the correct answer is obtained.

– AI's requirement that knowledge be made explicit will increase understanding of what biogeographers do, and possibly suggest ways of how it might be done better.

Viewing expert systems as intelligent assistants, associates and colleagues provides a positive, open-ended perspective on AI in phytogeography. AI's success in fields similar to phytogeography suggests that the time researchers devote to experimenting with knowledge-based systems will contribute significantly to phytogeographic research and understanding.

REFERENCES

Atkinson, W. D. and Gammerman, A. 1987. An application of expert systems technology to biological identification. Taxon, 36, 705–714.

Buchanan, B. G. Duda, R. O. 1983. Principles of rule-based expert systems. Advances in Computers, 22, 163–216.

Crovello, T. J. 1970. Analysis of character variation in ecology and systematics. Ann. Rev. Ecol. Syst., 1, 55–98.

Crovello, T. J. 1981. Quantitative biogeography: an overview. Taxon, 30, 563–575.

Crovello, T. J. 1982. Floristic similarities among 51 regions of the Soviet Union based on the Brassicaceae. Taxon, 31, 451–461.

Gardner, H. 1986. The Mind's New Science: A History of the Cognitive Revolution. Basic Books, Inc., New York.

Kuchler, A. W. 1974. Potential Natural Vegetation of the Conterminous United States. Amer. Geogr. Soc. Spec. Public. No. 36, New York.

Mascherpa, J. M. and Pellegrini, C. 1987 Approche par les systemes experts de la determination des familles de la flora de suisse. Cahiers de la Faculté des Sciences (Univ. de Geneve), 14, 7–21.

Pankhurst, R. (ed.). 1975. Biological Identification With Computers. Academic Press, New York.

Pankhurst, R. 1978. Biological Identification. Arnold, London.

Takhtajan, A. 1986. Floristic Regions of the World. Univ. of California Press, Berkeley.

Waterman, D. A. 1986. A Guide To Expert Systems. Addison-Wesley, Reading, Massachusetts.

Woolley, J. B. and Stone, N. D. 1987. Application of artificial intelligence to systematics: SYSTEX – a prototype expert system for species identification. Syst. Zool., 36, 248–267.

2. SOME QUANTITATIVE APPROACHES TO PROBLEMS OF COMPARATIVE FLORISTICS

LEONID I. MALYSHEV

Siberian Central Botanical Garden, Novosibirsk 630090, USSR

1. INTRODUCTION

Based on his comprehensive manual, "Geographie Botanique Raisonnée" (1855), Alphonse de Candolle, a prominent Swiss botanist, is the acknowledged founder of comparative floristics based on quantitative assessment. Somewhat similar ideas were formulated by Carl Claus (1851), a Russian chemist and botanist of German origin, four years prior to de Candolle. Unfortunately, the ideas of Claus were neglected until quite recently (Schmidt and Ilminskikh 1982), and now are of mostly historical significance.

The studies of de Candolle and Claus were made in the epoch of extensive assessment of data in botanical geography. But subsequent decades did not see further developments of this methodology, and a long period of only data accumulation followed. Meanwhile botanical geography finally separated as a distinct field from plant systematics. Botanical geography itself then divided into floristics and phytocoenology (geobotany in the stricter sense). Yet the affinity of floristics and phytocoenology facilitated the application of some primarily phytocoenological methods to quantitative floristic analyses. For example, coefficients of phytocoenological similarity on the species level, and the concept of species-area relationships, have proved useful for the assessment of local and larger floras.

A renaissance of comparative floristics began in the 1940s, 50s and 60s, stimulated by the remarkable advances in electronic computing. Analogous strides were made in computerization of herbaria (e.g. Crovello 1972) and in other areas of plant systematics (e.g. Crovello 1970; Crovello and MacDonald 1970). In the USSR Tolmachev (1986) was a pioneer in this comparative floristics renaissance along with other scientists, mostly from French- and English-speaking countries. The success of comparative floristics in the USSR is due partly to its vast area, which facilitates wide-range comparisons essential to botanical geography.

In 1931–32 Tolmachev (1986: 5–11) developed the method of "concrete", or "elementary", floras, primarily for the Arctic. Throughout this paper, "elementary" flora will refer to Tolmachev's concept. An elementary flora is the elementary unit of a flora. It is comparatively homogeneous as a whole, but still may include several vegetation types. This is possible since it is a floristic unit and not a vegetation unit. It remains popular among Soviet botanists (Yurtsev 1987) because it assesses the floristic situation of a large area based on detailed field study of a comparatively small area regarded as representative. Appropriate size of an elementary flora was estimated to be about 100 km^2 in the Arctic. Theoretically it would enlarge southward, reaching about 1000 km^2 in the tropics. The concept of an elementary flora is opposed to the notion of "collective" flora which should be of much larger size and comprise elementary floras of different taxonomic structure.

In European countries, with their generally more intense levels of floristic study, another method proved more effective. It may be designated as "species representation in a net-grid scale". The grid may be formed by political

P. L. Nimis and T. J. Crovello (eds): Quantitative Approaches to Phytogeography, pp. 15–33.

boundaries, by a regular grid of geographical latitude and longitude, etc. Originating in the USA, it was used widely in Europe after the publication of the "Atlas of the British Flora" (Perring and Walters 1962).

Both the method of elementary floras and of species representation on a grid are most valuable in comparative quantitative floristic analyses. Obviously, a thorough floristic check list is only the first stage of such research. But it places the study of floristic peculiarities and genesis in an objective framework which is an advance over unrestrained and ambiguous subjective speculations. For example, Malyshev (1976) used a comprehensive algorithmic scheme to study the flora Putorana in the Siberian Subarctic. Crovello (1981) reviewed several quantitative approaches to phytogeography. He also introduced the concept of an Operational Geographic Unit (OGU). Analogous to the Operational Taxonomic Unit (OTU) used widely in quantitative taxonomy, OGU is "any one of the set of geographic units to be analyzed in a study" (Crovello 1981: 563). Thus depending on the researcher's purpose, each OGU could be a political unit, such as a country, state or republic; or each OGU could be a unit in a uniform grid system; or each OGU could be defined as a geobotanical unit, e.g. tundra, taiga, etc., or the OGU could be based on other criteria.

2. ASSESSMENT OF FLORISTIC RICHNESS

Estimation of floristic origins benefits from quantitative estimates of floristic richness. Floristic richness is a balance of at least of two factors: heterogeneity (or floristic spatial diversity); and species abundance per unit area. Both characters may be computed on the basis of correlation between the species number (S) and size of an area (X). Arrhenius (1920) first expressed its regularity by an exponential curve of the function:

$$S = AX^z \quad \text{or} \quad \log S = \log A + z \log X, \qquad (1)$$

where the constant A assesses species abundance per unit area and the exponent z indicates the heterogeneity of a flora. This adequate and simple equation was applied primarily to phytocoenological studies of small areas.

To estimate values of A and z, it is necessary to know beforehand the observed number of species S_1 for the total area X_1 in comparison with a number of species S_2 for a part of area size X_2. Thus,

$$z = (\log S_1 - \log S_2) : (\log X_1 - \log X_2),$$

and

$$\log A = [(\log S_1 + \log S_2) - z (\log X_1 + \log X_2)] : 2,$$

or

$$A = Y/X^z.$$

More accurate estimates of the constants A and z may be derived by least squares regression analysis when the total area is divided into a series of parts of different size.

Gleason (1922) suggested a somewhat similar relationship expressed in the form:

$$S = A + b \log X, \qquad (2)$$

where A and b estimate accordingly the species abundance per area unit and the heterogeneity of a flora. This equation is an alternative to that suggested by Arrhenius and sometimes produces a better fit of observed and expected data (Rebristaya 1987). The value of the constant b varies considerably, perhaps from a few to several thousand units.

Evans et al. (1955) provide the simplest equation for the evaluation of species-area relationships:

$$S = A \log X, \qquad (3)$$

where A corresponds to the number of species per 10 area units, so an expected species number for one area unit ought to be zero. To avoid this unrealistic situation the authors arbitrarily suggested the addition of +1 in studies whose areas may be small. Consequently, this equation is not appropriate when extrapolation is needed.

Phytocoenologists have suggested more sophisticated equations, but they may have no greater goodness of fit in floristic data bases. Thus, Uranov's equation when compared with equations of Arrhenius and of Gleason would appear to be more accurate. It was used in a floristic analysis of the Württemberg flora in the Federal Republic of Germany (Makarova 1983). It has the form:

$$S = AX/(B + X^n) \qquad (4)$$

where S is the number of species intrinsic for an area of size X, B and n are constants characterizing the floristic heterogeneity of a plant community, and A corresponds to the maximum species number for the summarized hypothetical area of the whole phytocoenosis. *A priori* it is expected that use of an equation with more constants, such as this one, would ensure closer fitness of expected to observed data than equations with fewer constants. Nevertheless, fitting of additional parameters does not provide a closer fit (Malyshev 1987).

2.1. Floristic heterogeneity

Williams (1943, 1964) computed the heterogeneity (z) of floras for continental regions and islands of the world. For the world z was 0.26, for France it was 0.21, etc. Later it was found that for the British Isles z was 0.19 (Dony 1977). Using regression, Preston (1962) computed the magnitude of z to range from 0.22 to 0.30, being about 0.22 for the temperate zone and 0.25 for the Tropics. Isolated floras (oceanic islands, etc.) often have z values twice as high as unisolated floras. Thus, floras of the California coast have a z value of 0.16 compared to 0.37 for the adjacent islands. But when nonnative species are excluded from the assessment, z increases from 0.20 on the coast to 0.395 on the islands (Johnson et al. 1968).

On the basis of floristic and faunistic data MacArthur and Wilson (1967) concluded that for continental floras and for areas inside islands, z varies mostly between 0.12 and 0.17; whereas for small islands as a whole it is higher and usually varies between 0.20 and 0.27. They suggest this is due to the exclusion from small island biotas of species with smaller population sizes. Species equilibrium in nonisolated areas is maintained by an exchange with adjacent territories.

The phytogeographical importance of this phenomenon induced the author to recalculate floristic heterogeneity on a global scale. We assessed 404 areas classified as continential floras, parts of such floras and large islands (Table 1). Floristic heterogeneity (z) varies from 0.038 to 0.240 with a mean of 0.150, a coefficient of estimated variance equal to 25%, and a standard deviation of 0.037. Consequently z most often lies in the interval 0.11–0.19. Such values are somewhat lower than those of MacArthur and Wilson (1967). Furthermore, the sample mean value of z perhaps actually ought to be lower than 0.150 because in some cases a recorded species number for parts of a flora may be underestimated as compared with a whole flora estimated from incomplete field or herbarium data for parts of an area.

Floristic heterogeneity increases from the Arctic towards the tropics (see Table 1). Thus, for Devon Island in the Canadian Arctic archipelago, z equals 0.04, in the southern Arctic about 0.08, in the taiga zone 0.10–0.13, and in the tropical zone 0.20–0.22. Besides this general tendency, it is considerably higher for mountains than for nearby plains. So for the Putorana table land in the Siberian Subarctic zone and in the northern frontier of the taiga zone it is about 0.11 instead of 0.08–0.10. On the other hand, for deserts it falls to 0.07–0.15. Evidently due to historical events high floristic heterogeneity (0.22) is characteristic of the southern nondesert part of Africa (outside the tropical zone to the Cape region).

The statistics cited above were computed using both native and adventive vascular plants together. Data for microspecies were excluded as far as possible. Since they usually have small areas, microspecies increase estimates of floristic heterogeneity, while adventive species decrease it. Thus, in central Europe floristic heterogeneity is 0.127 for native plants only, versus 0.144 if

Table 1. Floristic heterogeneity z of the world (n indicates number of regional or local floras taken into account). (After Malyshev (1975) with minor changes.)

Localities of floras	z	n
Western and central Europe		
Scandinavia	0.130	16
British Isles	0.138	22
Middle and southern areas (except Switzerland)	0.150	62
Switzerland	0.178	9
Eastern Europe		
Arctic	0.084	2
Forest and steppe zones	0.119	60
Caucasus	0.206	2
Siberia, Soviet far east		
Putorana	0.106	13
Taiga region	0.102	16
Southern Soviet Far East	0.191	4
Soviet middle, or central Asia		
Desert and low mountainous areas	0.146	10
High mountainous areas	0.179	4
Asia Minor to India	0.109	9
Southeastern Asia	0.212	11
Canadian Arctic	0.038	4
USA		
Alaskan mountains	0.163	2
The main territory	0.162	65
Highland areas	0.205	4
Latin America	0.189	12
Africa		
Mediterranean areas	0.167	6
Sahara, Sudan, Senegal	0.075	15
Ethiopia	0.157	2
Tropical zone	0.201	33
Southern areas	0.224	14
Australia		
Desert areas	0.092	2
Savannah and forest areas	0.216	2

adventive species and microspecies are added. For native species including microspecies it is 0.157, and for native and adventive species without microspecies it is only 0.111 (Malyshev 1975).

For small and medium size islands the magnitude of z varies considerably. We have computed the data for 70 oceanic islands. For them the coefficient of estimated variance is 31%, the mean is 0.318, and the mean square deviation is 0.100. Consequently, z varies mostly from 0.22 to 0.42. It is higher for small islands than for larger ones, and in the case of equal areas it is higher for islands of the tropic zone than for islands of the temperate and especially frigid zones.

Some values of floristic diversity z for islands are cited below with corresponding zonal means in parentheses for a mainland (after Malyshev 1980):

Channel Islands 0.195 (0.15)
Greater Antilles 0.22 (0.21)
Bahama Islands 0.23 (0.20)
Japan (Hokkaido, Honshu, Kiushu and Shikoku Isles) 026 (0.19)
Islands of the White and Barents Seas (Velikiy and Charlamov Islands) 0.29 (0.10)
Ushkan Islands on the Lake Baikal 0.32 (0.10)
Islands near Toulon in France 0.33 (0.18)
Aldabra Island 0.34 (0.22)
Islands of Lower California 0.39 (0.16)
Islands of the Guinea Gulf 0.39 (0.22)
Small Islands near the western coast of Estonia 0.43 (0.12)
Islands of California 0.48 (0.18)
Lesser Antilles 0.49 (0.22)
Wolf Islands in the Bay of Funday, Canada 0.55 (0.15)

On the basis of quantitative estimates of bird species, Vuilleumier (1970) concluded that the isolated highland areas of the Andes within Venezuela, Columbia and Ecuador are similar to island biotas and have a z value of about 0.29. Our data confirm the assumption that the magnitude of z for highland floras is much larger than for nearby lowlands. For the Stanovoye Nagorye uplands in Siberia it is 0.24, and 0.23 for the East Sayan Mountains.

It is very important to remember that the constant, z, estimates the heterogeneity of a flora as a continuous character and assumes a more or less random species distributions. For discrete entities such as islands and other similar floras, z estimates the internal spatial diversity but also the heterogeneity between whole areas. Due to isolation this combined diversity may considerably exceed the heterogeneity within biotas.

In addition to floras of islands and isolated highlands, categories of the floristic hierarchy (floristic regions, provinces, districts, etc.) are more or less discrete, and therefore have high values of z. In this case z actually evaluates the similarity or difference of the categories. For example, in Siberia the difference between floristic districts in the southern part of the Krasnoyarsky Kray administrative region is 0.27, while the floristic heterogeneity of the same area as a whole is only 0.12, and within the floristic districts it is still lower, about 0.07. Similarly the difference between phytogeographical districts in the Altay Mountains is as high as 0.39.

Besides the Arrhenius equation (1), z can be estimated by the canonical equation of Preston (1962):

$$\left[\frac{S_1}{(S_{1-2})}\right]^{1/z} + \left[\frac{S_2}{(S_{1-2})}\right]^{1/z} = 1, \qquad (5)$$

where S_1 and S_2 are the number of species or other taxa in two compared floras, and S_{1-2} is the number of species in the combined flora of both areas. Instead of a direct computing of data, one can use a table in Preston's paper.

The values of z computed for the same floras using the formulae of Arrhenius and of Preston may not coincide. This is due to different approaches. The critical, or threshold magnitude, of the exponent z in the Preston equation is essential for the elucidation of floristic genesis. Usually it is 0.27, but for floras over 10,000 species it is 0.265. This magnitude was found theoretically by Preston. A z value higher than 0.265 or 0.27 indicates that the floras being compared are isolates or mutually discrete, and they may be regarded as separate units in the process of floristic genesis. On the contrary values less than 0.265 or 0.27 indicate a balanced equilibrium between floras. Thus they may be regarded as parts of a whole flora or at least the lesser of the two floras should be regarded only as a part of the larger one. In general terms, the formula $(1-z) \times 100\%$ may be used for evaluation of floristic similarity by the Preston equation.

In our opinion, criticism of the Preston equation expressed by Simberloff (1982), is invalid. It is based on the wrong assumption that the magnitude of z should be the same for separate islands and for a whole archipelago. In reality when estimating floristic heterogeneity within an island it ought to be less, and much larger for an archipelago composed of these islands.

2.2. Species abundance on mainlands and on larger islands

Floristic richness is the abundance of species (or other taxa) estimated on a comparative scale. It may be calculated directly for areas of equal size or computed by extrapolation. The constant A in the equations of Arrhenius (1) and of Gleason (2) designates the number of taxa per unit area, so it is a measure of specific floristic richness. On a graph with a log–log scale (in the case of the Arrhenius equation) or on a log–normal scale (for the Gleason equation) it corresponds to the initial point of the regression line on the ordinate. It is quite evident that the absolute value of constant, the A, depends not only upon the peculiarities of a flora but also upon the scale used for the area size measurement.

To use one square mile or square kilometre as the unit of area is inadequate to evaluate floristic richness because no flora of vascular plants can be so small. Therefore any estimates of floristic richness require that a larger area be used for the standard size.

Floristic representation of area size will be discussed later in this paper. For now we note that because of the logarithmic nature of the species–area relationship, the richness of floras with different spatial diversity would not differ in the same multiple for standard areas of different size. Thus floristic richness for 100 km^2 in the northern part of taiga zone with z equal to 0.10 and in the Tropics with z equal to 0.20 may be four times different, while the same floras for 100,000 km^2 should be eight times different in terms of floristic richness.

2.2.1. Previous estimates of species abundance

To estimate floristic richness, Cailleux (1953, 1961) chose 10,000 km^2 as the standard area size.

He used the phrase, "area richness" (*richesse areale*), for species number found for each such area. For extrapolation of data he probably used the mean value of z estimated by Williams (1943). This produces an overestimate for mainland floras. The original data then were put on a world map without generalization.

Later Lebrun (1960) created a map of comparative floristic richness in Africa, using equation (3) above with 10,000 km^2 chosen as the standard area size. Schmidt (1980: 137–166) developed a series of maps for the European part of the USSR. They show the observed number of species, genera and families recorded for numerous elementary and quasi-elementary floras. These data correspond to areas of quite different size, and therefore the isolines drawn for the generalization of floristic richness are ambiguous.

2.2.2. *Revised global estimates of species abundance*

For our analysis, we chose 100,000 km^2 as the standard area size for the comparative evaluation of floristic richness on a global scale because the observed data were recorded mostly from area of somewhat similar size. Thus it was possible to calculate data for 459 vascular plant floras with the areas ranging from 10,000 to 1,000,000 km^2. Especially abundant data exists for Europe, Africa and the USA. The Arrhenius equation was used for all calculations. The standard deviation of z values is 0.037, so the mean deviation of computed species numbers by extrapolation of possible actual values is only 2.6%. Actually it ought to be even less since the regional values of z were employed for mainland floras and large islands instead of the overall global mean value of 0.15.

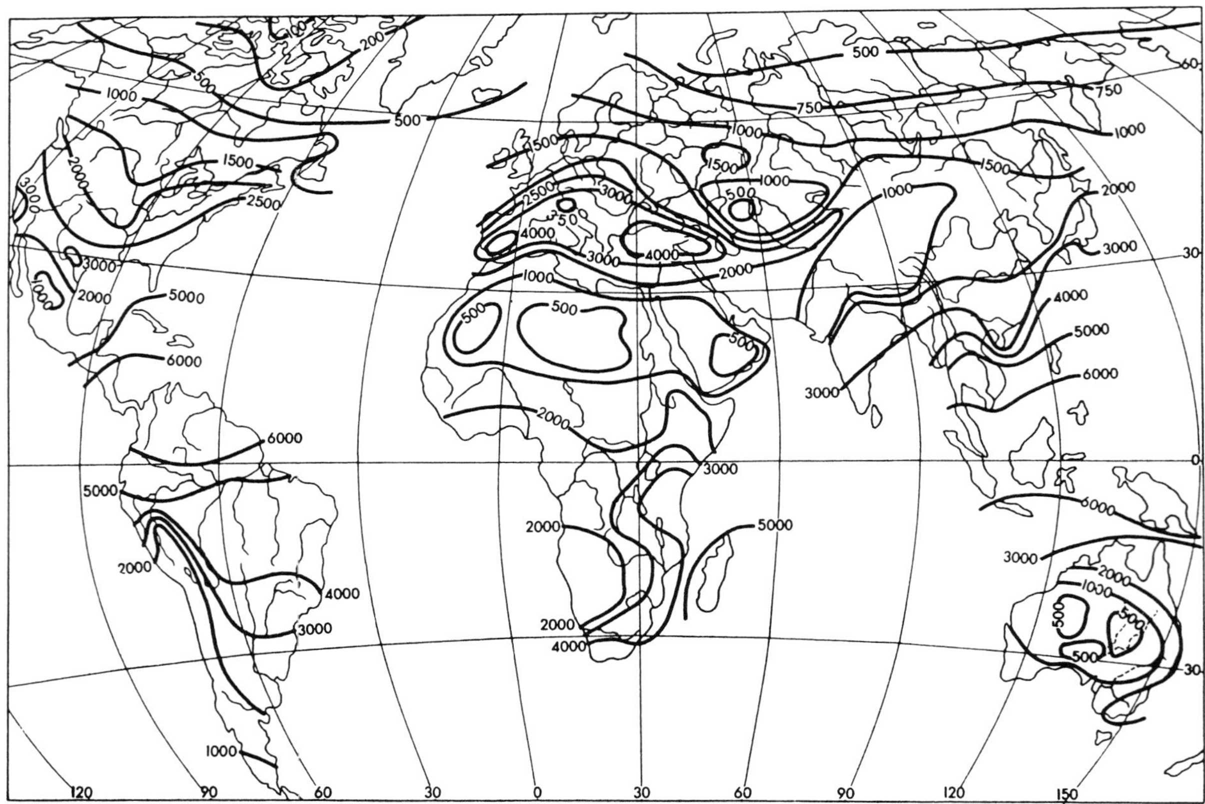

Fig. 1. Map of floristic richness of the world indicating the numbers of vascular plant species extrapolated for areas of 100,000 km^2 (after Malyshev 1975).

In the compilations we have endeavoured to use published data mainly for native and naturalized adventive plants, excluding species introduced in culture and excluding microspecies. In some cases the completeness of the check lists could not be ascertained. In addition, authors of floras most likely have somewhat different concepts of species abundance. This also affects the reliability of comparisons based on such data.

The comparative data for levels of species richness on a global scale (Fig. 1) confirm the general tendencies revealed by de Candolle (1855) and later scientists. These data illustrate in detail increasing floristics richness from the poles to the equator, with local decreases in the desert regions. Unfortunately, the smallest floristic richness in the tropics is still uncertain due to lack of reliable basic data.

Floristic richness may be regarded as a particular case of biological productivity. Therefore the map of floristic richness of the world is somewhat similar to maps of phytomass or primary productivity of vegetation (Lieth 1972, 1974, 1975). The similarity exists because floristic richness and productivity of vegetation are influenced largely by the same major climatic factors (amount of precipitation, level of evaporation and total annual days with mean temperature above a certain level).

Analysis of floristic richness for small areas has special importance for phytogeography when the Tolmachev concept of elementary floras is accepted. It was stressed earlier in this paper that Tolmachev opposes the concrete, or elementary floras in preference to the collective ones. His notion implies that concrete floras usually represent a continuum if their area surpasses the minimal size. In contrast, the collective floras, being composed of elementary ones are more heterogeneous, and so in extreme cases they may be divided into elementary units of the floristic subdivision.

We chose 100 km^2 as the standard area size for minor floras. Extrapolations based on the Arrhenius equation (1) was undertaken for 198 local floras of 10 to 1000 km^2 on mainlands and large islands of the world. Thus the mean deviation of computed species numbers from actual meanings should not exceed 2.6%. Unfortunately, absence of basic data at our disposal for most regions of the world (except Europe) prevents the generalization of computed data for plotting on a map.

In Europe the floras of the southern Arctic and of the Subarctic contain mostly 200–300 species per 100 km^2; in the main part of Scandinavia, northern and middle parts of the British Isles, 300–500 species; in southern England, Sweden, Denmark and in the Baltic region, 500–700 species; in middle and southern Europe, mostly 700–1000 species; but in the mountain regions (the Alps, Carpathians) sometimes up to 1130 species; and in the Lagodekhy nature reserve in the Caucasus in Soviet Georgia, Caucasus, about 1020 species.

In Asia the floras of the Subarctic and the northern part of the taiga zone have 260–300, and sometimes up to 360 species per 100 km^2; in the southern parts of the taiga zone; 460–730 species, in the southern parts of the steppe zone in Kazakstan, about 410 species; in the deserts of the Soviet middle, or central Asia, only 150–160 species; while in the nearby mountains, 500–1000 species; in the savannah of northern India, 720 species; on the Dekan peninsula, 635 species (also 357 and 387 species); in the Arabian desert near Aden, 232 species; while in Indo-Malaysia and on the Malay archipelago, 1000–1500 species.

Southeastern Australia in one area has 512 species per 100 km^2. In Africa, the Sahara desert is characterized by 156 species per 100 km^2; the basin of the Congo River, 644 species; Zimbabwe, 644 species; the Cape region, 1217 and 1825 species.

In the Western Hemisphere, areas of 100 km^2 contain the following: the Alaskan range (Ogotoruk-Creek), 297 species; the main expanse of the USA, 500 to 750 species; the southeastern United States and California, 750 to 1000–1100 species. Lagoa Santa in Brazil with its evergreen broadleaved forests has 2250 species.

To generalize, the estimated level of floristic richness for 100 km^2 areas is as follows: in the Arctic, a few dozen to about 200 species: in the

Sub-Arctic, 200 to 300 species; in the taiga zone, 300 to 500 species; in the zones of deciduous and mixed forests and in the steppe zone outside of high mountain ranges, 500 to 750 species; in the subtropical zone, 750 to 1000 species; in the tropical zone, 1000 to 2000 species; and in the deserts of temperate and warm climates, only 150–160 species.

In their main features these data correspond with the conclusion of Tolmachev (1986: 11–37) on the zonal changes of floristic richness in the Arctic and boreal zones. On the basis of elementary flora analysis Schmidt (1980) concluded that at least in the European part of the USSR the correlation of species, genus and family abundance with geographical latitude may be expressed by the reverse logistic function. But the estimates made by both authors are somewhat problematic due to their lack of knowledge (at least in some cases) of differences in areas of the floras being compared.

Later in this paper we will show the importance of floristic reliability assessments for relatively small areas. But, for now we state that the geographic tendencies of floristic richness for the standard area of 100 km^2 are similar to the general features for 100,000 km^2. But it is hardly possible to expect complete coincidence of these tendencies because floristic richness of small areas is more or less related to the specific diversity whereas floristic richness of larger areas becomes more and more dependent on spatial diversity as well. Thus, equatorial Africa, the Dekan peninsula, and islands of Japan and of southeastern Australia possess comparatively low levels of floristic richness for 100 km^2. But their paucity of species per unit area is counterbalanced by the high floristic spatial diversity due to historical factors.

The basic data for alpine floras are rather scanty. Nevertheless it seems evident that species abundance with 100 km^2 as the standard does not differ considerably in the temperate and warm zones. However, the altitudinal levels of alpine floras are quite different in both zones; their being correlated with geographical latitude and humidity of climate. Thus, for 100 km^2 alpine floristic richness is as follows: in the Siberian Subarctic about 172 species (the Putorana plateau); in the southern USSR's taiga subzone, 212, 247 and 262 species (Stanovoye Nagorye uplands); in the eastern Carpathians, 350 species; in the Rocky Mountains of Canada, 393 species (Sunshine National Park); in the Andes near the Tropic of Capricorn, 237 species; and in the Blue Mountains of Australia, 157 species.

2.3. Species abundance on islands

Basic floristic data for islands are rather numerous compared to that for mainland areas. Such an abundance of data allows more complete and comprehensive analyses, and thus provides more reliable and detailed estimates of floristic richness in relation to latitude. Thus using an area 100 km^2, Devon Island in the Arctic has only 102 species and the eastern coast of Greenland at 75°N has 152 species. In Japan floristic richness is estimated as being 500–583 species per 100 km^2, but it is higher in the mountains (e.g. Mount Yamizo in Honshu has 920 species). Barro-Colorado Island in the Tropics (Panama) has 1814 species per 100 km^2, while specific areas of New Zealand have only 364 and 452 species.

Usually the floras of small islands situated at great distances off the coast are much poorer than the local floras of the mainland. For example, the islands in the Gulf of Guinea at the equator have only *ca.* 303 species per 100 km^2, whereas the floras in the Congo River basin are at least twice as rich. The paucity of island floras is due mainly to their ecological isolation. Besides that, the isolation of many oceanic islands is perturbed by recent human activities, so that available ecological niches in many cases were occupied by adventive plants which colonized islands and have led to the extinction of native species.

The Galapagos archipelago has become the classical place for biogeographic studies since Charles Darwin's voyage around the world. Johnson and Raven (1973) concluded that the floristic richness on this archipelago depends on the size of an island, its maximum height, distance to a nearby island, distance to the center of the archipelago, and size of nearby islands. Distance between islands and the mainland affects dispersal of diaspores by natural agents, while island size predetermines the possibility of survival for rare species. On a larger time scale, the time an island has existed and the time since its separation from the mainland also are important factors. For example, the island of Great Britain differs from nearby small islands by the latter's continental type of floras (McCoy and Connor 1976).

We used the Arrhenius equation (1) with extrapolation to assess the floristic richness of islands for the standard area 100 km^2. The value of z was computed for individual archipelagos or island groups; otherwise the continental region means were used. For the remote oceanic islands of warm climates the mean value of 0.32 was employed (Malyshev 1980). Analysis of the data enabled us to conclude that many large islands (over 8000 km^2) have the continental type of floras at the level of floristic richness per 100 km^2. They are Great Britain, Ireland, Borneo, Corsica, Crete, Cuba, Luzon, Madagascar, both North Island and South Island of New Zealand, New Caledonia, Sakhalin, Sicily, Taiwan, Sri Lanka and Java. But other large islands (8900–226,000 km^2) have 1.2–1.5 times poorer floras than on the mainland. They are Hispaniola, Cyprus, Puerto Rico, Sardinia, Kyushu, Shikoku, Hokkaido and Honshu. At least some of them are more remote from the mainland when compared with richer neighboring large islands (e.g. Sardinia unlike Corsica or Sicily; Puerto Rico and Hispaniola unlike Cuba).

Small islands of 4 to 8000 km^2 may be similar in floristic richness to a mainland if they are found off a mainland or nearby island with the continental type of flora at a distance of less than 100 km or, in some rare cases, 200 km. In contrast, many other small islands at a distance off the mainland of 100–200 km have floras 1.2–1.5 times poorer at the standard of 100 km^2. Further, at distances of 200–500 km the floras are twice as poor, whereas additional significant remoteness from the source of plant dispersal produces 3–11 times the floristic impoverishment.

This general tendency can be modified by the different age of islands by the presence of intermediate islets facilitating the dispersal of diaspores, and to other conditions of isolation during periods of oceanic transgression.

3. EVALUATION OF FLORISTIC REPRESENTATIVITY

Evaluation of floristic representativity is most important for small areas when one wants to know whether an area size is sufficient for reliable phytogeographic comparisons. Knowledge of floristic representativity is essential when the method of elementary floras is used and one might have to work with an area of insufficient size for an elementary flora. In some other cases the assessment of floristic representativity may reveal whether a species checklist is incomplete. Also it seems quite vital for a biota conservation strategy.

3.1. The quantitative approach to floristic representativity

On the basis of the Gleason equation (2) it is possible to compute an area size X for cases when its doubling corresponds with an m-fold increase of species abundance in a flora:

$$\log X = \log 2/(m - 1) - A/b, \qquad (6)$$

where the constant A is the species number per unit area (floristic richness) and b is the measure of floristic heterogeneity (or floristic spatial diversity). It is evident that area size should be larger in cases of high floristic heterogeneity and, conversely, it ought to be smaller in cases of low floristic richness. So a doubling of area size should be more effective for floras of warmer zones and much less effective for floras of the temperate and especially of the frigid zone (Malyshev 1972a).

In that study, we suggested that a 20% increase in species number be regarded as a conventional criterion for minimal floristic area size, and a 14% increase be used for the optimal size. In cases where the numerical value of floristic heterogeneity is quite greater than floristic richness, the minimal area size ought to be less than 3200 km^2 and the optimal one less than 14,130 km^2 (Malyshev 1972a). But this situation is probably only possible for the richest tropical floras. Subsequently, we abandoned the idea of formalizing the notions of floristic minimal and optimal area size. Instead we suggested the more tenable concept of floristic representativity (Malyshev 1975).

The floristic representativity of an area corresponds conversely to the rate of increase in species abundance due to increase in area size. On the basis of equations (2) and (5) it is possible to calculate the comparative floristic representativity R or the expected m increase of a species abundance in the case of an area doubling in size:

$$m = \log 2/(\log X - A/b) + 1, \qquad (7)$$

or

$$R = [1 - \log 2/(\log X + A/b)] \times 100\%, \qquad (8)$$

where X is area size prior to doubling, the constant A is floristic richness per unit area, and the constant b represents floristic spatial diversity, or heterogeneity, computed by the Gleason equation. Consequently floristic representativity depends on area size and the ratio of species abundance and floristic heterogeneity (A/b).

Usually the ratio of species abundance and floristic heterogeneity (A/b) varies from zero to 10. For rich tropical floras it is minimal, and the heterogeneity is enormous, since their representativity is very sensitive to an area size. For an A/b ratio equal to 1, the representativity of 10 km^2 is only 85%, and for 1000 km^2 about 92.5%. In contrast, the northern boreal and especially arctic floras have high A/b ratios, so their representativity is considerable even for small areas and varies only slightly in response to area size differences. Thus, for an A/b ratio of

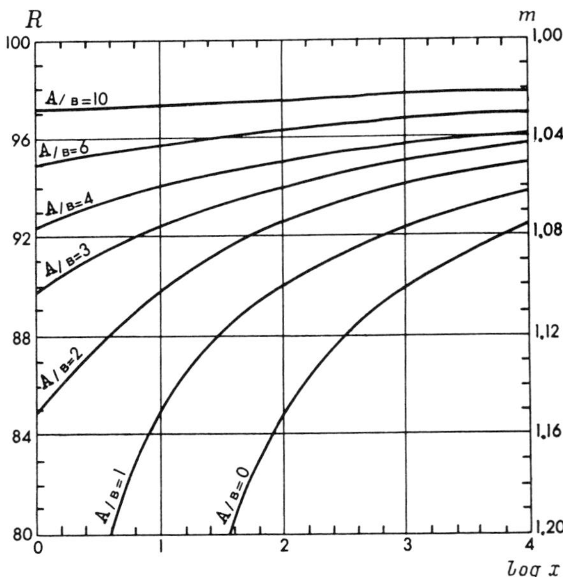

Fig. 2. Nomogram of floristic representativity R and corresponding m-fold increase of species abundance for floras of area size X and for ratios of species abundance A per area unit and of floristic heterogeneity b. (Computed by the Gleason equation, after Malyshev 1975.)

10, the floristic representativity of areas 10 to 10,000 km^2 would vary only from 97 to 98% (Fig. 2).

Data were obtained for seven elementary floras of the Yamal Peninsula in the Siberian Arctic (Rebristaya 1987). Floristic richness and spatial diversity only slightly decrease from 68°N to 72°N, while the A/b ratio decrease is even less. The increase of area size from 3.2 to 78.5 km^2 caused only very slight growth in floristic representativity: in the southern Hyparctic tundra from about 95.63 to 96.37%; in the northern Hyparctic tundra from about 96.70 to 97.15%; and in the Arctic tundra from about 97.10 to 97.45%. So it was concluded that areas about 30 km^2 are quite sufficient for studies by the method of elementary floras in the tundra plains of the Yamal Peninsula.

Sometimes any local boreal floras are thought appropriate for the direct comparison of taxonomic richness even when their area size is only approximately equal, and assuming they belong to the same phytogeographical groups (Schmidt 1987). This actually allows a rather wide range of size inequality without the need to any extrapolation from a standard area size. However, this approach is quite suspect. It is somewhat tenable only for the Arctic and even less so for the northern taiga regions with poor and rather homogeneous elementary floras. In other cases it is important to remember the warning by de Candolle long ago (1855: 1270), that it would be absurd to compare absolute species numbers for regions which are not more or less equal in area ("...il serait absurde de compare, au point de vue du nombre totale des espèces, des regions qui ne seraient pas sensiblement égales en surface").

The floristic representativity R may also be computed by the equation:

$$R = [(1 - \log 2)/(\log X + \frac{\log X}{X^2 - 1})] \times 100, \quad (9)$$

where X is area size, and z is an estimate of floristic spatial diversity corresponding to the constant in the Arrhenius equation (1). It differs from the alternative previous equation (8) by the substitution of A/b for $\log X/(X^z - 1)$. This simplifies the computing and allows us to plot a nomogram for different magnitudes of floristic spatial diversity (Fig. 3).

For example, it is obvious that any area of size 100 km^2 and of floristic heterogeneity equal to 0.08 should have floristic representativity of 95.36%, while an increase of heterogeneity to 0.27 for the same area size should diminish its floristic representativity to 89.17%. Likewise the same representativity of 93% would have an area of about 140 km^2 with a floristic heterogeneity of 0.14; an area of about 1470 km^2 would have a floristic heterogeneity of 0.18; and an area of about 9100 km^2 a floristic heterogeneity of 0.27.

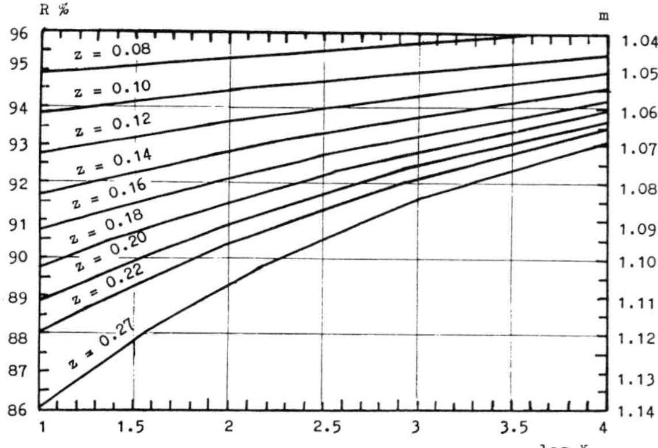

Fig. 3. The nomogram of floristic representativity R and corresponding m-fold increase of species abundance for floras of area size X and of spatial diversity z. (Computed by the Arrhenius equation, after Malyshev 1980.)

3.2. Implications of floristic representativity in the strategy of nature conservation

A crucial question is whether the nature reserves will ensure the long-term preservation of rare and endangered species of plants and animals. A positive answer to this question is not obvious when one considers all aspects of the long-term, lasting conservation of nature. It is just possible that at least some nature reserves sooner or later will be surrounded by cultural landscapes entirely transformed by human activities such as agriculture, industry and urbanization. These nature reserves then would inevitably be subject to ecological isolation, and actually would become isolates similar to oceanic islands, and thus at least some of the island biogeography concepts may be applied to them (Terborgh 1974; Sullivan and Shaffer 1975).

Ecological isolation of a nature reserve will diminish its floristic richness. These quasi-island biotas should have less species abundance and higher heterogeneity as compared with the initial state. The rare and endangered species would be the first victims of ecological isolation, although they were the main reason for the establishment of such nature reserves (Malyshev 1980). The decrease of floristic representativity in nature reserves due to ecological isolation may be evaluated by the above mathematical models (8) and (9) and nomograms (see Figs 2 and 3). Thus, the general trend in the dynamics of floristic or faunal improverishment seems quite obvious but the time scale is still obscure. To estimate the latter it is useful to consider the condition of land bridge islands.

It is postulated that, at least in the later period

of the Pleistocene epoch, the large islands (over 2500 km^2) of continental sea shelves were connected with the mainland, since at present they are floristically twice as rich in some cases than remote oceanic islands of equal area. Nevertheless, islands of less than 250 km^2 are much poorer in both cases (Diamond 1976). The real situation for islands was assessed by our computer analyses (see above in this paper). Consider the concept of paleogeographers that on a global scale enormous glaciation of the Earth's surface occurred from 70,000 to 10,000 years ago with the most frigid climate about 20,000 years ago. At that time the world's ocean level was 100 to 110 metres lower than it is today. The climatic warming and subsequent submergence of the sea shelf began about 15,000 years ago. So about 5000 to 6000 years ago the sea level was about 70 metres lower than at present (Pascoff 1987). In any case, due to ecological isolation floristic impoverishment correlates with area size, and endangered species should be first to disappear from the island biotas (Terborgh 1976).

The Barro-Colorado island presents perhaps a classic case for biogeography. Situated in the Panama Canal zone with an area about 17 km^2, it became isolated in 1914 due to the artificial flooding during canal construction. In 1923 the island became a nature reserve sponsored by the Smithsonian Institution. The amazing feature was the disappearance of 45 bird species of the initial 208 species in spite of 50 years of conservation enforcement in the area (Willis 1974).

The current lack of wide-scale experimental data does not allow accurate forecasting of the tempo of floristic or faunal impoverishment of natural reserves in cases of ecological isolation. A separate but related question is: "What is preferable for conservation strategy – one large territory or several smaller plots where total area equals that of the larger?" It would be unwise to recommend anything before careful examination of a concrete case. Nevertheless the island biogeography data may be helpful in making a comprehensive decision. The extrapolation of species abundance for a standard area has revealed that archipelagos are comparatively richer than the individual islands inside the archipelago. Thus, the archipelago near the French coast in the Mediterranean Sea has 895 species of vascular plants per 100 km^2, while its three main islands evaluated separately have only ca. 751 species (Porkerol, 860; Porkro, 530; Levan, 529), the Nicobar islands have 251 species while Car Nicobar has only 104 species, the Canaries have 820 species but Gran Canaria has only 622 species, the Aldabra islands have 179 species, while some of its larger islands taken separately have ca. 160 species, and the Galapagos archipelago has 306 species while its islands, separately, have only ca. 277 species.

The different levels of species abundance for an archipelago as a whole and for its islands separately are due to differences in their floristic or faunal structure. Islands far from the mainland (Aldabra, Hawaii, Galapagos, Canaries, etc.) are usually characterized by a high level of endemism, so some taxa are restricted to certain islands. As a result an entire archipelago is richer than its constituent islands taken separately (in comparisons where total areas are equalized). Besides that, comparison of islands of different size within the same archipelago may reveal the internal floristic heterogeneity of islands. Comparison of individual islands with a whole archipelago would show a much higher value of the exponent z for a whole archipelago. In this case it estimates the floristic heterogeneity of the archipelago as well as the difference in taxonomic structure between islands. The latter parameter may be evaluated directly by the Preston canonical equation (5).

If the analogy between the sea islands and conservation reserves is tenable, then a network of somewhat isolated reserve plots would ensure the preservation of major species abundance better than only one large plot of total area equal to the sum of the network. But plots that are too small cannot ensure the survival of certain particular species due to the ecological isolation among the small areas. In addition, the plots should not be too isolated from one another; otherwise there would be no archipelago. It has been shown above in this paper that sea islands should not be more than 100 to 200 km from the mainland. This will avoid pronounced floristic impoverishment due to ecological isolation. The distance ought to be considerably less for conservation reserves inside the cultural landscape because the ecological isolation between small areas should be much more consequential than between large and small areas. So it may be concluded that for the maintenance of initial floristic representativity, conservation of a large area is preferable. While a group of ecologically isolated small areas can give refuge to more species than one reserve of equal total size, the group is apt to increase the extinction of rare species.

Moreover the plots ought to be situated close to each other; otherwise the effects of ecological isolation will be emphasized. Because of this, buffer zones may be helpful.

4. AUTOCHTHONOUS AND ALLOCHTHONOUS TENDENCIES IN FLORISTIC GENESIS

Earlier in this paper it was stated that floristic richness may be regarded as a particular case of biological productivity that depends mostly on environmental factors. An analysis of faunal data allowed Terentyev (1963) to conclude that the abundance of vertebrate animal species depends 80–90% on external, or ecological factors, while historical factors mostly influence individual species assortment, so it is interesting to obtain similar assessments from floristic data.

4.1. Evaluation of floristic originality

Using the "Flora USSR" (Komarov 1934–64) as the basic source, the number of vascular plants species was calculated for the 51 regions. The main environmental factors chosen as influencing species numbers were area size, annual number of days with temperature above 0° C, aridity of climate, and the mountainous character of a region (Malyshev 1969). Among these characters aridity of climate depends mostly on the relation between thermal and water balance. So this has been estimated by the ratio of potential annual evaporation to actual evapotranspiration. Specific findings vary from 1.0 to 11.0. The number of altitudinal zones or belts was chosen as the indicator of the mountainous character of regions. It varies from zero to eight. Analysis using multiple curvilinear regression revealed a strong influence of the above ecological factors on species abundance. The coefficient of determination is 0.92, so the four indicated environmental factors account for 92% of the variation in floristic richness. Use of more sophisticated ecological criteria would perhaps reveal an even stronger environmental influence on species abundance in different regions of the world.

Differences between recorded and expected species abundance may be attributed to incomplete information and to historical factors. Meanwhile it is difficult, or perhaps it may be impossible, to separate these two possible causes. In any case it is evident just from the above data for the flora of the USSR that historical factors account for no more than 8% of variation in modern species abundance of the regions. Nevertheless evaluation of historical factors is of paramount importance for phytogeography. Thus, in addition to evaluation of endemism in the taxonomic structure of floras, the role of species immigration in floristic genesis also needs the consideration.

Speciation supposedly enriches the number of endemic taxa in a flora, and thus it leads to an autochthonous tendency in floristic genesis. This process is offset by the migration of species into a flora from other regions; i.e. the prevalence of immigration produces an allochthonous tendency in floristic genesis. Both tendencies, autochthonous and allochthonous, are opposed to each other because the level of species abundance in a flora is somewhat limited by various environmental factors as shown earlier in this paper. In other words, it seems that the richness of a flora is predetermined rigorously by ecological conditions.

Enhancement of autochthonous and allochthonous tendencies, alternatively or simultaneously, is possible in cases of floristic perturbations due to major climatic changes or to orogeny. Another reason for successful species immigration and perhaps for speciation is in response to the floristic vacuum that accompanies vegetation cover restoration after glaciation or sea regression.

Usually it is assumed that the high content of endemic taxa in a flora signifies its antiquity as well as the prevalence of an autochthonous tendency in its genesis. The rate of endemism depends on the ecological isolation of a flora, especially in the case of oceanic islands, and on ecological diversity which may be vigorous in mountains. Ultimately, the rate of endemism is due to the balance of allochthonous and autochthonous tendencies in a floristic genesis.

The prevalence of an autochthonous tendency in floristic genesis leads to the differentiation and consequent greater variability of some taxa more adapted to the available environments. As a result, the taxonomic structure of a whole flora would be homogeneous on the generic and higher taxonomic levels. In contrast, the prevalence of an allochthonous tendency would create a relatively heterogeneous taxonomic structure of a flora on the generic and sometimes higher taxonomic ranks. This is due to the somewhat random immigration of plant species from the outside, which usually belong to a wide variety of genera and in extreme case of families. So it may be expected that floras with a prevalent autochthonous tendency in their genesis should have a higher average mean number of species per genus or even per family than floras with an allochthonous tendency. Somewhat similar ratios may be revealed between genera and families. Furthermore, the ratio between species and genera should reveal relatively recent consequences of floristic genesis on the evolutionary scale, while the ratio between genera and families reveals mostly ancient consequences, and the ratio between species and families perhaps should have an intermediate significance.

Besides these historical factors, the ratio between species and genera or between species and families depends on general species abundance in a flora and at least partly correlates with geographical latitude (Candolle 1855). The regularity is stronger when only the richest taxa in a flora are taken into account. It was found that the ten richest families comprise 65–76% of species number in some Arctic floras, only 55–57% in some boreal floras, and even smaller percentages in some very rich tropical floras (Tolmachev 1986: 68–79).

Besides that it was revealed that percentage of species number per ten basic families (S/F_{10}) for distinct and quasi-distinct floras in the European part of the USSR somewhat *increases* towards the north, as expressed by the equation:

$$S/F_{10} = 328.0445 - 9.4389L + 0.0818L^2,$$

where L is north latitude, and the regression is curvilinear (Schmidt 1980). Meanwhile the regression becomes linear for ratios comprising entire sets of taxa. An average species number per family (S/F) is related to latitude (L) by the function: $S/F = 18.6536 - 0.1900L$. Likewise the average number of genera per family has the equation $G/F = 8.2098 - 0.0757L$, while the average number of species per genus may be expressed by the equation $S/G = 2.6838 - 0.0114L$. So it is evident that every additional degree of latitude *decreases* the ratio of species to families by about 0.2; the ratio of genera to families by about 0.1; and the ratio of species to genera only by 0.01. Thus, the ratio of species to genera only slightly correlates with geographic latitude, unlike the ratios of genera to families, and especially of species to families. So it seems that ecological factors do not significantly influence the ratio of species to genera, and that the major cause should be attributed to evolutionary reasons.

The relation between quantity of species and genera in a flora or fauna can be evaluated with the logarithmic series equation. The resulting regression has a concave slope somewhat similar to a hyperbola and has the equation:

$$S = \log_e(1 - G/\alpha), \tag{10}$$

where S is the expected species number, G is number of genera, and α is the index of generic diversity (Fisher et al. 1943; Williams 1964).

For 47 of the 51 floristic regions of the USSR, the actual numbers of species and of genera were recorded, and the regressions was computed by the method of least squares (Malyshev 1969) with the following result:

$$\hat{S} = 314.1 + 0.004538G^2. \quad (11)$$

In this equation $\hat{S}$ is the expected species number and G is the observed number of genera. The coefficient of correlation is 0.97, and thus the number of species in floristic regions of the USSR shares 94% of its variance with the number of genera.

The above empirical quadratic equation permits computation of the expected number of species on the basis of recorded number of genera. The estimate would be exact if the autochthonous and allochthonous tendencies are mutually balanced during floristic genesis; otherwise a difference would exist between actual (S) and expected ($\hat{S}$) species numbers. The difference indicates the prevalence of either an autochthonous or allochthonous tendency in floristic genesis, and thus may be used as an index of floristic originality (OR):

$$\mathrm{OR} = (S - \hat{S})/S. \quad (12)$$

Positive OR values indicate the prevalence of an autochthonous tendency, while negative values suggest an allochthonous tendency in floristic genesis. Zero magnitude indicates the counterbalanced state of both tendencies (Malyshev 1976).

4.2. Some consequences of the postglaciation floristic vacuum

In Europe the enhanced role of microspecies in floristic formation is evident in a trend from south to north. Thus, microspecies increase floras of some southern European countries by only 1–10%; in middle European floras by 10–20%; and in northern Europe (Norway, Sweden, Finland) and Great Britain by 20–30% (Malyshev 1981). In addition, the role of microspecies in a flora of the whole country usually is much higher than in its local floras. This is due to small areas occupied by most microspecies and to their pronounced "vicariation" as a result.

Apomixis is the most characteristic accepted process for the evolution of microspecies. The genera *Alchemilla, Hieracium* and *Taraxacum* are especially replete with microspecies in northern Eurasia, as is *Rubus* on the British Isles and in some other highly mountainous regions of Europe. In North America apomixis and microspecies speciation is usual in *Crataegus* and *Rubus*.

The enhancement of speciation on the microspecies level was stimulated by the end of the last (the Würm) glaciation about 11,000 years ago. It opened up vast areas in northern and partly in middle Europe to plant colonization. However, plant dispersion from glaciation refuges and immigration from southern Europe seemed inadequate to entirely fill the available floristic vacuum. So genera capable of prompt speciation by apomixis had an advantage in the colonization of free postglacial areas.

The flora of Iceland displays the widespread occupation of an area by microspecies where they constitute over 40% of the whole flora. The modern floristic structure of this subarctic island is due to immigration of plants in postglacial times. Yet the possibility of preservation during glaciation of some species on the island at specific sites cannot be dismissed entirely (Fridriksson 1962). The flora of Greenland presents another remarkable illustration. The main surface of the island is covered even today by the ice shield, and most endemic species along its coasts are apomictics (Böcher 1963). Investigations in North America perhaps would reveal an abundance of microspecies predominantly in areas formerly subjected to the Wisconsin glaciation.

In the flora USSR *Hieracium* constitutes 4.46% of the species, *Taraxacum* 1.15% and *Alchemilla* 0.86%. Meanwhile the role of *Taraxacum* increases and reaches 2.2–4.4% in the subarctic and arctic local floras. *Hieracium* expands in the northwestern regions to 9–12%, and the abundance of *Alchemilla* species is also highest (0.4–1.6%) in the northwestern regions (Malyshev

1972b). These northern and especially northwestern regions were subjected to the most powerful glaciation at the last stage of Pleistocene when compared with other USSR regions. This area also was the last to be freed from the retreating Scandinavian ice shield.

Speciation success in such available vacant areas is due to low competition levels during formation of the floras or faunas. In the long run microspecies survive if they can adapt to these new ecological niches. Furthermore, during species radiation there seems to be a strong correlation between chorological (phytogeographical) categories and the types of habitat, as Nimis (1984) revealed for the flora of Sicily. Chorology is the study of the geographic, or arealogical, distribution of organisms or taxa.

Besides microspecies, the chorology of colonizing plants seems to be related in many cases to their ability to expand into open or unstable natural habitats. The macabre fate of many native floras in oceanic islands is well known. They are overrun by species introduced mostly involuntarily by human activities. Continental floras reveal a somewhat different situation, but nevertheless one that is strongly influenced by human activities. Anthropochores (i.e. naturalized adventitious, or alien) species constitute about 7.3% of the European flora. But in many regional floras of Europe their percentage usually is much higher. Thus, in middle Europe the anthropochores (hemerochores in other terminology) constitute more than 16% of the total flora, with 7% of them being archeophytes, and 9% neophytes. The latter are species that invaded the area only since the sixteenth century or later due to development of worldwide transportation routes (Sukopp 1972; Sukopp and Trepl 1987). Furthermore, the intrusion of neophytes has caused intensive speciation due to hybridization between species with formerly isolated areas, as was ascertained by Haeupler (1976) for the flora of Lower Saxony. In addition, the percentage of anthropochores increases northward. Southern European floras usually contain 3–10% alien species, while the floras of middle and northern Europe are characterized by 11–23% and about 25% in the British Isles, and perhaps 23% or even 27–29% by some estimates (Webb 1978; Malyshev 1981).

The comparative abundance of adventitious species in middle and northern Europe may be due to the last Pleistocene glaciation. Using circumstantial evidence, the local native floras seem not to have entirely reconstructed themselves after their catastrophic destruction during the last glacial epoch. So perhaps they have not achieved the state of stable dynamic equilibrium characteristic of closed ecosystems, since their habitats today seem to encourage favorable access for the anthropochores.

Disturbance of natural vegetation cover by modern human activities also creates favorable conditions for the intrusion of alien species. Nevertheless, there is no reason to assume that the ecosystems of southern Europe were subjected to human activities any less than those of the northern regions, and perhaps it is the reverse. Furthermore, it is quite possible that the complex of synanthropic plants in southern European floras includes at least some apophytes among native species which became anthropochores in middle and northern Europe. This idea seems to be compatible with the common notion that the southern European countries belong to or border to the zone of ancient agriculture which includes the Mediterranean area and the Middle East as the cradle of European civilization.

Unlike Europe, the Pleistocene glaciations in Siberia, at least in later epochs, were less vigorous. Corresponding to the Würm glaciation, the Zyriansk glaciation is characterized by a very cold and dry climate. It was not catastrophic in most Siberian regions and mainly transformed their vegetation cover from one type to another. This common idea finds indirect confirmation in alpine floras of the Baikal region. Thus, comparative data do not support any catastrophic consequences of late glaciation for the Alpine flora on the main expanse of the Stanovoye Nagorye uplands. The only exception is a humid and steep western slope in the northern area of the Baikalsky range near the Ogneva river. There the Alpine zone contains only about 94 species of vascular plants in an area of 38 km^2. So its floristic richness is about half that of other sites of the uplands when normalized for equal area (Malyshev 1972c).

The uniqueness of the Ogneva flora merits a brief description. Its modern climate is very humid, and the relief is characterized by sharp Alpine forms with a predominance of granites. *Abies sibirica* is found at timber line at about 1000 m. Higher up, the landscape is one of bare granite cliffs, meadows, dense bushes of several subalpine species of *Salix* and of *Pinus pumila*. Many species have not been found there although they are rather usual on the Stanovoye Nagorye uplands as a whole. At the same time, other species that are widespread on the uplands form almost pure aggregations near the Ogneva River. Some *hygrids* (i.e. species characteristic for sites with humid climate) are especially abundant there. Other taxa either are *arids* (inhabitants of arid climate) or *mesids* (inhabitants of intermediate climate). Usually they belong to the complex of montane plants characteristic of several different altitudinal belts or to the adventive plants of the Alpine flora. They include forest-meadow shrubs and inhabitants characteristic of Alpine and Subalpine meadows. Besides that, other alpine species are very abundant locally (*Saxifraga melaleuca, Pyrethrum pulchellum*, etc.). So it is obvious that the Ogneva flora is not yet complete. It has not attained stable genetic equilibrium after the destruction caused by the late Pleistocene glaciation.

Only two centers of floristic destruction were revealed in the East Sayan mountains (Malyshev 1965). One of them occupies the eastern part of the Udinsky range and the neighboring area of the Okinsky range; the other center is in the eastern extremity of the Tunkinsky range. The local climate is characterized by high humidity. In both centers the destructive effects exist mostly in the alpine floras.

The remarkable feature of the first center is the disjunction of 100–200 km in the distribution of Alpine and montane species like *Carex bipartita, Dryas grandis, Gentiana falcata, Saxifraga androsacea*, etc. Such species belong to the hygrids and mesids, but mostly to the arids. Modern ecological conditions do not satisfactorily explain the phenomenon, so it may be attributed to extinction caused by the last glacial events.

In the second center some Alpine species are totally absent while they are conspicuous in the western and central parts of the Tunkinsky range. They belong to the arids (for example, *Nardosmia gmelinii, Primula nivalis*), mesids (*Tofieldia pusilla*) and hygrids (*Carex bipartita, Salix lanata*). Besides that, at the base of the southern slope in the eastern extremity of the range one finds *Rhododendron parvifolium*. This montane arid species is more characteristic of the upper part of forest belt and for the Alpine belt in subarid regions of East Sayan. So at the base of Tunkinsky range it may be regarded as a late Pleistocene relict that marks the former periglaciation zone. The last glaciation in the East Sayan Mountains seems in general to have had transformation consequences for the Flora, which stimulated long-distance migrations of Alpine and Arctic–Alpine species.

5. CONCLUSION

Besides ecology and phytocoenology, comparative floristics constitute the most substantial part of phytogeography. They are the cornerstones of phytogeography, or geobotany in a wider sense. Like two sides of the same coin, comparative floristics may be approached from two opposite directions. One is based essentially on qualitative characters, while the other is based on quantitative ones. The approaches are not in conflict, and actually they supplement each other in the elucidation of peculiarities and evolutionary tendencies in the genesis of a flora.

The evaluation of qualitative characters in floristic studies implies a mostly superficial perception at the visual or emotional background. It is highly subjective and depends on good judgment and the erudition of the researcher. In contrast, the quantitative approach implies analysis on a statistical or formal basis; it diminishes or even eliminates excessive subjectivity.

Historically, comparative floristics formerly was based mainly on the evaluation of qualitative characters, but later on it became enhanced by quantitative analyses. At present quantitative methods are essential for the elucidation of a flora's characteristics and its origins.

Quantitative assessment of floristic richness and originality reveals the most important characters of a flora. Furthermore, floristic richness incorporates two essentially different basic parameters: taxonomic abundance per unit area; and floristic spatial diversity (floristic heterogeneity). Likewise, floristic originality quantitatively sums up the issue of two different evolutionary tendencies in floristic genesis: autochthonous and allochthonous trends.

Quantitative evaluations constitute the core of comparative floristics. They help reveal evolutionary tendencies in intricate floristic genesis. Thus, they elucidate some consequences of the postglaciation floristic vacuum. Besides that, assessment of floristic representatively seems to be very valuable in choosing strategies for the conservation of nature.

Other quantitative approaches to the problems of comparative floristics exist but are outside the scope of this paper. Perhaps most important among them is the comparison of taxonomic structure in different floras; this in turn is especially valuable for procedures to produce floristic classifications.

REFERENCES

Arrhenius, O. 1920. Distribution of species over area. Meddel Vetens Nobelinst 4, No. 1, 1–6.

Böcher, T. W. 1963. Phytogeography of Greenland in the light of recent investigations. In: North Atlantic Biota and Their History, pp. 285–295, Pergamon Press, London.

Cailleux, A. 1953 (1e ed.), 1961 (2e ed.). Biogeographie Mondiale. Presse Universitaire, Paris.

Candolle, A. de. 1855. Geographie Botanique Raisonnée, Volumes 1–2. Masson, Paris.

Claus, C. F. 1851. Localfloren der Wolgagegenden. St Petersbourg.

Crovello, T. J. 1970. Analysis of character variation in ecology and systematics. Ann. Rev. Ecol. Syst., 1, 55–98.

Crovello, T. J. 1972. Computerization of the Edward Lee Greene Herbarium at Notre Dame. Brittonia, 24, 131–141.

Crovello, T. J. 1981. Quantitative biogeography: an overview. Taxon, 30, 563–575.

Crovello, T. J. and MacDonald, R. D. 1970. Index of EDP-IR projects in systematics. Taxon., 19, 63–79.

Diamond, J. M. 1976. Island biogeography and conservation: strategy and limitations. Science, 193, 1027–1029.

Dony, J. G. 1977. Species–area relationships in an area. Ecology, 65, 485–484.

Evans, F. C., Clark, P. J. and Brandt, R. H. 1955. Estimation of the number of species present in a given area. Ecology (Durham), 36, 342–343.

Fisher, R. A., Corbet, A. S. and Williams, C. B. 1943. The relation between the number of species and the number of individuals in a random sample of an animal population. Anim. Ecol., 12, 42–58.

Fridriksson, S. 1962. Um adflutning islenzke florunnar. Natturufraedingurinn, 32, 175–189.

Gleason, H. A. 1922. On the relation between species and area. Ecology, 3, 158–162.

Haeupler, H. 1976. Die verschollenen und gefährdeten Gefässpflanzen Niedersachsens, Ursachen ihres Rückganges und zeitliche Fluktuation der Flora. Schriftenr. Vegetationsk, 10, 125–131.

Johnson, M. P., Mason, L. G. and Raven, P. H. 1968. Ecological parameters and plant species diversity. Amer. Nat., 102, 297–306.

Johnson, M. P. and Raven, P. H. 1973. Species number and endemism: the Galapagos Archipelago revisited. Science, 179, 893–895.

Komarov, V. L. (ed). 1934–64, Flora USSR, Vols 1–30. Moscow and Leningrad.

Lebrun, J. P. 1060. Sur la richesse de la flore de divers territories africains. Bull. Seances Acad. Roy. Sci. Outre-Mer, 6, 669–690.

Lieth, H. 1972. Über die Primärproduktion der Pflanzendecke der Erde. Angew. Bot., 46, 1–37.

Lieth, H. 1974. Primary productivity modelling of the world. Ekologia (USSR), 2, 13–23.

Leith, H. 1975. Primary production of the major vegetation units of the world. Ecol. Stud., 14, 203–215.

MacArthur, R. H. and Wilson, E. O. 1967. The Theory of Island Biogeography. Princeton Univ. Press, New York.

Makarova, S. I. 1983. Comparison of some mathematical models describing the dependency of species number in a flora on the size of the area. Botanicheskiy Zhurn., 68, 376–381 (in Russian, English summary).

Malyshev, L. I. 1965. Flora Alpina Montium Sajanensium Orientalium. "Nauka", Moscow (in Russian).

Malyshev, L. I. 1969. Dependence of species abundance of a flora on environmental and historical factors. Botanicheskiy Zhurn., 54, 1137–1147 (in Russian, English summary).

Malyshev, L. I. 1972a. Representative areas of a flora in comparative floristic studies. Botanicheskiy Zhurn, 57, 182–197 (in Russian, English summary).

Malyshev, L. I. 1972b. Floristic spectra of the Soviet Union. In: Istoria Flory i Rastitelnosti Evrasii, pp. 17–40. "Nauka", Leningrad (in Russian).

Malyshev, L. I. 1972c. Peculiarities and genesis of the flora. In: Malyshev, L. I. (ed.), Alpine Flora of the Stanovoye Nogorye Uplands, pp. 150–189. "Nauka", Novosibirsk (in Russian).

Malyshev, L. I. 1975. The quantitative analysis of flora: spatial diversity, level of specific richness, and representativity of sampling areas. Botanicheskiy Zhurn., 60, 1537–1550 (in Russian, English summary).

Malyshev, L. I. 1976. Quantitative characteristics of the Flora Putorana, pp 163–186. "Nauka", Novosibirsk (in Russian).

Malyshev, L. I. 1980. Isolated reservation areas as pseudo-insular biotas. Zhurn. Obchey Biol., 41, 338–349 (in Russian, English summary).

Malyshev, L. I. 1981. Changes of the world floras due to anthropogenetic causes. Biologicheskie Nauki, 3, 5–10 (in Russian)

Malyshev, L. I. 1987. Modern approaches to quantitative analyses and comparisons of floras. In: Yurtsev, B. A. (ed.), Theoreticheskie i Methodicheskie Problemy Sravnitelnoy Floristiki, pp. 142–148. "Nauka", Leningrad (in Russian).

McCoy, E. D. and Connor, E. F. 1976. Environmental determinants of island species number in the British Isles: a reconsideration. J. Biogeogr., 3, 381–382.

Nimis, P. L. 1984. Contributions to quantitative phytogeography of Sicily, 1: Correlation between phytogeographical categories and environmental types. Webbia, 38, 123–137.

Paskoff, R. 1987. Les variations du niveau de la mer. La Recherche, 191, 1010–1019.

Perring, F. H. and Walters, S. M. (eds). 1962. Atlas of the British Flora. Nelson, Norwich, UK.

Preston, F. W. 1962. The canonical distribution of commonness and rarity. Ecology, 43, 185–215, 410–431.

Rebristaya, O. V. 1987. Use of the elementary flora method in the West Siberian Arctic (Yamal Peninsula). In: Yurtsev, B. A. (ed), Theoreticheskie i Methodicheskie Problemy Sravnitelnoy Floristiki, pp. 67–90. "Nauka", Leningrad (in Russian).

Schmidt, V. M. 1980. Statistical Methods in Comparative Floristics. Leningrad University Press, Leningrad (in Russian).

Schmidt, V. M. 1987. On the role of A. I. Tolmachev in the development of comparative floristics. In: Yurtsev, B. A. (ed.), Theoreticheskie i Methodicheskie Problemy Sravnitelnoy Floristiki, pp. 43–46. "Nauka", Novosibirsk (in Russian).

Schmidt, V. M. and Ilminskikh, N. G. 1982. On the role of C. C. Claus in the elaboration of the methods of comparative floristics. Botanicheskiy Zhurn., 67, 462–470 (in Russian).

Simberloff, D. S. 1982. Island biogeography theory and organization of conservation territories. Ekologia (USSR), 4, 3–13 (in Russian).

Sukopp, H. 1972. Wandel von Flora und Vegetation in Mitteleuropa unter dem Einfluss des Menschen. Ber. Landwirt., 50, 112–139.

Sukopp, H. and Trepl, L. 1987. Extinction and naturalization of plant species as related to ecosystem structure and function. In: Schulze, E. D. and Zwölfer, H. (eds), Ecological Studies, 61, pp. 245–276. Springer Verlag, Berlin.

Sullivan, A. L. and Shaffer, M. L. 1975. Biogeography of Megazoo: Biogeographic studies suggest organizing principles for a future system of wild lands. Science, 189, 13–17.

Terborgh, J. 1974. Preservation of natural diversity: The problem of exinction prone species. BioScience, 24, 715–722.

Terborgh, J. 1976. Island biogeography and conservation: Strategy and limitations. Science, 193, 1029–1030.

Terentyev, P. V. 1963. Experience with the application of variance analysis on the quantitative richness of USSR terrestrial vertebrates. Vestnik Leningrad Univ., 21, 19–26 (in Russian).

Tolmachev, A. I. 1986. The Methods of Comparative Floristics and the Problems of Floristic Genesis. "Nauka" Novosibirsk (collected papers, in Russian)

Vuilleumier, F. 1970. Insular biogeography in continental regions, 1: The northern Andes of South America. Amer. Nat., 104, 373–388.

Webb, D. A. 1978. Flora Europaea – a retrospect. Taxon, 27, 3–14.

Williams, C. B. 1943. Area and number of species. Nature (London), 152, 264–267.

Williams, C. B. 1964. Patterns in the Balance of Nature. Academic Press, New York.

Willis, E. O. 1974 Populations and local extinctions of birds on Barro Colorado Island, Panama. Ecol. Monog., 44, 153–169.

Yurtsev, B. A. 1987. Preface. In: Yurtsev, B. A. (ed.), Theoretical and Methodological Problems in Comparative Floristics, pp. 3–10. "Nauka", Leningrad (in Russian).

3. ECOLOGICAL PHYTOGEOGRAPHY OF THE SOUTHERN YUKON TERRITORY (CANADA)

DUILIO LAUSI AND PIER LUIGI NIMIS

Department of Biology, The University of Trieste, I-34127 Trieste, Italy

1. INTRODUCTION

This paper analyzes the relations between the present distribution of boreal plants and their ecological requirements. Based on a survey carried out in the southern Yukon (Canada), a large, phytogeographically interesting region whose vegetation was rather poorly known, the paper includes a classification of the vegetation into community types, their ecological characterization on the basis of direct data, and a phytogeographic analysis based on the world ranges of all relevant vascular plant species. Multivariate methods of classification and ordination were used in extensive vegetational, ecological and phytogeographic coordinated analyses to study the correlation between some major ecological factors and the frequencies within communities of species with similar world distribution.

Correlation between the distribution and ecology of plant species is an important research area, since phytogeographic theories derive from two main explanatory principles: actualistic and historical. Our approach, based on present ranges and present ecological requirements of the species, is purely actualistic. This approach also is useful for the solution of historical problems since a series of climatic changes is often represented in floristic evolution time. Therefore, the study of the relations between present species ranges and present ecological conditions is valuable in itself, but also provides important factual evidence on which historical inferences can be based.

Walter (1954) and Walter and Straka (1970) established a theoretical basis for a causally oriented, ecological phytogeography. On the other hand, the works of Meusel (1943) and his collaborators (Meusel et al. 1965; Jaeger 1968, 1970, 1972) were mostly based on a descriptive–comparative approach. The two approaches sometimes are seen as antithetical (Walter 1954; Haeupler 1974). We think they are complementary: the descriptive–comparative approach represents the classificatory stage of phytogeography; its main aim is to describe and order the great variety of plant ranges into a phytogeographic classification system. Its results are important also for an ecologically oriented, causal approach to phytogeography, since they provide a first operational level of its conceptual generalization. In this study the ecological interpretation of the species ranges is preceded by a classification of species into phytogeographic elements.

> According to Ritchie (1984) "plant geographers have a traditional predilection for compiling distribution maps of species ranges, grouping them rather casually, comparing them, usually simply by cartographic inspection, with the limits of various environmental factors, and then developing explanatory generalizations; ... the interest of this process is transitory if the hypotheses cannot be tested by one or other of the only two available methods: the factual data of palaeobotany, or the results of experimental ecology and experimental taxonomy."

At least in part Ritchie's considerations may apply also to the phytogeographic approach de-

P. L. Nimis and T. J. Crovello (eds): Quantitative Approaches to Phytogeography, pp. 35–122.

veloped by Hultén (1937), i.e. to the "Method of the Equiformal Progressive Areas". This is based on the analysis of concentric ranges of species, assuming that they had spread after glaciation from the same area. It is suprising that although Hultén's work is generally appreciated as a milestone in the development of modern phytogeography, his method is rarely referred to in the literature. Probably there are two main reasons for this: (1) the method can be employed only in areas where distribution maps are available for the whole flora; (2) the method contains strong elements of subjectivity, of which Hultén himself was fully aware. Its main weakness is that the selection of the species to be jointly mapped is completely arbitrary.

Recent advances in the fields of numerical taxonomy and quantitative plant ecology produced several new methods of numerical analysis, which in many cases may be useful also in phytogeography. These developments are bringing about a radical change in the methodological and logical pathways followed by biogeographers. "Qualitative Biogeography", where the formulation of hypotheses is based mostly on intuitive thinking, is moving towards "Quantitative Biogeography" (Crovello 1981), where data analysis and inference must be based on a strict formal logical consistency. We think that the developments of computerized database analysis allow us to modify Hultén's approach in such a way that its elements of subjectivity are reduced to a minimum. In this study the selection of species whose ranges are jointly mapped is not subjective, since it derives from the numerical treatment of two data sources: (1) classifications based on vegetational and ecological data (species with similar ecology); (2) classifications based on distributional data (species with similar distribution). If significant correlations are established between these two data sets, these correlations cannot be considered as simple "hypotheses". They constitute complex phytogeographic facts requiring a causal explanation; it is in the search for such explanations that working hypotheses may be formulated. In other words, we do not think that the main aim of an ecologically oriented phytogeography should be the formulation of hypotheses; the main aim is to discover phytogeographic facts.

The causal explanation of the degree of "phytogeographic order" in a given area requires several lines of experimental research. The correlations between the bioclimatic conditions in a given habitat and those prevailing over the ranges of the species occurring in that habitat could be specified in terms of direct ecological data; autecological studies could be of great phytogeographic relevance if interpreted in connection with species ranges, and in terms of adaptive radiation; finally, the results of present ecological phytogeography should be taken into consideration by Quaternary palaeobotanists, since these results provide the only factual evidence on the relations between ecology and distribution of the species.

A quantitative approach to ecological phytogeography may help to disengage phytogeography from the hitherto prevalent descriptive–speculative phase, and opens a promising field of research to plant ecology, plant taxonomy and vegetation science by discovering new phytogeographic facts and generating working hypotheses which can be tested on the basis of experimental research.

2. SURVEY AREA

Most of the phytosociological relevés were obtained along the Alaska Highway, in southern Yukon, from Watson Lake to the US border (Fig. 1) in the summer of 1978. A minor set of releves came from the Klondyke and Dempster Highways, from Whitehorse to the Richardson Mountains, in the central and northern Yukon. The Alaska Highway represents a transect, approximately oriented in a E–W direction, which crosses the entire southern Yukon, passing through five main ecoregions (see Oswald and Senyk 1977), i.e. regions with different climate, geomorphology and vegetation. From east to west they are as follows:

(1) Liard River Ecoregion: mostly below

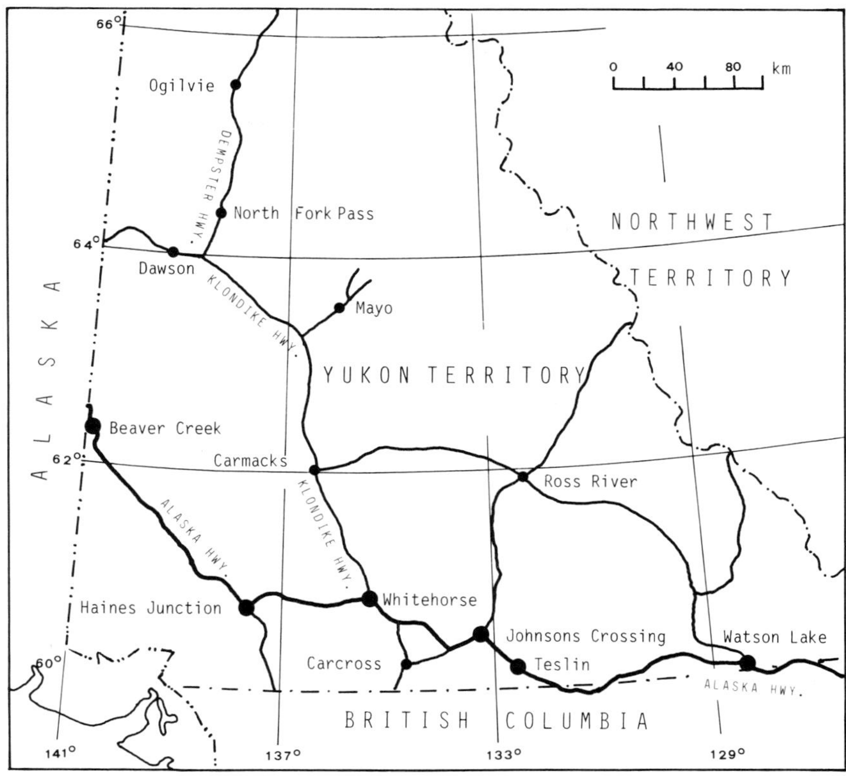

Fig. 1. Survey area with main roads and main localities.

900 m, annual precipitation ca. 430 mm, mean annual temperature −3 °C at an elevation of 685 m, January mean ca. −25 °C, July mean 15 °C. Substrate: mostly glacial and fluvial deposits; it lies in the discontinuous, widespread permafrost subzone. Alaska Highway intercept: 93 km (10%).

(2) Pelly Mountains Ecoregion: mostly between 600 and 1500 m, annual precipitation up to 625 mm, mean annual temperatures between −4 °C and −6 °C (Burns 1973). Substrate: glaciofluvial deposits and fine-textured lacustrine sediments in the valley bottoms, siliceous rocks at higher elevations. Alaska Highway intercept: 89 km (10%).

(3) Lake Laberge Ecoregion: mostly between 600 and 900 m, annual precipitation from 250 mm at Whitehorse to 325 mm at Johnson's Crossing. Mean annual temperature of Whitehorse and Teslin −1 °C, average temperatures in January and July −20 °C and 14 °C, respectively. Substrate: extensive deposits of glacial drift, and large proglacial Lake deposits. Permafrost: discontinuous and scattered. Alaska Highway intercept: 279 km (30%).

(4) Ruby Range Ecoregion: mostly between 600 and 900 m, mean annual precipitation from 190 to 285 mm. Mean annual temperatures ca. −4 °C, with a range of −3 °C to −5 °C. Winters are very cold, average January temperatures of −30 °C are common. Mean July temperature ca. 12 °C. Winds are frequent and strong, clouds are rare in summer, which results in high evapotranspiration rates. Substrate: mostly fine-textured deposits, often of loess; volcanic ash deposits are frequent. Alaska Highway intercept: 375 km (41%).

(5) Wellesley Lake Ecoregion: mostly between 600 and 900 m. Mean annual precipitation ca. 412 mm, more than half coming during June to August. Mean annual temperature ca. −7 °C. January mean temperatures of ca. −34 °C and July mean temperatures of ca. 13 °C. Substrate: volcanic rocks (greenstone and tuff breccia),

often overlain by glacial deposits. Permafrost discontinuous but widespread. Loess and volcanic ash deposits are frequent. Alaska Highway intercept: 81 km (9%).

The Alaska Highway avoids mountain ranges: most of the transect lies between 500 and 900 m. The highest points, above 1000 m, occur in the Pelly Mountains and Ruby Range Ecoregions; consequently, the upper montane belt is poorly represented.

Geologically, the survey area has been classified as a part of the Cordilleran Orogenic Province (Fremlin 1973), which has a great physiographic diversity (Bostock 1965). The area lies within the Yukon, Mackenzie and the much smaller Alsek drainage systems. The dominant rock types are sedimentary, metamorphic and intrusive. Volcanic rocks also occur, but rarely. The geological pattern of the southern Yukon is very complex (Douglas 1970); the area lies outside the Canadian Shield, so that the variety of rocks is much higher here than in the rest of North America at corresponding latitudes.

Most of the area was heavily glaciated at least once during quaternary times (Hughes et al. 1969). The glaciers have altered the surface, leaving behind smoothed plateaus, mountains and valleys; the bedrock is often overlain by glacial, fluvial or lacustrine deposits. In southern and central Yukon volcanic ash has been abundantly deposited in the upper soil horizons from 1220 to 1900 B.P. (Hughes et al. 1972). Given the complex geological pattern, the pedogenesis is correspondingly complex. True podzols are scarcely developed, even on acid parent material, because of the low rainfall. The prevailing soil types under the boreal forest vegetation are more or less leached brown soils with acid to subneutral reactions. Low pH values prevail under coniferous woods where loess deposition does not occur, whereas subneutral reactions prevail in zones with active loess deposition, and under broadleaved tree stands. Cryosols occur where permafrost is present; they are most frequent in the Wellesley Lake Ecoregion. Cryoturbational phenomena are rare, and limited to areas at higher elevation. Alkaline Regosols, supporting a steppe-like vegetation, occur in areas with strong loess deposition, chiefly in the Ruby Range Ecoregion.

The climate of the southern Yukon is continental, with long cold winters and short warm summers. The mean yearly temperatures are always below zero. The southwestern part of the region lies in the leeside of the St Elias Mountain Range, which effects a partitioning of the waterloaden clouds coming from the Pacific Ocean; as a result, precipitation is low, particularly around Kluane Lake, where strong dry winds of the Foehn type frequently blow from the mountains; here the climate is strongly continental, winter temperatures are lower and summer temperatures higher. The western part, under moderate marine influence, has higher precipitation.

As far as vegetation is concerned, the main formation types are:

(1) *Picea mariana* muskeg: a taiga-like open woodland, dominated by mosses (*Aulacomnium*, and sometimes *Sphagnum* species), with scattered dwarf trees (*Picea mariana*, sometimes, on limestone soils, *Larix laricina*), always occurring on Cryosols with acid to subacid reaction and high organic matter content.

(2) *Picea glauca* forest: including communities dominated by *Picea glauca*, in closed stands on relatively well-drained Brunisols with acid or subneutral reaction. The early successional stages after fire, leading towards a *Picea glauca* forest are dominated by *Pinus contorta* in the Ecoregions 1–3, by *Populus tremuloides* in the Ecoregions 4–6.

(3) Shrub vegetation: communities dominated by *Salix* are frequent after burning of the muskeg or of the boreal forest, and along rivers. Thickets dominated by *Alnus crispa* occur along creeks in the mountains.

(4) Grasslands: common in the most continental parts of southern Yukon, on loess soils or sand deposits. They are dominated by *Artemisia* or *Agropyron* species and occur on subacid to strongly alkaline Regosols.

(5) Alpine tundra: restricted at higher elevations in southern Yukon, and never intercepted by the Alaska Highway.

(6) Bogs, riverbanks and lakeshores: underrepresented in our data.

The Yukon Territory (Canada) and the state of Alaska (USA) are of great biogeographical importance, for two main reasons:

(1) Past connections with northeastern Siberia. A land bridge connecting Siberia and North American existed during the Pleistocene, permitting intermingling of the Siberian flora with the flora of Alaska–Yukon (Hultén 1937; Hopkins 1967; Gjaerevoll 1980).

(2) Large areas in this region were ice-free during the glacial period. A connection with the regions located south of the North American ice-sheet was ensured by a narrow deglaciated corridor that separated the Cordilleran glaciers from the continental ice-sheet during xeric interglacials (Douglas 1970).

Alaska–Yukon is therefore an important refugial area with intensive migration, chiefly through the Bering Bridge and through the Cordilleran Corridor. This is true for plants (Hultén 1937), animals (Youngman 1975; Dillon 1956) and man (Mueller-Beck 1967; Laughlin 1967).

The survey area is relatively well known from the floristic point of view. Floras have been published by Hultén (1968), Porsild and Cody (1980), Scoggan (1978–79) and Welsh (1974). Important floristic research has been carried out by Porsild (1945, 1951, 1966, 1974), Loeve and Freedman (1956), Jeffrey (1959), Argus (1984). Some aspects of the natural vegetation have been studied by Hanson (1953), Spetzman (1959), La Roi (1967), Larsen (1970), La Roi and Stringer (1976), Douglas (1974), Hoefs et al. (1975), Orlóci and Stanek (1979), Murray et al. (1983), Nimis (1981, 1982, 1989), Lausi and Nimis (1985a, b). Ritchie (1984) provides an excellent synthesis of palynological and palaeobotanical data from this area.

3. DATA AND METHODS

The data on which this paper is based are largely shared with another monograph by Orlóci and Stanek (1979). The two papers, however, have very different goals. Orlóci and Stanek's (1979) goal was mainly to reduce the great complexity of the data to scales manageable by the practitioner interested in quick referencing of vegetation types and gradients. For this reason these authors partitioned their data set according to Ecoregions, and laid great stress on data reduction. In the present monograph our focus is on community structure, ecology and phytogeography. For this reason we completely avoided data reduction by species ranking, and we partitioned our data set using formation types and not Ecoregions as the first level strata in data analysis.

3.1. Sampling design

The data have been collected in the summer of 1978. The survey was limited to the immediate vicinity of the Alaska Highway within the Yukon Territory; 323 sampling points were established via stratified random sampling. At the end of this survey 19 additional relevés were taken also in northern and central Yukon. From the total of 342 sampling points, 37 relevés were discarded because the sampling points were heavily disturbed. The vegetational analysis is thus based on a total of 305 relevés.

The sampling design kept the points at least 0.16 km apart to prevent chance clumping. Given 917 km as the total length of the transect, the number of 0.16 km segments which could receive random sampling points is 5731. On this basis, with 323 plots in the sample, a sampling intensity of 5.6% is achieved. Since the sampling area is vast and heterogeneous, a nested stratification was used. The first level strata are the Ecoregions of Oswald and Senyk (1977), the second level strata are terrain types (Foothills 1978). To improve the chances of the sample capturing the entire breadth of the vegetational and environmental variation in the survey area, the sampling design included at least one sampling point in each terrain type compartment. The sampling points were located via random numbers. Further details on the sampling strategy are in Orlóci and Stanek (1979).

Table 1. Percent frequency of the main soil types and main soil orders within the relevé groups obtained by numerical classification of the relevés (see Fig. 2).

RELEVE GROUP No.		PM1	PM2	PM3	PM4	PG1	PG2	PG3	PG4	PC1	PC2	BL3	S2	S3	G2	G3	G4
Brunisolic Order																	
O.MB	(Orthic Melanic Brunisol)						5					3					
O.EB	(Orthic Eutric Brunisol)							8		5	10	29					
GL.EB	(Gleyed Eutric Brunisol)						17	19				3					
O.SB	(Orthic Sombric Brunisol)							4	22			10			14		
E.SB	(Eluviated Sombric Brunisol)								22								
GL.SB	(Gleyed Sombric Brunisol)						6		12			3		12			
O.DYB	(Orthic Dystric Brunisol)				25		11	15	11	30	55	13					
E.DYB	(Eluviated Dystric Brunisol)				25			11		30	15	3					
GL.DYB	(Gleyed Dystric Brunisol)			9	50		5	31		25		5					
Cryosolic Order																	
R.TC	(Regosolic Turbic Cryosol)	4															
GL.TC	(Gleysolic Turbic Cryosol)	3															
BR.SC	(Brunisolic Static Cryosol)						11						8				
R.SC	(Regosolic Static Cryosol)						6										
GL.SC	(Gleysolic Static Cryosol)	43	50	33		72	11						8				
FI.OC	(Fibric Organic Cryosol)	18	17	8									16				
HU.OC	(Humic Organic Cryosol)			9			6										
THU.OC	(Terric Humic Organic Cryosol)			17			11										
TME.OC	(Terric Mesic Organic Cryosol)	4		8									8				
Gleysolic Order																	
O.HG	(Orthic Humic Gleysol)	7	16			14		4						33			
R.HG	(Rego Humic Gleysol)	4				14							25	11			
O.G.	(Orthic Gleysol)											8	9	11			
Regosolic Order																	
GL.R	(Gleyed Regosol)													11			
O.R.	(Orthic Regosol)			8					11	5	5	8				50	50
CU.R	(Cumulic Regosol)								11		15				86	50	50
GLCU.R	(Gleyed Cumulic Regosol)		17				11	8	11			15	17	11			
GL.HR	(Gleyed Humic Regosol)									5				11			
CU.HR	(Cumulic Humic Regosol)			8													
Organic Order																	
TY.F	(Typic Fibrisol)	3															
T.H	(Terric Humisol)	11											9				
TME.H	(Terric Mesic Humisol)	3															
BRUNISOLS		-	-	9	100	-	44	88	66	90	80	60	-	12	14	-	-
CRYOSOLS		72	67	75	-	72	45	-	-	-	-	-	40	-	-	-	-
GLEYSOLS		11	16	-	-	28	-	4	-	-	-	8	34	55	-	-	-
REGOSOLS		-	17	16	-	-	11	8	34	10	20	23	17	33	86	100	100
ORGANIC SOILS		17	-	-	-	-	-	-	-	-	-	-	9	-	-	-	-

At all sampling plots an area of uniform vegetation was described according to the Braun-Blanquet relevé method. Depending on the formation, plot sizes vary from 200 m^2 to 2 m^2. For each relevé, cover/abundance data of all occurring species were recorded. Eleven environmental variables were measured or otherwise described within the plots (Tables 1 and 2): (1) soil type, in accordance with the Canadian system of Soil Classification (Canada Soil Survey Committee 1978); (2) soil texture (four variables: gravel/stone, sand, silt and clay content); (3) organic matter content; (4) thickness of organic horizon; (5) depth to frozen soil horizon in June; and (6) in September (permafrost); (7) soil pH; (8) exposure; (9) slope; (10) drainage (wet,

Table 2. Average values of some main soil variables in the relevé groups obtained by classification of the relevés (see Fig. 2).

RELEVE GROUP No.	PM1	PM2	PM3	PM4	PG1	PG2	PG3	PG4	PC1	PC2	BL3	S2	S3	G2	G3	G4
Depth of organic layer (cm)	42	30	35	7	30	13	17	4	4	3	5	27	11	2	4	1
Permafrost conditions																
Plots with frozen ground in June (%)	100	100	80	50	100	40	70	-	30	-	-	100	40	-	-	-
Depth of ice in June (cm)	11	14	22	24	16	23	16	-	30	-	-	20	42	-	-	-
Plots with frozen ground in Sept.(%)	70	80	70	60	50	-	40	-	-	-	-	-	-	-	-	-
Depth of ice in Sept.(cm)	48	41	55	48	61	-	62	-	-	-	-	-	-	-	-	-
Drainage classes																
Wet	33	17	-	-	-	-	-	-	-	-	-	46	-	-	-	-
Impeded	67	83	66	-	44	20	40	-	25	-	7	54	50	-	-	-
Moderate	-	34	44	50	66	56	58	50	30	20	54	-	50	-	-	-
Excessive	-	-	-	50	-	24	2	50	45	80	39	-	-	100	100	100
pH																
Average	5.8	6.5	6.2	5.8	7.0	6.5	5.7	7.3	5.3	5.9	7.0	6.7	6.9	7.7	6.9	6.5
Standard deviation	0.6	0.9	0.6	0.7	0.3	0.6	0.7	0.4	0.4	0.5	0.6	0.8	0.7	0.3	0.5	0.3

impeded, moderate, excessive); and (11) erosion potential (high, medium, low). The determinations followed the usual standard techniques, including replicate sampling within the plots. In each stand the age of the tallest tree was measured by ring counting.

3.2. Data analysis

The data analysis is subdivided into three main steps: (A) Analysis of the relevés; (B) analysis of the vegetation types; (C) phytogeographic analysis.

The objective of the first analysis of relevés (A) is to determine the multivariate structure of the vegetation of the survey area, and to search for compositional gradients. To reduce the computing load, numerical classifications were carried out on six subgroups of relevés, stratified according to physiognomical features, as follows:

(1) Open woods dominated by *Picea mariana* (PM).
(2) Forests dominated by *Picea glauca* (PG).
(3) Forests dominated by *Pinus contorta* (PC).
(4) Stands dominated by broadleaved trees (BL).
(5) Stands dominated by low shrubs, mainly *Salix* species (S).
(6) Grassland vegetation (G).

The resulting classifications allowed us to distinguish several main vegetation types. Ordination programs also were applied to the six subsets of relevés to reveal possible compositional gradients, whose interpretation was made on the basis of the ecological data.

The analysis of vegetation types (step B) permitted a higher level of synthesis, on which the phytogeographic analysis could be based. The data set is the contingency table of the species and of all vegetation types obtained in step A. Also, ordination programs were used to reveal the main compositional gradients and to interpret them on the basis of direct ecological data.

The phytogeographic analysis (step C), has two main objectives: (1) to define, by numerical classification, groups of species with similar distribution (phytogeographic elements); and (2) to study the relations between phytogeographic elements and the main ecological factors in the survey area. The basic data are the distributional ranges of each species in the Northern Hemisphere. The ranges were obtained from the maps of Hultén (1968), taking into account also more recent data published by Meusel et al. (1965), Porsild and Cody (1980), and Hultén and Fries (1986). The ranges have been indicated on the same type of map used by Hultén (1968) and Hulten and Fries (1986). The maps are based on the azimuthal equal-area projection from the polar aspect, and are limited to the Tropic of Cancer. On equal-area projections, the areal scale is constant from point to point. The quantitative data on plant distributions have been obtained subdividing the maps with a network of squares. The squares were used in the analysis as Operational Geographic Units, i.e. geographic units chosen for analysis (OGUs, Crovello 1981). The size of the OGUs is based on a compromise between degree of detail and computing load; each OGU covers an area of ca. 100,000 km^2. The data were organized into a taxa by OGU matrix, which was submitted to a classification program, to reveal groups of taxa (at specific or infraspecific level) with similar distribution (phytogeographic elements). The graphic representation of the phytogeographic affinities of the vegetation types and of the phytogeographic elements is based on the overlapping of the ranges of a given set of species, i.e. on the percent of species present in each OGU. For such maps (e.g. Fig. 4) we propose the term "Chorogram".

The relations between phytogeographic elements and the variation of ecological factors were studied on the basis of the results of the ordination of vegetation types obtained in step B. The frequencies of each phytogeographic element within the vegetation types have been plotted against the sequence of the vegetation types obtained by ordination. Since this ordination was previously interpreted on the basis of ecological data, this procedure permits one to relate phytogeographic with ecological data.

Data analysis was carried out mainly using the program package of Wildi and Orlóci (1980). It includes numerical classification and different types of ordination techniques, mainly Principal Component Analysis (PCA) and Analysis of Concentration (AOC). The clustering and ordination algorithms will be specified at due place in the text.

Taxonomic nomenclature mainly follows Hultén (1968).

4. RESULTS: ANALYSIS OF VEGETATION

Numerical classification of the relevés was carried out on the six phytosociological tables of species and relevés by which we subdivided our data set:

(1) *Picea mariana* muskegs (PM), Table 3;
(2) *Picea glauca* forests (PG), Table 5;
(3) *Pinus contorta* woods (PC) Table 7;
(4) Broadleaved tree stands (BL) Table 9;
(5) *Salix* thickets (S) Table 13;
(6) Grasslands (G), Table 15.

The clustering algorithm is the Average Linkage Clustering (Anderberg 1973), with the correlation

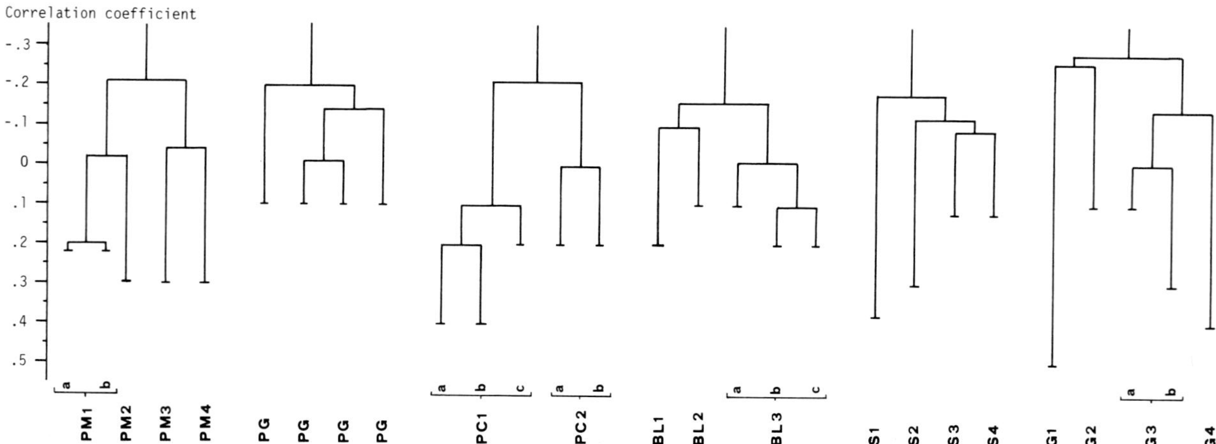

Fig. 2. Dendrograms of the relevé groups. The six dendrograms were obtained by numerical classification of the following data sets: PM, Table 3; PG, Table 5: PC, Table 7; BL, Table 9; S, Table 13; G, Table 15. For further explanations see text.

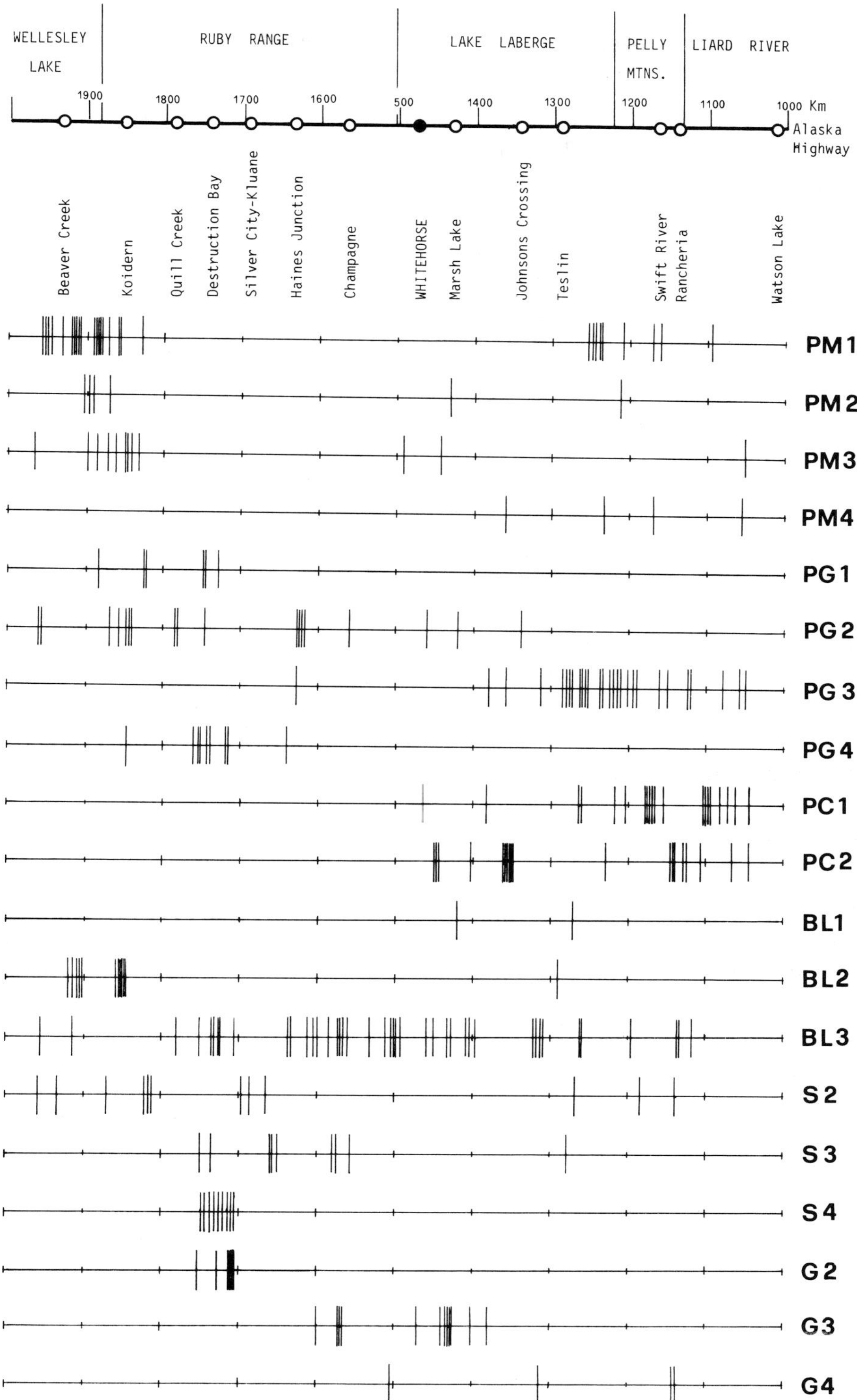

Fig. 3. Location of the relevé groups obtained by classification (see Fig. 2) along the Alaska Highway.

Table 3. Phytosociological table of the *Picea mariana* muskegs. The releves are arranged according to the results of numerical classification (see Fig. 2).

RELEVÉ No.	1	2	3	4	5	6	7	8	9	10	11	12	13	14	15	16	17	18	19	20	21	22	23	24	25	26	27	28	29	30	31	32	33	34	35	36	37	38	39	40	41	42	43	44	45	46	47	48	49	50	51	52	53	54	55	56	57	1a	1b	2	3	4
	PM1 a																							PM1 b									PM2								PM3													PM4								
Sphagnum warnstorfii	7	9	7	5	3	3	7	2	5	3	5	5	2	7	3	3	7	3		7	2	2	3	5	3	3	3	5	7	3	7	2																2										5	5		1	
Oxycoccus microcarpus	3	3	3	3			3		5	3	3	3		3	5	5	3			3		3			3	3	3	3		3	3			3																								4	4	1		
Sphagnum fuscum	5							2							2	5	3		3	3			2							3																												2	1			
Sphagnum magellanicum					2	3	3			2				2	3							2																							3													2			1	
Drepanocladus revolvens							2													2			2					2	2	3															2													1	2		1	
Stereocaulon tomentosum			2	2					2					2										2		2																	2															1	2		1	
Ledum palustre	5	3	3	5	3	3	3	3	3	3	5			5	2	2		2			3	3	5	3	3				3	3	5	3	2		3	3	3	5	3	3																		4	4	5		
Rubus chamaemorus	3	3	2	2	2		2	2	2	2	2	2	2	2											2			2	2	3	2							2	2	2		3																3	3	2	1	
Cetraria nivalis				2	2								2	2										2			2	2	2	2		2		2					2	2																		1	4	2		
Cetraria cucullata	2	3												2				2						2	2		2	2		2			2		2				2	2																		1	3	3		
Andromeda polifolia															2					2					2	2	2	2	2	3			2				3		2																			1	4	2		
Dryas integrifolia																								3	3	2	3	2		2		2						2	3				2																4	2	1	
Chamaedaphne calyculata	3		2				2											2							2				3	2	2					2				2	2																	1	3	2	1	
Equisetum arvense				2																	2					2	2	2			2	2	2					2		2																		1	3	2		
Vaccinium microphyllum			3											2				3							2				3	2								3																				1	2	1		
Carex dioica																3				2	2	2	2				2							2	2																							2	1	2		
Picea mariana (pl.)			5	5	5	5	2	2	5	5	5	7	2								2			3	3	3	3		5	3	5	3				3	3	3	2	5	2	5	2	2			2	3	2									3	5	4	3	
Potentilla fruticosa			2	2		2	2	2		2	2					3		2		2		2	2	2	2	2	2	2	2	2	2	2			2	2	2	3	2	3	2	3	2		2	3	2	2	2									3	5	4	4	
Eriophorum vaginatum	7	5	5	5	3	2	5	7	2	5	3	3		2	2			3						5	5		2	2	5	3	5	2	2	3		2	3	2		3	2	2	3															4	5	4	2	
Senecio lugens			2	2		2						2												2	2	2	2		2	2		2				2	2	2	2		2	2	2		2		2	3	3									1	4	3	3	
Rhododendron lapponicum			3	3		3	3	3									7								3	2	3		3	2		3							3	2		3	2															2	4	2	1	
Carex aquatilis			2								2					3	5		2	2			3					2							3			2			2	2	2															2	1	2	2	
Hedysarum americanum						2																		2	2	2	2	2				2				2		2									2	2	2	2								1	4	2	2	
Festuca altaica						2			2	2	2	2									2	2	2													2			2												2							2		2	1	
Carex capillaris		5														2	2	2									2			2			3							3			2				2	2					2					1	2	2	2	
Petasites sagittatus				2	2					2	2	2		2														2	2		2					2							2		2													2	2	1	1	
Mertensia paniculata																			2	2	2		2		2															2				2			2			2	2	2	2					1	1	1	3	
Tofieldia pusilla																							2	2	2	2	2			2			2	2					2				2						2									1	3	2	1	
Tomenthypnum nitens																								2									2	3							3							3	3										1	1	2	
Pinus contorta															2							2																																2	3	3	5	1				5
Cornus canadensis																			2																																	2			2	2	3	1			1	4
Cladina stellaris																							2																		2														2	2	2	1			1	4
Linnaea borealis																																														2								2	2		3				1	4
Lycopodium complanatum																																									2														2		2				1	3
Cladonia ecmocyna																																																								2						2
Picea mariana	5	5	5	5	5	5			3	5	5	3	3	5	3	5	2	2	7	5	5	5	5	5	5	2	5	3	2	5	5	5	5	3	3	5	5	5	3	3	5	5	3	3	5	2	3	3	3	3	3	7	7	3	5	3	3	5	5	5	5	5
Aulacomnium palustre		3	7	7	7	7	7	3	7	7	7		5	3	7	5	5	5	7	5	5	5	7	7	3	7	5	7	5	7	7	7	3		7	3	7	7	7	3	3	7	7	2		7	2	3	3	5	3	2	3		2	2		5	5	5	5	4
Hylocomium splendens			3	2	2	2	2		2	3	2	5	7	2				3			7		2	3	2	3			3	2	3	2	5	7	2	5	3	3	2	3	7	3	3	8	2	2	5	5	5	5	5	7	7	5	5	3		4	4	5	5	4
Peltigera aphtosa	2		3	3	2	2	2		2	3	2		2		2	2	2	2	3	2	3	2	2		2		2	2	3			2		2	2			2	2	2	3	2	2	3	3		2					2	2	3	3	3	2	5	3	4	4	5
Vaccinium vitis - idaea	2	2	2	2	2	3	3	2	2	2	3	3	3	2	2									3	2							2				3	3	2	3	3	5	2	2	3	2	2	3	3		3	3	5	3	3		5		4	2	4	5	3
Equisetum scirpoides			2	2	2	3			2	2	3		2					2	3	2		2	2	3	3	3		3	3	2	2	3			2	2	2	2	3	3	3	2			3	2	2	2		2	2						3	3	5	4	4	2
Empetrum nigrum			3	2	2	3			3	2	3	2	3	5	3		1		2	3	2	3	2				2		2	2	3			2		2	2			2	2	3	3		2			2		3			2	3	2		2	4	3	3	3	4
Vaccinium uliginosum			5	3	5		2	2	5	3	3	2	2	3	3	2	3	3	2	3	3		2	3	3	3	3	2	3		3	3	2		3	3	3	3	3	2	2	3	3	2				3	3	3	2	2	2		2			5	5	5	4	2
Arctostaphylos rubra			2	3	3	3		3						3	2		3	2	3	2	2	2	2	3	3		3		3	3	2	3	2	3	2	2	3	2	3	2	2	3	3		3	3	5	2	3	3			5			2		4	4	5	4	2
Salix glauca			3	2	2	2	3	5	2	2	3			3	2			2	2	2	3	2		2	2	2	2	3	2	2	3		2		3	2	2	3		2	2	2	2	2	2	2					2		2	2				4	5	4	4	2
Betula glandulosa			2	2	2		5	7	2		3		3	5	5	2	2	5	2	3	5	5	7		3			3	3	3	3		2	3	3		3			2	2		2	2	2	2			2					2	2			4	3	4	3	3
Cladonia amaurocraea			2	2				2	2	2				2					2	2	2	2	2	2	2			2		2	2	2		2		2		2			2	2	2	2		2	2		2	2					2			3	4	2	4	2
Pedicularis labradorica		2	2	2	2	2	2		2	2	2	2	2	2	2	2	2	2		2		2	2		2	2	2	2	2	2	2	2	2		2	2	2	2	2	2		2			2	2			2						2		2	5	5	5	2	3
Thamnolia subuliformis						2		2		2	2		2		2			2				2	2	2	2	2	2	2	2	2	2	2			2				2	2	2		2		2		2	2						2			2	2	5	2	2	3
Dicranum bergeri		2				2		2	2		2	3	2	2	2			2	2	2	2	2		3					2	3		2		3		2	3			2	2						2	2	2			2		2		2		4	3	3	2	3
Salix myrtyllifolia								2			2					5			5		3	3	2		2		2	2			2	2		3	5		2	3			2				5	3								2	2			2	3	3	2	3
Ledum groenlandicum				2			3					5	5				2	3	5	5	3	3				3		3						3	3	3						3	3	2	3	2	3	3	3	3	5	5	5	2	2	5	3	3	2	2	5	5
Rhytidium rugosum					3						2						3							2	2	2		2		3						3	3	2		3	3	2	3				3	3	3	3	3					3		1	3	3	4	2
Abietinella abietina										2		3					2												3	2	2				3	5	3	2	5	5		2							3	5	3	3	3	3		5		1	2	4	3	3

Picea glauca																	2			2	2		2				2					2		3	2			2	3				3		3	3	3	2	2	3	3			3	2	1	2	1	2	3	4	5
Cladina mitis			2						2	2											2	2	2	2			2		2	2						2	2				2	2					2	2	2			2	2	2				2	3	2	3	2
Cladina rangiferina		3								2			2				2	3												2	2					2			2		2	2	2	2			2	2	2							2		2	2	2	3	2
Cladina arbuscula			2	2					2	2	2		2	2					2	2											2		2				2	2																	2	2	2	2	1	2		4
Rosa acicularis																																			2	2					2			2		2	2			2	2		2	2	2		2			2	3	4
Pleurozium schreberi												3	3		2						3	2																													3	3	3					1	2		1	1
Cladonia gracilis		3												3	3						2		2																																2	2		2				3
Geocaulon lividum																				2											2										2		2							2	2		2	2		2		1	2		2	3
Carex concinna																						2		2		2														2									2	2		2	2					1	2	1	2	
Salix reticulata																							2		2					2			3	2		2	2		2																			1	2	4		
Carex scirpoidea																							2					2		2					2			2		3		2	2															1	2	2	1	
Alnus crispa														2											1	3												3														3				2		1	2	1	1	2
Salix planifolia																2	2											2		2										2												2			2			1	2	1	1	2
Salix alaxensis								3								2					2							3	2																							2						1	2		1	
Carex membranacea				2		2		2												2																					2	2																1			1	
Cetraria islandica	2	2												2																		2					2		2																			1	1	2		
Salix rostrata							2								5						2		2														2																					1		1		
Equisetum sylvaticum									2	2	2	2																											3																			1		1		
Ditrichum flexicaule																		2																	2								2					2				2							1	1	2	
Drepanocladus uncinatus																		3											2				3							2																			1	2	1	
Pyrola asarifolia							2																													2				2											2						2	1		2	1	2
Pyrola secunda													2						2																																	2	2				2	1			1	2
Rubus arcticus																							2											2	2											2												1		2	1	
Stellaria longipes					2																															2											2						2					1		1	1	
Dactylina arctica	5	1																															2																									1		1		
Carex lugens	3	2																															2																									1		1		

Table 4. Ecological data relative to the relevés of *Picea mariana* muskegs reported in Table 3. Soil type abbreviations are in Table 1. Other abbreviations: GS gravel; SA, sand; SI, silt; CL, clay. Org: thickness of organic horizon. Erosion potential: high (hi), medium (me), low. Drainage classes: wet, impeded (imp), moderate (mod), excessive (ex).

Releve No.	1	2	3	4	5	6	7	8	9	10	11	12	13	14	15	16	17	18	19	20	21	22	23	24	25	26	27	28	29
Exposure	-	-	-	NW	SW	-	-	-	-	-	-	-	-	-	S	SSW	-	-	-	E	NE	NNE	-	-	-	NE	NW	-	NNE
Slope	-	-	-	5	5	-	-	-	-	-	-	-	-	-	5	5	-	-	-	5	5	5	-	-	-	5	5	-	5
Total cover	100	100	100	100	100	100	100	100	100	100	100	100	100	100	100	100	100	100	100	100	100	100	100	100	100	100	100	100	100
Soil type	. .	. .	GL. SC	TFI OC	GL. SC	GL. SC	FI. OC	TME OC	GL. SC	GL. SC	R. TC	GL. SC	TFI OC	. .	TME H	T. H	TY. F	. .	O. HG	T. H	O. HG	T. H	R. HG	GL. SC	GL. SC	GL. SC	TFI OC	GL. SC	FI. OC
GS %	.	.	-	30	-	-	-	-	-	15	10	-	-	.	80	-	-	.	75	80	10	80	80	10	-	-	30	20	-
SA %	.	.	-	30	-	-	-	20	2	10	50	-	-	.	14	15	-	.	23	20	50	30	18	20	-	20	30	30	-
SI %	.	.	50	30	50	85		30	48	30	40	40	-	.	5	35	-	.	2	-	39	-	1	30	95	45	30	45	-
CL %	.	.	50	40	50	15	-	-	50	45	-	60	-	.	1	50	-	.	-	-	1	-	1	40	5	35	10	5	-
Org. (cm)	.	.	23	45	5	30	100	50	20	13	15	15	100	.	45	60	100	.	18	100	30	40	30	35	30	15	45	30	100
pH	.	.	5.5	6.3	6.4	6.3	5.8	6.2	5.6	5.8	6.7	5.2	4.6	.	5.8	5.1	6.0	.	5.9	6.0	6.7	5.8	5.5	.	4.5	5.6	6.3	6.5	.
Erosion pot.	.	.	med	hi	med	med	low	med	med	med	med	med	low	.	low	med	low	.	low	low	med	low	low	med	med	med	hi	med	low
Drainage	.	.	imp	imp	imp	imp	imp	imp	imp	imp	imp	imp	imp	.	imp	wet	wet	.	wet	wet	imp	wet	imp	wet	wet	imp	wet	imp	imp
Depth of ice June (cm)	.	.	22	10	10	9	18	10	30	25	20	20	18	.	.	.	1	.	18	18	30	20	6	10	9	12	1	18	7
Depth of ice Sept.(cm)	.	.	45	36	38	50	36	50	60	66	65	58	30	.	.	.	.	.	-	-	-	-	-	30	40	60	36	60	38

Releve No.	30	31	32	33	34	35	36	37	38	39	40	41	42	43	44	45	46	47	48	49	50	51	52	53	54	55	56	57
Exposure	-	N	-	W	SE	SSE	-	-	SW	E	NE	-	SSW	SW	SW	WSW	-	S	-	-	SW	SSW	N	SSW	N	SSE	NW	S
Slope	-	5	-	20	5	5	-	-	5	5	5	-	5	8	5	5	-	5	-	-	5	20	10	5	7	5	10	45
Total cover	100	100	100	100	100	100	100	100	100	100	100	100	100	100	100	100	100	100	100	100	100	100	100	100	100	80	100	80
Soil type	GL. SC	GL. TC	GL. SC	. .	GL. CUR	O. HG	GL. SC	TFI OC	GL. SC	. .	GL. SC	. .	GL. SC	HU. OC	O. R	GL. SC	CU HR	FI. OC	THU OC	THU OC	GL. SC	TME OC	GL. DYB	GL. SC	GL. DYB	E. DYB	O. DYB	GL. DYB
GS %	20	-	-	.	3	5	-	-	60	.	-	.	50	-	50	1	15	-	17	17	-	-	50	-	20	30	22	4
SA %	30	50	-	.	50	10	-	-	60	.	20	.	-	90	48	29	50	-	40	40	100	10	30	50	60	70	49	90
SI %	45	50	85	.	22	30	50	60	20	.	60	.	42	10	2	10	33	-	8	8	-	40	17	50	20	-	27	5
CL %	5	-	15	.	15	55	50	40	-	.	20	.	8	-	-	25	2	-	5	5	-	50	3	-	-	-	2	1
Org. (cm)	30	25	24	.	25	30	30	45	25	.	24	.	15	45	15	25	15	100	40	40	18	45	30	35	10	7	6	5
pH	6.0	4.7	6.2	.	7.7	7.5	6.5	5.4	5.8	.	6.4	.	6.4	6.6	5.2	6.7	7.3	5.9	6.1	6.3	6.3	5.4	6.0	6.7	5.5	6.6	5.0	6.0
Erosion pot.	med	med	med	.	med	med	med	med	hi	.	med	.	med	med	med	med	low	low	med	med	med	hi	med	med	med	low	low	med
Drainage	imp	imp	imp	.	wet	imp	imp	imp	imp	.	imp	.	imp	imp	mod	imp	mod	imp	imp	imp	imp	mod	mod	imp	mod	ex	mod	mod
Depth of ice June (cm)	18	13	8	.	2	13	15	10	10	.	20	.	10	15	10	13	-	24	30	30	18	10	37	10	10	-	23	-
Depth of ice Sept.(cm)	60	45	60	.	10	-	25	60	50	.	60	.	60	45	-	60	-	55	120	60	45	50	-	50	25	-	-	-

coefficient used as the resemblance measure, with binary data.

For the purpose of classification of the vegetation into community types we considered as pertaining to the same cluster all the relevés fusing together below a resemblance threshold of 0.0. Further subgroups have been distinguished below this threshold when this appeared to be justified by the data structure, and are treated as variants within the main relevé groups.

The results of the classifications are shown in Figure 2: a total of 21 main clusters of relevés has been obtained. The location of the relevés of each main cluster along the Alaska Highway is shown in Figure 3. This figure does not report the distribution of clusters S1 and G1. The former includes relevés taken in northern Yukon, the latter refers to a community of weeds along roadsides which has been the object of a much more detailed study by Lausi and Nimis (1985b).

In the following, each cluster will be treated as a community type, and will be briefly discussed. The phytogeographic affinities of the community types will be shown by means of chorograms. A synthesis of the ecological relations of the community types within each data set has been obtained on the basis of ordination methods applied to the six matrices of species and relevés. A general synthesis of the floristic and ecological data relative to the 21 community types is provided in the next section.

In this paper we have avoided a formal description of associations and other syntaxa according to the principles of the phytosociological school of Braun-Blanquet. The main reason for this is our conviction that the establishment of new syntaxa is a fruitful endeavor only if it is based on a very large number of relevés, representative of a comparatively broad area. Otherwise, as has all too frequently happened, the description of new syntaxa becomes a serious source of nomenclatural confusion. Although our survey area encompasses the whole of southern Yukon, we think that it is not large enough to represent the main trends of floristic variation in the whole of northwestern North America. The data which are available from interior Alaska and the Northwest Territories are still too scarce to allow phytosociological generalizations. For this reason, our community types will be designated only with a brief English name, referring to the dominant species.

4.1. Picea mariana muskegs

4.1.1. Classification

In the dendrogram of the relevés (Fig. 2), based on the data of Table 3, four main groups are formed (PM1, PM2, PM3, PM4). The three last clusters have the same degree of within cluster homogeneity. PM1 may be subdivided further into two subclusters at correlation coefficient 0.2. They are PM1a (Rel. 1–23) and PM1b (Rel. 24–32).

The relevés of PM1 and PM2 represent a formation with pronounced ground structuring in hollows and hummocks and sparse trees at the top of them. The relevés of PM3 and PM4 are transitional towards the closed boreal forest dominated by *Picea glauca*. The ecological data of each relevé are in Table 4. The chorograms of the relevé groups are shown in Figure 4.

PM1 *Picea mariana-Sphagnum* community

This is an open-canopied formation dominated by dwarf *Picea mariana* trees (muskeg). The most frequent life forms are thallochamaephytes (37%, mostly mosses) and chamaephytes (27%). The high incidence of mosses agrees with the dependence of this vegetation type on rain water. According to Walter (1979), muskegs can be considered as ombrogenous bogs with a sparse tree cover due to a strongly continental climate that slows down the growth of mosses.

Frozen ground persists throughout the year, so that Cryosols are the most frequent soil type. Regosolic Static Cryosols occur in 66% of the plots. The organic horizon is thick (average 42 cm), and peat formation is frequent. Soil texture has a high content of silt and clay. Fine-textured soils and permafrost cause impeded drainage in all plots. An open water table was

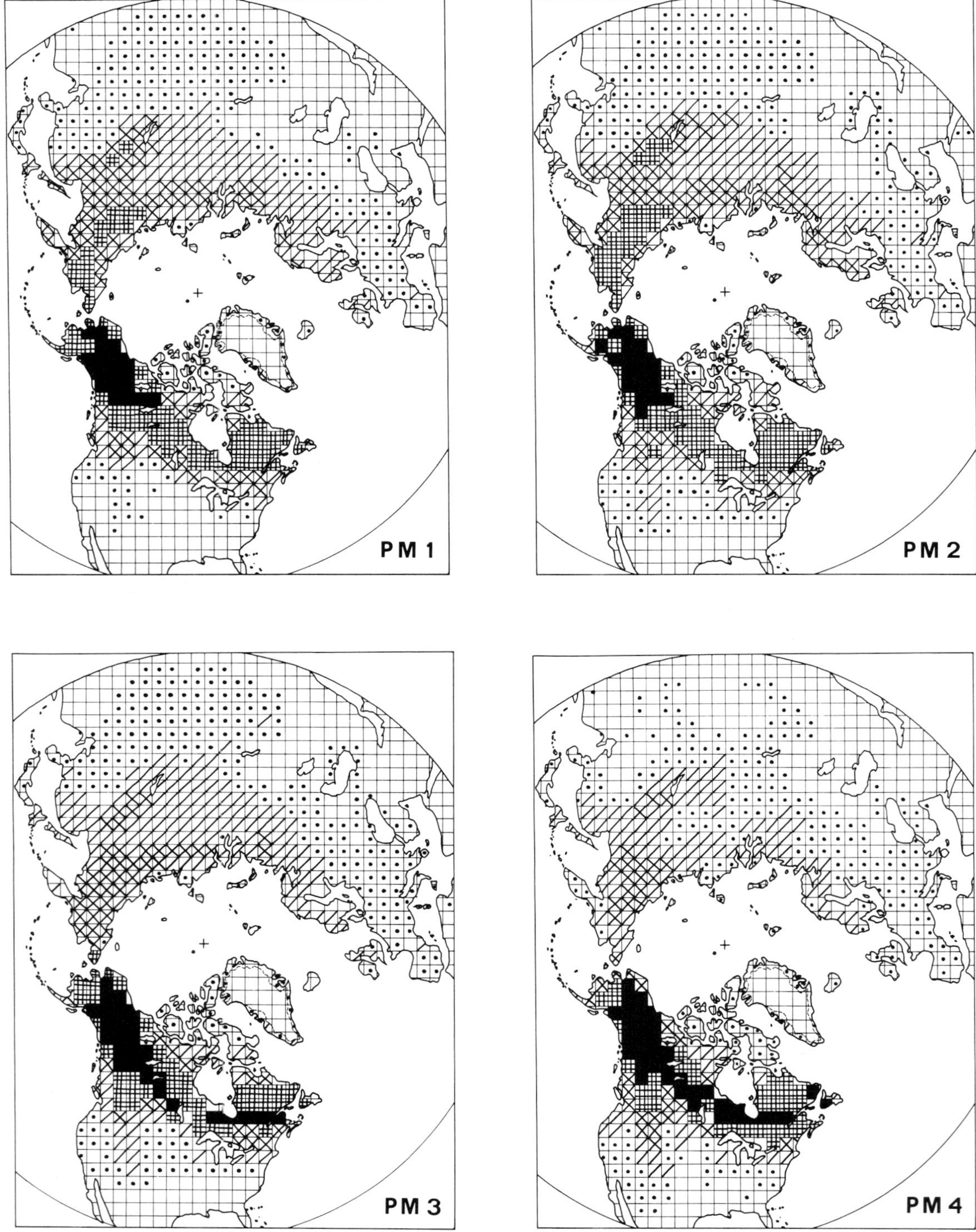

Fig. 4. Chorograms of the four relevé groups obtained by classification of the data in Table 3 (*Picea mariana* muskegs). Different shadings refer to the percent of species occurring in the OGUs, on a 5-class scale with intervals of 20%.

observed in 33% of the plots. Soil pH ranges from 4.5 to 6.7 with an average of 5.8. The relatively high pH values are a probable consequence of scarce rainfall and loess deposition.

The ground is strongly patterned, with alternating hummocks and hollows. Water conditions range from extreme aridity on the top of the hummocks to water stagnation in the hollows. The hummocks are occupied by a lichen synusia dominated by *Cladonia amaurocraea* and *Thamnolia subuliformis* (Nimis 1981), the hollows host a moss community dominated by *Drepanocladus revolvens*. Therefore, this type is a highly structured vegetation mosaic composed of at least three synusiae: (a) Lichen synusia, top of hummocks; (b) *Drepanocladus* synusia, in the hollows, submerged; (c) *Sphagnum* synusia, areas in between.

The releves (Table 3) may be subdivided into two subgroups: PM1a and PM1b. Those of the latter subgroup were always on slopes on calcareous substrates. PM1b is an edaphic vicariant replacing PM1a on limestone soils.

In the Yukon this community is montane–Subalpine (Subarctic). It is most frequent along the northwestern section of the Alaska Highway, from Koidern to the Alaskan border, and in the Pelly Mountains section between Rancheria and Teslin Lake (Fig. 3). PM1b is restricted to the northwestern section, where limestone outcrops occur. Along other portions of the Highway, scattered occurrences of relevés included in PM1a were observed in depressions with permafrost. Relevés 1 and 2 were taken along the Dempster Highway, near Ogilvie. They appear as impoverished stands of this community, whose distribution seems to extend all the way to northern Yukon, with little floristic variation.

PM2 *Picea mariana–Ledum palustre* community

This community differs from PM1 by the absence of *Oxycoccus microcarpus* and *Sphagnum* species. Soil type and organic matter content are similar to those of PM1. Soil texture, however, has a higher silt and clay content, which may be due to local accumulation of loess. Although soil drainage is always impeded, an open water table was observed only in 17% of the plots. The pH range is 5.8–7.7, with an average of 6.5. This is the main ecological difference with respect to PM1, and it agrees with the scarcity of acidophytic species and the higher frequency of neutrophytic species such as *Thuidium abietinum* and *Rhytidium rugosum*.

This community replaces PM1 on eutrophic Cryosols. The two communities have a similar distribution. PM2, however, is much less frequent in the survey area (Fig. 3).

PM3 *Picea mariana–Ledum groenlandicum* community

This is an open-canopied mixed coniferous forest, with *Picea mariana* as the dominant tree and *Picea glauca* as occasional codominant.

Chamaephytes and thallochamaephytes are the most frequent life forms (21 and 27% resp.), although less frequent than in PM1 and PM2. The main phytogeographic elements are the circumboreal and the boreal North American (see Fig. 4).

This type lacks true differential species, but it is well defined by the absence or low frequency of *Oxycoccus microcarpus, Sphagnum* species, *Rubus chamaemorus, Dryas integrifolia, Chamaedaphne calyculata* and *Cetraria* species. *Ledum palustre* is replaced here by *Ledum groenlandicum*.

Frozen ground is present in 80% of the plots in June and in 70% of the plots in September. The active layer is thicker than in the previous communities, both in June and in September, with an average of 22 and 45 cm respectively. True Cryosols occur in 65% of the plots. The organic horizon is thick (average 35 cm), with frequent peat formation. The soils are mostly sandy or sandy-loamy. The amounts of clay are considerably lower than in the previous communities; 75% of the plots were on gentle slopes, with an inclination range from 4 to 24°. Slope and soil texture improve drainage conditions, impeded drainage occurring only in 66% of the plots. Stagnant water is always absent. The pH ranges from 5.4 to 6.7, with an average of 6.2.

PM3 is a vicariant of PM1 on better drained

soils, and represents a transition towards closed-canopied forests dominated by *Picea glauca*. It frequently occurs on slopes (not on calcareous talus and scree slopes, which are colonized by PM1b) mainly in the northwestern section of the Alaska Highway (Fig. 3).

PM4 *Picea mariana–Pinus contorta* community

This is a closed-canopied mixed coniferous forest dominated by *Picea mariana, Picea glauca* and *Pinus contorta*. Thallochamaephytes and chamaephytes are the most frequent life forms (33 and 16% resp.) phanerophytes are more frequent, hemicryptophytes less frequent than in the previous types.

Frozen ground is sometimes present in June. Permafrost was noticed in one plot only. The soils are moderately well-drained Brunisols with high sand and low organic matter content. The average thickness of the organic horizon is of only 7 cm. All plots were on slopes, with an average inclination of 16.5°. The pH ranges from 5.0 to 6.2 (average 5.7). This community is transitional between muskegs and closed boreal forests. Traces of past fires have been observed in all plots, which explains the presence of *Pinus contorta*. This type is restricted to the south-eastern section of the Alaska Highway, within the range of *Pinus contorta* in the Yukon (Fig. 3).

4.1.2. Ordinations

The results of PCA on the binary data of Table 3 are in Figure 5. The first Principal Component has been rotated to enhance interpretation. The clusters obtained by classification are still recognizable in the ordination. In Figure 6 the centroid of each group has been projected on the rotated first axis. The following variables have been plotted against the axis (averages for each relevé group): thickness of the organic horizon, sand content, active layer thickness in June, frequency of stands occurring on flat ground, and average slope. In Figure 6 the

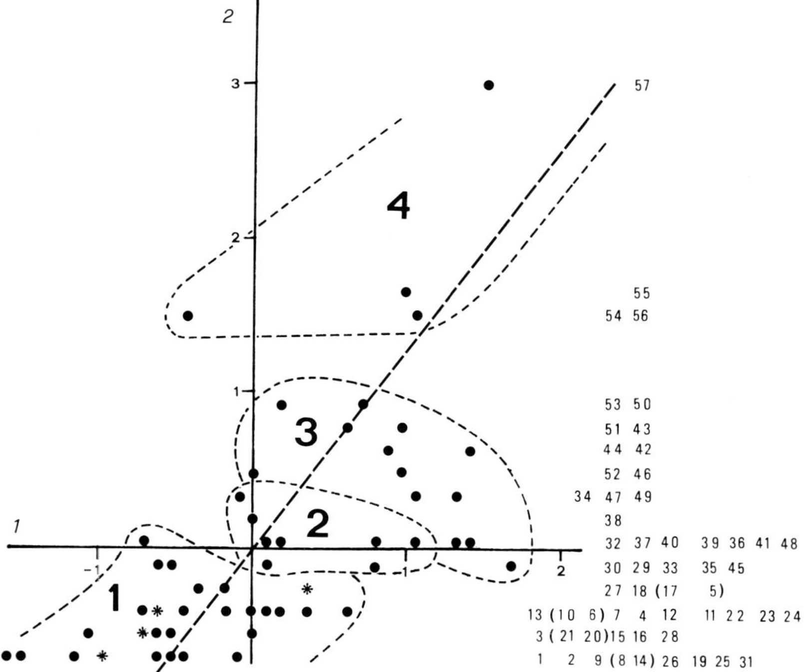

Fig. 5. Ordination (by PCA) of the relevés, based on the data of Table 3 (*Picea mariana* muskegs), Principal Components I and II. Large type numbers locate the centroids of the relevé groups obtained by classification, small numbers refer to the releve numbers in Table 3.

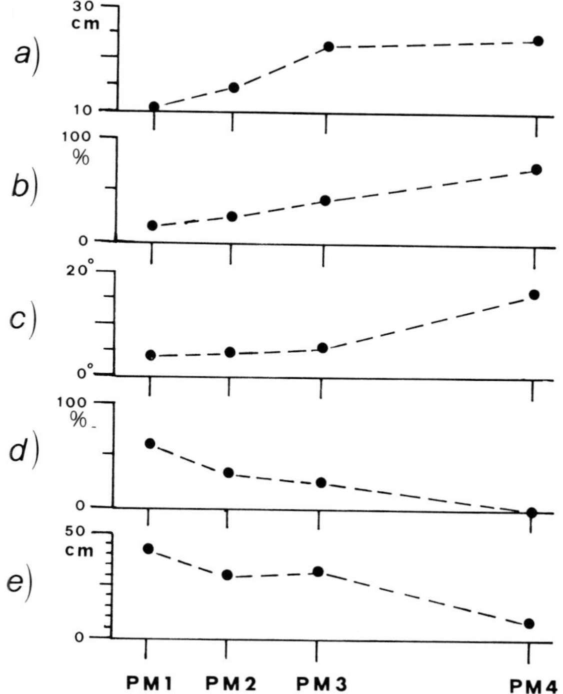

Fig. 6. Trends of ecological variation within *Picea mariana* muskegs. The *x*-axis reports the scores of the centroids of the relevé groups in the ordination of Figure 5; on the *y*-axes the following variables are plotted: (a) active layer thickness, (b) sand content, (c) average slope, (d) percent occurrence of stands located on flat ground, (e) thickness of the organic horizon.

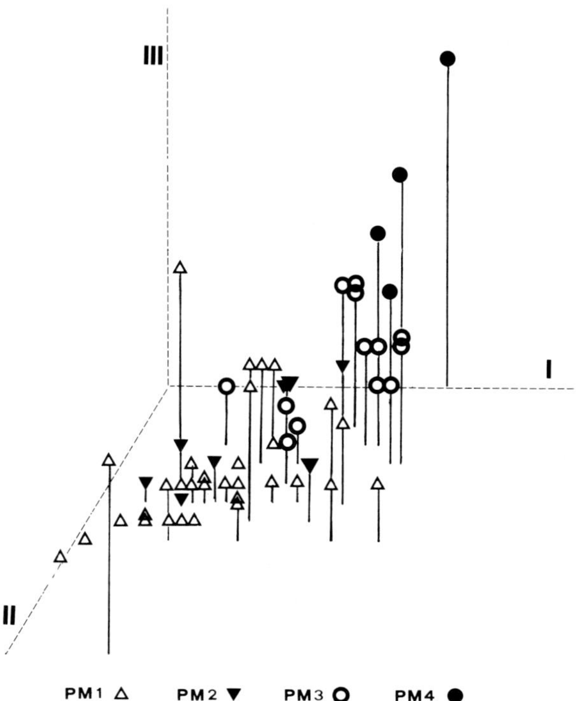

Fig. 7. Ordination (by PCA) of the relevés, based on the data of Table 3 (*Picea mariana* muskegs), Principal Components I, II and III. Symbols refer to relevé groups as in Figure 2.

following trends are recognizable from PM1 to PM4: increase of active layer thickness, increase of sand content in the soils, and increase of the average slope. The thickness of the organic horizon decreases from PM1 to PM4, as does the frequency of stands located on flat ground. All of these factors directly influence drainage conditions, the former by increasing, the latter by decreasing water percolation through the soil. The rotated first axis reflects a gradient in water availability. Along this gradient a clear trend of phytogeographic variation is evident (Fig. 4), which results in a progressive decrease of circumboreal species that are replaced by species with boreal North American distribution. The arrangement of the relevé groups on the third Principal Component (Fig. 7) maintains the sequence PM1–PM4.

The percentages of soil types and soil orders in the four vegetation types are in Table 1. Cryosols decrease from PM1 to PM3, and are absent in PM4. The most frequent soil type is Regosolic Static Cryosol. Brunisols are mostly present in PM4, with Orthic Dystric Brunisols as the most frequent type. Gleyed Dystric Brunisols are also present in PM3: this fact confirms the transitional character of this vegetation type towards the true closed boreal forest, which in the survey area normally occurs on Brunisols. The frequency of Regosols increases from PM1 to PM3. They are all of the Gleyed Cumulic Humic subgroup, a fact that is understandable considering the poor drainage conditions prevailing in these communities. True Gleysols occur only in PM1, where stagnant water in depressions is most frequent.

Permafrost conditions are summarized in Table 2. The frequency of frozen ground decreases from PM1 to PM4, both in June and in September. The thickness of the active layer in

June increases from PM1 to PM4, whereas in September the values tend to be constant between 40 and 55 cm. Plots with frozen ground are most frequent in June than in September. The values relative to the month of September correspond to the occurrence of permafrost, since temperatures under freezing are common in October, and prevent further ice melting. The values relative to ground ice in June are the most important for the vegetation, since they refer to conditions falling in the middle of the vegetative season. The higher frequency of *Picea glauca* and *Pinus contorta* in PM3 and PM4 depends on the relatively high thickness of the active layer in these two types, where true permafrost is less frequent.

The thin active layer in PM1 (only 11 cm) is chiefly due to the insulating effect of the *Sphagnum* carpet, that prevents ice melting and causes a continuous rising of the permafrost table.

The correlations between the four vegetation types and soil conditions have been investigated by submitting the data sets of Table 1 and Table 2, separately, to AOC analyses. The strongest correlation is given in the analysis of drainage data (Table 2). Its scatter diagram (Fig. 8) shows a clear gradient on the first canonical variate, with a good correlation between the four vegetation types and the four drainage classes. Water availability seems to be the most important factor explaining floristic variation in the *Picea mariana* stands. This conclusion is supported by the results of both PCA and AOC.

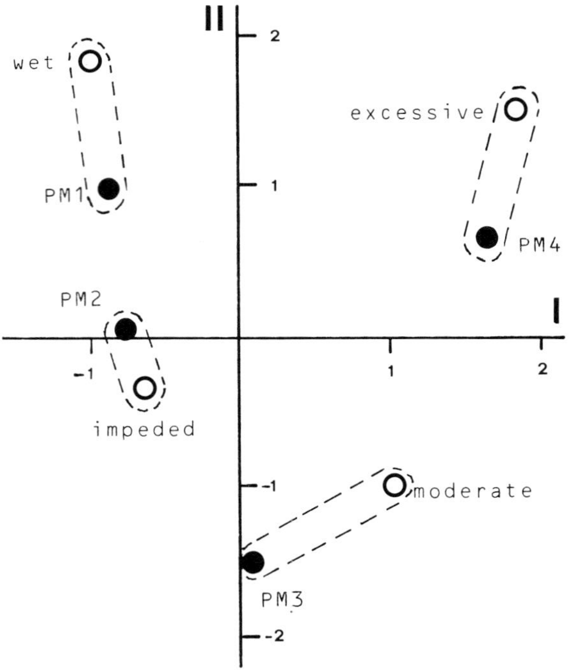

Fig. 8. Scatter diagram obtained by AOC on the drainage data of Table 2, for the four relevé groups obtained by classification of Table 3 (*Picea mariana* muskegs, see Fig. 2).

4.1.3. Discussion

In the survey area *Picea mariana* stands are confined to poorly drained soils; poor drainage is mostly due to ground ice. Brown (1967, 1973), divided permafrost in the Yukon into the continuous zone in the north and the discontinuous zone in the south. The latter is subdivided into the widespread subzone, where permafrost predominates, and the scattered subzone, where permafrost is rare. Most of the Alaska Highway lies within the scattered subzone (Oswald and Senyk 1977). Permafrost is most common at high elevation, on flat ground, and in depressions or on gentle north exposed slopes at lower elevation. The distribution of *Picea mariana* muskegs (above all of PM1 and PM2), corresponds with the areas where permafrost is most common, i.e. the northwestern section of the Highway and the Pelly Mountains section.

Picea mariana stands are old, mature communities. Viereck (1970) described a successional series on a flood plain near Fairbanks, Alaska. The succession starts with *Salix* thickets, followed by a *Populus balsamifera* wood and by a *Picea glauca* forest. As the white spruce stand develops, a thick moss layer forms on the forest floor. Soil temperatures are lower beneath this thick insulating layer, and ice starts forming in the soil, until drainage becomes impeded, and a *Picea mariana–Sphagnum* community is formed. The results of Viereck (1970) show that in the boreal zone permafrost may be due to vegetation dynamics, and that *Picea mariana* stands may represent the final stage of a successional series. According to Zoltai and Tarnocai (1974), *Picea mariana* stands in the Mackenzie River Valley may be up to 250 years old.

This accords with the evidence presented by Drury (1956) and Benninghoff (1952), and corresponds to Sjoer's claim (1963) that paludification is a general process in boreal regions. The development of the characteristic hummocks by cryoturbational phenomena has been described by Zoltai and Tarnocai (1974) and Zoltai and Pettapiece (1973). Although signs of recent fires were observed by us in *Picea mariana* stands, it seems that burning has little effect on their composition. Black and Bliss (1978) showed that most of the phanerogamic species in the muskegs can sprout vegetatively after burning, so that the major changes regard their size and relative abundance. Cryptogams, on the contrary, shows a well-defined succession of species (see also Maikawa and Kershaw 1976; Scotter 1964). In old stands, *Picea mariana* chiefly reproduces by layering, but burning must occur for stand regeneration by seeds (Black and Bliss 1980). The semiserotinous cones of *Picea mariana* (Vincent 1965) and the small seed size require burning to open the cones and to remove accumulated duff and interfering vegetation for effective establishment (Black and Bliss 1980). Considering the relative stability of *Picea mariana* stands with respect to fire, the process of permafrost formation in the boreal zone, and the old age of *Picea mariana* in the survey area, we think that *Picea mariana* stands should be considered as the climax vegetation of the survey area where permafrost may develop under a moss carpet on flat ground. These areas correspond well with the Wellesley Lake and Pelly Mountains ecoregions of Oswald and Senyk (1977).

Both *Picea mariana* and *Picea glauca* have boreal North American, transcontinental ranges. *Picea mariana* is a low concurrence species with *Picea glauca*. It is most common on poorly drained soils or in rock outcrops (Knapp 1965; Collinwood and Brush 1947). In the Yukon, *Picea mariana* is more common on poorly drained soils (Krajina 1975).

Since our community types have a prevalence of circumboreal and boreal North American species, their main affinities are to be looked for in boreal North America and throughout the boreal zone.

Picea mariana stands on rock outcrops are most frequent in the Canadian Shield area, with open lichen woodlands differing floristically and ecologically from our communities (Ahti 1977). The composition of *Picea mariana* stands on moist soil is not constant throughout boreal North America. A vegetation resembling the *Picea mariana–Sphagnum* community (PM1) is reported by Viereck (1970) for the Chena River area in interior Alaska. The stands of eastern North America (east of Lake Superior) differ due to the presence of suboceanic species such as *Epigaea repens, Coptis groenlandica, Gaultheria procumbens, G. hispidula* etc., and due to the high frequency of *Kalmia angustifolia* and *Kalmia polifolia* (La Roi and Stringer 1976; La Roi 1967; Carleton and Maycock 1980). This resulted in the inclusion of part of North American Boreal coniferous stands in an endemic Order, *Gaultherio–Piceetalia* (Braun-Blanquet et al. 1939). The stands of the Yukon, however, due to their chiefly circumboreal affinities, belong to the class *Oxycocco–Sphagnetea* (PM1) or to the class *Vaccinio–Piceetea* (PM2, PM3, PM4), two syntaxa with a circumboreal distribution.

4.2. *Picea glauca forests*

4.2.1. Classification

The data include 75 relevés (Table 5) of coniferous stands dominated by *Picea glauca*. The ecological data of the relevés are in Table 6 and Figure 9 presents the chorograms of the relevé groups. In the dendrogram of the relevés (Fig. 2), four main groups formed at a fusion level of 0.0 (PG1, PG2, PG3, PG4). All relevé groups have the same within cluster homogeneity. PG1 includes intermediate stands between muskegs and closed boreal forest, the other groups belong to the true boreal forest formation.

Table 5. Phytosociological table of *Picea glauca* forests. The relevés are arranged according to the results of numerical classification (see Fig. 2)

	PG 1													PG 2																	PG 3																											PG 4																	
RELEVE No.	1	2	3	4	5	6	7	8	9	10	11	12	13	14	15	16	17	18	19	20	21	22	23	24	25	26	27	28	29	30	31	32	33	34	35	36	37	38	39	40	41	42	43	44	45	46	47	48	49	50	51	52	53	54	55	56	57	58	59	60	61	62	63	64	65	66	67	68	69	70	71	72	73	74	75	1	2	3	4
Aulacomnium palustre	3	3	5	5	7	5	7	2	5	5	7	3	7	8					2								2															3	2																2								2																					5	1	1	
Vaccinium uliginosum	2	3	3		3	3	3	2	3	3	3	3	3	3					2																								3																																						5	1	1	
Potentilla fruticosa	3		2	3	2	2	2	2	2	2			2	2				2								2																								2																					2													4	1	1	1
Pedicularis labradorica	2	2	2	2	2	2	2			2	2	2															2											2			2	2		2																																				4	1	1	
Rhododendron lapponicum	3	3	5	3	3	3	2	2	3	3	2		2																																																																	5			
Dryas integrifolia	3	3	5	3	3	3	5			3	2		3	3																																																																4			
Picea mariana					3		2	3					2	2					3	2																	3			2					7					3										3																				2	1	1	
Equisetum palustre	2		2			2				2	2	2	2																																																										2				3			1
Tofieldia pusilla	2	2			2	2	2						2	2																																																														3			
Carex membranacea	2		2	3	2	2	2																																																																				3			
Ditrichum flexicaule	2			2	2	2	2							2																																																														3			
Cladonia cenotea	2	2	2	2	2	2																																																																					3			
Cetraria cucullata				3	2	2	3					2																	2				2																2																										2	1	1	
Stellaria longifolia	2	2		2	2	2														2																																																							2	1		
Carex scirpoidea	2	3	3			2								2																																																													2			
Salix reticulata		2		2	2	2	2																																																																				2			
Cladina stellaris	2	2			2				2																																								2																										2		1	
Eriophorum vaginatum				5		3	5						3																																																														2			
Andromeda polifolia					2	2	3						2																																																														2			
Carex lugens				3	2	3	2																																																																				2			
Cassiope tetragona					2	5	2																																																																				2			
Juniperus nana	3	2	3																																												7					3																									2		1	
Cypripedium passerinum								2		2	2		2																																																													2			
Rhytidium rugosum	2		2		2		2	2		2		2		3	2	2	3	3	2	5	3	2	3	3	3	2	2	3	3	2	2	2																																													3	5		
Hedysarum americanum															2		2	2		2									2																																				2													2		1
Cladonia ecmocyna																												2				2						2		2	2	2	2	2	2	2	2	2	2	2	2	2	2	2	2	2	2	2	2	2	2																	1	5	
Cornus canadensis																																	2	3	2	2	3	3	3	3		3	3	3	2		2	3					3			2		2	2																			4	
Festuca altaica			2		2				2					2																		2	2	2	3	2	2	2	2	2					2	2	2	2	2	2	2												2														2	1	3	1
Cladina arbuscula	2	2		2								2																2						2	2	2	2	2	2	2	2	2	2					2								2		2		2	2		2																	2	1	3	
Ptilium crista-castrensis											2																		3					3	3	2				2	3		2	3	2	2	2	3	2									2	2	2	2	3																		1	1	3	
Nephroma arcticum																																				3		2	2	2		2					2	2	2									2																						2	
Calypso bulbosa														2																		2		2					2															2		2	2	2																	1		2	
Lycopodium annotinum																																	2				2					2	2				2	2																																	2	
Pyrola asarifolia				2																										2				2	2			2	2		2	2	2	2	2				3										3																		1	1	2	
Pinus contorta																																	2		2		2	2	2	3	3	3				2		3	3		3																												3	
Salix scouleriana															2										2							2	2	2	2												2			2	2												2			2											1	2	1
Cladonia multiformis																2																		2					2			2	2	2	2			2			2				2				2																			1	2	
Pleurozium schreberi		3														3	3				3			2		7			7			2	5	7	5	5	5	3	3	3	5	7	2	3		5	5		3			3	3	2			3	3	5	3																	1	2	4	
Dicranum fuscescens																2						2					3	3	2	2	2	3	2	2			2	2		2	2	2				2	2	2	2				2	2	2	2	3	2		2						2											3	4	1
Pyrola secunda																2				2	2	2								2	2			2	2						2	2	2	2	2	2				2		2		2	2	2												2	2			2	2			2		2	3	2
Dicranum acutifolium																					2			3	2		2	3	2												2													3						2		2																			2	1	
Hylocomium splendens	5	3	3	3	5	3		5		2	3	5			8	3	7	7	8	5	5	5	3	5	3	5	8	5	5	8	8	8	7	8	8	5	7	5	7	8	5	7	8	5	7	8	5	7	7	5	3	8	8	8	7	8	5	8	7	8						2											4	5	5	1
Peltigera leucophlebia						2		2				3			2	3	3	3	3	2	3	3		3	3	3	3	3	3	3	3	3	3	3	3	3	3	3	2	3	3	3	3	2	2	2	3	3	3	2	2	2	3	3	3	3	3	3	3	2	2																2	5	5	
Arctostaphylos rubra	2	3	3	3	3	5	3	2	3	3	3	5	3	3		2		3	2	2	2	2		2	2									3	2		2		2	2									2					2		3	3	2	2	3													2	3	3	5	3	3	1
Vaccinium vitis-idaea					3	2	2	2		2				2	2	2	3	3	3	3									2		2		3	5	3	2		5	3	3	3	5	3	3	3	3	3	3					3		3		3	5		2															3	3	3	4	1
Betula glandulosa					3	2	2	2		2				2	2	2	3	3	3	3			3								2		2		2	2	2	2	2	2																																				3	3	2	
Habenaria obtusata								2			2				2						2				2																										2	2	2	2				2	2																		1	1	2	
Alnus crispa												3			2	2	2														3										2	2								2				2																								1	2	1	
Viburnum edule															2	2	2	2	2												3										2		2	2	2											2		2																					2	2	
Hypnum procerrimum															2	5					2	3	3	3	3		3	2																																					3	3	3	3	3	3	7	7	7	7	5	5	3	2	3		3		5
Peltigera canina															2	2										2			2																							2	2		2							2				3	3	3	3	3	3	3	3	3	3	3	3	3	3	3		2	1	5
Cladonia pocillum																									2																																						2	2	2	2	2	2	2	2	2	2	2	2	2	2		1			5
Shepherdia canadensis																										2								2											2							3	3			2	2						2	2	2		2	3		2									2			1	2	2
Hedysarum mackenzii																															2																																	2			2	2	2	2	3					1		2
Tortula ruralis																																																														2							2					2	2	2	2				2
Arctostaphylos uva-ursi	2	5																																																											5	5		2							3					1			2
Ceratodon purpureum																																																	2													3		3			2			2	2							1	2
Oxytropis deflexa																																																											2															2	2	2			1	1
Anemone multifida																																																																	2			2	2	2	2										2
Astragalus alpinus																																																																2	2	2					2										2
Picea glauca	7	5	5	5	5	5	5	3	5	5	3	3	5	5	7	5	5	5	3	5	8	7	5	7	5	7	8	7	7	8	8	8	8	7	7	3	5	8	5	5	5	7	7	3	7	8	7	5	7	5	7	8	8	7	7	8	5	8	5	5	5	3	5	3	5	5	5	5	7	5	5	7	5	5	5	5	5	5	5
Linnaea borealis														2	3	3	2	3			2	3		2		3	2	2	2	2	2	2	2	2		2			3	2	2	2	2	2	3	2	2	3	3	3	2	2	3	2	3	2		2	2	2					3	2	3	2	2		3	2			2	1	4	5	3
Thuidium abietinum	2	2						2	2	2		5		3	3	7	3	3		2			5	5	5	3				2			2				2			3	3	5		2	3		2	2	2	5	3		5				2		3		2	2	8	2	3	3	7	7	7	7	5	3	3	3	7	3	3	3	5
Mertensia paniculata			2					2							2	2	2	2	2	2	2		2						2		2	2	2	2				2							2	2	2		2				2	2		2		2	2			2	2				2	2	2	2		2					2	1	4	3	2
Ledum groenlandicum	5	2	2		3	2	3	3		5	3	3	2	3				5	5	3	3																2	3	5	3				3			2			3		2						3		5.		5	5											2		2				5	2	3	1
Rosa acicularis																		2	2	2	2						2		2	2	2	2	2	2	2	2									2	2	2	2	2	2		2		2		2		2	2			2	2				2	2									2			3	3	1

Species	Relevé values (in printed order)				
Geocaulon lividum	2 2 2 2 2 2 2 2 2 2 2 2 2 2 2 2 2 2 2 2 2 2 3 2 2 2 3 2 2	2	2	3	1
Carex concinna	2 2 2 2 2 2 2 2 2 2 2 2 2 2 2 2 2 2 2 2 2 2 2 2 2 2 2 2 2	2	2	2	4
Salix glauca	2 2 2 2 3 3 2 2 2 2 2 2 2 2 2 2 2 2 2 2 2 2 2 3 2 3 3	3	3	2	2
Equisetum scirpoides	2 2 3 3 2 3 2 2 3 3 2 2 3 2 2 2 2 2 3 2 2 2 3	4	2	2	1
Lupinus arcticus	2 2 2 2 2 2 2 2 2 2 2 2 2 2 2 2 2 3 2 2 2 2 2 2	1	1	3	3
Empetrum nigrum	2 3 3 3 2 3 2 3 3 2 2 2 2 3 2 2 2 2 2		2	3	1
Epilobium angustifolium	2 2 2 2 2 2 2 2 2 2 2 2 2 2 2 2 2 2 2 2	1	1	3	1
Drepanocladus uncinatus	2 2 3 3 2 3 2 2 2 2 2 2 2 3 2 3 7 3 3	1	2	1	3
Mitella nuda	2 2 2 2 2 3 2 2 2 2 2 2 2 2 2 2 2 2	1	1	3	1
Tomenthypnum nitens	2 2 2 2 2 2 2 2 2 2 2 2	2	2		1
Cladonia chlorophaea	2 2 2 2 2 2 2 2 2 2	1	1	1	1
Salix alaxensis	2 3 2 2 3 2 2 2	2		1	1
Moneses uniflora	2 2 2 2 2 2 2 2		2	1	
Lophozia hatcheri	1 1 1 1 1 1 1 1 2	1	1	2	
Salix myrtillifolia	2 3 2 2 3 2 3 2	2		1	1
Corallorhiza trifida	2 2 2 2 2 2 2	1	2	1	1
Populus tremuloides	2 2 2 3 2 3			1	1
Populus balsamifera	2 3 2 2 2 3			1	1
Salix planifolia	2 1 3 2 2 3			2	
Oxytropis campestris	2 2 2 2 2 2	2			1
Dicranum elongatum	2 2 2 2 2 2	2			1
Betula papyrifera	2 3 3 2 2		1	1	
Arnica cordifolia	2 2 2 2 2			1	
Calamagrostis canadensis	2 2 2 2 2		1	1	1
Ribes triste	2 2 2 2			1	
Artemisia borealis	2 2 2 2			1	
Cladonia gracilis	2 2 2 2			1	
Goodyera repens	2 2 2			1	

PG1 *Picea glauca–Rhododendron lapponicum* community

This is an open canopied Subalpine–Subarctic coniferous forest dominated by *Picea glauca*. The most frequent life forms are chamaephytes (22%) and thallochamaephytes (23.9%); the latter are mainly cushion forming mosses. Circumboreal and boreal northwestern North American species are the most frequent phytogeographic elements (Fig. 9). This community resembles PM3, except for the constancy of *Picea glauca* and *Rhododendron lapponicum*.

The soils are mainly of the Cryosolic Order. Frozen ground is present in 100% of the plots in June, and in 50% of the plots in September. The active layer has an average thickness of 16 cm in June and of 61 cm in September. Soil texture is similar to the one of PM3, drainage conditions are slightly better (more sandy to sandy-loamy and moderately well-drained soils). The pH,

Table 6. Ecological data relative to the relevés of *Picea glauca* forests reported in Table 5. For the abbreviations, see caption of Table 4.

Releve No.	1	2	3	4	5	6	7	8	9	10	11	12	13	14	15	16	17	18	19	20	21	22	23	24	25
Exposure	S	SW	S	E	E	S	-	SSW	NNE	NE	NE	NE	-	SW	-	SW	SSW	SSW	-	SSW	-	N	NNE	-	-
Slope	20	35	30	15	5	5	-	2	2	2	2	5	-	5	-	40	4	10	-	5	-	5	5	-	-
Total cover	100	100	100	100	100	100	100	100	100	100	100	100	100	100	100	100	90	100	100	100	90	100	100	100	100
Soil type	. .	. .	. .	. .	. .	. .	. .	O. HG	GL. SC	R. HG	GL. SC	GL. SC	GL. SC	GL. SC	GL. CUR	R. SC	HU. OC	THU OC	BR. SC	THU OC	BR. SC	GL. EB	GL. SC	GL. SB	GL. SC
GS %	.	.	.	.	.	.	.	33	-	30	10	-	-	-	80	15	-	5	15	20	30	2	2	5	2
SA %	.	.	.	.	.	.	.	46	60	40	20	70	60	90	20	40	20	55	20	60	40	25	55	25	55
SI %	.	.	.	.	.	.	.	20	40	30	30	30	40	10	-	40	-	5	30	20	20	38	43	55	43
CL %	.	.	.	.	.	.	.	1	-	-	40	-	-	-	-	5	-	35	35	-	10	35	-	15	-
Org. (cm)	.	.	.	.	.	.	.	20	15	25	18	80	20	30	9	15	80	95	15	35	10	7	15	10	15
pH	.	.	.	.	.	.	.	6.6	7.0	7.3	7.5	7.2	6.9	6.6	6.6	6.1	7.3	7.3	5.5	6.4	6.3	6.3	6.3	6.6	6.0
Erosion pot.	.	.	.	.	.	.	.	low	med	med	med	med	med	med	med	med	med	hi	med	med	med	med	med	med	med
Drainage	.	.	.	.	.	.	.	imp	imp	imp	imp	imp	imp	imp	imp	mod	imp	mod	mod	mod	mod	mod	mod	mod	imp
Depth of ice June (cm)	.	.	.	.	.	.	.	20	-	15	15	15	8	7	-	30	11	13	15	13	10	18	15	-	15
Depth of ice Sept.(cm)	.	.	.	.	.	.	.	-	-	-	45	25	60	60	-	75	50	40	60	40	45	-	45	-	90

Releve No.	26	27	28	29	30	31	32	33	34	35	36	37	38	39	40	41	42	43	44	45	46	47	48	49	50
Exposure	SW	E	ESE	NE	SW	S	-	SW	ENE	ESE	SSW	NNW	S	NW	SSW	W	SSW	ESE	-	S	S	NNW	-	E	S
Slope	5	5	2	2	10	15	-	10	5	5	5	5	5	5	5	10	10	35	-	45	10	5	-	10	45
Total cover	90	90	80	100	100	100	100	100	100	100	100	100	100	100	100	90	100	100	100	100	100	100	100	100	80
Soil type	O. DYB	GL. EB	GL. EB	GL. DYB	GL. CUR	OH. B	O. DYB	GL. EB	GL. EB	GL. EB	O. HG	E. DYB	O. EB	GL. DYB	GL. EB	O. DYB	GL. DYB	O. EB	GL. DYB	E. DYB	O. DYB	E. DYB	GL. DYB	O. DYB	GL. DYB
GS %	15	5	30	1	14	25	20	30	20	30	33	70	8	20	33	65	10	10	10	40	50	70	40	50	4
SA %	30	45	35	90	26	45	40	40	50	40	46	25	82	60	46	30	75	60	30	45	40	25	50	35	90
SI %	54	40	30	9	58	15	35	29	28	29	20	5	9	20	20	5	15	30	45	12	9	5	9	13	5
CL %	1	10	5	-	2	15	12	1	2	1	1	-	1	-	1	-	-	-	15	3	1	-	1	2	1
Org. (cm)	9	5	8	8	9	9	23	9	5	9	15	13	4	9	15	9	9	4	11	9	6	13	9	4	5
pH	6.9	5.7	6.5	5.4	6.6	7.0	7.8	5.9	5.9	5.6	5.9	4.4	5.9	5.7	5.1	5.0	5.0	5.5	5.5	6.4	4.9	4.4	4.2	5.4	6.0
Erosion pot.	low	med	low	low	med	med	med	med	med	med	med	low	med	low	low	low	med	med	low	med	low	low	low	low	med
Drainage	ex	mod	imp	mod	imp	mod	mod	mod	mod	mod	imp	ex	mod	mod	imp	ex	mod	mod	mod	mod	ex	ex	imp	mod	mod
Depth of ice June (cm)	-	-	-	18	15	20	23	20	-	-	-	-	13	28	-	-	-	8	40	-	-	-	-	-	-
Depth of ice Sept.(cm)	-	-	-	-	145	-	-	-	-	-	-	-	-	-	-	-	-	-	-	-	-	-	-	-	-

Releve No.	51	52	53	54	55	56	57	58	59	60	61	62	63	64	65	66	67	68	69	70	71	72	73	74	75
Exposure	SE	ESE	S	NNW	SE	SE	N	-	SSE	SSW	NE	NE	NW	NE	SSW	NE	ENE	ENE	S	ENE	S	ENE	NE	E	ESE
Slope	30	10	30	3	13	10	5	-	5	5	15	20	15	10	5	10	10	24	5	20	7	11	5	5	5
Total cover	100	100	100	90	90	90	100	90	100	100	90	80	90	50	60	70	.	.	.	.	.	.	80	70	90
Soil type	GL. CUR	GL. CUR	. .	GL. EB	O. SB	GL. DYB	O. DYB	GL. DYB	GL. DYB	GL. EB	O. SB	O. DYB	O. SB	E. SB	C. UR	GL. CUR	. .	. .	. .	. .	. .	. .	GL. SB	O. R	E. SB
GS %	40	30	.	30	10	5	10	5	10	33	20	15	15	10	5	1	.	.	.	.	.	.	-	40	1
SA %	32	40	.	40	60	50	20	75	45	46	40	60	45	60	80	90	.	.	.	.	.	.	20	40	70
SI %	20	29	.	25	29	44	64	19	44	20	40	25	40	30	13	9	.	.	.	.	.	.	79	18	29
CL %	8	1	.	5	1	1	6	1	1	1	-	-	-	-	7	-	.	.	.	.	.	.	1	2	1
Org. (cm)	14	3	.	9	13	15	15	9	15	10	-	5	4	9	6	8	.	.	.	.	.	.	-	2	4
pH	7.0	7.1	.	6.0	6.1	6.5	6.7	5.9	6.0	5.8	7.8	7.6	7.6	7.5	6.8	6.4	.	.	.	.	.	.	7.3	7.5	7.3
Erosion pot.	med	low	.	low	med	med	hi	low	med	med	hi	med	med	med	med	low	.	.	.	.	.	.	med	med	med
Drainage	imp	mod	.	mod	mod	mod	imp	imp	mod	mod	ex	ex	mod	ex	mod	mod	.	.	.	.	.	.	mod	mod	ex
Depth of ice June (cm)	-	25	.	50	-	20	18	25	25	-	-	-	-	-	-	-	.	.	.	.	.	.	-	-	-
Depth of ice Sept.(cm)	-	-	.	-	-	-	-	-	-	-	-	-	-	-	-	-	.	.	.	.	.	.	-	-	-

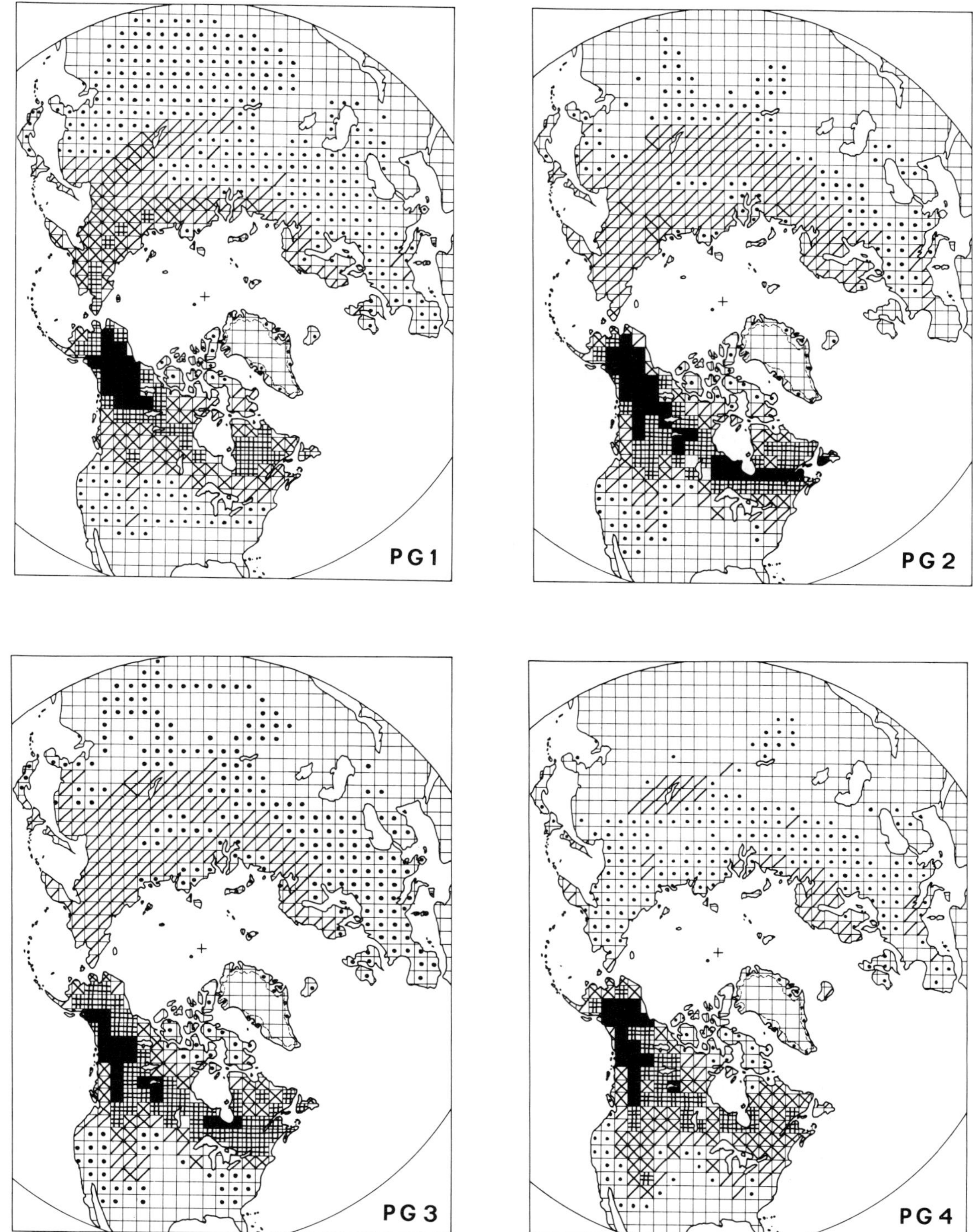

Fig. 9. Chorograms of the four relevé groups obtained by classification of Table 5 (*Picea glauca* forests, see Fig. 2). Different shadings refer to the percent of species occurring in the OGUs, on a 5-class scale with intervals of 20%.

however, is consistently higher, with and average of 7.0.

Relevés 1–7 are from northern Yukon, along the Dempster Highway, near Ogilvie, on calcareous scree slopes. They contain a set of differential acidophytic species with circumboreal distributions. Relevés 8–14 are from southern Yukon, Kluane region. In this area, accumulation of loess is frequent. These relevés have a set of neutro- to basiphytic species, such as *Thuidium abietinum, Tortula ruralis* and *Carex concinna*.

This community replaces *Picea mariana* stands on neutral to weakly alkaline, moderately well-drained Cryosols, in areas with limestone, or with consistent loess deposition, mainly in the upper montane and Subalpine vegetation belts. The dominance of *Picea glauca* on soils with ground ice suggests that this species can successfully compete with *Picea mariana* on minerotrophic Cryosols.

PG2 *Picea glauca–Rhytidium rugosum* community

This community has no characteristic species. The moss *Rhytidium rugosum* is a good differential species towards PG3. The chorogram (Fig. 9) shows a clear prevalence of boreal North American species.

Two facies may be distinguished: relevés 15–20 include *Betula glandulosa, Viburnum edule, Alnus crispa, Vaccinium vitis-idaea, Ledum groenlandicum* and *Mertensia paniculata*. Relevés 16–32 are characterized by the constant presence of the xerophytic moss *Hypnum procerrimum*. They represent a transition towards PM4 (see later) and hence towards more xeric conditions.

The soils are mostly sandy-loamy to sandy Brunisols. The organic horizon has an average thickness of 13 cm. Mull is the most frequent humus type. Frozen ground is present in 40% of the plots in June, always absent in September. Most of the stands were on slopes, with an average inclination of 11.4°. Drainage is moderate in 56% and excessive in 24% of the plots. Impeded drainage in 20% of the plots is a seasonal phenomenon due to the presence of ground ice in spring and early summer. The pH tends to be subneutral, with an average of 6.1. The differential species *Rhytidium rugosum* is a neutrophytic moss, whereas the differential species of PG3 lacking in this community are strongly acidophytic.

This community occurs along the northwestern section of the Alaska Highway (Fig. 3). Its distribution coincides with the one of the secondary stands dominated by *Populus tremuloides*, outside the range of *Pinus contorta* in the Yukon. The relatively high pH values with respect to PG3 most probably reflect the influence of *Populus* litter, in contrast with the acidifying effect of *Pinus* litter in the secondary stages preceding PG3 in the successional series.

PG3 *Picea glauca–Hylocomium splendens* community

This community represents the typical boreal forest formation in the survey area. The stands are closed canopied and *Picea glauca* is dominant.

Thallochamaephytes, chamaephytes and hemicryptophytes are the most frequent life forms (resp. 27%, 20% and 18%), boreal North American species are the most frequent phytogeographic element (Fig. 9). The lichen synusia of the understory grows on dead trunks covered by humus, and is dominated by *Cladonia ecmocyna* and *Cladonia multiformis* (Nimis 1981).

A floristic variation trend is recognizable in Table 5, from the releves at the right to those at the left side. The former are characterized by the high frequency of *Carex concinna*, a rather xero- and heliophytic species, the latter by a set of mesophytic species. This trend reflects a gradient in soil moisture.

While frozen ground is present in 40% of the plots in June, at an average depth of 23 cm, it is always absent in September. The community occurs on a wide variety of soils (Table 2), with a prevalence of Brunisols, or other soil types evolving towards the Brunisolic Order. They have mostly a sandy-loamy texture. The average thickness of the organic horizon is 17 cm, with Moder as the main humus type. Drainage is moderate in 56%, impeded in 40% and excessive in 4% of the plots. Impeded drainage is due to frozen ground at the start of the vegetative season, drainage tending to improve during summer. 74% of the plots were on slopes, with average inclination of 4.5°. The pH is always acid, with an average of 5.6.

This community represents the mature coniferous forests on permafrost-free soils in south-eastern Yukon. It is replaced by PG2 and PG4 along the central section of the Highway, owing to a drier climate and higher pH of the soils (Fig. 3). Its geographic distribution extends to northern Yukon: we recorded it along the Dempster Highway, where it is confined to permafrost-free soils along creeks and rivers. Its distribution along the Alaska Highway corresponds with the limits of *Pinus contorta* (Fig. 3). The western limit of this tree corresponds with the isoiet of 375 mm. The low soil pH is probably due to the joined effect of relatively high precipitation and the acidifying effect of *Pinus* litter on soil development. Furthermore, in the area of this community acid sands and granite are frequent and no active deposition of loess has been observed. Precipitation, however, is still too scarce to produce strong podzolidation and true podzols are absent.

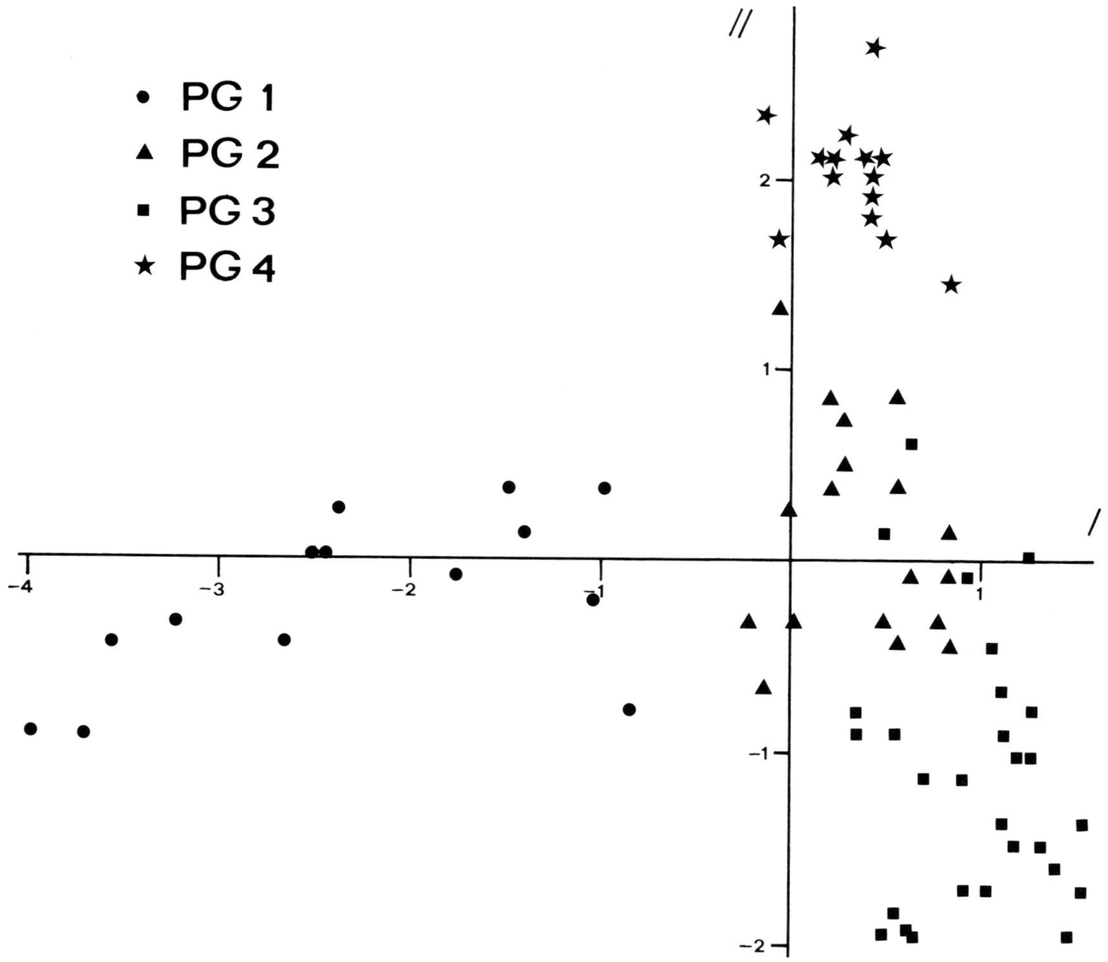

Fig. 10. Ordination of the relevés of *Picea glauca* forests (by PCA with correlation coefficient) of the relevés, based on the data of Table 5. Principal Components I and II. The symbols refer to the releve groups obtained by classification of the same data set, as in the legend (see also Fig. 2).

PG4 *Picea glauca–Hypnum procerrimum* community

This community was described by Hoefs et al. (1975) from Sheep Mountain in the Kluane Lake region of southern Yukon. It is a closed- or open-canopied forest dominated by *Picea glauca*. The life form spectrum differs from the previous communities for the high frequency of hemicryptophytes (27%) and thallohemicryptophytes (18%), and for the lower incidence of both chamaephytes (22%) and thallochamaephytes (15%).

Circumboreal species are relatively rare, whereas boreal North American species have the highest frequency among all of the community types of the survey area, and the Cordilleran element is well represented (Fig. 9). These structural and phytogeographic features indicate an affinity with the secondary *Populus tremuloides* stands (BL3).

The soils are sandy-silty Brunisols, poor in organic matter, derived from loess deposits. The pH values are high, with an average of 7.2. Most of the plots were on slopes, with an average

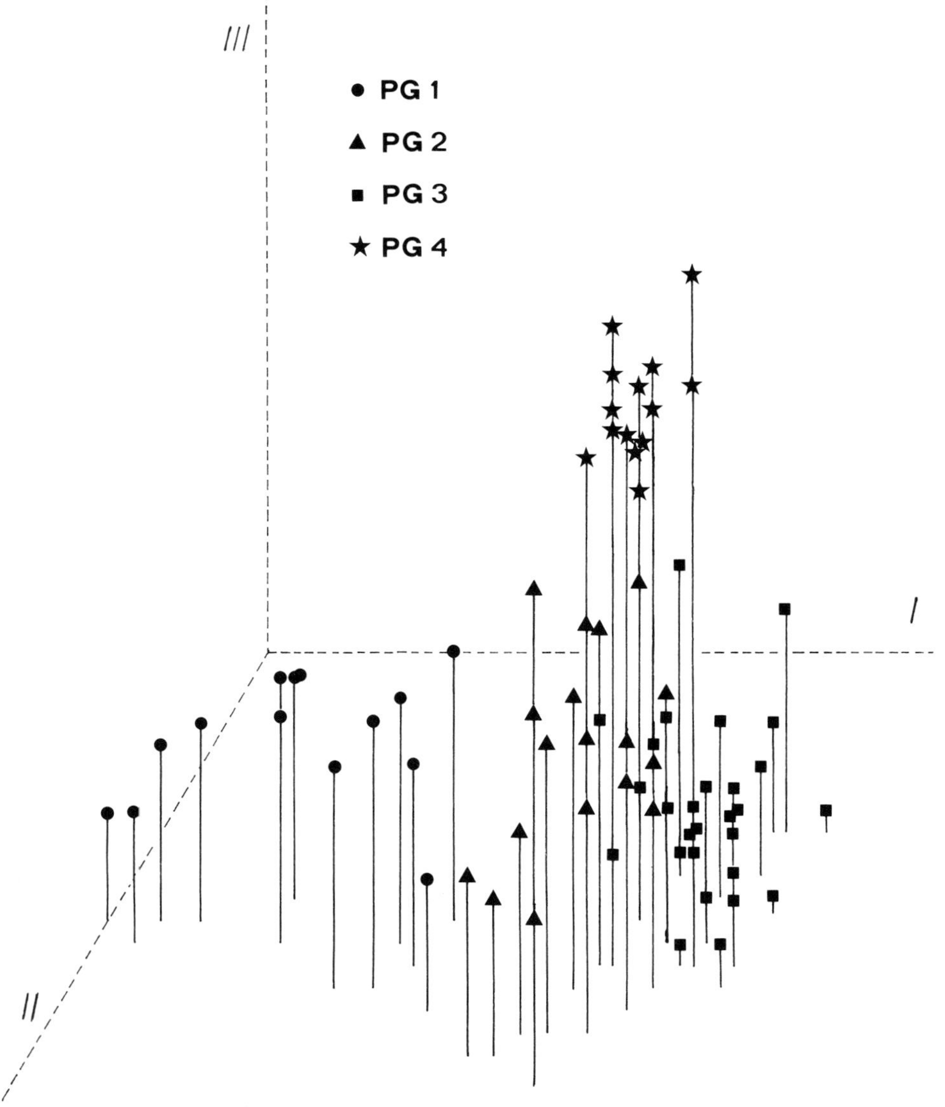

Fig. 11. Ordination (by PCA with correlation coefficient) of the relevés of *Picea glauca* forests, based on the data of Table 5. Principal Components I, II and III. The symbols refer to the relevé groups obtained by classification (see also Fig. 2).

inclination of 14.5°. Frozen ground was always absent and drainage is excessive to moderate.

This community is restricted to the Kluane Lake area, at the leeside of the St Elias Mountain Range, a region with very low precipitations and a strongly continental climate (Fig. 3). It is mostly confined to the principal components of the hydrographic net, the slopes inbetween being occupied by steppe-like grasslands (G2). This is the most xerophytic coniferous forest community in the Yukon Territory.

4.2.2. Ordinations

PCA was performed on the binary data of Table 5 both with the correlation coefficient and Similarity Ratio. The results using the correlation coefficient are in Figures 10 and 11.

In Figure 10 the relevé points are disposed according to the first and second Principal Components. The first component separates the relevés of PG1. The relevés of PG4, PG3 and PG2 are disposed along the second component in a sequence such that the clusters obtained by classification are still recognizable. In Figure 11 the third component separates the relevés of PG4 from the others. The ecological interpretation of this ordination is that the first and third axes respectively separate relevés of moist and xeric stands. The second axis clearly reflects a pH gradient, as illustrated in Figure 12. This figure reports the position of the relevés on the second component and the corresponding pH values. A regression line obtained by the least squares method fits the centroids of the groups, showing a high correlation between pH and the arrangement of the relevés, from the subacidophytic PG3 to the basiphytic PG4.

The ordination of relevés on the basis of the Similarity Ratio (Fig. 13) shows them dispersed around a curved line that revealed trends in the variation of some ecological factors. The sequence of PG1, PG2, PG3, PG4, obtained by connecting the centroids of each group (dotted line in Fig. 13) reveals the following trends, illustrated in Figure 14: an almost linear decrease of the frequency of frozen ground in June, and of the thickness of the organic horizon, tendency towards improved drainage conditions, decrease in the frequency of Cryosols and corresponding increase in the frequency of Brunisols.

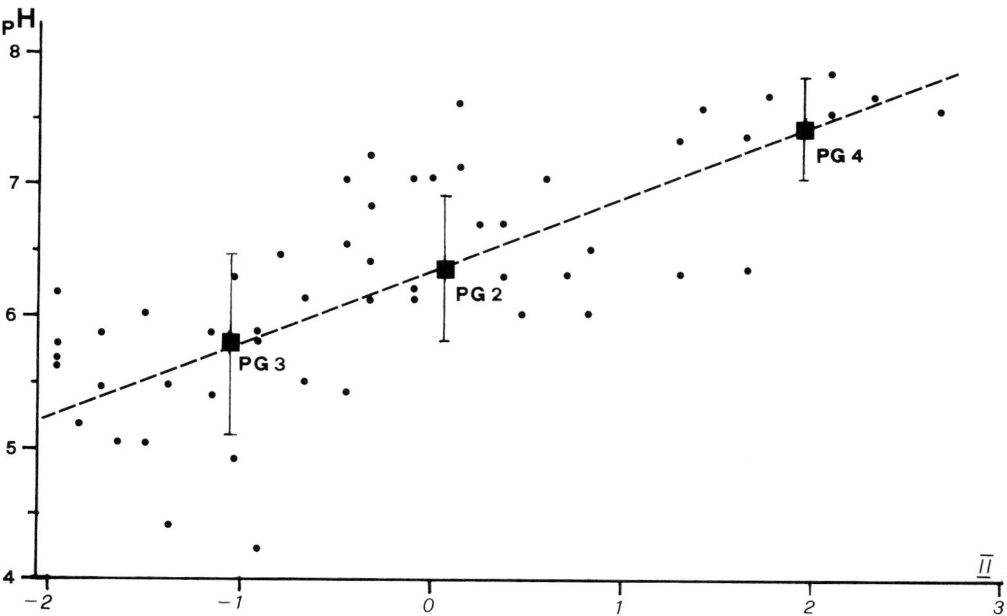

Fig. 12. Interpretation of the ordination of Figure 10. The x-axis reports the scores of the relevés of the groups PG2, PG3 and PG4 on Principal Component II of Figure 10; the y-axis reports the soil pH measured in each relevé (data in Table 6).

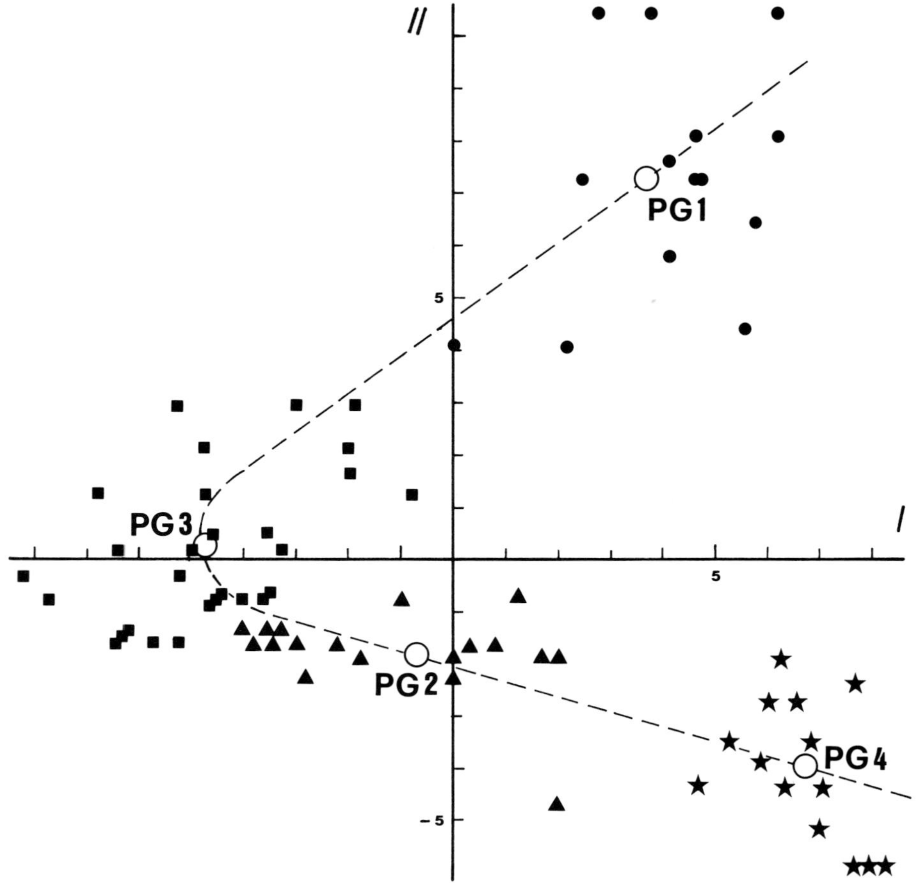

Fig. 13. Ordination of the relevés (by PCA, with Similarity Ratio) on the data in Table 6 (*Picea glauca* forests). Symbols refer to the relevé groups, as in the legend to Figure 10. Open circles locate the position of the centroids of the groups. The dotted line represent an axis utilized for the construction of Figure 14.

The ordinations obtained by the correlation coefficient and the Similarity Ratio proved to be complementary in detecting trends of ecological variation that underlie the floristic variation. The two most important factors seem to be drainage and pH.

4.2.3. *Discussion*

Picea glauca and *Picea mariana* have very similar ranges. The former, however, is most common in the southern and western parts of the boreal zone (Halliday and Brown 1943; Nienstaedt 1957; Hart 1959; Rowe 1977), where it often forms almost pure stands on well-drained soils, whereas in the north it is mostly confined to permafrost-free soils along water courses, or on calcareous slopes. The two species were reported to hybridize in several parts of their ranges (Larsen 1965; Roche 1969; Dugle and Bols 1971). But a more recent study suggests that most reports of natural hybridization apparently have resulted from underestimation of the variation present in both species, and that no hybridization occurs between them (Parker and McLachlan 1978). *Picea glauca* is generally assumed to be a more mesophytic species than *Picea mariana*: it grows best on nutrient-rich, permafrost-free, well-drained soils (Carleton and

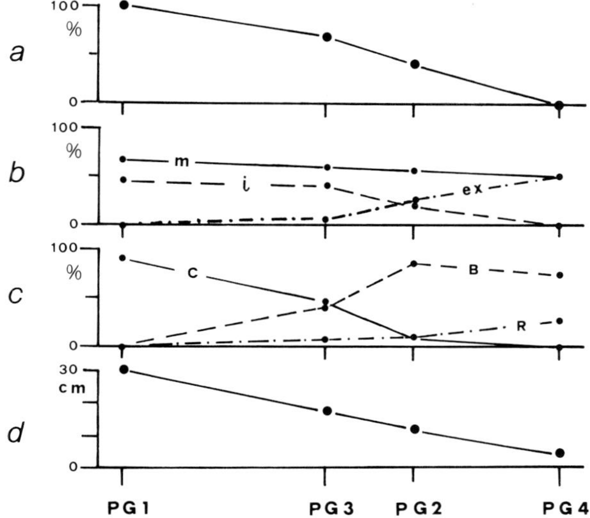

Fig. 14. Interpretation of the ordination in Figure 13 on the basis of the ecological data of Tables 1 and 2. The *x*-axis reports the position of the centroids of the relevé groups on the dotted line in Figure 13, the *y*-axes indicate: (a) percent frequency of the occurrence of frozen ground in June, (b) average values of the four drainage classes, (c) percent frequencies of the four main soil orders (C: Cryosols, B: Brunisols, C: Regosols), (d) average thickness of the organic horizon.

Maycock 1978, 1980; Dyrness and Grigal 1979), where it often forms a closed boreal woodland. Of the four vegetation types dominated by *Picea glauca* that are described in this paper, only two (PG2 and PG3) represent a true closed boreal forest. PG1 includes open stands near the northern distributional limit of the species. PG4 represents an open woodland in an area with a continental, dry climate, where the prevalent vegetation is of grasslands and aspen thickets.

Douglas (1974) reports a vegetation closely resembling PG2 from the Alsek River region (SW Yukon). His *Picea glauca–Salix glauca* community is considered as a climax stage on poorly to well-drained, moderately moist to mesic, glacial, alluvial and lacustrine parent materials. A type most similar to PG3 is reported by Viereck (1970) from the Chena River, in interior Alaska. According to this author, the white spruce stands are replaced by black spruce and bog as permafrost develops under the old white spruce stands. This contradicts the claim of Baxter and Wadsworth (1939) who considered the white spruce/birch stands along the lower Yukon to be a climax type.

A preliminary classification of white spruce stands throughout boreal North America, based on bryofloristic and vascular floristic criteria has been provided by La Roi (1967) and La Roi and Stringer (1976). They show that considerable floristic variation exists from east to west. There is a good correspondence between PG2 and PG3, and the *Populus–Salix–Shepherdia* group of La Roi (1967), whose distribution extends from interior Alaska to the western part of the Northwest Territories. Further east, *Picea glauca* woods have a large set of species absent in the Yukon flora. These include *Gaultheria hispidula, Dryopteris austriaca, Coptis groenlandica, Oxalis montana, Clintonia borealis, Streptopus roseus, Majanthemum canadense*, etc., and other bryophytes such as *Rhinchostegium serrulatum, Tetraphis pellucida, Bazzania trilobata, Cephalozia media*, etc. Our communities, compared with those of eastern North America, have a much lower incidence of ferns, and a high frequency of rather xerophytic plants with subcontinental or Cordilleran distributions. This reflects the continental climate in the survey area.

4.3. *Pinus contorta* woods

4.3.1. Classification

This data set includes 40 relevés (Table 7) of coniferous forests dominated by *Pinus contorta*. Table 8 contains the ecological data and Figure 2 presents the dendrogram of the relevés, with two main relevé groups, PC1 and PC2. Both can be subdivided further into subgroups, which are interpreted here as facies of the two main community types. PC1 represents closed, rather mature stands leading towards a forest dominated by *Picea glauca*. PC2 is an open lichen woodland occurring on particularly dry and poor soils. The chorograms of the two relevé groups are shown in Figure 15.

Table 7. Phytosociological table of *Pinus contorta* woods. The relevés are arranged according to the results of numerical classification (see Fig. 2).

	PC1																				PC2																					
	a					b							c								a											b										
RELEVÉ No.	1	2	3	4	5	6	7	8	9	10	11	12	13	14	15	16	17	18	19	20	21	22	23	24	25	26	27	28	29	30	31	32	33	34	35	36	37	38	39	40		
Hylocomium splendens	5	7	2	5	5	8	8	3	2	3	3	3	5	8	7	3	3	5	2	7			2																2		5	1
Pleurozium schreberi		7	5	7	5	3	3	2	2			2	3	5	2	5	3	5		3	2		2	2						2											4	1
Empetrum nigrum	2				2	3	3		5	3	5	3	2	2		4		2		2	2						2				2							2			4	1
Ledum groenlandicum	2	2			2	7	7	2	3	2	2	5	2	7	5	2									2									2	2						4	1
Cornus canadensis		2				3	3	3	3		2		2	2		2	2	2													5	2		2					2		4	1
Polytrichum juniperinum	2	2	2	2				3	2	2	2	2			2					2			2				2			2							2				3	1
Dicranum fuscescens	2	3	2	2	2			2	2	2			2	2	2		2	2		2																					4	
Cladonia ecmocyna	2	2		2	2			2	2	2	2	2	2	2	2						2										2										3	1
Drepanocladus uncinatus	2	2	2					3	2	2	2		3				3		3				2						2											2	3	1
Lycopodium complanatum				2		2	2	2	3	2	2	2	2		2	2															2										3	1
Thuidium abietinum	2	5	5	5	2								3		2		2	5	2				2							2											3	1
Alnus crispa	5	3	7	5	5			2						2	3																				2					5	2	1
Lophotia barbata			2	2		2	2	2	2	2		2								2	2									2											3	1
Ptilium crista-castrensis			5	3			3	2	2		2	2			2	2		3																							3	
Lycopodium annotinum	2				2		2	2	3	2	2	2			2																										3	
Nephroma arcticum	3		2			3			2			2			2	2																2									2	1
Vaccinium uliginosum						3	2	2	3	2	2	3																													2	
Betula glandulosa								3	3		5	2																													1	
Picea mariana		2											2	2	3	3							2												2						2	1
Arctostaphylos uva-ursi				2				2					2	2		2		2	3		3	2	3	5	3	7	3	3	5	5	2	5	3	2	3	2	5	5	3	2	2	5
Peltigera canina		2	2							2	2			2								3	2	2			3	2	2	2	2	2	2	2	2	3	2	2	2	2	2	5
Polytrichum piliferum	2	2						2													2	5		3		3	3	2	3	3	2	5	2		3	3	3	3	2	5	1	5
Cladina mitis																				2	2	2	2	2	2	2	2	2	2	2	2	2	2		2	2	2	2		2	1	5
Cladonia gracilis																						2			2	2	2	2	2		2	2	2	2	3	2	2	2		2		4
Cladonia clorophaea			2		2						2		2											2		2								2	2	2	2	2	2	2	1	3
Pulsatilla patens																				2		2		2	2	2	2	2	2	2						2	2	2			1	3
Gentiana propinqua																					2	2										2	2	2	2	2	2	2	2	2		3
Cladonia multiformis																		2		2	2	3	2	2					2	2		2				2				2	1	3
Cladonia cornuta		2																			3	2		2	2		2	2		2						3		2			1	2
Cetraria laevigata														2	2						2	2		2				2								2	2	2			1	2
Cetraria cucullata												2					2			2				2		2					2	2				2	2				1	2
Cladonia uncialis											2				2	2						5				3				3						3	3	3			1	2
Cladonia coccifera	2											2						2			2	2	2		3				3						3						1	2
Cladonia cariosa																										2			2			5	5	5				3		5		2
Geocaulon lividum																									2	2			2	2		2			2							2
Cetraria nivalis																														2		2				2	2	2	2			2
Cladonia verticillata																													2	2						2	2	2				2
Zygadenus elegans																			2								2	2	2		2						2	2			1	2
Pinus contorta	3	7	7	7	5	5	7	5	7	5	3	7	8	7	3	5	7	7	7	7	7	5	7	5	7	7	7	7	7	7	7	5	7	7	7	3	7	5	5	7	5	5
Linnaea borealis	2	3	2	3	2	3		2	2		2	3	3	2	2	3	3	2	3	2	3	3	2	3	3		5	7	3	3	3	3		2	3	2	2	2	3	2	5	5
Peltigera aphtosa		3	3	2	2	3	2	3	2	2	2	3	2	5	2	3	3	3	3	3	2	2	2		2	2				2	3	2				2	2	1	2	2	5	4
Vaccinium vitis-idaea	3	3	2	3		5	3	3	3	3	2		3	3	5	7	7	7		2	5	5		3			5	3	5	5	7	3		2	2		5	2	3	2	5	4
Festuca altaica						2		5	2	5	3	2	2	2	2	2	2	2	3	2	2	2				3	3	3		3	2	2	5	2		2	5	3	3	2	4	4
Cladina arbuscula	2	2			2	2		2	2	2	2	2	2	2	3	2	2	2	2		3	5	2	2			2	2	2	2	3	2	2			5	2				4	4
Cetraria pinastri (epiph.)		2	2	2	2	2	2		2		2	2	2	2	2	2	2	2		2	2	2						2	2	2	2	2									4	2
Lupinus arcticus						2	2	2	2	2		2	2				2	2	3	2									2	2		2		2	2		2	2	2		3	2
Pyrola secunda			2	2	2				2	2		2	2	2			2	2	2	2	2											2	2	2	2						3	2
Ceratodon purpureus	2				2						2		2		2						2	2	2	2			2	2	2	2					3	3	2	2			2	3
Stereocaulon tomentosum	2												2	2	2	2	2				3	2	2	2		2		2				2		2	2	2	2				2	3
Picea glauca			2			2		2			2		2	2	2	2	2	2		2				2		3				2	2				2						3	2
Rosa acicularis														2	2		2	2	2	2		2				2	2	2	2	2		2	2	2				2			2	3
Epilobium angustifolium			2					2			2	2	2	2		2		2	2		2	2												2	2	2				2	2	3	2
Salix scouleriana										3	3	3			3			2	2				2								2	2	2		2		2				2	2
Shepherdia canadensis																	5	2	2	2	5		2		2	3						3	5	2				2			1	2
Pyrola asarifolia	2	2	2	3	2															2							2		2	2	2			2							2	2
Cladonia deformis		2						2	2	2	2										2	2										2		2			2	2			2	2
Cladina rangiferina	2										2	2			2	2					2	2									2						3	2			2	2
Viburnum edule										2			2	2	2			2											2			2						2	2		2	1
Calamagrostis purpurascens																2				2	2		2	2	3	2										2	2				1	2
Cladina stellaris								2			2			2	2	2					2	2														2		2			2	1
Mertensia paniculata											2						2		2													2	2						2		1	1
Populus tremuloides															2	3														2	3		5								1	1
Abies lasiocarpa					2					2																				2		2									1	1
Equisetum scirpoides					2																						2	2	2												1	1
Pedicularis labradorica																	2							2							2			2							1	1
Dicranum scoparium				2															2				2														2				1	1
Polytrichum commune						2			2																			2									2				1	1
Populus balsamifera																																		2								1
Lycopodium clavatum								2	2	2																															1	
Achillea borealis												2																					2	2							1	1
Stereocaulon alpinum																											3			2									2			1
Cornicularia aculeata																										3	3				2											1
Solidago decumbens																	2											2			2										1	1

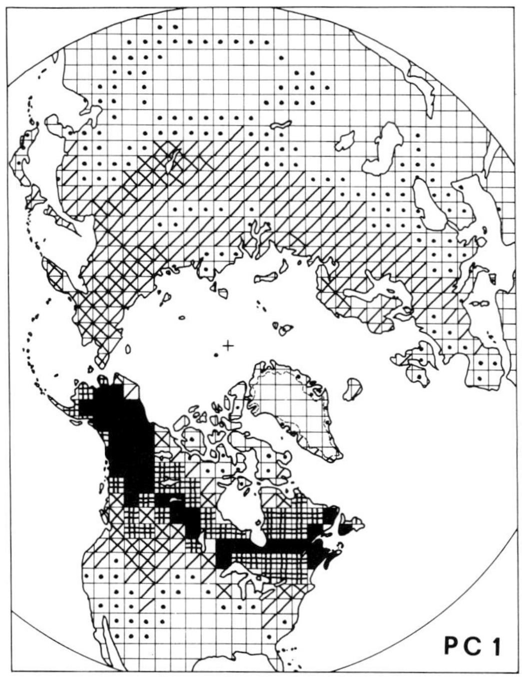

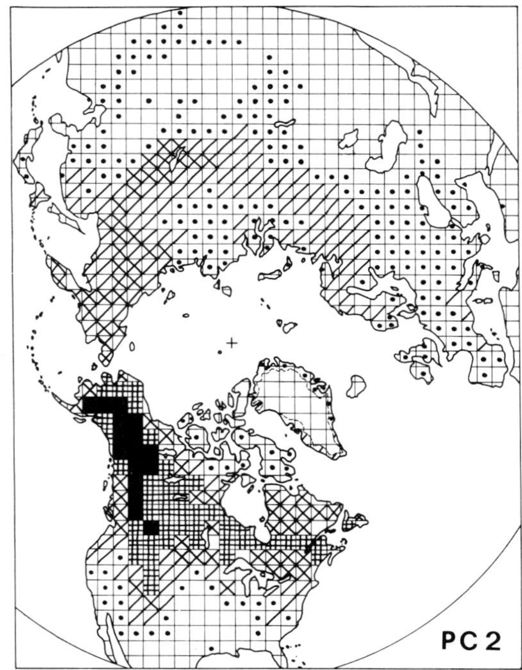

Fig. 15. Chorograms of the two relevé groups obtained by numerical classification of the data in Table 9 (*Pinus contorta* woods). Different shadings refer to the percent of species occurring in the OGUs, on a 5-class scale with intervals of 20%.

PC1 *Pinus contorta–Hylocomium splendens* community

This is a closed-canopied coniferous forest dominated by *Pinus contorta*, in whose understory mosses and ericaceous chamaephytes are prevalent. The chorogram (Fig. 15) shows a prevalence of boreal North American species, and is similar to the one of PG3.

The soils are mostly sandy Brunisols with acid pH (average 5.3). The organic horizon is 3–7 cm deep. The main humus form is Moder. Drainage is moderate (55% of the plots) to excessive (45% of the plots). Frozen ground occurs in 30% of the plots in June at an average depth of 30 cm, and is always absent in September.

In Table 7 three subgroups of relevés may be distinguished: the first (Rel. 1–5) is characterized by the constant presence and high cover of *Alnus crispa, Thuidium abietinum, Pyrola asarifolia*, and by the absence of *Festuca altaica* and *Lupinus arcticus*. The soils are sandy-gravelly and the stands mostly occur on flat ground, so that drainage is moderate.

The second subgroup is characterized by *Betula glandulosa, Lycopodium complanatum, Lophotia barbata, Lycopodium annotinum, Vaccinium uliginosum, Lycopodium clavatum*. The soils have higher amounts of silt and clay, and the drainage is moderate to seasonally impeded for the presence of frozen ground in early summer, a fact that explains the presence of mesophytic–hygrophytic species.

The third subgroup is characterized by *Stereocaulon tomentosum* and *Arctostaphylos uva-ursi*, and by the higher frequency of fruticose lichens. The soils are similar to those of the first subgroup, but most stands are on rather steep slopes, so that drainage is always excessive.

This is the most xeric variant of the community, and represents a transitional stage towards the *Pinus*–lichen woodland community (PC2).

This community is restricted to the southeastern part of the Alaska Highway, within the range of *Pinus contorta* in the Yukon (Fig. 3).

PC2 *Pinus contorta* – lichen woodland community

This is an open-canopied secondary forest dominated by *Pinus contorta* and by fruticose lichens. Macrolichens are responsible for the

Table 8. Ecological data relative to the relevés of *Pinus contorta* woods. For the abbreviations see the caption to Table 4.

Releve No.	1	2	3	4	5	6	7	8	9	10	11	12	13	14	15	16	17	18	19	20
Total cover	90	100	100	100	90	100	100	70	90	70	80	100	100	100	100	100	90	100	90	100
Cover of trees	70	65	60	60	70	35	60	40	70	50	20	70	70	70	30	50	50	70	50	50
Cover of shrubs	50	10	50	30	10	65	60	20	30	40	40	20	10	40	40	50	10	10	5	10
Cover of grasses	5	15	5	5	5	15	30	30	5	30	40	30	5	5	5	5	70	60	60	25
Cover of lichens	5	5	10	10	5	10	5	20	5	5	20	10	5	5	45	25	10	10	5	5
Cover of mosses	10	90	90	80	30	70	50	30	60	30	30	70	30	70	50	30	30	60	5	80
Exposure	E	-	-	SSE	S	S	S	SSW	S	SE	S	NNE	E	-	NNE	N	WSW	SSE	-	E
Slope	10	-	-	10	10	10	10	20	5	15	5	15	10	-	40	10	40	5	-	5
Soil type	E. DYB	O. DYB	GL. DYB	O. DYB	E. DYB	E. DYB	O. EB	GL. DYB	GL. DYB	O. DYB	GL. DYB	GL. HR	E. DYB	O. DYB	E. DYB	E. DYB	O. R	GL. DYB	O. DYB	O. DYB
GS %	40	75	35	30	75	33	5	20	15	20	15	20	46	4	35	50	40	75	50	25
SA %	55	20	58	50	23	46	85	20	30	70	30	60	55	72	60	35	40	20	45	60
SI %	3	4	5	16	2	20	10	30	40	9	40	19	3	22	2	13	19	5	4	14
CL %	2	1	2	4	-	1	-	35	15	1	15	-	2	2	2	2	1	-	1	1
Org.(cm)	4	4	4	4	6	9	4	2	3	7	9	7	4	3	4	4	1	3	4	2
pH	5.4	4.9	5.0	6.0	5.5	5.3	5.4	5.1	5.0	5.1	5.0	5.6	4.9	4.9	4.5	4.9	5.7	5.1	6.2	6.1
Erosion pot.	low	low	low	low	low	med	med	med	med	med	med	hi	low	low	med	low	hi	low	low	low
Drainage	mod	mod	imp	mod	mod	mod	mod	imp	imp	mod	mod	imp	mod	ex	mod	ex	ex	ex	ex	ex
Depth of ice June (cm)	-	20	-	50	-	18	10	7	-	-	-	13	-	-	7	-	-	-	-	-

Releve No.	21	22	23	24	25	26	27	28	29	30	31	32	33	34	35	36	37	38	39	40
Total cover	100	100	80	60	60	70	85	80	85	80	100	50	70	90	70	80	80	80	70	100
Cover of trees	70	25	60	30	50	40	70	70	60	70	50	20	-	-	60	20	30	20	40	-
Cover of shrubs	30	15	5	10	5	10	5	5	5	5	5	30	60	60	5	10	10	10	50	80
Cover of grasses	5	-	10	35	5	50	50	50	10	60	35	30	40	30	5	10	10	50	30	5
Cover of lichens	40	70	25	20	20	10	10	10	10	10	15	20	40	10	60	50	40	30	40	10
Cover of mosses	20	10	10	5	5	5	5	5	5	15	60	10	10	20	10	20	10	10	10	20
Exposure	SW	SW	SE	-	S	SW	SSE	S	SSE	ESE	S	ESE	SE	S	SE	NE	S	SSE	SW	-
Slope	5	5	5	-	30	5	5	20	30	10	15	35	40	40	20	5	5	5	5	-
Soil type	O. DYB	E. DYB	O. EB	O. DYB	CU. R	O. DYB	O. DYB	O. DYB	O. DYB	O. DYB	O. DYB	E. DYB	CU. R	O. R	CU. R	O. EB	O. DYB	O. DYB	O. DYB	E. DYB
GS %	60	20	1	5	1	20	5	5	5	5	75	70	90	90	40	1	5	5	50	40
SA %	30	75	50	93	50	79	93	93	93	93	20	25	10	5	35	50	93	93	47	50
SI %	10	4	48	2	48	1	2	2	2	2	4	4	-	5	25	48	2	2	3	9
CL %	-	1	1	-	1	-	-	-	-	-	1	1	-	-	-	1	-	-	-	1
Org. (cm)	4	9	2	-	2	4	2	4	3	3	3	4	-	-	2	2	2	2	5	5
pH	5.5	5.4	6.2	5.7	6.7	5.8	6.0	5.5	6.7	6.1	4.9	4.9	6.9	6.4	.	6.0	5.6	5.7	5.8	5.8
Erosion pot.	low	low	med	low	hi	low	low	med	med	low	med	med	hi	hi	hi	med	low	low	low	low
Drainage	mod	mod	mod	ex	ex	ex	ex	ex	ex	ex	ex	ex	ex	ex	ex	ex	ex	mod	ex	ex

high incidence of thallochamaephytes (40%). Boreal North American species are less frequent than in PC1 (Fig. 15). The high frequency and cover of lichens is the most striking feature of this community. The lichen synusia corresponds to the *Cladonietum mitis* Krieger, and it has already been discussed by Nimis (1981).

Two subgroups can be distinguished in Table 7. The first is characterized by the high frequency of *Cladonia cornuta, C.coccifera, C.arbuscula, C.multiformis*; the second by *Cladonia cariosa, C.chlorophaea* and *Gentiana propinqua*. No significant pedological differences exists between the two subgroups. They can be interpreted as two successional stages following fire: the second subgroup is probably an early stage of the community (presence of *Gentiana propinqua*, one of the few therophytes of the Yukon flora, and of the rapidly growing *Cladonia cariosa*, a good fire indicator in the survey area). The first subgroup represents a more mature stage (presence of slow-growing *Cladoniae*, chiefly *Cladonia arbuscula*).

Brunisols occur in 63%, Regosols in 27% of the plots. The soils are always sandy, with an organic horizon of 2–6 cm. The community often occurs at the top of old, stabilized sand dunes. Moder is the main humus form. Frozen ground is always absent and drainage is excessive in 80% of the plots, mainly because of the high sand content in the soils. The pH is generally acid, with an average of 5.8. Although soil conditions

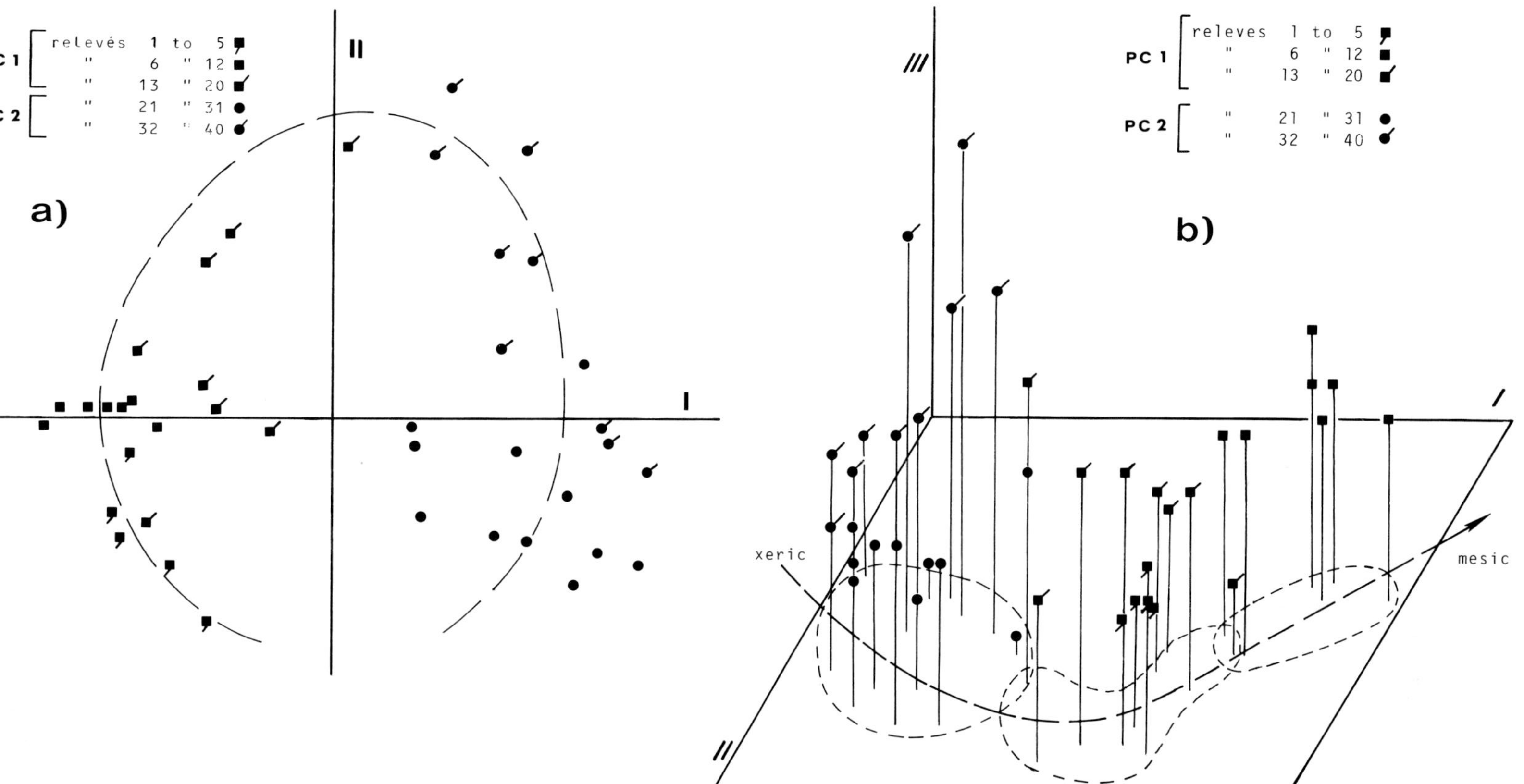

Figure. 16a–b. Ordination (by PCA) of the relevés of *Pinus contorta* woods on the basis of the data in Table 7. Principal Components I and II; Principal Components I, II and III. Symbols refer to subdivisions within each relevé group, as in the legend.

are good for podzolization, true podzols are very rare in the area, because of low precipitation. Excessive drainage causes the open-canopied appearance of this community: the trees are widely spaced because of root competition for water. This is particularly evident on old sand dunes.

This community is restricted to the south-eastern portion of the Alaska Highway (Fig. 3).

4.3.2. Ordinations

PCA, using the correlation coefficient on binary data, was performed on the data of Table 7. The scatter diagram according to the first three components is given in Figure 16. On the first and third Principal Components three main clusters can be distinguished: the first (left) includes all the relevés of PC2, the second (right) includes relevés 6–12 of PC1, and the third (center), includes relevés 13–20 and 1–5 of PC1. The third Principal Component separates relevés 32–40 from the others belonging to PC2. At the left side of the scatter diagram are relevés of stands whose soils have moderate drainage and higher content of silt and clay. From left to right there is also a trend towards a slight acidification in the soils: in PC2 (left) the average pH is 6.0 ± 0.5, whereas in relevés 6–12 of PC1 the average pH is 5.2 ± 0.2. Considering the ecological requirements of the differential species, the arrangement of the relevés in the ordination seems to reflect mainly a gradient from xeric (left) to mesic (right) conditions.

4.3.3. Discussion

Pinus contorta occurs at elevations ranging from sea level along the Pacific coasts to over 3350 m in the Sierra Nevada and in the Rocky Mountains. It extends from 64° in the Yukon Territory to the S. Pedro Martir Mountains of Baja California, latitude approximately 31°. Longitudinally, it extends from the Pacific coasts to the Black Hills of South Dakota (Mirov 1967). The extensive range of *Pinus contorta* overlaps the range of the largely trans-Canadian *Pinus banksiana* in Alberta (Moss 1949) and in the Mackenzie region (Raup 1947). Where the two come in contact, they cross, forming extensive hybrid areas (Liddicoet and Righter 1947; Moss 1949). According to Mirov (1967) before glaciation *Pinus contorta* was the only *Pinus* species growing in the northwest. Hultén (1937) claims that it migrated southwards before the Wisconsin glaciation from refugia in the upper Yukon valley. A study by Jeffers and Black (1963) provides evidence for the existence of refugial stations also along the Pacific coasts. Succession and establishment of pines in the post-Pleistocene were sometimes determined not by glaciation, but by edaphic factors or by the influence of fire (Hansen 1947, 1950). According to Mirov (1967) the occurrence of *Pinus contorta* in excessively dry or wet sites is due to its low competitive power compared to other tree species.

In the Yukon, *Pinus contorta* is clearly favoured by fires; its more or less closed forests are a secondary successional stage leading towards a *Picea glauca* forest. This is particularly evident for the community whose understory is dominated by mosses (PC1). However, already Clements (1916) stated that *Pinus contorta* may establish itself as a permanent climax in a given site. This consideration chiefly applies to the open *Pinus*–lichen woodlands (PC2). Such woodlands, dominated by different *Pinus* species, or by other conifers (Larsen 1980), are widespread throughout the boreal zone. The succession of boreal lichen woodlands after fire has been studied in Europe (Jalas 1953; Uggla 1958), Asia (Rabotnov 1936) and North America (Ahti 1957, 1959; Scotter 1970; Rouse and Kershaw 1971). According to Ahti (1977) the successional pattern can be subdivided into five main stages, culminating in a mature reindeer stage dominated by *Cladonia stellaris* ca. 80 (120) years after fire. Because of the widespread occurrence in the boreal zone of glacial sandy deposits and rock outcrops with very oligotrophic and dry soils, the occurrence of open tree stands with understory dominated by lichens is frequent. It is obvious that in such conditions the successional development towards more mature stages is extremely slow, which would explain the climax-role attributed by Clements (1916) to *Pinus contorta*. In the survey area, the PC2 community can be considered an indicator of old, stabilized sand dunes or, in general, of dry, oligotrophic soils.

Pinus contorta, owing to its broad distribution and ecological tolerance, cannot be considered a characteristic species of any association. However, it is noteworthy that the ground layer of the *Pinus*–lichen woodlands is very uniform throughout the boreal zone (Kornas 1972). According to Ahti (1977) it includes 25–40 species of lichens and 5–15 bryophytes. The lichen synusia corresponding to the more mature stages (*Cladonietum alpestris*; Klement 1955; Looman 1964; Nimis 1981; or *Cladonia alpestris* sociation; Trass 1968) was described from Europe, and has a circumboreal distribution.

4.4. Broadleaved tree stands

4.4.1. Classification

This data set includes 62 relevés (Table 9) of open stands dominated by broadleaved low trees (chiefly *Alnus, Populus* and *Betula*). The ecological data are in Table 10 and the dendrogram of the relevés is in Figure 2. Three main clusters are formed: BL1, BL2 and BL3. The cluster BL3 can be further subdivided into three subclusters: BL3a, BL3b and BL3c, which are considered here as facies of BL3. The chorograms of the three relevé groups are shown in Figure 17.

BL1 *Salix arbusculoides–Alnus incana* community

These are stands dominated by broadleaved trees and shrubs, occurring along rivers and creeks. The most frequent dominant tree species is *Alnus incana*, followed by *Populus balsamifera* and *Populus tremuloides*. A good characteristic species for this relevé group is *Salix arbusculoides*. The chorogram (Fig. 17) reveals a high incidence of boreal northwestern North American species.

Quantitative soil data are not available for most of the relevés. The soils are mostly gravelly Regosols, sometimes with an upper layer of finer fluvial deposits. The stands are often flooded and water availability is good, a fact that is reflected in the presence of rather hygrophytic species such as *Ranunculus macounii, Anemone richardsonii, Arctostaphylos rubra, Potentilla fruticosa, Polygonum viviparum* and *Carex aquatilis*. The first four relevés were taken along the Dempster Highway, in the Subalpine–Subarctic zone. In this area the narrow strip of vegetation along rivers is one of the few sites with permafrost-free soils, where permafrost-intolerant species, such as the geophytes *Cypripedium passerinum* and *Habenaria obtusata*, find the northernmost refugial stations. The remaining two relevés (Nos. 5, 6) were taken along the Alaska Highway (Fig. 3), were this community type is also present, apparently with little floristic variation.

BL2 *Betula papyrifera–Alnus crispa* community

These are thickets dominated by *Alnus crispa* occurring along meltwater channels on slopes in the Subalpine and upper montane belts of the southern Yukon, on siliceous parent material. Most of the specimens of *Alnus crispa* belong to subsp. *crispa*, but intergradation with subsp. *sinuata* was sometimes observed. Among the differential species, *Boschniackia rossica* could be considered as characteristic of this community type, being an obligate parasite on *Alnus*.

The stands are closed canopied, with a well-developed layer of high shrubs and sometimes a tree layer dominated by *Betula papyrifera*. The chorogram (Fig. 17) is similar to the one of the previous community, with a higher incidence of wide ranging boreal species also occurring in eastern Siberia.

Notwithstanding the accumulation of litter in the upper soil horizon, a moss layer is almost always present, dominated by *Hylocomium splendens*. No quantitative soil data are available, except for relevé No. 17. In general, the soils are distric or eutric Brunisols with a thick surficial organic layer formed by decaying leaves.

Table 9. Phytosociological table of broadleaved tree stands. The relevés are arranged according to the results of numerical classification (see Fig. 2).

	BL1						BL2												BL3																																																
																		a													b																c																			
RELEVÉ No.	1	2	3	4	5	6	7	8	9	10	11	12	13	14	15	16	17	18	19	20	21	22	23	24	25	26	27	28	29	30	31	32	33	34	35	36	37	38	39	40	41	42	43	44	45	46	47	48	49	50	51	52	53	54	55	56	57	58	59	60	61	62					
Salix arbusculoides	2	3	3	2	2	2																																																													5				
Salix alaxensis			2	2	3	2																															2																								3	2		5			1	1
Pyrola grandiflora		2	3	3	3								2																																			2												2			4	1			1
Equisetum arvense	2	2	2		3	2																																																												2			5				1
Anemone richardsonii	2	2		2	2	2																																																													5				
Ribes hudsonianum	2	2				2			2																										2																									2					3			1	1
Alnus incana ssp.tenuifolia		5	5	3	3																																																														4				
Habenaria obtusata	2	2	2																																																															3				
Rubus arcticus	2	2				2																																																															3				
Ranunculus macounii				2		2																																																															2				
Cypripedium passerinum			3		2																																																														2				
Sanguisorba officinalis	2		2																																																															2				
Polemonium coeruleum	2	2																																																														2				
Hylocomium splendens					2		5	5	5	3		3	2	2	3	3		5																				2	2			2								2																1	4		1	1
Alnus crispa							8	8	8	8	5	5	5	2	8	3	5	3																																														5			
Betula papyrifera							3	2		3	3	7	7	2													3								5	3											2																		3	1	1	
Ledum groenlandicum							2	2	2			2		2		2																																		2	2	3	2																	3		1	1
Ribes triste	2						3	3	3	5		2	3		3	2	3																																	2																		1	4		1	
Polytrichum juniperinum							2	2	2		2		2	2			3																		2																			2													3		1	1
Peltigera aphtosa								2	2					2	2			2	2																									3	2																										3		1	
Equisetum scirpoides	2	2			2					2	2			2	2	2		5																															2			2														3	3		1	1
Boschniackia rossica							2	2	2	2	2		2					2									2																																								3	1		
Gymnocarpion dryopteris							2	2	2	2					2	2																																																			3			
Lycopodium annotinum											5			2		2											2																																								2	1		
Aconitum delphinifolium						2	3	3	3	3					2																																																1	3			
Solidago decumbens																				2	2	2			2	3	3	2	3	2	2												2								1				2	2		2	2	2			2			4	1	2
Galium boreale	2		2																2	2		2	2		2	2	2	3	2																				2										2									2		4	1	1
Pulsatilla patens																		5	2	2	2			2						2																																				1	2		
Ribes oxyacanthoides																2									2			2	2																																						1	2		
Penstemon procerus																											2		2	3																													2										2		1
Polemonium pulcherrimum																												2	3																													2										1		1
Aquilegia brevistyla																												2	2											3																													1	1	
Oxytropis campestris																												2			2		2																																			1	1	
Hierochloe odorata																												2	2																																							1		
Geum perincisum																												2	2																																							1		
Calamagrostis purpurascens																								2	2	2																																									2		
Poa glauca																									2	2																																	2						1		1
Saxifraga tricuspidata																									2	5																					2																				1	1	
Juniperus horizontalis																						2	2																																													1		
Drepanocladus uncinatus			2																2	3			3									3			2	2	2	2	2	2	2	2					3	3							2		3				2			2	3	2
Lupinus arcticus																			2	3													2	2		2		3	3	3	2	2	2	2	2	2	2																	2	3	1		1	5	1
Peltigera rufescens																																2													2	3	2	2	3		2	2	2	3	2	2	2		2	2		2						2	4
Ceratodon purpureum																																		2									2	2	2	2	2	2	2	3	2	3		3		2		3	2		2				2	4
Populus balsamifera	2		8	8										5		3												2	5			5		3			2	2	2			2			2							2				3			3	3	7	3	1	1	3	2
Tortula ruralis																																3							2							2	3	2							2	2		2	2	2							2	2
Carex concinna																																						2	2							2		2	2		2		2						2	2	2				1	3
Salix scouleriana																	3																2	3																	2			2	2						5		2			2		1		1	2
Cladonia pyxidata																															2	2								3		2																	3	2	2					1	1	1
Pohlia nutans																																					2		2					2						2							2	2							1	1
Oxytropis deflexa																					2															2																								2	2	2	2				1	1	2
Gentiana propinqua																								2																			2	2	2						2		2											1	1	1
Cladonia cariosa																																		2											4	3											3		2							1	1
Cornus canadensis																																			2						2	2		2	2																					2	
Pedicularis labradorica																																						2		2	2			2		2																				2	
Arctostaphylos uva-ursi																			5	3	8	3	7	7	3	7		5	2	7	7	2	5	5	5	3	5	3	3	3	7	2	3	5	2	3		3	3	3	2	3	3	5	2	8	2	7	3	2	8				5	5	5
Shepherdia canadensis			5	7	7								5					3	2			5		3			5	2		2	2	3	3	2	2		2	2	3	2	3	5	2	5		3	2	5		2		2	2	2	2	2	2	2	2	5	7	3	1	3	5	5
Salix glauca		3													2			2		2		2		2			3	2		3					3	2		2	2	2	3						2	2	3	3	3	2	2	2	3		2	2	5			1	1	3	2	4

Species																																																																				
Festuca altaica							2	2											2	3	2	3	3	2				2			2				2		2		2	2		3	2	5	2	2	2					2			3				2			2		1	4	4	2
Anemone multifida																							2		2		2	2	2	2	2	2	2																	2			2		2	2	2			2			2			4	1	3
Achillea borealis																				2	2	2	2	2				2	2			2	2											2		2							2			2	2	2		2							3	2	2
Fragaria virginiana	2																				2	3					2									2															3	2		2	2	2		2						1		2	1	2
Zygadenus elegans															3					2	2	2					2								2	2																			2	2			2						1	2	1	2
Hedysarum alpinum			2	2	3																2	2	2	2									2																			2	2			2	2	2						3		2	1	2
Pinus contorta																	2			2																		3					2	2			3	3																	1	1	2	1
Arnica cordifolia																	2		3																				2			2	2	2	2																			1	1	2	
Cladonia ecmocyna																	2																							3	2	2		2																			1	1	2	
Populus tremuloides	2				7	7						3	3	7		3			5	5	7	5	7	5	8	2	2	8	3	3	3	3	7	7	3	5	5	7	5	7	7	5	7	7	5	7	7	5	3	7	5	5	5	7	5	7	2	5	3				3	2	5	5	4
Epilobium angustifolium	2		2	2	2	5	2	2	2					5		3	2		2	2	2	3	2	2	2	2	2	3				3		3	2	2	2	2	2	2	2	2	2	2	2	2	2	2	2	2	3	2	3	2	2		2	2		2	2	3	5	3	4	5	5
Rosa acicularis	2		3	3	3	3					2	2	2	3		2	2		2	2	2	2	2		2	2	2		2	2	2	2	2	3	2	2		3	2	2	3	2	3		2		2			2		2	2	3	2	2						3	5	3	5	5	3
Picea glauca	3		2	2	2		2	2	2			3	2	3		2	2	7		2	2	2	3	2				2	2	2	2	2		2			3		2	2		2				1		2	3			2	1				2	3	2	2	2	3	4	4	4	3	4
Mertensia paniculata	2	2	2	2	2	3	2	2	2	2	2	2	3	5	2	3	2					3						2	2			2	2	2	2	2	2	2	2	2	2		2	2	2										2	2			2	2	3	5	5	2	5	2
Linnaea borealis				2			2	2	2	2	2			5		3	2			2												3		5		3	3	2	3	2	3	2	2		5	2		3		3	2	3		3		3								1	4	1	4	2
Calamagrostis canadensis	2	2	2	2	2		2	2	2	3	1	2	3	2						2		2		2				2		2	2	2		2	2				3				2		2		2		2					3				2				5	4	3	2	2
Pyrola secunda													2	2	2			2			2																		2	2	2	2	2	2	2	2	2											2							2	1	3	1
Pyrola asarifolia							2	2	2		2		2	2	2		3																					2	2	2		2	2									2			2								4		2	1
Vaccinium vitis-idaea							2	3	3	3		2	2	3										3													7	2				3			3		2		2	3														3	1	2	1
Viburnum edule				5	2		2				2	3	3	3	2	2									3									3	2					2	5	3	5																			2	3	1	2	
Geocaulon lividum												2		2	2			2														2	2	2	2			2	2			2																						2		3	
Juniperus communis																			3															3												2		2																				1	1	
Delphinium glaucum			2			3														2	2	2																										2																			2		2	1	
Senecio atropurpureus																							2																																	2				2		2								1		1
Arctostaphylos rubra	2	3		2	2		2							2																																				3	2			2											4	1			1
Stellaria longipes								2		2	2											2																																															2	1		
Rhytidium rugosum									2						2	2																					2																															2		1	
Dicranum bergeri							2					2	2																																					2																	2			1
Valeriana capitata					2		2	2		2																																																									1	2			
Potentilla fruticosa				3	3										2																																																								2	1			

Table 10. Ecological data relative to the relevés of broadleaved tree stands. For the abbreviations see the caption to Table 4.

Releve No.	1	2	3	4	5	6	7	8	9	10	11	12	13	14	15	16	17	18	19	20	21
Total cover	80	65	100	100	100	100	90	90	90	100	90	100	95	100	100	90	90	100	85	70	95
Cover of trees	10	-	90	90	95	85	5	5	-	-	5	70	80	90	-	40	-	50	70	70	70
Cover of shrubs	60	65	85	85	90	40	80	80	80	80	80	60	70	70	80	40	70	30	50	20	20
Cover of herbs	10	5	40	40	55	80	5	5	50	50	5	50	50	50	50	30	40	10	50	50	90
Cover of lichens	-	-	-	-	-	-	-	-	-	-	-	-	1	1	-	-	5	-	-	-	-
Cover of mosses	1	5	-	1	1	-	50	50	20	10	-	25	10	-	10	20	20	20	1	5	-
Exposure	W	SE	-	-	-	-	N	N	N	NE	N	SW	NE	NE	N	N	SW	N	SW	-	-
Slope	20	15	-	-	-	-	25	35	25	15	35	25	10	25	35	30	7	30	26	-	-
Soil type	GL.	O.	.	.	.	.	.	.	.	.	.	.	.	.	.	.	GL.	.	O.	E.	GL.
	CUR	HG	.	.	.	.	.	.	.	.	.	.	.	.	.	.	DYB	.	DYB	DYB	DYB
GS %	70	30	.	.	.	.	.	.	.	.	.	.	.	.	.	.	25	.	.	1	5
SA %	20	55	.	.	.	.	.	.	.	.	.	.	.	.	.	.	20	.	.	90	75
SI %	9	14	.	.	.	.	.	.	.	.	.	.	.	.	.	.	24	.	.	9	18
CL %	1	1	.	.	.	.	.	.	.	.	.	.	.	.	.	.	1	.	.	-	2
Org. (cm)	-	25	.	.	.	.	.	.	.	.	.	.	.	.	.	.	15	.	4	3	6
pH	7.1	7.5	.	.	.	.	.	.	.	.	.	.	.	.	.	.	5.4	.	6.8	6.2	7.0
Erosion pot.	hi	hi	.	.	.	.	.	.	.	.	.	.	.	.	.	.	med	.	low	low	low
Drainage	imp	imp	.	.	.	.	.	.	.	.	.	.	.	.	.	.	imp	.	mod	mod	mod
Depth of ice June (cm)	-	15	.	.	.	.	.	.	.	.	.	.	.	.	.	.	13	.	-	-	-

Releve No.	22	23	24	25	26	27	28	29	30	31	32	33	34	35	36	37	38	39	40	41	42
Total cover	90	80	75	100	90	50	85	70	85	100	70	65	100	100	80	75	70	60	70	80	100
Cover of trees	80	80	40	-	5	-	80	70	-	-	60	-	60	25	25	20	40	60	45	60	50
Cover of shrubs	10	15	10	90	5	5	35	5	20	30	5	45	35	20	20	65	25	5	30	60	60
Cover of herbs	60	15	70	20	90	50	60	20	80	80	40	40	40	80	60	40	30	5	30	35	60
Cover of lichens	-	-	-	-	-	-	-	-	1	1	1	-	-	-	-	-	-	5	5	1	1
Cover of mosses	-	5	-	-	-	-	-	-	-	-	1	5	-	20	1	1	1	5	5	1	1
Exposure	NW	SE	SE	S	S	S	S	S	SW	-	E	E	S	S	S	-	S	NE	SW	SW	E
Slope	5	5	5	45	30	30	5	5	5	-	5	10	20	20	25	-	10	5	10	5	10
Soil type	O.	O.	O.	O	O	O	.	CU.	GL	O.	O.	CU.	.	.	.	O.	O.	O.	O.	O.	O.
	EB	EB	EB	N	N	N	.	R	EB	EB	G	R	.	.	.	SB	EB	EB	DYB	R	DYB
GS %	5	5	-				.	2	5	5	15	80	.	.	.	15	10	5	50	5	20
SA %	35	20	20	R	R	R	.	83	75	75	85	20	.	.	.	30	60	20	30	60	50
SI %	30	40	45	O	O	O	.	13	18	18	-	-	.	.	.	54	28	40	19	34	28
CL %	30	35	35	C	C	C	.	2	2	2	-	-	.	.	.	1	2	35	1	1	2
Org. (cm)	7	5		K	K	K	.	2	5	-	9	3	.	.	.	4	9	-	9	1	5
pH	6.9	6.4	.	.	.	.	.	7.0	7.1	6.2	7.7	7.3	.	.	.	6.8	6.9	7.6	6.2	7.4	5.9
Erosion pot.	med	low	med	.	.	.	.	low	med	med	low	low	.	.	low	low	low	med	low	hi	med
Drainage	mod	mod	mod	.	.	.	.	mod	mod	ex	mod	ex	.	.	ex	mod	ex	ex	mod	mod	mod
Depth of ice June (cm)	-	30	-	.	.	.	.	-	-	-	-	-	.	.	60	-	-	-	-	-	-

Releve No.	43	44	45	46	47	48	49	50	51	52	53	54	55	56	57	58	59	60	61	62
Total cover	100	90	90	85	60	70	60	80	45	70	60	80	60	100	60	80	70	60	100	70
Cover of trees	70	90	-	-	-	50	25	80	35	60	60	65	50	65	5	70	-	5	15	40
Cover of shrubs	60	20	65	70	60	5	30	5	5	10	5	15	10	10	40	15	40	50	25	50
Cover of herbs	30	20	30	20	40	45	5	25	2	25	5	50	40	80	10	25	60	15	90	25
Cover of lichens	-	1	10	5	5	10	5	1	15	1	5	1	1	1	15	1	5	-	5	-
Cover of mosses	1	1	5	15	5	10	10	5	15	1	5	1	5	1	5	5	10	5	5	5
Exposure	S	E	S	-	S	NE	N	NE	-	N	NE	S	NE	SE	SE	-	S	-	N	SW
Slope	10	10	20	-	30	20	15	20	-	5	5	5	5	5	5	-	10	-	5	13
Soil type	O.	O.	O.	O.	CU.	CU.	O.	O.	CU.	CU.	O.	O.	O.	O.	O.	O.	O.	GL.	O.	O.
	DYB	DYB	R	R	R	R	SB	EB	R	R	EB	EB	MB	EB	G	G	EB	SB	SB	SB
GS %	50	60	90	4	98	80	15	2	2	5	5	-	5	5	85	10	5	15	50	5
SA %	45	25	5	50	1	5	45	30	30	20	20	60	70	35	15	30	15	60	30	25
SI %	5	13	5	45	1	14	30	40	40	40	40	35	40	30	-	30	40	20	20	45
CL %	-	2	-	1	-	1	10	28	28	35	35	5	35	30	-	30	40	5	-	25
Org. (cm)	5	3	-	3	2	4	15	2	5	-	-	5	2	2	4	-	8	10	3	10
pH	6.2	6.7	6.4	7.5	6.7	6.8	7.5	7.6	7.3	8.3	7.6	7.5	7.4	7.8	6.4	6.3	7.4	7.3	7.8	7.2
Erosion pot.	low	low	hi	med	hi	med	med	med	med	med	med	med	med	med	low	low	hi	med	low	med
Drainage	ex	ex	ex	ex	ex	mod	ex	ex	mod	mod	ex	mod	mod	mod	imp	imp	ex	mod	ex	mod
Depth of ice June (cm)	-	-	-	-	-	-	-	-	-	-	-	-	-	-	30	25	-	-	-	-

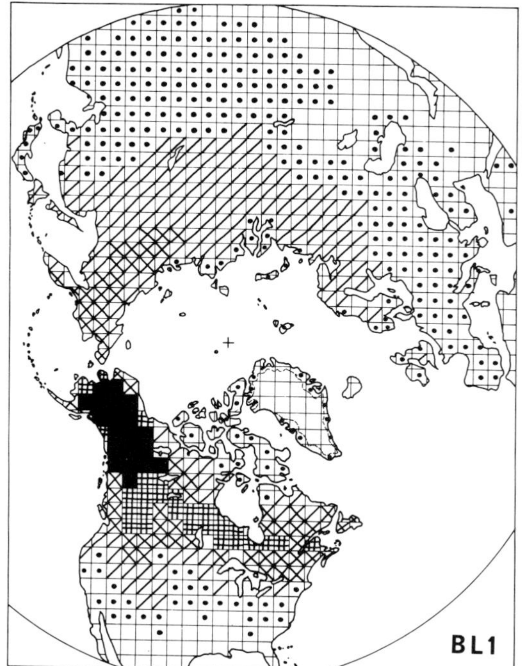

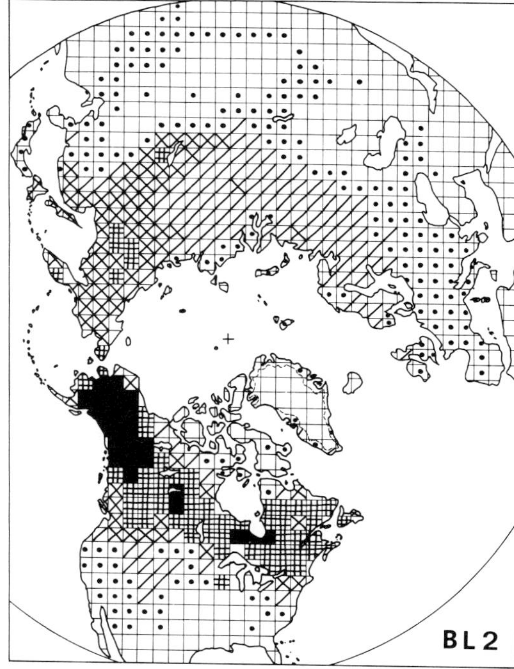

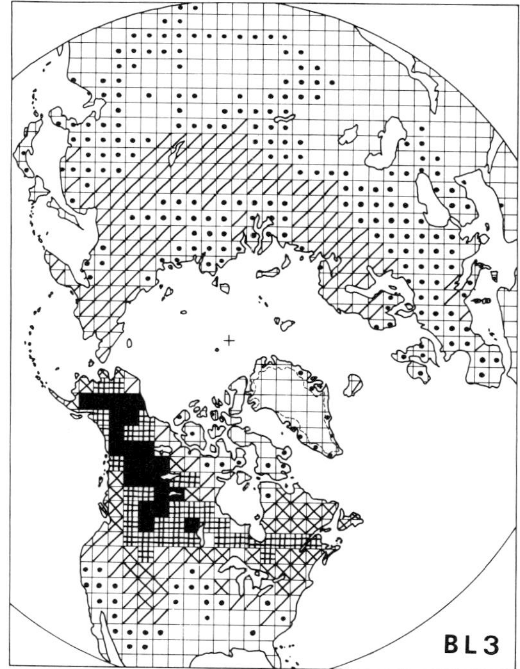

Fig. 17. Chorograms of the three relevé groups obtained by numerical classification of the data in Table 9 (broadleaved tree stands, see Fig. 2) Different shadings refer to the percent of species occurring in the OGUs, on a 5-class scale with intervals of 20%.

The lower horizons have signs of gleyzation. This is due to the fact that, although all of the stands occurred on slopes (inclination range from 5° to 35°, see data in Table 10), they were mostly northern exposed and located in drainage channels, so that water availability is good. Although direct pH data are not available, the stands of this group probably occur on soils with the lowest pH among all broadleaved stands.

This can be deduced by the presence of acidophytic species such as *Hylocomium splendens, Polytrichum juniperinum, Peltigera aphtosa* and *Lycopodium annotinum*, and by the absence of a whole set of neutro- to basiphytic species that are frequent in the relevés of the next group (BL3). In relevé 17 the pH is 5.4. This community is most frequent in the western section of the Alaska Highway, were precipitations are highest, from Koidern to the Alaskan border (Fig. 3).

BL3 *Shepherdia canadensis–Populus tremuloides* community

This is the most frequent deciduous broad-leaved forest type in the Yukon Territory. It is dominated by *Populus tremuloides*, with *Populus balsamifera* as occasional dominant or codominant. Most of the stands are secondary woods following fire. The principal structural difference with the secondary woods dominated by *Pinus contorta* is the scarcity of bryophytes and lichens in the understory and the higher frequency of hemicryptophytes. Circumboreal species have a relatively low incidence, whereas boreal North American and Cordilleran species are rather frequent (see Fig. 17).

The soils are mainly orthic and eutric Brunisols, with high sand content. Regosols occur in 32% of the plots. The organic horizon is poorly developed, with an average thickness of 4 cm. The humus is always of the Mull type. Most of the releves were on slopes, with moderate to excessive drainage. The pH is always subneutral to alkaline, with an average of 7.1. This is the main ecological difference with the secondary *Pinus contorta* woods, that always occur on acid soil.

The high pH values do not necessarily reflect a difference in parent materials, mostly being a consequence of the accumulation of *Populus* litter and of low precipitations.

This community is the main secondary forest type on well-drained Brunisols in the central section of the Alaska Highway. It also occurs within the distributional range of *Pinus contorta*, in the southwestern section of the Highway, where, however, it is confined to very steep, mostly southern exposed rocky slopes (Fig. 3).

The results of the numerical classification allow the distinction of three variants within the community (see Table 9). They are:

(a) Variant with *Galium boreale* (BL3a).
(b) Variant with *Mertensia paniculata* (BL3b).
(c) Variant with *Ceratodon purpureus* and *Peltigera rufescens* (BL3c).

The first variant includes stands that are mostly located on south-facing rocky slopes with strong erosion. They were affected by fire about 52 years ago (± 15 years), and are relatively mature. Floristically, this variant is quite heterogeneous, being characterized by the high frequency of *Galium boreale* and *Solidago decumbens*, and by the occasional presence of species of xeric grasslands. This variant depends on slope and exposure. The xeric topoclimate and the strong soil erosion do not allow the development of a moss carpet or of species typical of mature *Populus* stands, despite the relatively old age of the dominant trees. It is noteworthy that in 5 of the 13 relevés included in this variant no traces of past fires have been observed. They probably represent primary stands of *Populus tremuloides* on south exposed rocky slopes, where the development of a boreal forest dominated by *Picea glauca* is very slow because of the unfavorable ecological conditions.

Variants (b) and (c) appear to be dynamically linked. The former includes relevés of open stands recently affected by fire (29 years ago ± 10), and is characterized by the high frequency of two heliophytic cryptogams: *Peltigera rufescens* and *Ceratodon purpureus*. In contrast, variant (b) includes mature stands (average distance of fire 63 years ago), with a moss layer dominated by *Drepanocladus uncinatus* and with a set of mesophytic phanerogams. A successional trend is thus evident from the variant with *Ceratodon* to the one with *Mertensia*, the latter already including species of mature *Picea glauca* forests. Summarizing, the three variants respectively represent: (a) primary stands on south exposed slopes; (b) young secondary stands following fire; and (c) old secondary stands evolving towards a *Picea glauca* forest.

4.4.2. Ordination

The relevés of Table 9 were subjected to PCA using the correlation coefficient and binary data. The results are shown in Figure 18. In the space defined by the first three Principal Components, the relevés belonging to BL1, BL2 and BL3 are positioned at different levels along the 3rd axis. Note that the relevés of BL1, BL2 and BL3b are disposed within horseshoe-shaped portions of the three-dimensional space. The sequencies of the relevés within each horseshoe reveal three distinct gradients, one for each vegetation type.

The releves of BL2 have the highest scores on the third axis. In Table 11 the relevés are arranged from left to right according to their sequence along the horseshoe. The presence of those species that showed trends along the gradient has been reported. The relevés on the left side of Table 11 are characterized by several shrub or tree species, while those at the right side are the typical Subalpine *Alnus crispa* stands near treeline. It seems therefore that the main factor underlying the ordination of the relevés of BL1 within the horseshoe is elevation.

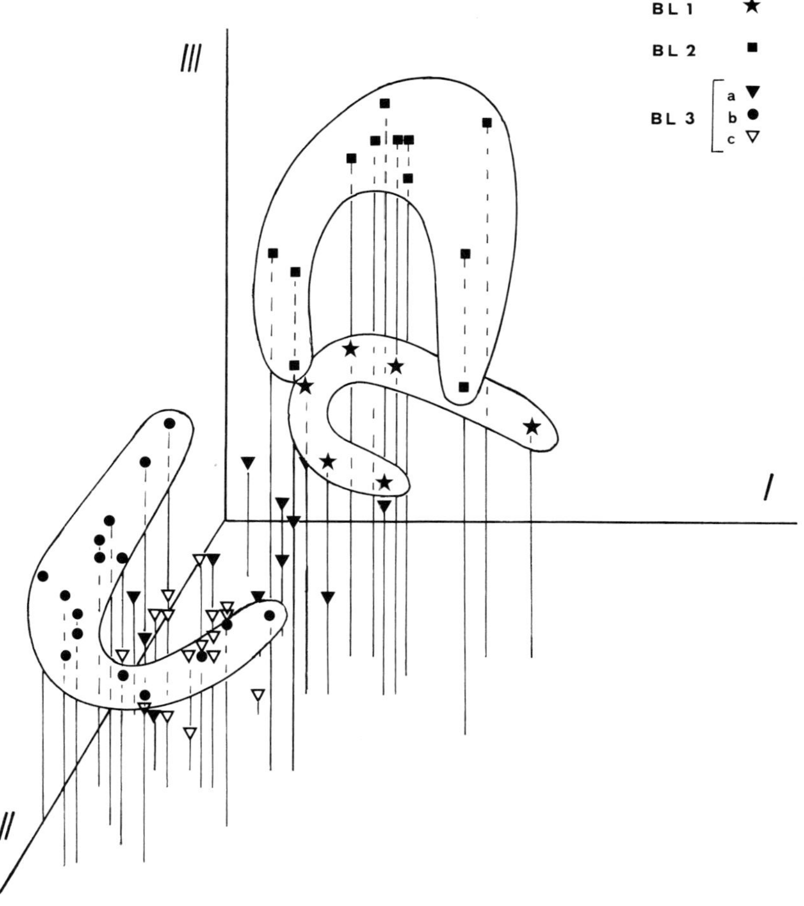

Fig. 18. Ordination (by PCA) of the relevés, based in the data of Table 9 (broadleaved tree stands). Principal Components I, II and III. Symbols refer to the relevé groups and subgroups, as in the legend (see also Fig. 2).

Table 11. Relevés of BL2, arranged according to their sequence in the horseshoe shown in the ordination of Figure 18. Only those species have been retained that reveal compositional variation trends along the gradient.

Releve No.	17	16	14	12	13	7	9	8	11	10	15
Populus tremuloides		3	7	3	3						
Rosa acicularis	2	2	3	2	2				2		
Picea glauca	2	2	3	3	2	2	2	2			
Viburnum edule		2	3	3	3	2			2		2
Populus balsamifera		3	5								
Dicranum undulatum				2	2	2					
Boschniakia rossiea					2	2	2	2	2	2	
Aconitum delphinifolium					3	3	3			3	2
Gymnocarpion dryopteris					2	2	2			2	2
Stellaria longipes						2	2	2			
Alnus crispa	5	3	2	5	5	8	8	8	5	8	8

The relevés of BL1 have lower scores on the third axis (Fig. 18). Also in this case their arrangement within the horseshoe reveals a gradient. The first four relevés from the left were taken along the Dempster Highway, between 1,034 and 900 m, whereas the last two relevés were taken along the Alaska Highway at elevations of 750 and 700 m. Notwithstanding the considerable differences in latitude, no subarctic species characterizes the relevés of the Dampster Highway. This is because in the north the narrow strip of ground along rivers and creeks is one of the few biotopes where permafrost is absent, so that these biotopes are refugial stations for many southern, permafrost-intolerant forbes and herbs. This is reflected in the structure of the stands, which varies along the gradient from closed stands with a well-developed herb layer (left) to less closed stands with a scarcely developed herb layer at the right side of the horseshoe.

The relevés of BL3 have the lowest scores on the third axis. The relevés of BL3b (variant with *Mertensia paniculata*) are arranged within a horseshoe. From left to right, the following factors show regular trends: decrease of the age of trees, increase of the percent of gravel in the soil, and decrease of the percent of sand (the soils tend to become more finely textured). The floristic variation along the gradient is shown in Table 12, where the relevés are ordered according to their sequence along the horseshoe. Those at the left side are characterized by species of more mature and closed stands, whereas those at the right side contain a set of xero- and helio-

Table 12. Relevés of BL3b, arranged according to their sequence in the horseshoe shown in Figure 18. Only those species have been retained that reveal compositional variation trends along the gradient.

Releve No	34	35	40	36	32	37	39	43	41	42	38	45	44	33	46	47
Picea glauca	2		2		2	3	2			2					1	
Salix glauca			2	3		2	2		2	3						
Populus balsamifera	3				5	2	2			2	2					
Drepanocladus uncinatus			2	2		2	2	2	2	2	2			3		
Viburnum edule	3	2	2					5	5	3						
Pyrola asarifolia			2				2	2	2	2						
Hylocomium splendens			2	2		2										
Vaccinium vitis-idaea		7	3	2				3			2					2
Geocaulon lividum	2	2			2		2			2	2			2		
Pinus contorta								2			3	7	2			3
Ceratodon purpureum												2	2		2	2
Peltigera rufescens												3	2		2	2
Salix alaxensis														3	2	
Ledum groenlandicum												2			2	3
Gentiana propinqua												2	2		2	
Polytrichun piliferum												2	2		2	

phytic species such as *Gentiana propinqua, Ledum groenlandicum, Peltigera rufescens* and *Ceratodon purpureus*. They constitute a transition towards the variant with *Peltigera rufescens* and *Ceratodon purpureus* (BL3c). This is also evident from the position of these last relevés (45–47) in the ordination of Figure 18, where they are located among the releves of BL3a and BL3c, i.e. among the stands of the most xeric variants of BL3.

4.5. *Salix* thickets

4.5.1. Classification

This data set includes relevés of vegetation dominated by low shrubs, mainly *Salix* species. The numerical classification (Fig. 2) of the relevés (Table 13) produced four main relevé groups. The dendrogram reveals a strong inhomogeneity in the data structure, both within and among groups: the most homogenous groups are S1 and S2. The ecological data of the relevés are in Table 14, and the chorograms of the relevé groups are shown in Figure 19.

S1 *Salix pulchra* community

This is a Subalpine–Subarctic community dominated by *Salix pulchra*. Hemicryptophytes are the most frequent life forms (47%), followed by thallochamaephytes, chiefly lichens (30%). The chorogram (Fig. 19) shows a prevalence of species with amphiberingian distribution.

No quantitative soil data are available. The soils are organic Cryosols derived from Precambrian quartzite and schists. All relevés were on slopes, with solifluction phenomena. The community forms extensive thickets in the subalpine vegetation belt near North Fork Pass, along the Dempster Highway, in northern Yukon. Most of this area was unglaciated, which explains the high incidence of amphiberingian species. The lichen synusia has been described by Nimis (1981) as *Cetrario-Masonhaleetum richardsonii*, and is dominated by *Masonhalea richardsonii*, an amphiberingian lichen. Further details on the phytogeography of this community are in Nimis (1989).

S2 *Betula glandulosa–Salix myrtyllifolia* community

This is a low scrub community dominated by *Salix myrtyllifolia* and *Betula glandulosa* and by mosses. The most frequent life forms are phanerophytes (31%) and hemicrytophytes (27%). Most of the species have an incompletely circumboreal or boreal North American distribution (Fig. 19).

The soils are mostly Cryosols (42%) or Gleysols (33%) with high sand and organic matter content. In June frozen ground was present at an average depth of 20 cm. In September, 38% of the plots had permafrost at an average depth of 56 cm. Drainage is always impeded, and water stagnation frequent. An open-water table was observed in 46% of the plots. The pH ranges from 5.6 to 7.8, with an average of 6.7. The relatively broad pH range may be due to the development of the community after the burning of a *Picea mariana* muskeg (ephemerous ash accumulation).

This community is often a secondary stage of a succession leading towards a *Picea mariana* muskeg. In some cases (e.g. around ponds and small lakes) it is a long-lasting stage, since permafrost may hinder the ecesis of trees. Most of the stands were in shallow depressions created by subsidence after burning in soils underlain by permafrost. The community occurs throughout the boreal (montane) vegetation belt of southern Yukon (Fig. 3).

S3 *Arctostaphylos–Salix glauca* community

These are secondary *Salix* thickets developing after the burning of *Picea glauca* forests. Hemicryptophytes (26%) and nanophanerophytes (26%) are the most frequent life forms. Boreal North American species are the most frequent phytogeographic elements, followed by species with boreal North American ranges extending southwards along the Cordilleras; the circumboreal element is scarcely represented (see Fig. 19).

The soils are either Gleysols (55%, mostly of

Table 13. Phytosociological table of *Salix* thickets. The relevés are arranged according to the results of numerical classification (see Fig. 2).

				S1									S2										S3									S4								
RELEVÉ No.	1	2	3	4	5	6	7	8	9	10	11	12	13	14	15	16	17	18	19	20	21	22	23	24	25	26	27	28	29	30	31	32	33	34	35	36	37				
Salix pulchra	5	5	8	7	5	2	8					3					2											2										5	1	1	
Polemonium acutiflorum		2	2	2	2	2	2												2																			5	1		
Carex lugens	2	2	2	3	3	2	2					2																										5	1		
Petasites hyperboreus	2	2	7	3	3	3	5																															5			
Luzula parviflora	2	2	3	3	2	3	2																															5			
Polytrichum juniperinum	2	7	2	2	2	2	2																															5			
Empetrum nigrum	3		3	5	5		5	2																														5	1		
Artemisia arctica	2	3		3	3	2	2																															5			
Trisetum spicatum		2	2	2	2	3	2																															5			
Poa arctica	2	2	2	3		3	2																															5			
Senecio lugens			2	2	2	2	2																															4			
Cladina mitis	5	5		3	2	3																																4			
Masonhalea richardsonii	5	2		2	3	2																																4			
Cetraria nivalis	2	2	2	2																					2													3		1	
Cetraria cucullata	2		2	2	2	2																																4			
Cladina rangiferina	5	3	2	2	2																																	4			
Polygonum bistorta			3	2	2	2	3																										2			2		4			1
Veronia wormskioldii			2	2	2		2																															4	1		
Cetraria islandica	2	3		3	2																																	3			
Dactylina arctica	2	2		3	2																																	3			
Carex microchaeta	2					5	2																															3			
Cetraria laevigata		2	2			3	2																															3			
Gentiana algida					2	3	2																															3			
Salix myrtillifolia								2	5			5	3	3	2	3	5	5	2					7		3													5	1	
Arctostaphylos rubra								3	3	3		2	2	3	3	3	2	2	2					3															5	1	
Potentilla fruticosa								3	3	2	2	2	2	2	2	2	2		2																				5		
Salix planifolia								2	2	2	2	3	3					3																					3		
Equisetum scirpoides								2				2	2			2	2																						3		
Ledum groenlandicum											3			2			2																						2		
Eriophorum vaginatum											3			2	7	2	3																						3		
Vaccinium vitis-idaea											3			2	2																								2		
Oxycoccus microcarpus								2		2	2																												2		
Equisetum variegatum											2		2				2	2																					2		
Rhododendron lapponicum											3				2	2																							2		
Picea glauca																		2		2	2	3	2	2		2													1	4	
Carex concinna												2				2						2	2	2	2			2											1	3	
Ceratodon purpureum																				2		2	3		2	2	2	5												4	
Arctostaphylos uva-ursi																						7		2	5	5	2	2												4	
Artemisia borealis																			2	2	2				2														1	2	
Lupinus arcticus																				2			2	2			2													3	
Peltigera canina																							2		2	2		2												3	
Shepherdia canadensis																		2						2			2	2											1	2	
Achillea borealis																					2			2		2	2													3	
Pyrola secunda																						2		2			2													2	

Anemone multifida																									2	2	2												2	
Pyrola grandiflora																								2		2	2												2	
Salix setchelliana																													5	5	3	3	2	2	5	3	3				5
Deschampsia caespitosa																													1	2	3	5	3	2	2	2	2				5
Epilobium latifolium																													2	2	4	3	6	2	2	2	3				5
Carex maritima																													3	5	3	3	3	5	2						4
Castilleja yukonis																															2	2	2								2
Aulacomnium palustre		5	3	7	2	2		7	5	5	7	7	2	8	5	5	3							3									5					4	5	1	1
Betula glandulosa	5	5	2	5	3	2	3	5	3		7	3	3	2	3	5									7	3							5					5	4	2	1
Valeriana capitata			2	2		2	2		2		2	2	2																									3	2		
Equisetum arvense			2				3			2			2	3	2	2		2	7		3									2			2					2	3	1	2
Stellaria longifolia	2	2	2	2	2	2	2							2		2																						5	1		
Rubus arcticus		2					2	2	2	2		2											2														2	2	1	
Pedicularis sudetica ssp.	2	2		2			2					2	2	2		2																						3	2		
Peltigera aphtosa	3	2	2			2	2	2	2		2												2														4	2	1	
Ledum palustre	3		2	2				3		2	3				3	2																						3	3		
Vaccinium uliginosum	2						2	3			3			3	3	3	3																					2	2		
Calamagrostis canadensis				2			2		2		2							2	2																			2	3		
Hylocomium splendens		3		5	3		5				2																											3	1		
Rubus chamaemorus			2								2			2	2	2																						1	3		
Salix reticulata							2					2	2	3																							1	2		
Salix glauca									5		3	5	5	3	5	2	3	3	5	5	8	7	5	5	3	5	5	5												5	5	
Salix scouleriana									3				3	3	3			2	3	2	2		3	3			2	3		2	2	2								3	4	2
Mertensia paniculata		2		2	2	2	2	2	2	2									2	2	3			2			2	2										4	2	3	
Populus balsamifera																			2									3			2	2			2					1	1	2
Festuca altaica	2	3		2			2														2				2	5			2	3	2	3						3		3	4
Salix alaxensis								2	2				3					2	3	2	2	2																		3	3	
Polygonum viviparum												2	2			2	2														2	2								2		2
Salix bebbiana													2	2		2				2		2	3										2						2	3	1
Hedysarum alpinum													2	2		2		2					2				2						2						2	2	1
Gentiana propinqua									2			1												2	2	2														1	2	
Anemone parviflora				2		2	2									2																						3	1		
Drepanocladus uncinatus													2			2						2	3															1	2	
Delphinium glaucum						2														3	2						2											1		2	
Dicranum bergeri	2					3									2	2																						2	1		
Cladonia chlorophaea			2	3		2		2																													3	1		
Juncus arcticus																		2	2											2							2		1		2
Fragaria virginiana										2											2				2	2												1	2	
Astragalus umbellatus														2		2		2		2		2																2	1	
Picea mariana											2				2																								1		
Pedicularis labradorica								2			2																												1		
Sphagnum magellanicum										5	8																												1		
Gentiana glauca		3			2																																2			

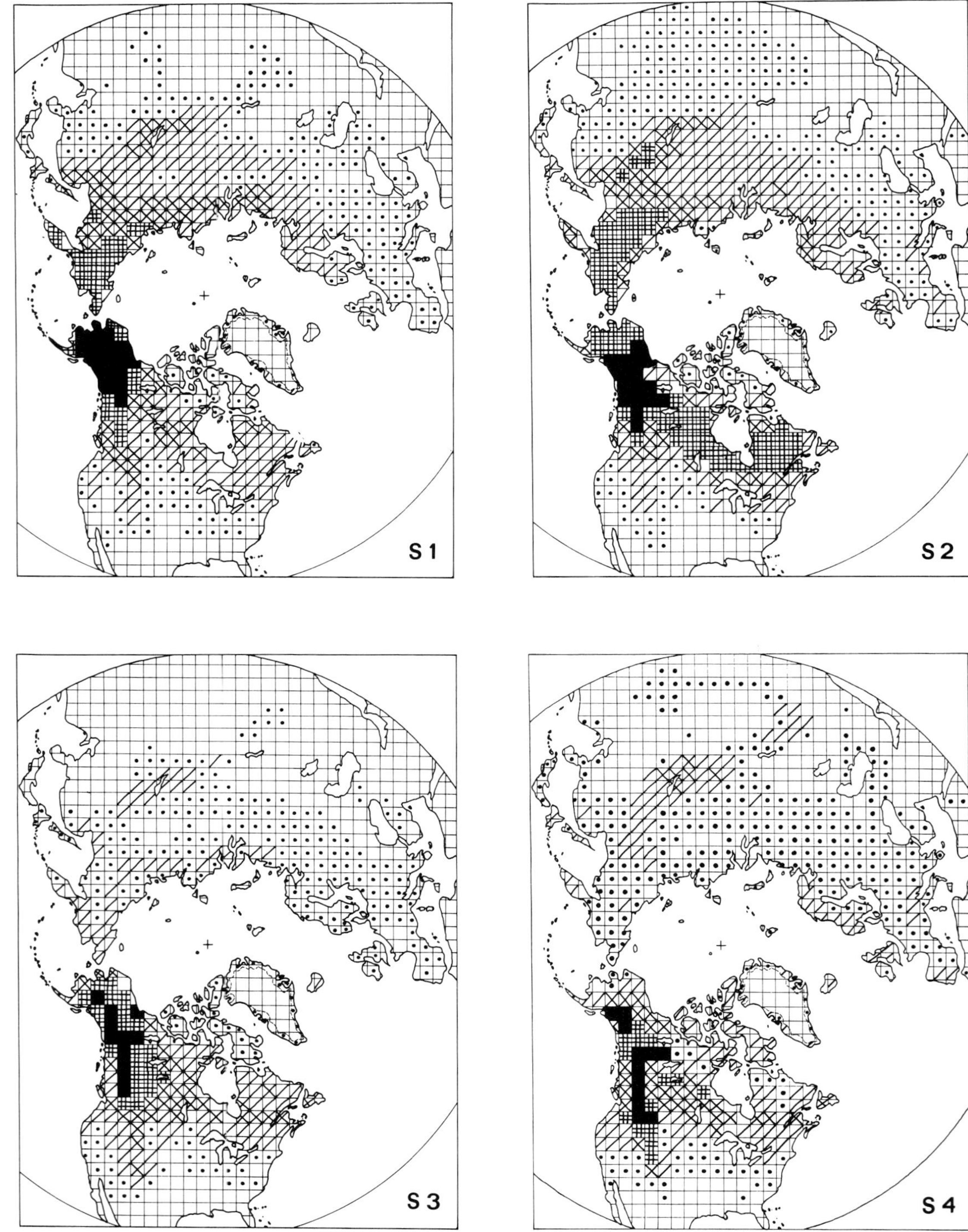

Fig. 19. Chorograms of the four relevé groups obtained by classification of the data in Table 13 (*Salix* thickets). Different shadings refer to the percent of species occurring in the OGUs, on a 5-class scale with intervals of 20%.

Table 14. Ecological data relative to the relevés of *Salix* thickets. For the abbreviations see the caption to Table 4.

Releve No.	8	9	10	11	12	13	14	15	16	17	18	19	20	21	22	23	24	25	26	27	28
Total cover	100	100	100	100	100	100	100	100	100	100	80	100	80	80	80	85	100	60	100	70	60
Cover of shrubs	60	80	40	40	80	75	80	70	40	40	65	70	75	70	80	75	80	50	60	70	50
Cover of grasses	5	20	80	50	20	65	20	60	40	50	30	40	30	30	10	20	20	30	50	25	15
Cover of lichens	5	5	-	5	-	-	-	-	-	-	-	-	-	-	-	-	-	5	-	-	-
Cover of mosses	40	60	40	70	70	25	60	20	50	40	10	5	-	-	5	15	10	-	-	-	5
Exposure	SSW	S	-	-	-	-	-	-	-	-	-	SE	-	-	-	-	WSW	-	NNW	NNE	-
Slope	20	4	-	-	-	-	-	-	-	-	-	10	-	-	-	-	10	-	5	5	-
Soil type	R. HG	O. HG	T. H	FI. OC	GL. SC	GL. CUR	R. HG	TME OC	BR. SC	R. HG	GL. CUR	TFI OC	O. HG	O. HG	R. HG	GL. R	GL. HR	GL. CUR	O. HG	O. G	GL. SB
GS %	70	40	15	-	5	3	10	-	10	-	40	-	10	15	5	10	50	-	-	70	15
SA %	12	30	20	-	30	77	50	20	55	50	40	60	30	31	20	60	40	20	25	25	45
SI %	7	29	25	-	55	20	30	30	30	50	18	20	20	50	40	30	10	40	45	5	35
CL %	1	1	10	-	10	-	10	-	5	-	2	20	40	4	35	-	-	40	30	-	9
Org. (cm)	10	30	30	100	20	15	15	50	12	10	4	60	5	10	20	-	10	20	-	9	10
pH	6.2	7.2	6.8	5.8	5.7	7.7	5.7	6.2	6.0	7.8	7.7	7.6	5.8	5.9	7.6	7.6	6.4	7.7	7.0	6.6	7.3
Erosion pot.	low	med	med	low	med	med	low	med	med	med	med	med	med	med	hi	med	low	med	med	low	med
Drainage	imp	wet	wet	imp	imp	wet	wet	imp	imp	wet	imp	wet	imp	imp	imp	mod	imp	imp	imp	imp	mod
Depth of ice June (cm)	-	-	-	18	10	13	13	10	18	15	-	12	60	-	-	-	-	45	50	-	-
Depth of ice Sept.(cm)	-	-	-	36	30	-	-	50	70	-	-	60	-	-	-	-	-	-	-	-	-

the Orthic group) or Regosols (33%), mostly of the gleyed group. Frozen ground was present in 44% of the stands in June, at an average depth of 42 cm, and was always absent in September. Drainage is moderate to impeded, and improves after ice melting in late summer. All stands were on flat ground or on very gentle slopes. Soil type is the main difference with the secondary *Populus tremuloides* stands: the latter mainly occur on steep arid slopes with good drainage conditions, so that the soils rarely present traces of gleyzation. The pH values are high, with an average of 7.1 and a deviation of 0.7. In a single relevé the pH reached 8.2, because of ash accumulation following fire. This community is most common in the Ruby Range Ecoregion (Fig. 3).

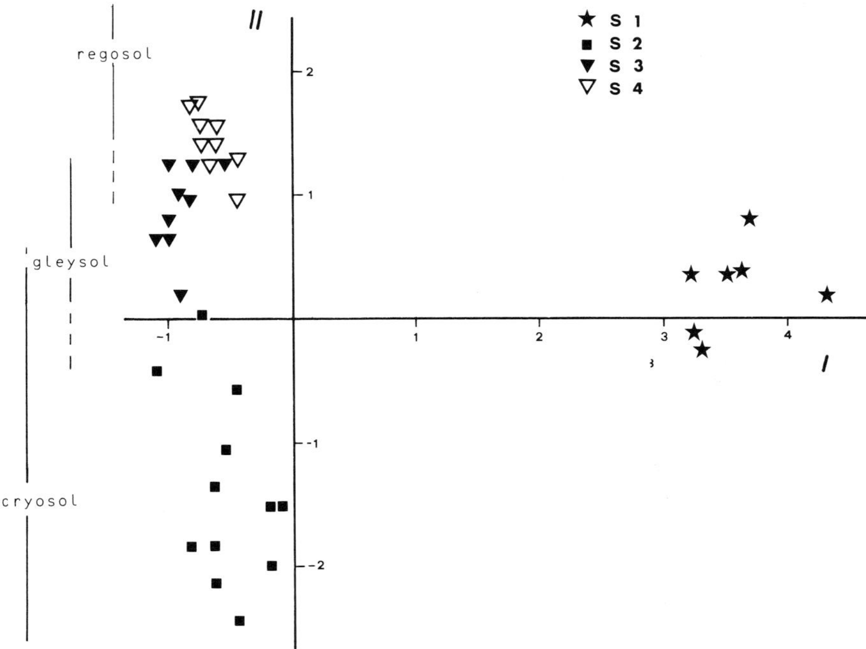

Fig. 20. Ordination (by PCA) of the relevés of *Salix* thickets, based on data data of Table 13. Principal Components I and II. Symbols refer to relevé groups, as in the legend (see Fig. 2).

S4 *Salix setchelliana* community

This is a very open pioneer community on gravel dominated by the small shrub *Salix setchelliana*. Hemicryptophytes are the most frequent life forms (50%), followed by chamaephytes (26%). The chorogram (Fig. 19) shows a prevalence of Cordilleran species, followed by endemics of Alaska–Yukon. The high incidence of endemics is the most striking feature of the community; *Salix setchelliana* itself is endemic of central Alaska and southern Yukon. No quantitative soil data are available for this community, which occurs on gravelly Regosols with varying amounts of silt and sand, along the shores of Kluane Lake (Fig. 3).

4.5.2. Ordination

The data of Table 13 have been submitted to PCA (binary data), whose scatter diagrams are shown in Fig. 20 (first and second axes) and Fig. 21 (second and third axes). In Fig. 20 the relevés of S1 are separated by the first axis; they are the only ones taken in the Subalpine–Subarctic zone of northern Yukon. The second axis is correlated with soil types. The sequence of the relevé groups (S2, S1, S3, S4) corresponds well with a transition from Cryosols (most relevés of S2, all releves of S1) to gleyed soils (S3) and Regosols (S4), which represents a gradient in increasing soil aridity. In Figure 21 the peculiarity of the *Salix setchelliana* community (S4), in terms of ecology (pioneer community on Regosols) and floristic composition (richness in endemics) is revealed by its separation on the third axis.

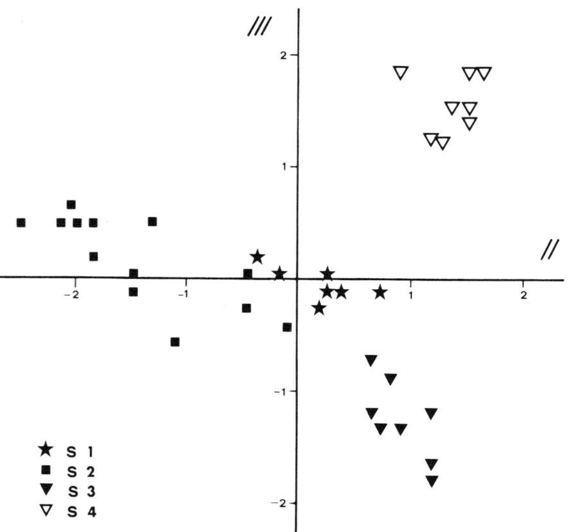

Fig. 21. Ordination (by PCA) of the relevés of *Salix* thickets, based on the data of Table 13. Principal Components II and III. Symbols refer to relevé groups, as in the legend (see Fig. 2).

4.6. Grassland vegetation

4.6.1. Classification

This data set includes relevés of vegetation dominated by grasses and low shrubs, with no or few trees. All plots were below the treeline. The dendrogram resulting from the classification of the relevés (Table 15, Fig. 2) produced four main clusters (G1, G2, G3 and G4). The less homogeneous cluster is G3, which can be divided in two subclusters: G3a (rel. 16–27, Table 15) and G3b (rel. 28–30, Table 15). The ecological data of the relevés are in Table 16. The chorograms of the relevé groups are shown in Figure 22.

G1 Weed vegetation of roadsides

This is a very open, pioneer grassland community occurring along roadsides. Its ephemeral character is reflected by the high frequency of therophytes (23%), which is somewhat unusual for a vegetation type of the boreal zone. This, together with the high incidence of broad-ranging species (Fig. 22), indicates the anthropogenous nature of the community. However, the chorogram (Fig. 22) indicates that the main phytogeographic affinities of this community are with the Cordilleras and with the sagebrush prairies of western North America.

No quantitative soil data are available. The community occurs on regosols created by man for the construction of the Alaska Highway; it is best developed on flat ground periodically disturbed by vehicles, being more fragmentary on road shoulders.

Table 15. Phytosociological table of grassland vegetation. The relevés are disposed according to the results of numerical classification (see Fig. 2).

	G1							G2								G3 a												G3 b			G4							
RELEVÉ No.	1	2	3	4	5	6	7	8	9	10	11	12	13	14	15	16	17	18	19	20	21	22	23	24	25	26	27	28	29	30	31	32	33	34				
Hordeum jubatum	2	3	2	5	2	3	3																												5			
Taraxacum officinale s.l.	5	3	7	3	2	5																													5			
Polemonium pulcherrimum	2	2	2	3	2	3										2									2										5		1	
Epilobium angustifolium	2	2	3	2	3	2	2																											2	5			2
Agropyron violaceum		2	3	3	2	2																													4			
Fragaria virginiana	2	2		2	3	5	2									2	2										2								5		1	
Rubus idaeus		2	2	2	2	2																													4			
Astragalus alpinus	2	2		2	3	2																													4			
Aira caryophyllea	2	2	2	2		2	2																												5			
Equisetum arvense	3	2	2			2																													4			
Lupinus arcticus	2	2			2	2																													4			
Bromus pumpellianus		2	2	3																															4			
Plantago major	2	2	2				2																												4			
Trifolium repens	2		2		2		2																												4			
Agropyron yukonense								2	2	3	3	2	2	5	3			2																		5	1	
Erigeron caespitosus								2	2		2	2		2	2						2				2											4	1	
Linum perenne ssp.lewisii									2	2	2	2							2											2						3	1	
Psora rubiformis								2	2	2				2	2																					4		
Caloplaca tominii									2	2	2		2	2	2																					4		
Squamarina lentigera								2			2		2	2	2																					3		
Hypnum vaucheri								2	2				2	2																						3		
Psora decipiens													2	2	2																					2		
Arctostaphylos uva-ursi								7	7		3																									2		
Oxytropis viscida								2		2	2																									2		
Buellia elegans									2			2	2																							2		
Erigeron compositus															2		2	3	2	2	2	2	2		2											1	3	
Potentilla hookeriana																2	2	2		2	2	2	2	2	2	2	3										4	
Sedum lanceolatum																2		2	2	2	2	2	2	2	2	2											4	
Cetraria nivalis																2			2	3	3	2	3	3	3	2			2								4	
Oxytropis coampestris																				2		2		2	2	2											2	
Galium boreale																	2										2	2	2	2							2	
Carex obtusata																							3					7	2	3							2	
Dryopteris fragrans																															2	5	3	5				5
Stereocaulon tomentosum																															2	2	3	5				5
Cladina stellaris																															2	2	2	2				5
Cladina arbuscula																															3	3	2	3				5
Racomitrium canescens																															3	3		3				4
Parmelia wyomingica																2	2			3	2	3	5	3	3	3		3	3	7	5	2					4	3
Cladonia chlorophaea																2		2		2	3		2					2	2	2	2	2					3	3
Saxifraga tricuspidata																5	5			2	2			3	2			2	2		4	3	5	5			3	5
Tortula ruralis													2			2		3	2	2	2	3	3	2	3	2	2				2	2	2			1	4	4
Anemone multifida												2		2					5	5	3	3	2	2		2	2		2		2					2	3	2
Peltigera rufescens					2													2	2	2	2									2	2	2	2		1		2	4
Cladonia cariosa																				3	3		2						2	2		2	2				2	3
Cornicularia aculeata																				3	3	3	3						2		2	2					2	2
Arabis hoelbelii				2														2				2	2			2	2	2					2		1		2	2
Plantago canescens																				2	2	2	2					2	2								2	
Potentilla pennsylvanica							2										2	2						2	2			2						3	1		2	2
Draba aurea																								2		2	2		2					2			2	2
Calamagrostis purpurascens	2	2	2	2	2					2	3	2	3		2	2	2	2	2	3	3	3	3	3	3		3	3	5	5	3	2	5	2	4	4	5	5
Artemisia frigida								2	2	5	5	5	2	3	3		3	3	3	2	2	3			5	3	5		3					2		5	4	2
Carex filifolia										5	2		2	3	3			3	5	3	3	3			2	3		2			3					4	3	2
Penstemon gormani										3	2	2		2				2		2	2	2	3	2			3	3								3	3	
Antennaria rosea var. nitida	2	2		2							2				2	2	2	2	2	2	2				2			2							3	2	3	
Poa glauca	2	5	5	2			2				2				2					3	2				3			2	2			2			4	2	2	2
Achillea borealis	2	2	3	2	2	2	2								2	2	2							2					2		2				5	1	2	2
Rosa acicularis						2				2	2	2	2	2	2	2	2												2		2	2	2	2	1	4	1	5
Solidago decumbens	2	2				2	2			2					3		2							2				3	2			2			3	2	2	2
Penstemon procerus									2		2	2		2	2			2						2	2	3	2			2						3	2	
Festuca brachyphylla													3		2	2		3	3					2	3		3				3	2	2			2	2	4
Artemisia alaskana								2	2									2					3		2			5	2	3						2	2	
Ceratodon purpureus								2						2				2		2	2				2	2					2		2			2	2	3
Pulsatilla patens									2		2	2	2	2		2		2	2					2												3	2	
Juniperus nana								3	5							2	2							3	2									5		2	2	2
Juniperus horizontalis												5			2	2	3							3		2			5							2	2	
Populus tremuloides					2			2											2					2			2								1	1	1	
Melandrium apetalum								3											2	2	2								2							1	2	
Potentilla norvegica	2	2	2				3									2												2		2					3		1	
Gentiana propinqua						2		2	2																										1	2		
Catopyrenium lachneum													2	2																						2		
Shepherdia canadensis								2	3																											2		
Hedysarum americanum								2	2																											2		
Cladonia pocillum																		2	2																		1	
Chamaerhodos erecta																						2	2														1	
Geum perincisum																														2		2					1	2

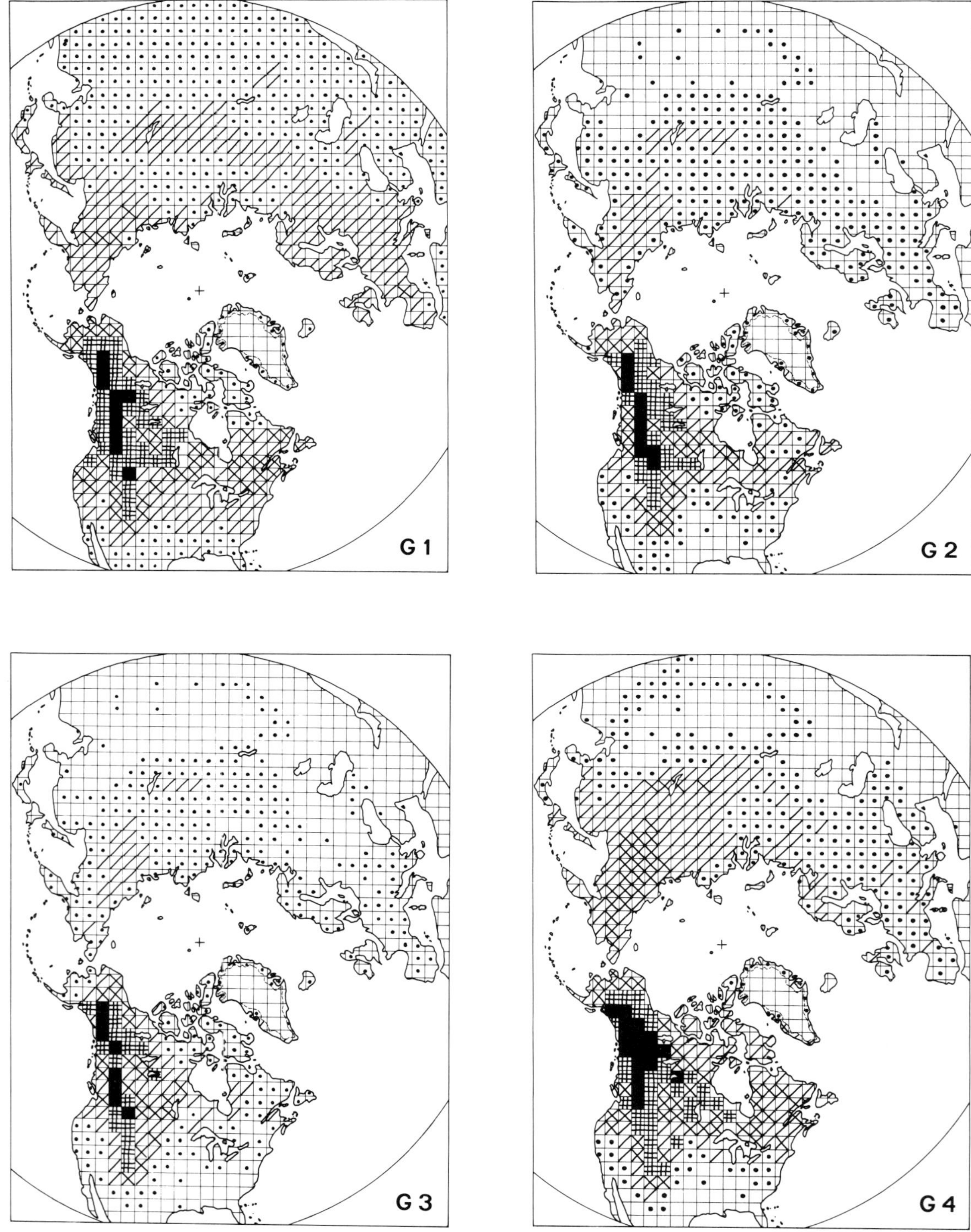

Fig. 22. Chorograms of the relevé groups obtained by classification of the data in Table 15 (grasslands). Different shadings refer to the percent of species occurring in the OGUs, on a 5-class scale with intervals of 20%.

Table 16. Ecological data relative to grassland vegetation. For the abbreviations see the caption to Table 4.

Releve No.	8	9	10	11	12	13	14	15	16	17	18	19	20
Exposure	SW	NE	S	SE	S	W	SE	SW	SW	S	S	.	W
Slope	35	15	30	25	20	25	15	35	15	20	30	.	20
Total Cover	40	80	30	65	30	60	60	40	60	60	35	.	70
Soil type	CU.	CU.	O.	.	CU.	CU.	CU.	CU.	.	B	CU.	CU.	.
	R	R	SB		R	R	R	R	.	E	R	R	.
GS %	30	50	15	.	20	-	40	50	.	D	30	5	.
SA %	45	30	30	.	40	70	35	30	.	R	40	75	.
SI %	23	20	30	.	35	30	25	20	.	O	28	18	.
CL %	2	-	25	.	5	-	-	-	.	C	2	2	.
Org.(cm.)	2	2	3	.	2	2	2	2	.	K	5	5	.
pH	7.7	7.8	7.1	.	7.8	7.5	8.0	7.8	.	.	6.9	7.6	.
Erosion pot.	hi	low	hi	.	hi	hi	hi	low	.	.	hi	med	.
Drainage	ex	ex	ex	.	ex	ex	ex	ex	.	.	ex	ex	.

Releve No.	21	22	23	24	25	26	27	28	29	30	31	32	33	34
Exposure	W	SW	W	SE	S	S	S	SW	SW	W	N	S	S	S
Slope	15	15	15	8	20	35	30	25	35	25	25	15	30	15
Total Cover	70	70	70	60	70	50	60	80	40	75	100	20	40	40
Soil type	.	.	.	O.	O.	.	B	.	.	.	CU.	O.	.	.
	.	.	.	R	R	.	E	.	.	.	R	R	.	.
GS %	.	.	.	70	30	.	D	.	.	.	90	80	.	.
SA %	.	.	.	15	60	.	R	.	.	.	5	15	.	.
SI %	.	.	.	14	10	.	O	.	.	.	5	4	.	.
CL %	.	.	.	1	-	.	C	.	.	.	-	1	.	.
Org.(cm.)	.	.	.	10	-	.	K	.	.	.	1	2	.	.
pH	.	.	.	6.3	6.7	.	6.8	.	.	.	6.3	6.7	.	.
Erosion pot.	.	.	.	low	hi	.	med	.	.	.	med	med	.	.
Drainage	.	.	.	ex	ex	.	ex	.	.	.	ex	ex	.	.

Our relevés were obtained in 1978 in the southeastern section of the Alaska Highway, from Teslin to Watson Lake. In 1983 we devoted an entire summer to the study of weed vegetation in Alaska–Yukon. For further details on weed vegetation of the survey area consult Lausi and Nimis (1985b).

G2 *Artemisia frigida–Agropyron yukonense* community

This is an open sagebrush grassland on alkaline soils, dominated by hemicrytophytes (31%) and chamaephytes (22%). Of note is the high frequency of crustose lichens, a distinctive feature of this community. Cordilleran and endemic species are the most frequent phytogeographic element (see Fig. 22).

The lichens belong to a group of basiphytic species of the alliance *Toninion coeruleonigricantis*, hitherto known only from Europe and Asia. The lichen synusia has been described by Nimis (1981a, b) as *Fulgensio-Caloplacetum tominii*.

The soils are alkaline Regosols (average pH: 7.7) derived from loess accumulation; in most cases the organic layer could be best described as chernozemic, since it is underlain by a Ca-horizon. Organic matter, however, is scarce, because the stands are on steep slopes with strong erosion and excessive drainage; frozen ground is always absent.

This community is restricted to the Kluane Lake area (Fig. 3), which has a very dry, continental climate; it behaves as a long-lasting stage of a succession leading, through an intermediate stage with *Populus tremuloides* and *Arctostaphylos uva-ursi*, to a xerophytic *Picea glauca* forest (PG4). Hoefs et al. (1975) described a very similar community from the same area.

G3 *Sedum lanceolatum–Parmelia wyomingica* community

This is an open grassland on neutral to subacid sandy soils. Hemicryptophytes are the most frequent life forms (44%). Cordilleran species are the main phytogeographic element (Fig. 22).

In contrast with the previous community, in G3 crustose lichens are replaced by neutro- to acidophytic fruticose and foliose lichens (see Nimis 1981).

The relevés of G3 are floristically heterogeneous; most of the differential species present in relevés 16–27 (G3a), are lacking in relevés 28–30 (G3b), which are characterized by *Carex obtusata* and *Galium boreale*. All the relevés of subgroup G3a were taken along the Alaska Highway (Fig. 3), those of G3b along the Klondyke Highway, near Carmacks. The latter subgroup probably represents a distinct community, whose formal description should be based on more relevés.

The soils are sandy-gravelly Regosols. Siliceous sand grains and loess particles are equally present, causing subneutral pH values (average 6.9). A thin acid layer at the soil surface allows the presence of acidophytic lichen species. Most of the stands were on slopes, often on sand dunes; the community is relatively frequent in the Lake Laberge Ecoregion of southern Yukon (Fig. 3).

G4 *Dryopteris fragrans* community

This is a very open grassland on siliceous rocky

slopes. It has a relatively higher incidence of thallochamaephytes (mostly lichens, 26%) and of circumboreal species (see Fig. 22).

The soils are sandy-gravelly Regosols occurring in pockets among granitic or schistose boulders. They could be considered as protorankers, according to the classification of Kubiena (1958). The pH tends to be lower than in the other grasslands, with an average of 6.5.

This community occurs mainly in the central and southeastern part of the Alaska Highway (Fig. 3). It also has been recorded along the Klondyke Highway, between Koidern and Whitehorse.

4.6.2. *Ordination*

The ordination of the relevés of Table 15 is shown in Figure 23. The clusters obtained by classification are still recognizable, and occupy three distinct quadrants. The first axis separates the relevés of G1 (weed vegetation) while the arrangement of the other relevé groups on the second axis reflects a gradient in soil pH. This is shown in Figure 24, where the pH values of the relevés have been plotted against the relevé scores on the second axis. A parallel trend concerns soil texture, which tends to become coarser from G2 to G4. The content of silt and clay ranges from 20 to 55% in G2 (loess deposition), and from 5 to 30% in G3 (sand dunes) and G4 (rocky slopes).

Along this gradient the phytogeographic affinities of the communities (Fig. 22) show a trend from the prevalence of narrow-ranging, Cordilleran or endemic species (G2), to the prevalence of species with a broader distribution, sometimes extending to northeastern Asia (G4).

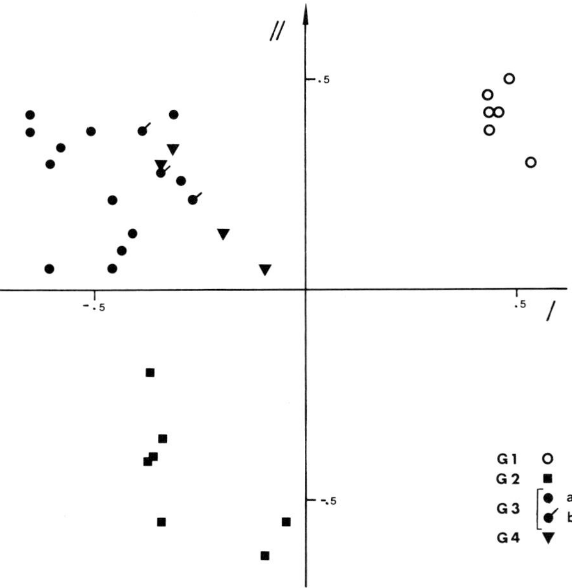

Fig. 23. Ordination of the relevés (by PCA), based on the data of Table 15 (grasslands). Principal Components I and II. Symbols refer to relevé groups, as in the legend (see Fig. 2).

4.6.3. *Discussion*

The existence of a steppe-like vegetation in Alaska and Yukon has been the object of much debate since the publication by Guthrie (1968) of data concerning four fossil assemblages of large mammals from the late Pleistocene sediments near Fairbanks. Guthrie (1968) claimed that "the high percentage of grazers in the fossil community suggests that interior Alaska was a grassland environment during the late Pleistocene". This hypothesis was quickly embraced by other zoologists (Hoffmann 1974; Harington 1978), archaeologists (Morlan 1980) and plant ecologists (Bliss 1975; Young 1976). Maxima of *Artemisia* in Pleistocene pollen spectra of Alaska seemed to confirm Guthrie's hypothesis. Some

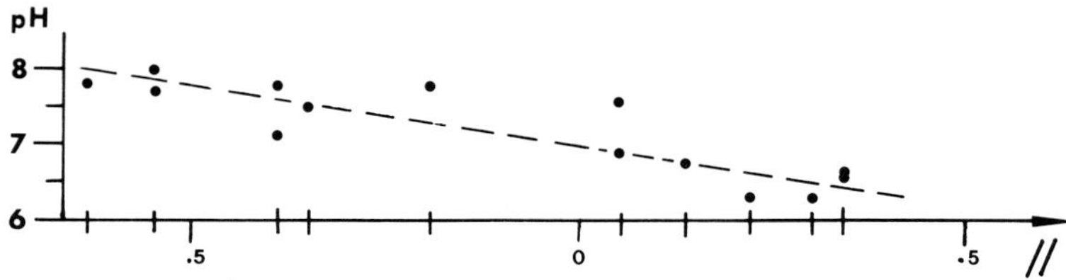

Fig. 24. Interpretation of the ordination of Figure 23. The *x*-axis reports the scores of the relevés of G2, G3 and G4 on the second Principal Component (Fig. 23), the *y*-axis indicates the soil pH in each relevé (data in Table 16).

authors (Matthews 1976; Hopkins et al. 1982) assumed that both the plant and animal communities were reduced to extinction in the Holocene. Others, such as Young (1976), Yurtsev (1981), Murray et al. (1983), claimed that remnants of the ancient steppes were still extant in Alaska and the Yukon. Ritchie (1984) denies the presence of steppe vegetation in the Yukon, claiming that the term "steppe" should be left for grass- and herb-dominated zonal communities of continental zones with warm summers, homologous with the grasslands of the western interior of North America. According to Ritchie (1984), the "steppes" of the Yukon are just grassy tundras. We think that this statement is not correct. In the northern Yukon, the area studied by Ritchie, true steppes are probably absent. We have also observed there several examples of "grassy tundras" with *Oxytropics* and *Astragalus* species. However, in southern Yukon, and particularly in the Kluane region, the grasslands, corresponding to type G2, should be considered of steppe character, for the following reasons:

(1) They occur in an area with a strongly continental, dry climate.

(2) The soils have a weak Ca-horizon (chernozemic), and can be considered as steppe-soils.

(3) They are dominated by *Artemisia* and *Agropyron*, species (and *Artemisia arctica*, which is not a steppe plant, is absent in these grasslands).

(4) Tundra species are extremely rare; the few species occurring also in the tundra vegetation are broad ranging plants with a wide ecological tolerance.

(5) The main phytogeographic affinities of these grasslands (Fig. 22) are with the sagebrush grassland regions of the Cordilleras and of western and central North America.

(6) The terricolous lichens are typical of steppe vegetation throughout the Holarctic (Nimis 1981a).

Nimis (1981b) reported as new to North America a crustose soil lichen, *Caloplaca tominii*, that is common in the vegetation type G2, and was hitherto known only from steppe areas of Eurasia. Nimis (1981b) suggested, on the basic of ecological and phytogeographic considerations, that this lichen should be looked for in the western interior American grasslands. This prediction proved to be right, and now *Caloplaca tominii* is known from almost all parts of Canada and the United States with natural sagebrush grasslands (Thomson, 1982). The species has never been recorded from the herb-tundras of northern Yukon.

The existence of a true steppe vegetation in the Yukon is beyond any doubt. The problem of its connections with the steppes of Asia will be treated in the section devoted to the phytogeographic analysis of vegetation.

5. ECOLOGICAL SYNTHESIS

This section provides an ecological synthesis wherein the 21 community types described in the previous section will be compared on the basis of floristic and ecological data. Its results will be used as a basis for the phytogeographic analysis.

A first evaluation of the floristic affinities of the 21 community types is provided by the dendrogram based on the contingency table of species and community types (Table 17). The dendrogram was obtained using Complete Linkage Clustering on binary data, with the Similarity Ratio (Westhoff and van der Maarel 1973) as resemblance measure (Fig. 25). Four main clusters emerged (a, b, c, d) and cluster "c" can be further subdivided into two subclusters (c1, c2). This grouping correlates with soil variables, and reflects the three major vegetation units in the survey area:

(1) Clusters "a" and "b": 97% of the relevés are on Regosols. The group includes relevés of grassland vegetation (G1–4) and the *Salix setchelliana* community (S4) occurring on gravelly soils along lake shores.

(2) Cluster "c": 51% of the relevés are on Brunisols, and the group includes the boreal forest types dominated by *Picea glauca* and their secondary stages The soils in c1 are mostly neutral (pH: 6.5–7.3), and in c2 are subacid (pH 5.3–5.9). In c1 the closed coniferous forest is

Table 17. Contingency table of the species and of the relevé groups. Species frequency within each group is expressed on a 5-class scale with intervals of 20%. The relevé groups are designed by numbers, as follows: (1) S1; (2) S2; (3) PM1; (4) PM2; (5) PM3; (6) PG1; (7) BL1; (8) PM4; (9) PG2; (10) BL2; (11) PG3; (12) PC1; (13) S3; (14) PG4; (15) PC2; (16) BL3; (17) G4; (18) G2; (19) G1; (20) G3; (21) S4.

RELEVÉ GROUPS	1	2	3	4	5	6	7	8	9	10	11	12	13	14	15	16	17	18	19	20	21
1 Salix pulchra	5	1					1						1			1					
2 Polemonium acutiflorum	5	1																			
3 Stellaria longifolia	5																				
4 Petasites hyperboreus	5																				
5 Luzula parviflora	5																				
6 Artemisia arctica	5																				
7 Trisetum spicatum	5																				
8 Poa arctica	5																				
9 Masonhalea richardsonii	5																				
10 Polygonum bistorta	4																				
11 Veronica wormskjoldii	4																				
12 Carex microchaeta	3																				
13 Gentiana algida	3																				
14 Gentiana glauca	2																				
15 Salix bebbiana		3									1				1						
16 Sphagnum magellanicum		1	2		1																
17 Sphagnum warnstorfii			5		1																
18 Sphagnum fuscum			2																		
19 Equisetum palustre			1		1	3								1							
20 Carex membranacea			1	1		3															
21 Cladonia cenotea						3									1						
22 Ditrichum flexicaule						3															
23 Cassiope tetragona						2															
24 Salix arbusculoides			1	1			5														
25 Anemone richardsonii			1				5														
26 Pyrola grandiflora				1	1		4			1			2		1	1					
27 Habenaria obtusata			1		1	1	3		1		2										
28 Ribes hudsonianum							3			1						1					
29 Alnus incana subsp. tenuifolia							4														
30 Ranunculus macounii							2														
31 Sanguisorba officinalis							2														
32 Polemonium coeruleum							2														
33 Cornus canadensis		1	1		1			4			4	3			1	1					
34 Polytrichum commune			1	1		1		3			1	1			1	1					
35 Peltigera leucophlebia						2			5		5										
36 Moneses uniflora									2		1										
37 Ribes triste							1			4	1	1				1	2				
38 Boschniakia rossica									1	3						1					
39 Aconitum delphinifolium							1			3											
40 Gymnocarpion dryopteris										3											
41 Pleurozium schreberi						1			2	1	4	4			1						
42 Ptilium crista-castrensis						1			1		3	3					2				
43 Dicranum fuscescens									3		4	4		1							
44 Lophozia barbata												3			1						
45 Mitella nuda						1			1		3			1							
46 Salix scouleriana		2							1	1	2		3	1		1					1
47 Cladonia pocillum									1					5		1				1	
48 Hypnum procerrimum									3					5							
49 Hedysarum mackenzii									1					2							
50 Gentiana propinqua		1		1	1										3	1		2	1		
51 Cladonia gracilis			1					3		1	1				4	1					
52 Cladonia multiformis									1		2	1			3						
53 Polytrichum piliferum												1			5	1					
54 Cladonia cornuta			1									1			3	1					
55 Cladonia uncialis												1			2						
56 Cladonia coccifera												1			2						
57 Cladonia verticillata															2						
58 Populus tremuloides			1				3	2		2	1	1		1	1	5		1	1	1	
59 Dryopteris fragrans																	5				
60 Racomitrium canescens																	4				
61 Agropyron yukonense														1		1		5		1	
62 Linum perenne																1		3		1	

	1	2	3	4	5	6	7	8	9	10	11	12	13	14	15	16	17	18	19	20	21
63 Erigeron caespitosum																		4		1	
64 Squamarina lentigera																1		3			
65 Psora decipiens																		2	1		
66 Caloplaca tominii																		4			
67 Hypnum vaucheri																		3			
68 Psora rubiformis																		4			
69 Dermatocarpon hepaticum																		2			
70 Polemonium pulcherrimum																1			5	1	
71 Aira cariophyllea																			5		
72 Hordeum jubatum																			5		
73 Taraxacum officinale																			5		
74 Agropyron violaceum																			4		
75 Plantago major																			4		
76 Trifolium repens																			4		
77 Bromus pumpellianus																1			3		
78 Potentilla norvegica																			3	1	
79 Erigeron elatus																			2		
80 Erigeron compositus																		1		3	
81 Potentilla hookeriana																				4	
82 Sedum lanceolatum																				4	
83 Carex obtusata																				2	
84 Plantago canescens																				2	
85 Salix setchelliana																					5
86 Deschampsia caespitosa																					5
87 Epilobium latifolium																					5
88 Carex maritima																					4
89 Castilleja yukonis																					1
90 Anemone parviflora	3	1	1								1										
91 Valeriana capitata	3	2	1				1			2											
92 Cetraria islandica	3		1	2																	
93 Ledum palustre	3	3	4	5																	
94 Salix reticulata	1	2	1	4		2															
95 Andromeda polifolia	5	1	2	2		2															
96 Rubus chamaemorus	1	3	3	2	1																
97 Senecio lugens	4		2	3	3																
98 Oxycoccus microcarpus		2	4	1																	
99 Astragalus umbellatus		2	1		1																
100 Petasites sagittatus		2	2	1	1																
101 Polygonum viviparum		2	1	1	1		2														
102 Rhododendron lapponicum		2	2	2	1	5															
103 Eriophorum vaginatum		3	4	4	2	2															
104 Carex dioica		2	1	2		1															
105 Chamaedaphne calyculata			2	2	1																
106 Carex capillaris			1	2	2																
107 Dryas integrifolia			2	2	1	4															
108 Tofieldia pusilla			1	2	1	3															
109 Carex scirpoidea			1	2	1	2															
110 Carex lugens			1	1		2															
111 Salix rostrata			1		1	2															
112 Drepanocladus revolvens			1		1	1															
113 Equisetum variegatum			1	1		1															
114 Cladonia amaurocraea			3	2	4			2													
115 Carex aquatilis			2	2	2	1	2														
116 Rubus arcticus	2	2	1	2	1		3														
117 Dicranum bergeri	2	1	3	3	2			3		2						1					
118 Dactylina arctica	3		1	1		1			1		1	1									
119 Cetraria cucullata	4		2	3		2			1		1	1			2						
120 Equisetum arvense	2	3	1	3	2	2	5		1				1			1			4		2
121 Vaccinium uliginosum	2	2	5	5	4	5		2	1		1	2									
122 Betula glandulosa	5	4	4	4	3	3		3	3		2	1	2								1
123 Aulacomnium palustre	4	5	5	5	5	5		3	1		1	1	1								
124 Cladina rangiferina	4		2	2	3	2		2			1	2			2						
125 Hylocomium splendens	3	1	4	5	5	4	1	4	5	4	5	5		1	1	1					
126 Empetrum nigrum	5	1	4	3	3			4	2	1	3	4		1	1	1					
127 Pedicularis labradorica		1	5	5	2	4		3	1		1	1			1	1					
128 Arctostaphylos rubra		5	4	5	4	5	4	2	3	1	3	1	1	1		1					
129 Equisetum scirpoides		3	4	4	4	4	3	2	2	3	2	1		1	1	1					
130 Potentilla fruticosa		5	4	4	4	4	2		1	1	1			1							

Table 17 (Continued)

	1	2	3	4	5	6	7	8	9	10	11	12	13	14	15	16	17	18	19	20	21
131 Picea mariana		1	5	5	5	2		5	1		1	2			1						
132 Salix myrtillifolia		5	2	3	2	2		3			1		1	1							
133 Ledum groenlandicum		2	2	2	5	5		5	2	3	3	4		1	1	1					
134 Rhytidium rugosum			2	3	4	3		2	5	2						1					
135 Thamnolia vermicularis			3	2	2			3	1					1							
136 Pyrola asarifolia			1	2	1	1		2	1	4	2	2			2	1					
137 Alnus crispa			1	1	1	1		3	2	5	1	2			1						
138 Thuidium abietinum			1	4	3	3		3	3		3	3		5	1						
139 Geocaulon lividum			1		2	2		3	2	2	3			1	2	1					
140 Carex concinna		1	1	1	2	2			2		2		3	4		2					
141 Salix planifolia		3	1	1	1			2			2				1						
142 Stellaria longipes			1	1	1	2			1	2						1					
143 Betula papyrifera			1		1				1	3	1	1				1					
144 Cetraria nivalis			2			1			1						2	1				4	
145 Galium boreale	2						2		1				1			2	2				
146 Picea glauca		1	1	3	4	5	4	5	5	4	5	3	4	5	2	3	2				
147 Vaccinium vitis-idaea		2	3	4	5	3		3	3	3	4	5		1	4	1					
148 Peltigera aphtosa	4		4	4	4			5		3		5	1		4	1					
149 Cladina mitis	4		2	2	3			2			1	1			5						
150 Mertensia paniculata	4	2	1	1	3	1	5		4	5	3	1	3	2	1	3					
151 Festuca altaica	3		2	2	1	2			1	1	2	4	3	1	4	3		1			4
152 Salix glauca		5	4	4	4	3	1	2	3	1	2	2	5	2	2	3					
153 Cladina arbuscula			2	2		2		4	1		3	4			4		5				
154 Calamagrostis canadensis	2	3					5		1	4	1			1		2					
155 Cladina stellaris			1		1	2		4			1	2			1		5				
156 Rosa acicularis		1		2	3		5	4	3	3	3	2	1	1	3	4	5	4	1	1	
157 Lupinus arcticus			1	1	1	1	1	3	1		3	3	3	3	2	2			4		
158 Shepherdia canadensis		1		1			3		1	1	2	1	2	2	2	4		2			
159 Pyrola secunda			1		1			2	2	2	3	3	2	2	2	2					
160 Linnaea borealis					1	1	1	4	4	4	5	5	1	3	5	3					
161 Viburnum edule					1		2		2	3	2	3	1		1	1					
162 Lycopodium complanatum					1			3			1	3			1						
163 Lycopodium annotinum										2	2	3				1					
164 Drepanocladus uncinatus		1				1			2		1	3	2	3	1	2					
165 Epilobium angustifolium						1	5		1	3	3	3		1	2	5	2		5		
166 Corallorhiza trifida						1			2	1	1			1		1					
167 Nephroma arcticum			1					2			2	2			1						
168 Cladonia ecmocyna								2	1	1	5	3			1	1					
169 Pinus contorta			1					5		1	3	5			5	1					
170 Dicranum scoparium									2		1	1			1						
171 Stereocaulon tomentosum					1				1		1	2	3		3		5				
172 Peltigera canina									2		1	2	3	5	5						
173 Populus balsamifera		1					3			1	1		1	1	1	2			2		2
174 Pulsatilla patens										1		1			3	1		3		2	
175 Ceratodon purpureus											1	2	4	2	3	2	3	2		2	
176 Juniperus nana						2		2			1				1	1	2	2		2	
177 Solidago decumbens		1									1		1	1	1	2	2	2	3	2	
178 Arctostaphylos uva-ursi						1						2	4	2	5	5		2			
179 Calamagrostis purpurascens												1		1	2	1	5	4	4	5	
180 Fragaria virginiana		1					1				1		2			2			4	1	
181 Achillea borealis												1			1	2	2	1	5	2	
182 Tortula ruralis						2								1		2	4	1		4	
183 Anemone multifida													2	2		2	2	2		3	
184 Astragalus alpinus														2		1			4		
185 Cladonia cariosa															2	1	3			2	
186 Rubus idaeus											1				1	1			4		
187 Cornicularia aculeata															1		2			2	
188 Poa glauca																1	2	2	4	2	
189 Peltigera rufescens																2	4		1	2	
190 Festuca brachyphylla																1	4	2		2	
191 Saxifraga tricuspidata																1	5			3	
192 Penstemon procerus																1		3		2	
193 Juniperus horizontalis																1		2		2	
194 Geum perincisum																1	2			1	
195 Artemisia frigida																	2	5		4	
196 Carex filifolia																	2	4		3	
197 Parmelia wyomingica																	3			4	

	1	2	3	4	5	6	7	8	9	10	11	12	13	14	15	16	17	18	19	20	21
198 Potentilla pennsylvanica																	2			2	
199 Draba aurea																	2			2	
200 Antennaria rosea var.nitida																		2	3	3	
201 Penstemon gormani																		3		3	
202 Artemisia alaskana																		2		2	
203 Melandrium apetalum																		1		2	
204 Hedysarum americanum		2	2	2	2		3		2				2	1		2		2			1
205 Cladonia chlorophaea	3	1				1			1		1	1		1	3		3			3	
206 Salix alaxensis		3	1		1	2	4				1		4	1		1					2
207 Delphinium glaucum	1						2				1		2		1	1					
208 Pedicularis sudetica	3	2							1		1			1							
209 Equisetum sylvaticum			1	1						1	1					1					
210 Calypso bulbosa						1					2	1				1					
211 Oxytropis campestris									2					1		1				2	
212 Arnica cordifolia										1	1	1				1					
213 Artemisia borealis		1									1		2			1					
214 Abies lasiocarpa			1								1	1			1						
215 Cladonia turgida	3			2	1				1		1		1								
216 Polytrichum juniperinum	5									3	1	3			1	1					
217 Zygadenus elegans									1			1		1	2						
218 Cladonia pyxidata			1		1										1	1					
219 Tomenthypnum nitens						2			2					1							
220 Barbilophozia hatcheri						1			1		2										
221 Oxytropis deflexa											1			1		1					
222 Cladonia deformis											1	2			2						
223 Carex vaginata						1				1						2					
224 Selaginella selaginoides			1		1			2													
225 Cetraria laevigata	3											1			2						
226 Cypripedium passerinum						2	2														
227 Dicranum elongatum									1					1							
228 Petasites palmatus			1								1										
229 Aquilegia brevistyla														1		1					
230 Vaccinium microphyllum			1	1																	
231 Stereocaulon alpinum															1	1					
232 Petasites frigidus			1	1																	
233 Rumex arcticus			1	1																	
234 Carex atratiformis				1	1																
235 Arabis hoelbelii																1	2				
236 Juncus arcticus		1																			2
237 Ribes oxyachanthoides										1						1					
238 Androsace septentrionalis										1						1					
239 Dicranum tenuifolium														2							
240 Goodyera repens											1										
241 Listera borealis						1			1												
242 Viola renifolia											1										
243 Saussurea angustifolia									1												
244 Betula occidentalis			1																		
245 Larix laricina			1																		
246 Amerorchis rotundifolia			1																		
247 Salix lanata			1																		
248 Spiraea beauverdiana			1																		
249 Lycopodium clavatum										1											
250 Antennaria rosea															1						
251 Chamaerhodos erecta																				1	
252 Amelanchier alnifolia																				1	
253 Stipa comata																				1	
254 Hierochloe alpina																				1	
255 Solidago multiradiata																		1			
256 Pohlia nutans																1					
257 Senecio atropurpureus																1					
258 Hierochloe odorata																1					
259 Stellaria crassifolia																1					
260 Potentilla diversifolia																1					
261 Dryas drummondii																1					
262 Erigeron sp.																1					
263 Cetraria tillesii																1					
264 Astragalus eucosmus																1					
265 Cladonia fimbriata																1					

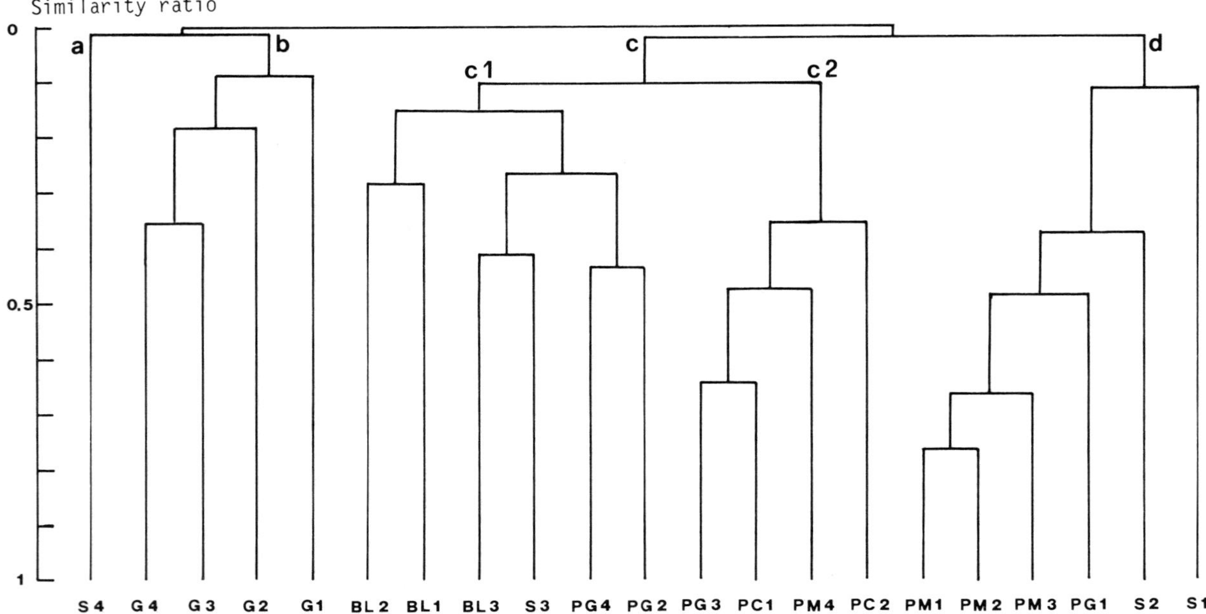

Fig. 25. Dendrogram showing the floristic affinities among the relevé groups, obtained by classification of the data in Table 17. Letters indicate the main clusters discussed in the text.

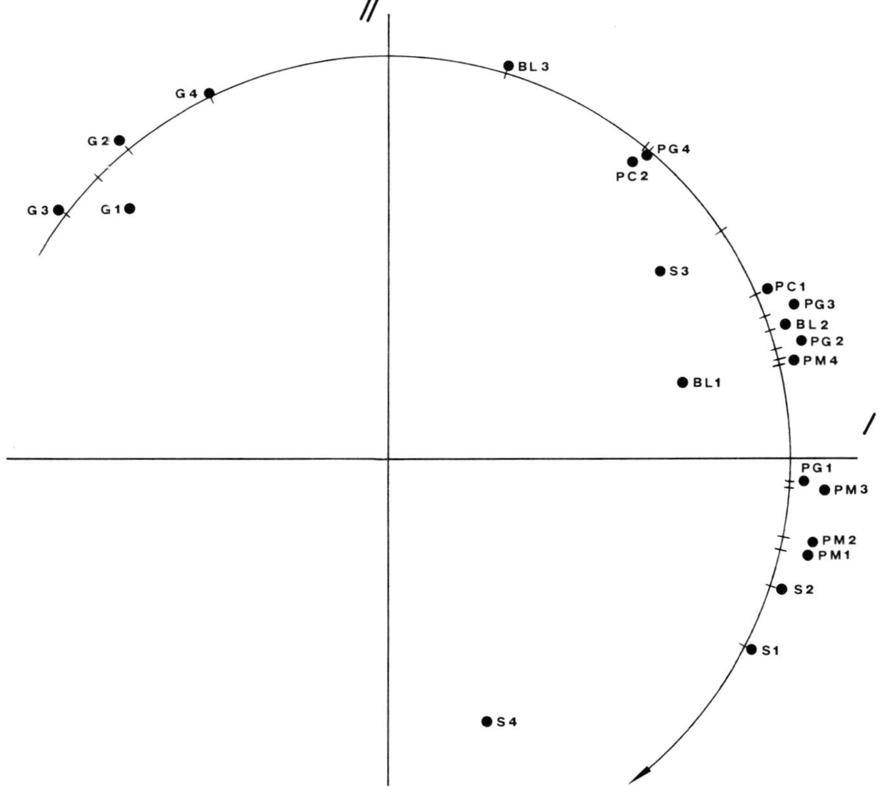

Fig. 26. Ordination (by SIPLO) of the relevé groups, obtained on the basis of the data in Table 17 (floristic composition).

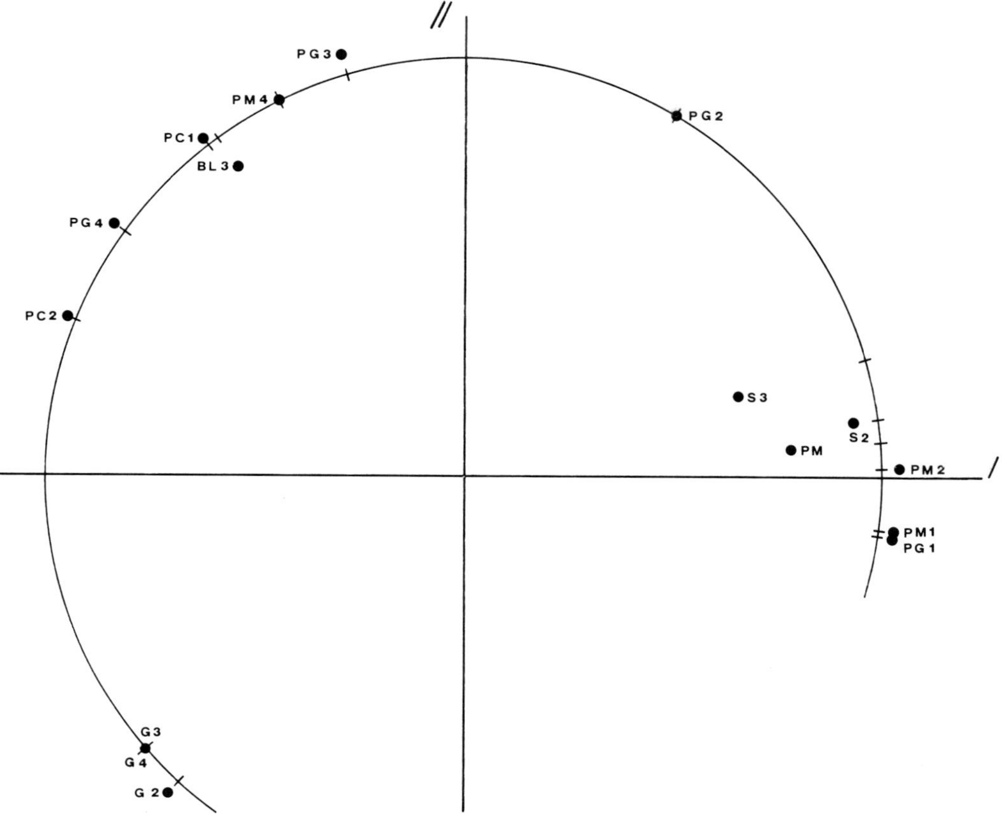

Fig. 27. Ordination (by SIPLO) of the relevé groups, obtained on the basis of the data in Table 1 (soil types).

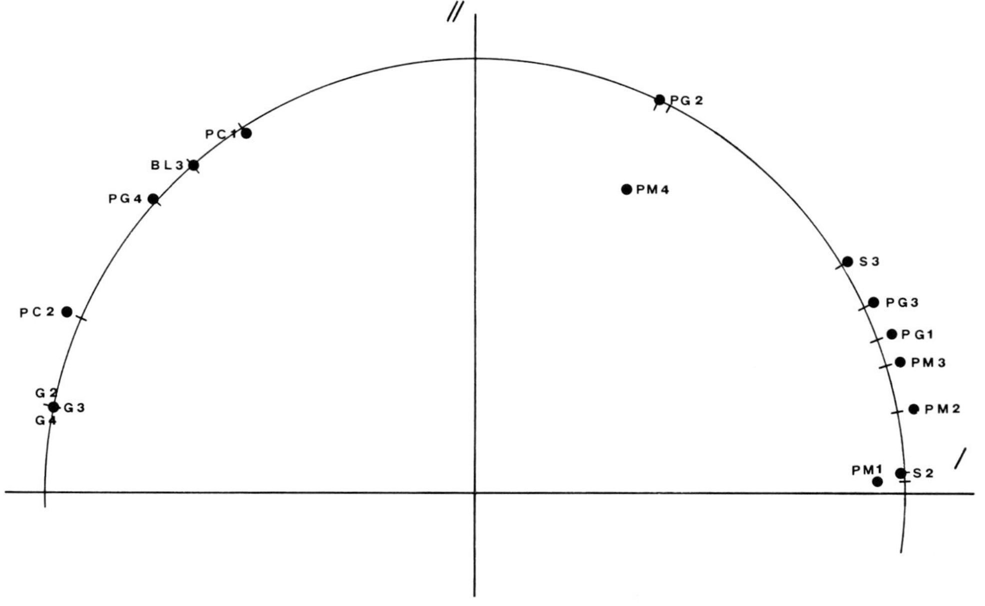

Fig. 28. Ordination (by SIPLO) of the relevé groups, obtained on the basis of the data in Table 2 (drainage and permafrost).

preceded by broadleaved *Populus* stands, and in c2 by *Pinus contorta* woods, whose litter has an acidifying influence.

(3) Cluster d: in 71% of the releves the soils were of the Cryosolic order. The cluster includes relevés of the muskeg formation (open taiga) and of its secondary stages.

An ordination of the vegetation types on the basis of floristic data (Table 17) was obtained with the program SIPLO (Feoli and Feoli Chiapella 1980). The results are shown in Figure 26. To quantify the correlation between vegetation and soil variables, the vegetation types were ordered using SIPLO, and on the basis of two further data sets: (1) percentages of soil types within the vegetation types (Table 1); (2) percentage occurrence of permafrost and of the different drainage classes within the vegetation types (water conditions, Table 2). The results of the two ordinations are given in Figs. 27 and 28, respectively.

The correlations between the ordinations obtained on the basis of (1) floristic composition, (2) soil types and (3) water conditions are given in Figures 29a–c. These figures are based on the angular seriations of the vegetation types in the space defined by the first two axes of the three ordinations.

A very strong correlation exists between water conditions and soil types (r = 0.94, Fig. 29a), between floristic composition and soil types (r = 0.88, Fig. 29b), and between floristic composition and water conditions (r = 0.86, Fig. 29c). These strong correlations demonstrate the high predictivity of the species combinations defining the vegetation types with respect to the variation of the main soil factors.

The relations between the community types and the soil types were studied by AOC performed on the soil data of Table 1. The results are shown in Figure 30a and b, which reveals the subdivision of the community types into three main groups, corresponding with the three main soil orders occurring in the survey area: Regosols, Brunisols and Cryosols. In this case the results of ordination based on ecological data correspond fairly well with those obtained by classification

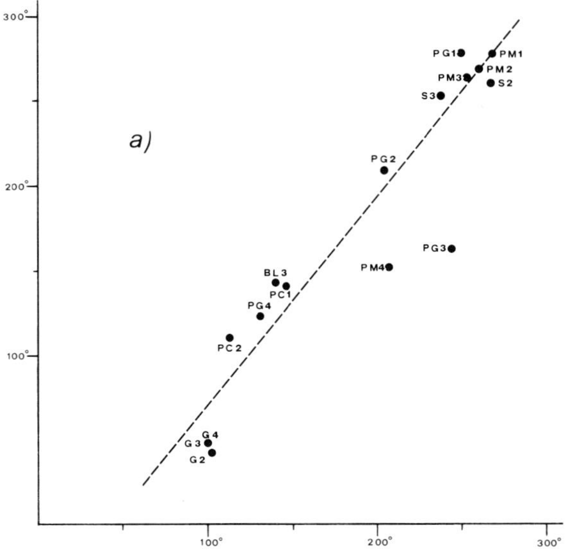

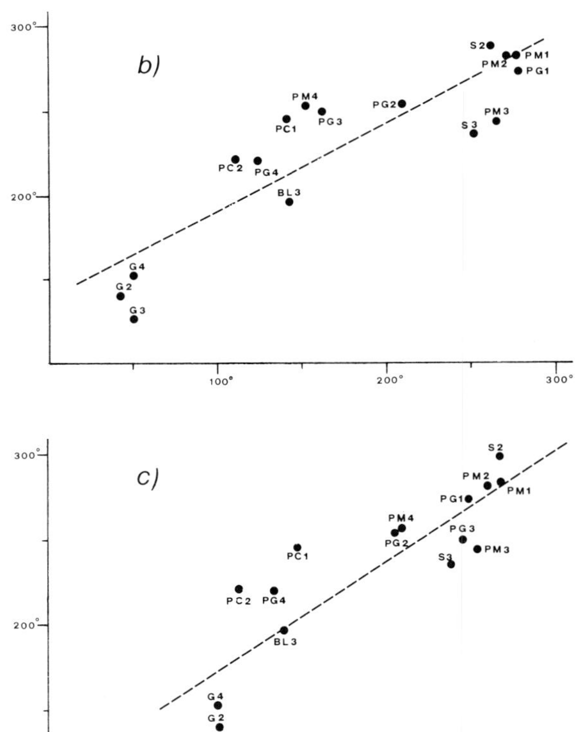

Fig. 29. Comparison between the three ordinations of Figures 26–28. The x- and y-axes report the angular scores of the relevé group points in the respective ordinations, as follows: (a) drainage and permafrost (Fig. 28) and soil types (Fig. 27); (b) floristic composition (Fig. 26) and soil types (Fig. 28); (c) floristic composition (Fig. 26) and drainage and permafrost (Fig. 28).

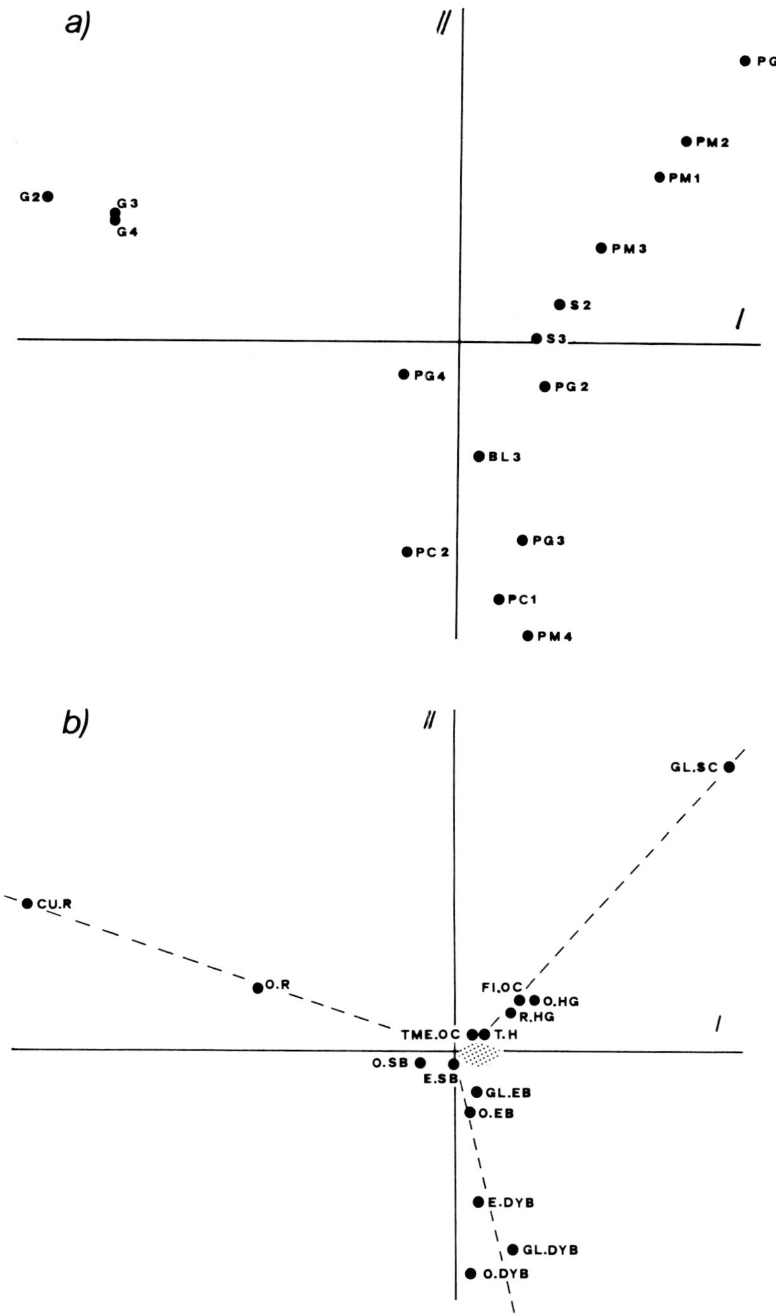

Fig. 30. Ordination (by AOC) of relevé groups (a) and soil types (b) obtained on the basis of the data in Table 1.

on the basis of floristic data (see dendrogram in Fig. 25).

These results are summarized in Fig. 31, where the community types are arranged on the *x*-axis according to their angular scores in the ordination obtained on floristic data; the following ecological variables, for each community type, are plotted on the *y*-axes: (1) percentage of Regosols; (2) percentage of Brunisols; (3) percentage of Gleysols; (4) percentage of Cryosols; (5) average thickness of the organic horizon; (6) percentage of drainage classes (wet, impeded,

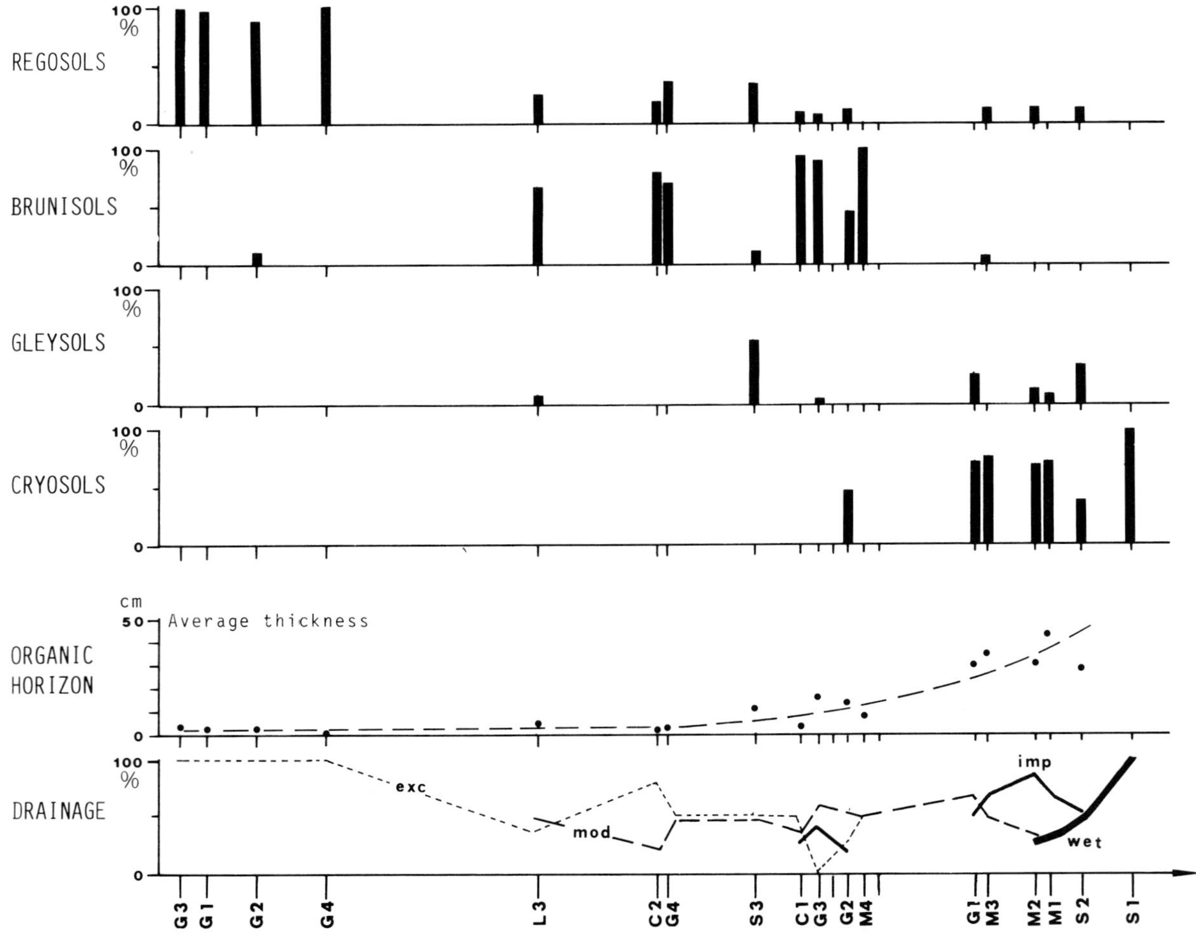

Fig. 31. Interpretation of the ordination of relevé groups based on floristic data (Fig. 26). The *x*-axis reports the angular scores of the relevé group points in the ordination of Figure 26.

moderate, excessive). The results shows a regular transition from Regosols to Brunisols to Cryosols, paralleled by a regular increase of the average thickness of the organic horizon and by a transition from excessive to moderate, impeded and wet drainage, and the progressive formation of permafrost. The floristic variation along this complex gradient is shown in Table 17, where the vegetation types are ordered according to their angular scores in Fig. 26. From Table 17 it is evident how the vegetation types obtained by classification represent noda along a compositional gradient which in general can be interpreted as a gradient in water availability in the soils.

Soil pH, in this context, seems to have only minor importance; Fig. 32 shows an ordination of the community types obtained by PCA on the basis of pH data. The pH ranges of the communities, ordered as in Fig. 32, are shown in Fig. 33. It is evident how the sequence of the types obtained on the basis of pH data does not correspond to any of the sequences obtained on the basis of floristical data, soil types, or water conditions. To test whether a pH gradient could be revealed by another ordination method, the vegetation types also were ordered by PCA on the basis of floristic data (Table 17). The results are shown in Fig. 34, where the sequence of the vegetation types fairly well corresponds with

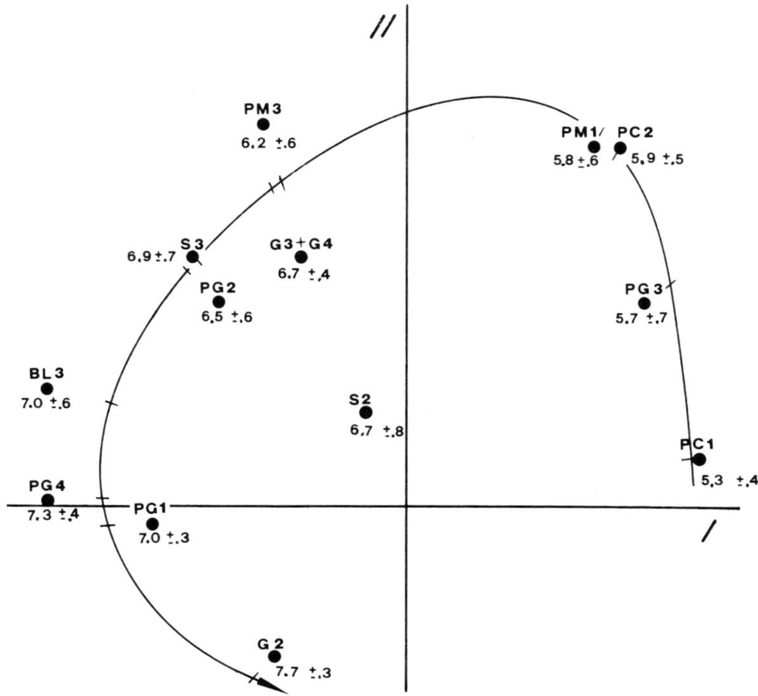

Fig. 32. Ordination of the relevé groups on the basis of pH data (see Fig. 33).

the one obtained by SIPLO (Fig. 26). In Fig. 34 some major trends concerning pH, however, are detectable. At the negative side of the first axis are communities growing on subacid to acid soils, whereas the communities whose points have positive scores have neutral to basic pH.

As far as the main successional phenomena are concerned, the only objective approach to their study is a series of long-term observations in permanent plots. Since these data are not available, we shall limit ourselves to some general considerations, which could lead to working hypotheses to be tested and enriched on the basis of further evidence. Four main factors should be considered:

(1) *Geographic distribution of the communities*: since some of the communities are restricted to limited portions of the survey area, the successional sequences will be different in different parts of the area. Three main regions have been distinguished (see Fig. 3):

(a) Southeastern portion (from Watson Lake to Whitehorse); it lies within the distributional limit of *Pinus contorta*. Communities restricted to it are PC1 and PC2.

(b) Central portion (from Whitehorse to Quill Creek): this is the driest part of the survey area. *Pinus contorta* is replaced by *Populus tremuloides*. Communities restricted to it are G2, S4, S3, PG4.

(c) Northwestern portion of the Alaska Highway (from Quill Creek to the Alaskan border): this is the most humid part of the transect. *Picea mariana* muskegs cover most of the ground, except the steepest slopes. Communities restricted to it are PG1 and BL2.

(2) *Age of the trees*: the average age of the tallest trees within each community can be considered as a rough estimate of their relative positions along successional series. The following values have been obtained:

Salix glauca (S3)	28 years
Pinus contorta (PC2)	56 years
Populus tremuloides (BL3)	60 years

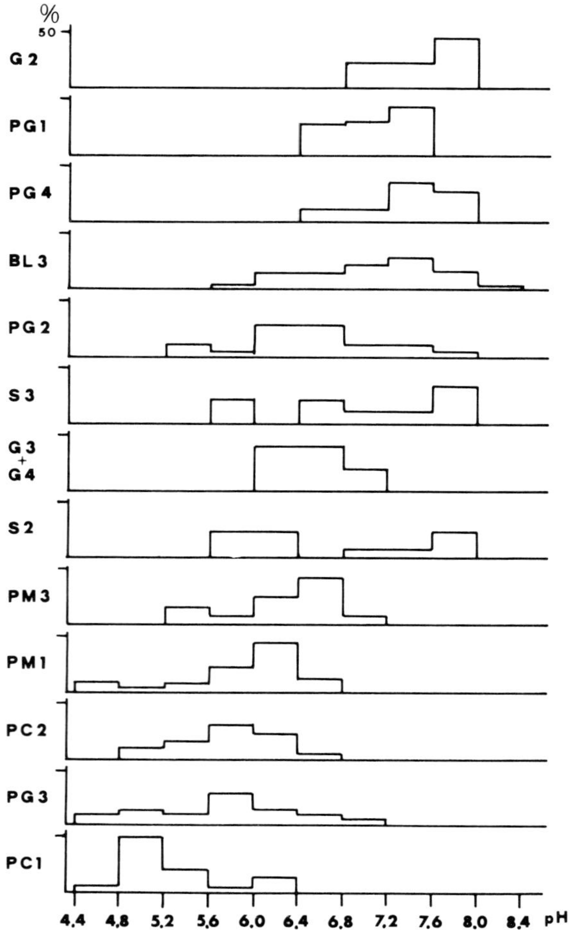

Fig. 33. Arrangement of relevé groups according to their pH ranges, following the sequence obtained by the ordination shown in Figure 32.

Pinus contorta (PC1)	75 years
Picea glauca (PG3)	109 years
Picea mariana (PM1)	130 years
Picea glauca (PG2)	175 years
Picea mariana (PM2)	177 years
Picea glauca (PG1)	180 years
Picea mariana (PM3)	194 years

From these data the following characterization may be given for the five main tree species occurring in the survey area:

Salix glauca: first pioneer species
Pinus contorta: species of secondary stands
Populus tremuloides: species of secondary stands
Picea glauca: mature forests (mainly on Brunisols)
Picea mariana: mature forests (mainly on Cryosols).

In general, trees growing on Cryosols tend to be older than trees growing on other soil types. An exception is *Picea glauca* in PG2: this community occurs in an area with a dry climate where Brunisols rarely evolve towards Cryosols.

(3) *Soil development*: a schematic representation of soil development in the study area can be summarized in the following two sequencies: (a) bedrock, Regosols, Brunisols, Cryosols (xeroseries); (b) Gleysols, Cryosols (hydroseries).

Since most of the area lies in the scattered permafrost subzone the development of Cryosols is not always reached. This is most frequently the case on south facing slopes or on sandy soils with excessive drainage. The successional sequence may stop at different stages, according to the local pedological and geomorphological conditions.

(4) *Edaphic vicarism among communities*: this regards sympatric communities with similar requirements for soil type, and different requirements for pH. An example is given by PM1, PM2 and PM3.

The following is a very broad successional scheme, valid for the entire survey area:

(a) Grasslands (on Regosols, throughout the area, different communities according to parent material).

(b) *Salix* thickets (following heavy fires of forest vegetation, mostly on Brunisols).

(c) Secondary tree stands dominated by *Pinus contorta* in the east, by *Populus tremuloides* in the west.

(d) *Picea glauca* forests (different communities according to climate and soil type, mostly on Brunisols).

(e) *Picea mariana* muskegs (widespread only in the western part of the area, and in the mountains, being bound to Cryosols and the formation of permafrost; different communities according to soil pH).

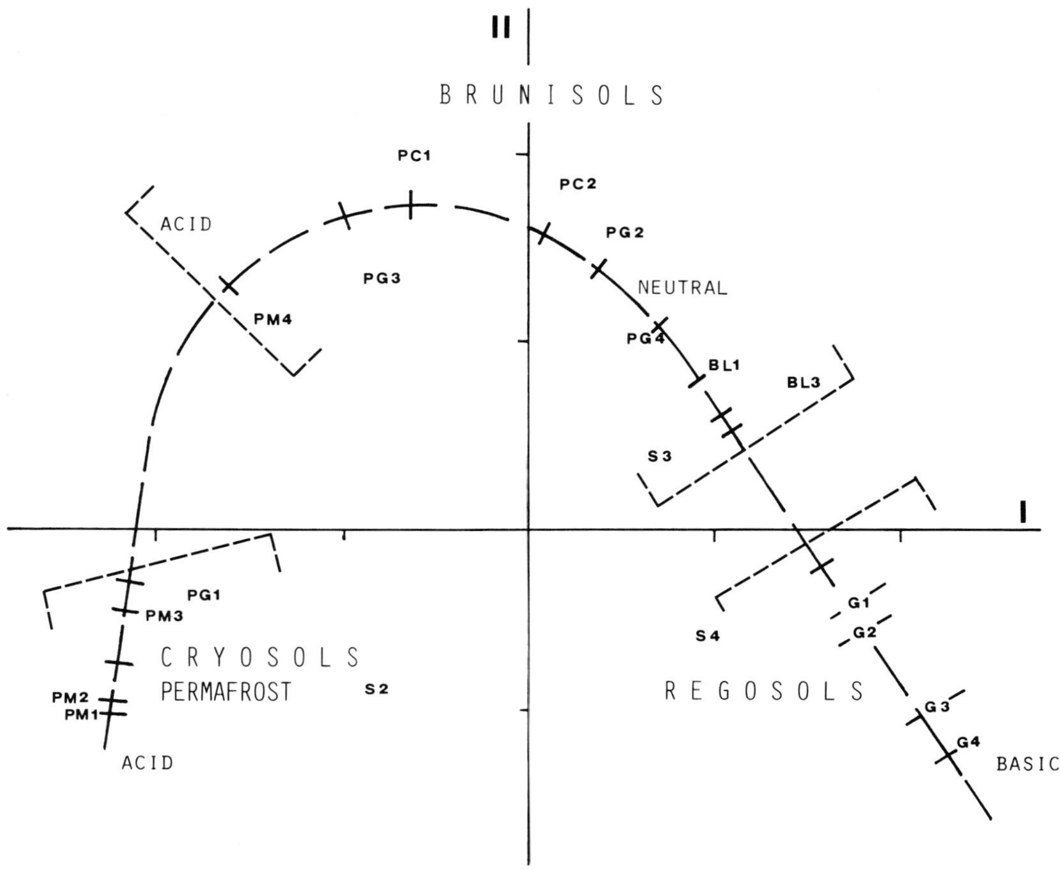

Fig. 34. Ordination (by PCA) of the relevé groups, based on floristic data (Table 17).

6. PHYTOGEOGRAPHIC ANALYSIS

6.1. Classification of species

A total of 163 species of vascular plants present in our data set (Table 17) have been classified on the basis of their world ranges, to obtain groups of species with similar distribution (phytogeographic elements). Classification programs have been applied to matrices of species and OGUs. To reduce the computing load, classification has been carried out on four separate data sets:

Data set A: *Salix pulchra* community (S1). This community was separated from the others, since it is the only one recorded in the northern Yukon. The classification (Fig. 35) produced four main species groups, A1, A2, A3, A4.

Data set B. Includes the muskeg formation and its secondary stages (cluster "d" in Fig. 25). The classification produced three main species groups, B1, B2, B3.

Data set C. Includes the boreal forest formation and its secondary stages (cluster "c" in Fig. 25). The classification produced six main species groups, C1, C2, C3, C4, C5, C6.

Data set D. Includes the grassland vegetation (clusters "a" and "b" in Fig. 25). The classification produced five main species groups, D1, D2, D3, D4, D5.

A higher level of synthesis was obtained by classification of the 18 species groups, based on the respective frequencies in the OGUs. As a result (see Fig. 35), seven main species groups were identified. In the following, each species group is briefly discussed. The list of the species, subdivided by groups, is reported in the appendix.

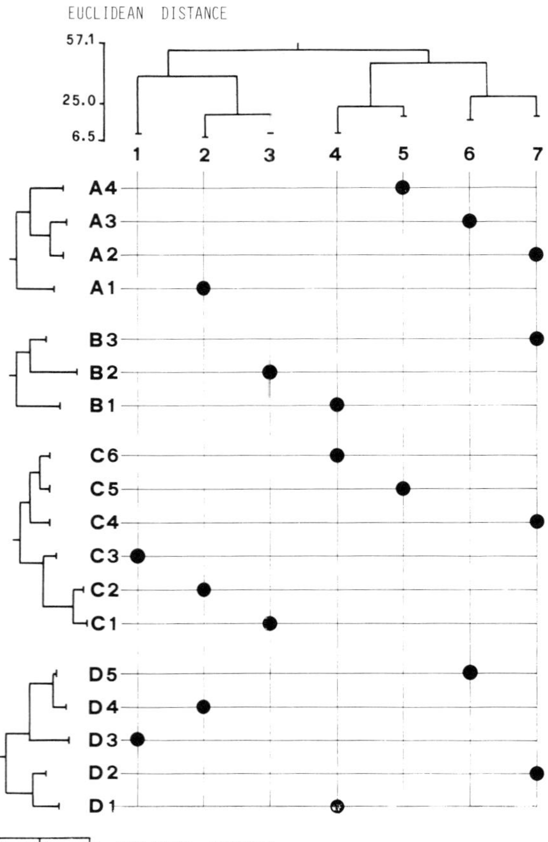

Fig. 35. Classification of the species on the basis of their occurrence in the OGUs. The species have been extracted from four eco-community data sets, which have been submitted to classification separately to obtain species groups with similar distribution: Subalpine *Salix* thickets of northern Yukon (A), *Picea mariana* muskegs and related communities (B, see cluster "d" in Fig. 25), *Picea glauca* forests and related communities (C, see cluster "c" in Fig. 25), grasslands (D, see clusters "a" and "b" in Fig. 25). The phytogeographic elements (designated with numbers 1–7) were obtained by classification of the contingency table of the 18 species groups and of their relative frequencies in the OGUs. For further explanations see the main text.

Species group 1 (*incompletely circumboreal–temperate species*)

This is a group of species with incompletely circumpolar, boreal-temperate ranges (chorograms in Fig. 36). They are absent in large parts of northeastern Siberia, an area with a strongly continental, cold climate. Some of these species occur within the boreal forest (C3), others are among the most widespread weeds in anthropogenous environments (D3). The boreal forest species have strongly disjoint stations south of the boreal zone, e.g. in the mountains of northern China, Japan and the Himalayas. This suggests that they had an ancient broad distribution in the mountains of the Northern Hemisphere before the Ice Age. They might have survived glaciations in refugia located mostly south of the ice-sheets, from which they later extended in connection with the expansion of the boreal forest during the postglacial periods. The harsh continental conditions in northeastern Siberia are probably the reason for their absence in this area. They probably reached Alaska–Yukon from the south, along the Cordilleras, from the refugial stations located south of the North American ice-sheet. Notwithstanding their relatively low number (12.3% of all species present in the boreal forest and its secondary stages), these species are among the most characteristic elements of the closed-canopied *Picea glauca* forests in the Yukon. Most of them are considered by Braun-Blanquet et al. (1939) as characteristic species of the circumboreal Order *Vaccinio–Piceetalia*.

The weed species have a different ecology and history. They belong to a widespread group of antropochore plants with Holarctic, or even worldwide distribution, centered in the temperate zone. They became part of the Yukon flora in recent times, as a consequence of increasing anthropization, and are now a frequent element along roadsides (see Lausi and Nimis 1985b).

Species group 2 (*Cordilleran species and endemics to Alaska-Yukon*)

This group includes species with distribution centered in the northern part of the Cordilleras of western North America. The chorograms are shown in Figure 37. They occur in the *Salix pulchra* community of northern Yukon (A1, 15.09% of the species), in the boreal forest vegetation (C2, 26.4% of the species) and in the grasslands (D4, 58.4% of the species). The Cordilleran species are an important element of the xerophytic vegetation in southern Yukon. They are most frequent in steppe vegetation on

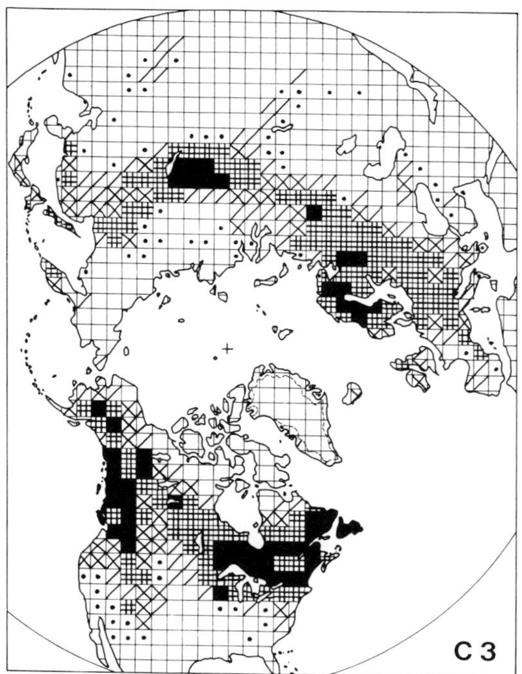

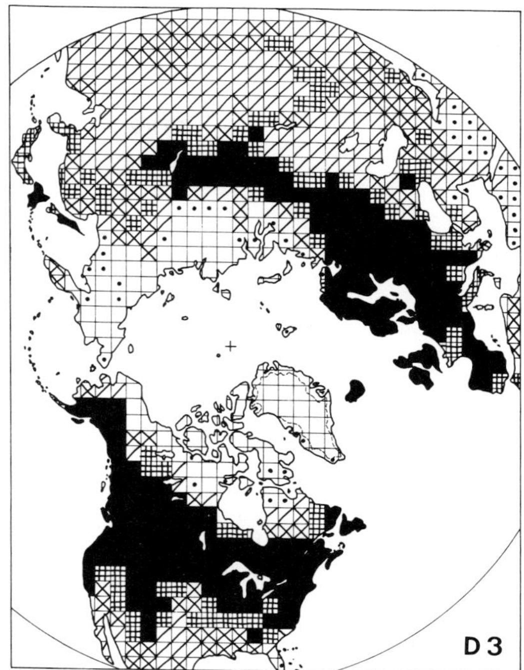

Fig. 36. Chorograms of the two species groups included in group 1 (see Fig. 35). Different shadings refer to the percent of species occurring in the OGUs, on a 5-class scale with intervals of 20%.

loess, but also penetrate into the secondary *Populus tremuloides* open woods.

According to Nimis (1982) and Lausi and Nimis (1985a) the history of the Cordilleran element in Alaska–Yukon is linked to two main palaeobotanical facts: (a) the existence during xeric interglacials of an ice-free corridor connecting Beringia with the regions of North America south of the ice-sheets; and (b) the persistence, in unglaciated portions of Alaska–Yukon, of a parasteppe vegetation throughout the Ice Age. The past ecological conditions along the corridor were such as to allow a relatively easy migration of grassland species from the periglacial steppes southwards, and from the central North American prairies northward.

Interestingly, this group also includes some grassland species which are endemic to Alaska and the Yukon. The presence of a relatively high number of endemics in this steppe vegetation is a further indication of the ancient origin of natural grasslands in the area. There is extensive palaeobotanical evidence (see Hopkins 1967) that the southeastern part of Beringia supported a parasteppe periglacial vegetation in which *Artemisia* species were particularly abundant. The existence of such a vegetation was mainly due to a strong continentality of the local climate, and there is reason to assume that the local conditions did not substantially change from the late Pliocene until now. After the retreat of the glaciers some of the grassland species were able to migrate through the Cordilleran corridor, others (the endemic grassland species) did not substantially change their distribution patterns. In other cases, the endemic status may be due to the fragmentation of a previously continuous population in two refugial areas, one located in Beringia, the other south of the ice-sheet. For example, the ranges of *Penstemon gormani* (endemic to Alaska–Yukon) and of the closely related *Penstemon eriantherus* (a prairie species) are separated by a rather wide area along the northern Cordilleras. For other species such geographic isolation did not result in differentiation. An example is the range of *Eurotia lanata*, with its main distribution in the North American interior plains and a disjunct station in southern Yukon.

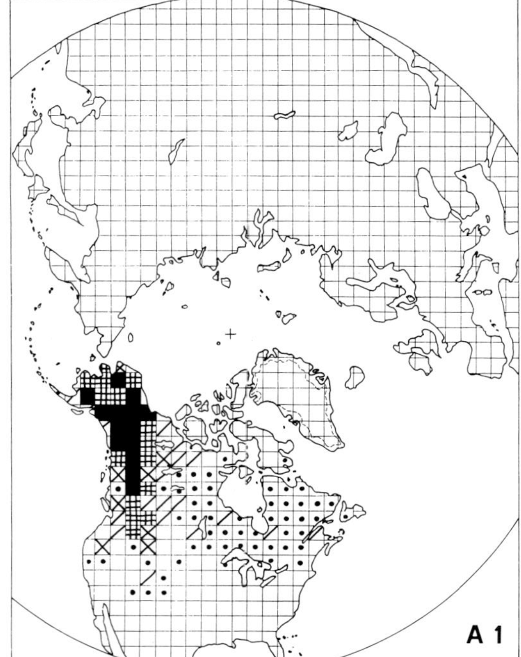

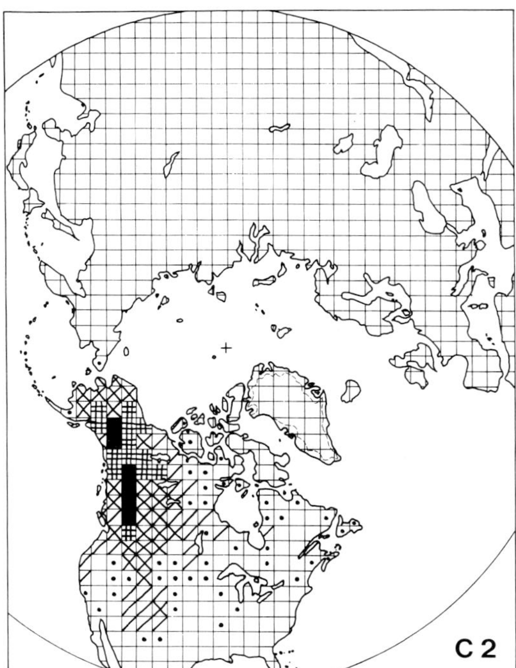

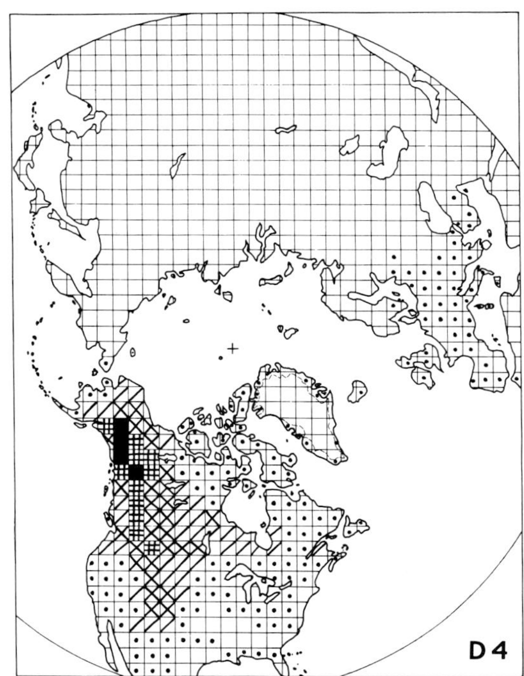

Fig. 37. Chorograms of the three species groups included in group 2 (see Fig. 35). Different shadings refer to the percent of species occurring in the OGUs, on a 5-class scale with intervals of 20%.

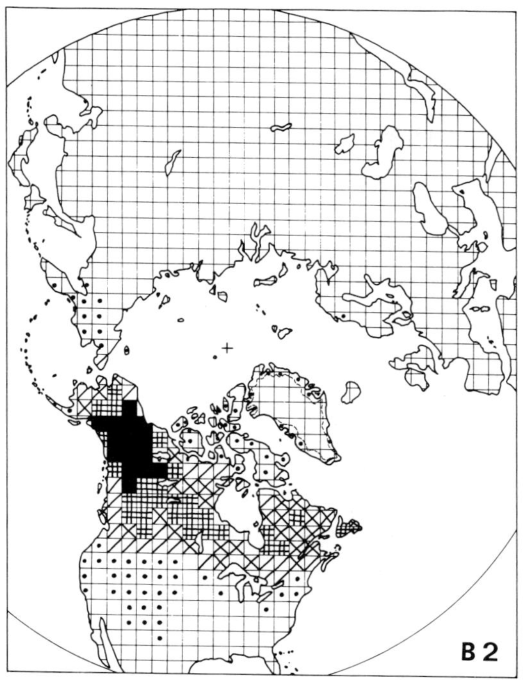

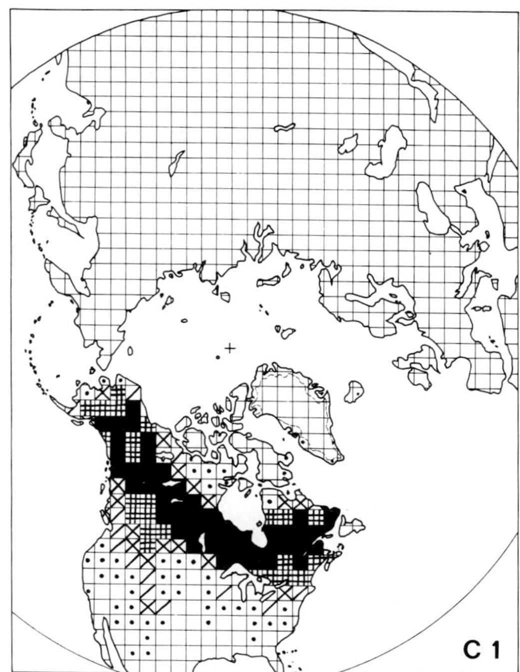

Fig. 38. Chorograms of the two species groups included in group 3 (see Fig. 35). Different shadings refer to the percent of species occurring in the OGUs, on a 5-class scale with intervals of 20%.

The fact that our classification places Cordilleran and endemic species in the same phytogeographic group is a further indication of their common history.

Species group 3 (*boreal North American species*)

This group includes species with boreal North American ranges, occurring in the muskeg vegetation (B2) and in the boreal forest formation and its secondary stages (C1). The chorograms are shown in Figure 38. Some of the species, such as *Carex concinna, Geocaulon lividum, Ledum groenlandicum, Picea glauca*, etc., have a coast to coast distribution. Other species, such as *Cypripedium passerinum, Juniperus horizontalis, Amerorchis rotundifolia*, etc., although present both in eastern and western North America, have a discontinuous range, with gaps located mainly south of the Hudson Bay. Their common absence from the Hudson Bay region can be explained by the much colder climate, compared to the rest of North America at corresponding latitudes. In the area between Hudson Bay and the Great Lakes, the boreal forest occurs at much lower latitudes, forming the narrowest belt in the whole of North America (between 48° and 50° latitude; Rowe, 1977).

A third subgroup of species, including *Pinus contorta*, is restricted to boreal northwestern North America. There is evidence that before glaciation *Pinus contorta* was the only *Pinus* species growing in the northwest (Mirov 1967). This species seems to have persisted in the Yukon in ice-free refugia during at least the last Wisconsin glaciation (Hansen 1943, 1947; Mirov 1967). Using multivariate methods, Jeffers and Black (1963) contributed to the elucidation of the infraspecific variation of *Pinus contorta*. Their results confirm the subdivision of the species into two main varieties, respectively found in the inland (var. *latifolia*) and in the coastal provinces (var. *contorta*). The existence of different refugial stations in northwestern North America can be postulated for most of the species which presently are restricted to this area. But this does not imply that other refugial stations, located

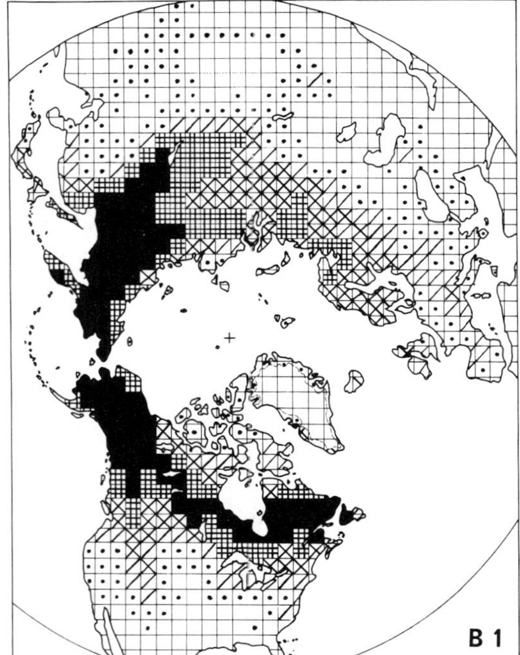

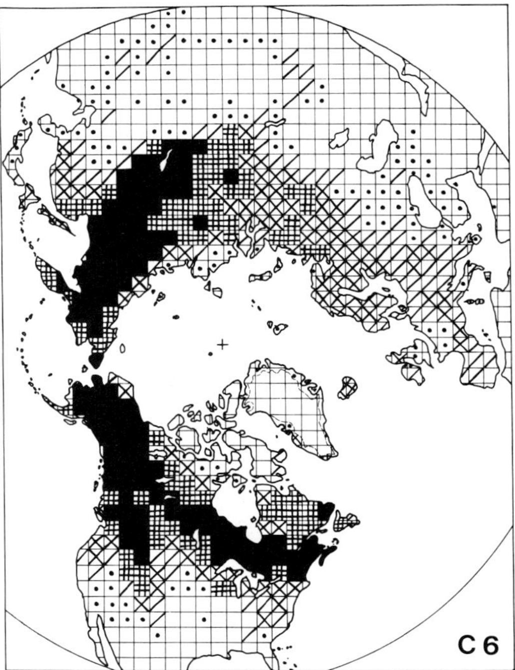

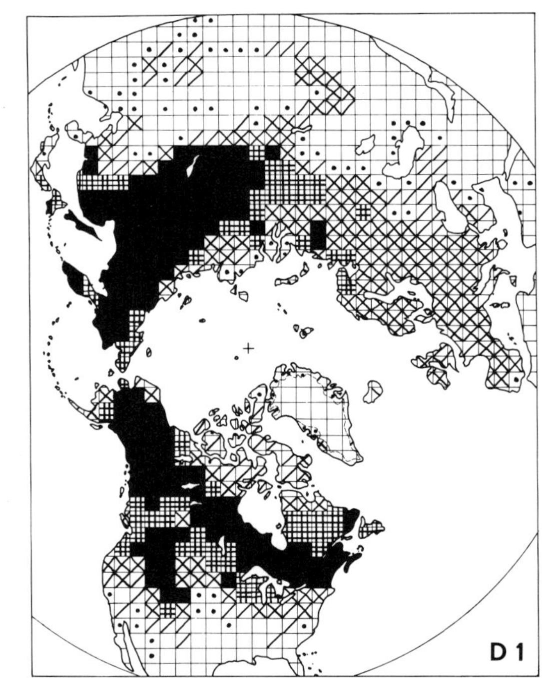

Fig. 39. Chorograms of the three species groups included in group 4 (see Fig. 35). Different shadings refer to the percent of species occurring in the OGUs, on a 5-class scale with intervals of 20%.

south of the ice-sheets, also did not exist. An example is given by *Picea glauca*; Ritchie (1984) summarizes the available palaeobotanical information concluding that this species probably entered the Yukon relatively late from refugia located south of the North American ice-sheets. However, the taxonomic differentiation of *Picea glauca* into a northwestern race (var. *porsildii*, Raup 1947) and the recent chemotaxonomic demonstration by Rudolf et al. (1981) that samples of this variety from northern Yukon display a richer gene pool than other populations, provide indirect support for the existence of refugial stations in the Yukon also, as assumed by Hopkins (1972) and Matthews (1976).

The close affinity between the boreal North American element and the boreal northwestern North American element of the Yukon flora, revealed by the results of our classifications, might reflect a similar history: both elements might have survived glaciation in ice-free refugia located in the northwest, but only the former group was able to spread westwards reaching the Atlantic coast. Perhaps this was due to the presence of further refugia south of the ice-sheets.

Species group 4 (*Arctic-montane species*)

This group includes species whose ranges (see chorograms in Fig. 39) are centered in boreal North America (mostly throughout the continent), and in the mountains of eastern Asia. Only a few of them reach Europe and have a truly circumboreal range. These species occur in the *Picea mariana* muskegs (B1), in the boreal forest (C6), in the grasslands (D1) and in the *Salix pulchra* vegetation of northern Yukon (A4). The majority of the species (50.9%) occur in the muskegs, and 23.5% of them are present in the boreal forest.

The chorograms of this species group (Fig. 39) correspond well with that of Plate 27 in Hultén (1937). According to Hultén (1937) these are Arctic-montane species, i.e. species which live not only on lowlands with Arctic or Subarctic vegetation, but also on mountains in far more southerly districts. The only places in Eurasia where mountain ranges reach the Arctic regions are in Scandinavia, the Urals and in northeastern Asia. As the two first mentioned areas were almost completely covered with ice, they can hardly be expected to be centres for the Arctic–montane plants; whereas the mountains in northeastern Asia were only partially glaciated, and form an excellent connection with the mountains of central Asia (Hultén 1937). It is interesting to note that most of these species in North America are bound to the northern boreal and Arctic zones; only a few steppe species (see chorogram of D1 in Fig. 39) have ranges extending south of the areas covered by the North American ice-sheet. According to Hultén (1937) the Arctic–montane plants spread from the mountains of northeastern Asia to America before the maximum glaciation, and did not reach the southern rim of the ice in that continent. Their further spreading westwards after the maximum glaciation mainly occurred along the Arctic coasts.

Species group 5 (*circumpolar (−boreal) species*)

This relatively small group includes species with a complete circumpolar range (see chorograms in Fig. 40). Given that most of the area presently occupied by these species was covered by ice during the glacial periods, and that the only extensive ice-free areas were those located in Beringia, one might conclude that most of these species survived glaciation in Beringia itself, and were able to expand their distribution eastward and westward in postglacial times. Interestingly, this group includes some of the most common and widespread species in the boreal–Subarctic vegetation throughout the Northern Hemisphere, such as *Carex vaginata, Empetrum nigrum, Vaccinium uliginosum, Vaccinium vitis-idaea, Andromeda polyfolia, Rubus arcticus*, and *Rubus chamaemorus*. Many are capable of vegetative reproduction and have a relatively broad ecological tolerance, being present in several community types. These features could have allowed a rapid migration from the Beringian refugia. A closer analysis of the taxonomy of this species throughout their ranges, however, indicates that their history might have

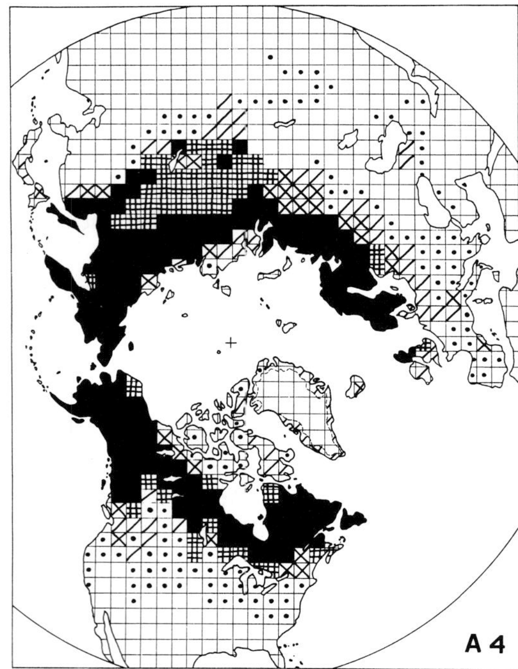

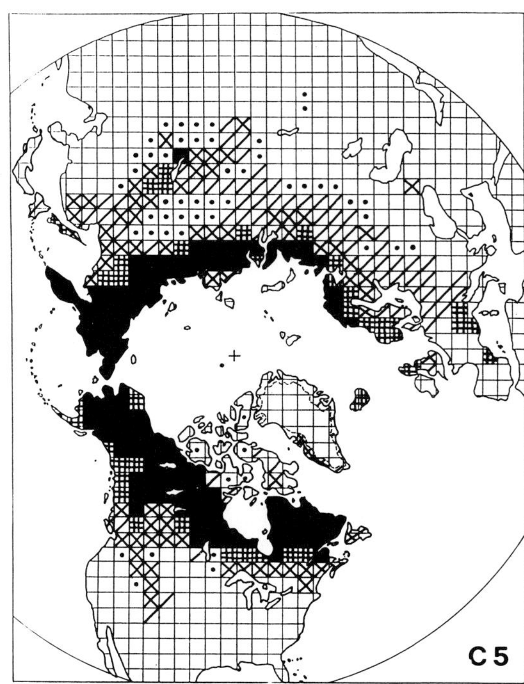

Fig. 40. Chorograms of the two species groups included in group 5 (see Fig. 35). Different shadings refer to the percent of species occurring in the OGUs, on a 5-class scale with intervals of 20%.

been much more complex. Within *Rubus arcticus* three subspecies are recognized: ssp. *arcticus*, with a mainly Eurasiatic range; ssp. *acaulis*, limited to boreal North America; and ssp. *stellatus*, ranging from southern Alaska to eastern Siberia. It should be noted that the ranges of the three subspecies overlap only in areas which were part of ancient Beringia. In contrast, *Astragalus alpinus* has three subspecies in the Yukon: subsp. *arcticus*, with an amphiberingian–Arctic range; subsp. *alaskanus*, endemic to Alaska–Yukon; and subsp. *alpinus*, which occurs also in the mountains of central Asia. *Vaccinium vitis-idaea* occurs in America with the subsp. *minor*, which has a circumpolar distribution. But its typical subspecies is restricted to Eurasia, which indicates the existence of other refugial stations in that continent. The same applies to *Vaccinium uliginosum*, which can be subdivided into several subspecific taxa, some of which are Eurasiatic. *Andromeda polifolia* has an incomplete circumpolar range, being replaced by *Andromeda glaucophylla* in most of eastern North America and west Greendland, which indicates the probable existence of other refugial centers along the Atlantic coasts. Without denying the great importance of Beringia as a refugial center for boreal and Arctic plants, it seems that several other refugias also were available throughout the Northern Hemisphere for current circumpolar taxa.

Species group 6 (*old preglacial relict species*)

This very small group includes species with rather narrow, disjunct ranges, centered in Alaska–Yukon, eastern Siberia, and in the mountains of southern Siberia. Their chorograms are shown in Figure 41. They occur in the *Salix pulchra* community of northern Yukon (A3) and in the grasslands of southern Yukon (D5). Most of them are restricted to rather xerophytic conditions. They can be considered as the remnants of an old, preglacial, parasteppe–montane vegetation. The strongly disjunct distribution patterns indicate that these species are presently restricted to their refugial centers, located in the unglaciated parts of Beringia or in the south Siberian mountains, and have spread very little.

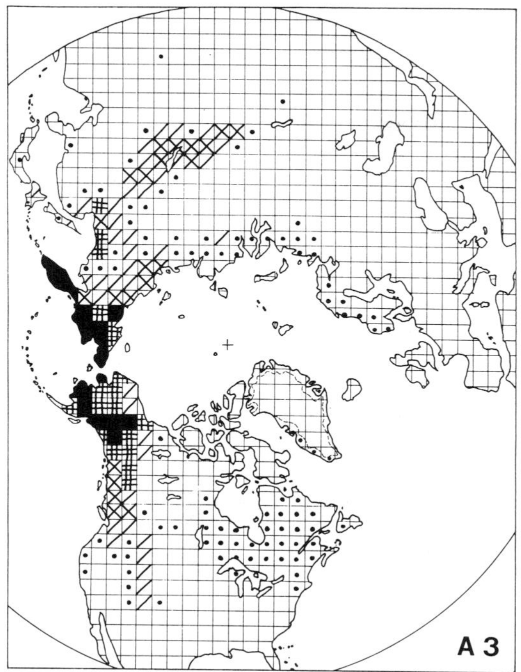

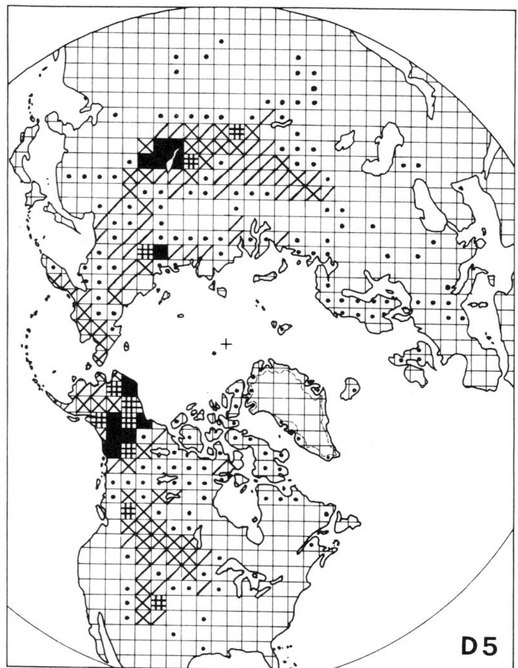

Fig. 41. Chorograms of the two species groups included in group 6. Different shadings refer to the percent of species occurring in the OGUs, on a 5-class scale with intervals of 20%.

Festuca altaica also might have survived glaciation south of the North America ice-sheet, but the southern populations underwent differentiation and are sometimes considered as a distinct species (*Festuca scabrella* Torr.). The most 'plastic' species of this group is *Luzula parviflora*, with a very narrow circumpolar extension to Scandinavia. The grassland species are the only ones which connect the steppes of central Asia with the American prairies, via the periglacial steppes of the Yukon Territory. They constitute less than 10% of all grassland species, which indicates a rather low phytogeographic affinity between American and Asiatic steppes (see Nimis 1982). Lausi and Nimis (1985a), on the basis of an analysis of the distribution of the species which are endemic in the Yukon, concluded that the grasslands of southern Yukon and the Subarctic–Alpine tundra of northern Yukon were the main centres for the differentiation and conservation of the endemic element. The presence of the species of this group in these two formation types is a further indication of the conservative role played by these environments in the history of the flora of the Yukon.

Species group 7 (*amphiberingian species*)

This rather large set of species has an amphiberingian distribution. The term "amphiberingian" is used here in the broad sense of Porsild and Cody (1980) and Yurtsev (1974) to encompass species' ranges that straddle the Bering Strait and extend eastwards as far as about longitude 100°W and as far into Eurasia as longitude 100°E (see chorograms in Fig. 42). They occur in all the formation types. The chorograms of Figure 42 underscore the great phytogeographic importance of Beringia as a refugial area for several boreal species during the glacial periods. This was already stressed by Hultén (1937) and by many subsequent authors (e.g. Hopkins 1967; Young 1976a; Gjaerevoll 1980; Murray 1981; Vitt et al. 1987). Hultén distinguished between species which radiated west- and eastward from Beringia after the retreat of the ice (his "plastic" species) and those with present ranges limited to parts of ancient Beringia (his "rigid" species). Within the group of "rigid" Beringian species we can now

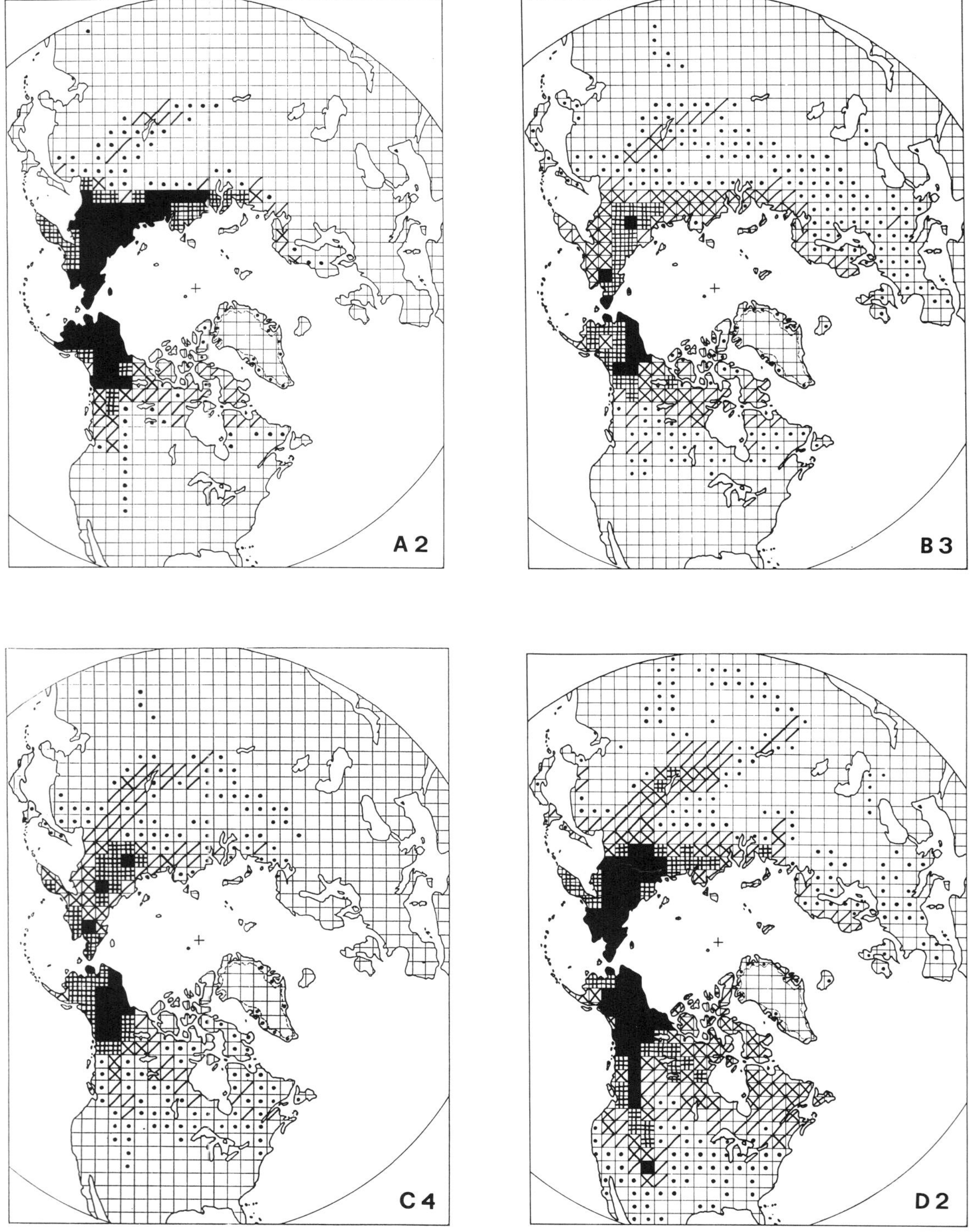

Fig. 42. Chorograms of the four species groups included in group 7. Different shadings refer to the percent of species occurring in the OGUs, on a 5-class scale with intervals of 20%.

distinguish two further groups: (a) truly amphiberingian species, occurring at both sides of the Bering Strait, and (b) species present only at one side of the Strait (endemic to eastern Siberia or to Alaska–Yukon). The species of our group belong to the former element. It is interesting to note that in different formation types different degrees of expansion from Beringia can be detected. In particular, the amphiberingian species present in grasslands (D2, Fig. 42) show a southward expansion of their ranges along the Cordilleras, which is probably related to the existence of the Cordilleran corridor. Those present in the Subalpine–Subarctic vegetation of northern Yukon (A2, Fig. 42), being adapted to a colder climate, have a broader extension along the Arctic coasts of Eurasia, and in a few cases attain an almost circumboreal range. The muskeg species (B3, Fig. 42) have more northern ranges, and thus differ from those of the boreal forest (C4, Fig. 42).

By comparing the chorograms of Figure 42 with those of Figure 36 one can understand how circumboreal ranges might have been attained in different ways. In the case of the species of group 1 the invasion of the ice-free areas was from refugial stations located south of the ice-sheet. In contrast, for those of group 7 the main refugial center was located in Beringia itself, and the process of east- and westward migration from Beringia is probably still in progress. In the Yukon flora the latter element is obviously the most common, and it is difficult to draw a sharp limit between amphiberingian and circumboreal species (see Fig. 41).

6.2. Summary

To summarize our phytogeographic analyses, the results of the numerical classification of the species on the basis of their distribution in the Northern Hemisphere allow us to subdivide them into seven main groups, as follows:

Group 1: *Incompletely circumboreal-temperate species*. These survived glaciation south of the ice-sheets. They are absent from northeastern Siberia because of the strongly continental climate; 11% of the total.

Group 2: *Cordilleran species and endemics to Alaska–Yukon*. These either survived glaciation in Beringia or reached Alaska–Yukon through a Cordilleran ice-free corridor which was open during xeric interglacials; mostly grassland species, 24% of the total.

Group 3: *Boreal North American species*. Most of them survived glaciation in Beringian refugia. Some were able to spread eastwards after the retreat of the ice, while others are still restricted to boreal northwestern North America; 18.4% of the total.

Group 4: *Arctic-montane species*. Often with incomplete circumboreal ranges; most are limited to boreal North America and to the mountains of northeastern Asia; 15.3% of the total.

Group 5: *Circumpolar (–boreal) species*. These probably survived glaciation in Beringia, with possible secondary refugial areas in other parts of Eurasia and/or North America; 6.7% of the total.

Group 6: *Old preglacial relict species*. Usually associated with a parasteppe vegetation, they are either endemic to Alaska–Yukon or have disjunct stations in the mountains of southern Siberia; 5.5% of the total.

Group 7: *Amphiberingian species*. They survived glaciation in Beringia itself; 19.0% of the total.

7. ECOLOGY AND PHYTOGEOGRAPHY

The relations between ecology and phytogeography have been analyzed on the basis of the ordination of vegetation types by PCA, based on floristic composition (see Fig. 34). The sequence of the vegetation types in Figure 34 according to their angular scores corresponds, as already discussed, to a gradient in water availability and soil development, from the humid Cryosols at the left side to the xeric Regosols at the right side. Furthermore, soil pH tends to be acid or subacid at the left side of the gradient, neutral or basic at the right side of the figure. In

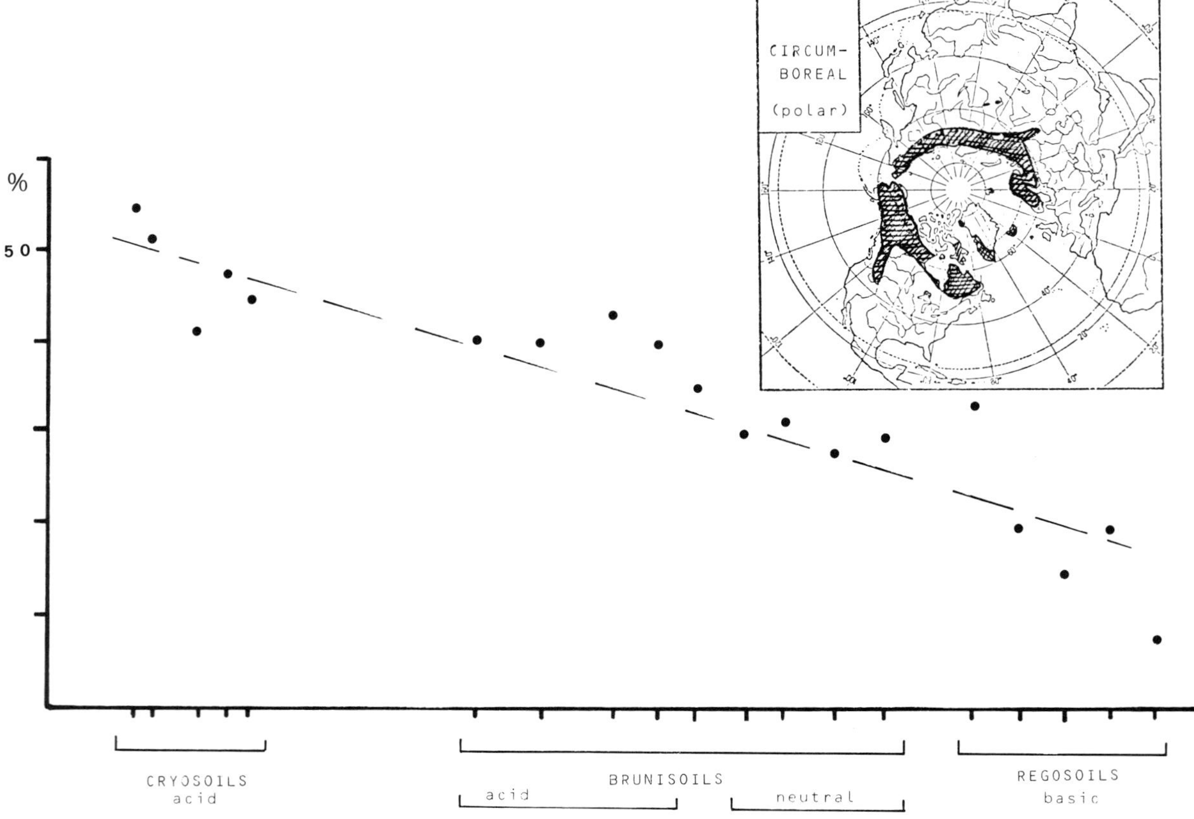

Fig. 43. Ecological behavior of Circumboreal species. The *x*-axis reports the positions of the relevé groups along the horseshoe shown in the ordination of relevé groups on the basis of floristic data of Figure 35. The *y*-axis indicates the percent occurrence of these species in each relevé group. The figure indicates that circumboreal species have a well-defined ecological response.

Figs. 43–48 the horseshoe-shaped curve of Fig. 34 has been stretched with unaltered positions of the vegetation types. The percentage of species with a given distribution pattern has been calculated for each vegetation type, and plotted on the *y*-axes in Figs. 43 to 48.

Figure 43 is based on all circumboreal and circumpolar species, i.e. those whose distribution extends through North America and Eurasia, within the Arctic, Subarctic and boreal vegetation belts, and reaching Europe. This category includes all species of groups 1 (Fig. 36) and 5 (Fig. 40), and the species of group 4 reaching northern Europe. This element has a clear maximum in the muskeg vegetation on Cryosols, and decreases linearly towards the grassland vegetation on Regosols. Circumboreal (circumpolar) species common both to muskeg vegetation and the boreal forest are relatively scarce. Most of the circumboreal (–polar) species occurring in the muskeg are common components of Alpine tundra throughout the survey area. The muskeg vegetation itself could be considered as a tundra-like vegetation with scattered dwarf trees. According to Johnson and Packer (1967), tundra climates persisted in the amphiberingian area throughout the Quaternary and the Bering Land Bridge was covered by a flat tundra. The high number of circumboreal species in the muskeg vegetation indicates that actually the Bering Strait has not been an effective barrier against tundra plant dispersal.

The distribution patterns of the species chiefly occurring in the boreal forest communities differ from those of most muskeg species in that they extend further south in the boreal zone. In this

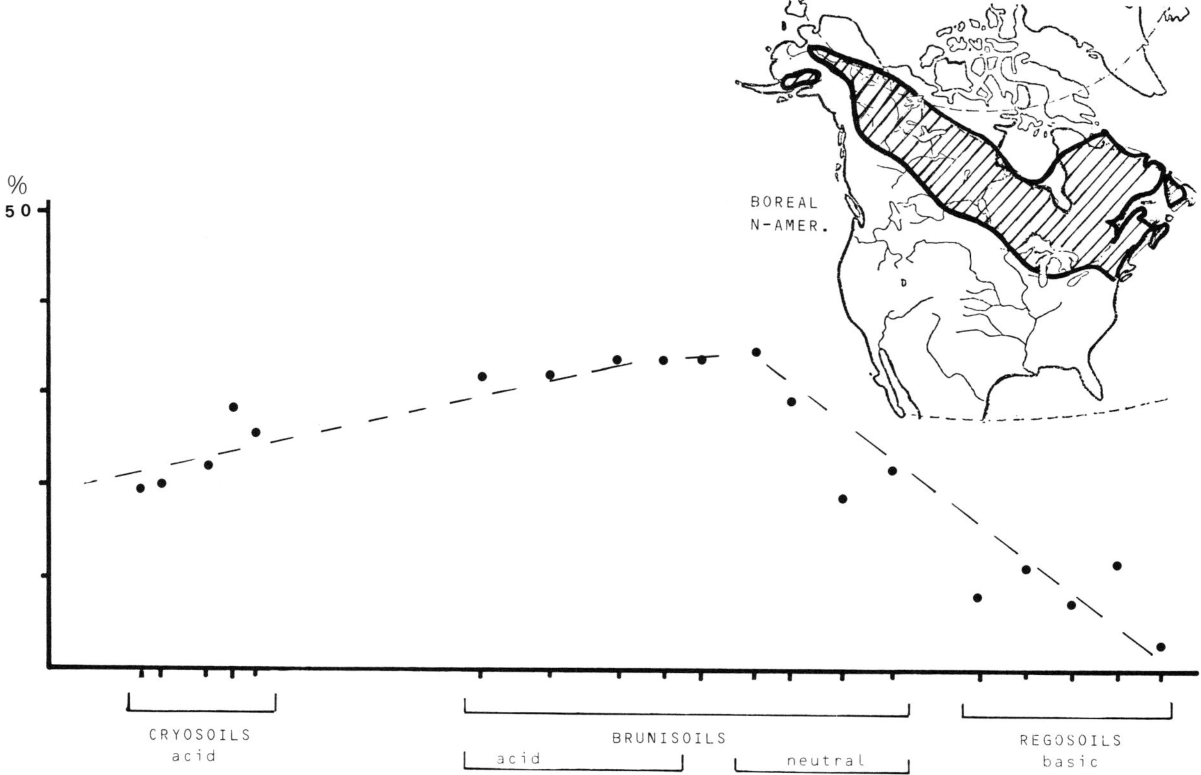

Fig. 44. Ecological behavior of boreal North American species. For explanation see caption to Figure 43.

sense, they are mostly true circumboreal species, whereas the circumpolar element prevails in the muskeg vegetation. None of these circumboreal species has its optimum in tundra-like vegetation on permafrost; some of them (those of group 1, Fig. 36) are even absent from eastern Siberia. This and the intolerance to permafrost of most of the boreal species make the hypothesis of their possible migration through the Bering Land Bridge less probable. The history of this species group is probably different from the one of the muskeg species. They seem to have survived glaciations by means of refugia located south of the North American ice-sheet, or in Alaska–Yukon itself, from which they spread later in connection with the expansion of the boreal forest during the postglacial periods.

The hypothesis that the Bering Land Bridge was mostly covered by tundra on Cryosols is further corroborated by the ecological behaviour of the boreal North American species, i.e. those species that are restricted to the boreal vegetation zone of North America (group 3, Fig. 38), which is shown in Fig. 44. These species have a clear maximum in the boreal forest on Brunisols, and in its secondary degradation stages following fire, being rare both on Cryosols and on Regosols. It should be remarked that all the major tree species present in the area belong to this category. They are: *Picea mariana, Larix laricina, Picea glauca, Pinus contorta, Populus tremuloides, Populus balsamifera*. The two former species are confined to Cryosols, the latter occur mainly on Brunisols. In eastern Siberia *Larix laricina* is replaced by *Larix dahurica*, whereas such species as *Picea obovata, Larix sibirica* and *Pinus sibirica* occur further west, in the less continental regions of western Siberia. Not a single coniferous tree species is shared by the two continents. These facts suggest that the boreal forests of North America and eastern Asia did not form a continuous belt during the Pleistocene, despite the

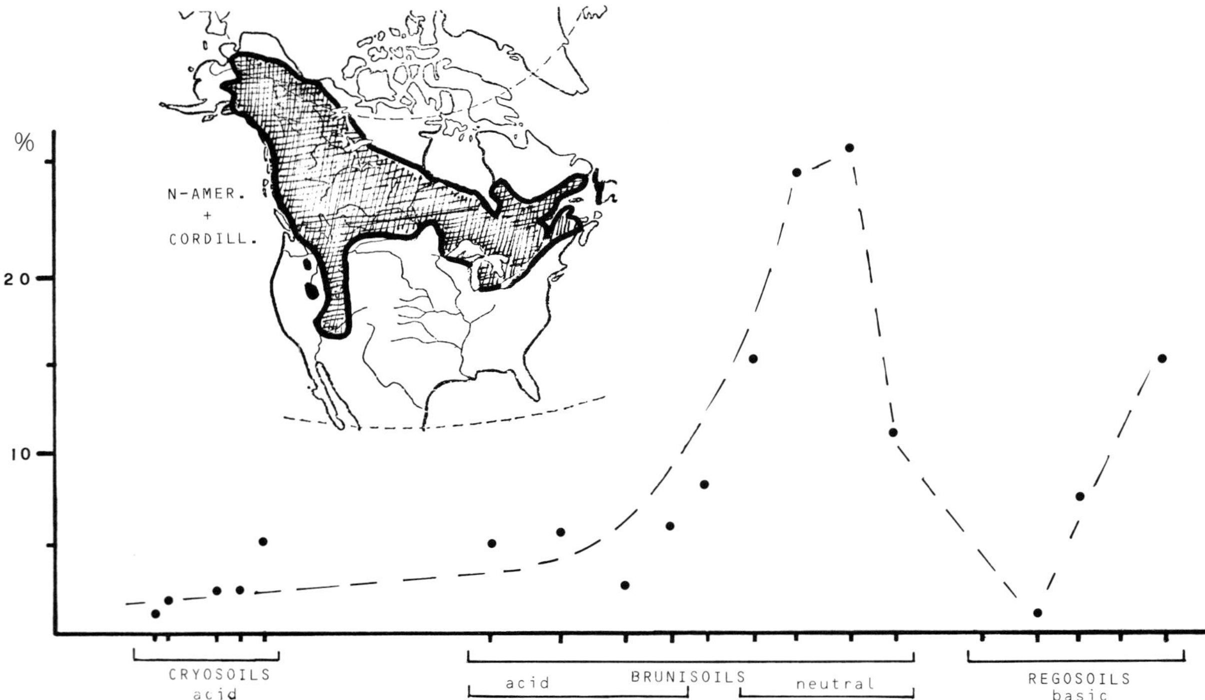

Fig. 45. Ecological behavior of boreal North American species extending along the Cordilleras. For explanations see caption to Figure 43.

presence of the Bering Land Bridge. This assumption is supported by palynological and palaeobotanical evidence. The separation of the forests of the two continents seems to date back to late Miocene or early Pliocene (Johnson and Packer 1967; Wolfe and Leopold 1967) and seems to have lasted throughout the Pleistocene (Colinvaux 1967; Giterman and Golubeva 1967).

Figure 45 shows the ecological behavior of a subgroup of boreal North American species, whose ranges are characterized by a pronounced southward extension along the Cordilleras. Among them are the two *Populus* species, *Shepherdia canadensis, Mertensia paniculata*, etc. Their frequencies in the various community types are generally low, with the exception of the degradation stages of the boreal forest on neutral Brunisols dominated by deciduous trees and shrubs, chiefly *Populus tremuloides* and *Salix* species. Compared with the entire group of boreal North American species, this subgroup has a frequency maximum more towards the right along the pedological gradient, i.e. within more xerophytic vegetation types. It seems that a whole set of species with optima in the *Populus tremuloides* stands moved as a whole from their probable refugial stations along the Cordilleras, a fact that is now reflected in the high number of species with similar distributions that occur in these communities. In this example it is interesting how a relatively small difference in species ranges (a southward extension along the Cordilleras) is reflected in a clear ecological difference. This is still more evident in Figure 46 where two subgroups taken from the species of groups 4 (Fig. 39) and 7 (Fig. 42) are plotted on the same graph. The two subgroups were selected as follows: (a) broad-ranging amphiberingian species, present both in North America and in north-eastern Asia, but not reaching Europe, not extending along the Cordilleras; (b) as the previous, but with a pronounced extension along the Cordilleras. The difference in distribution patterns is reflected in a very different behavior

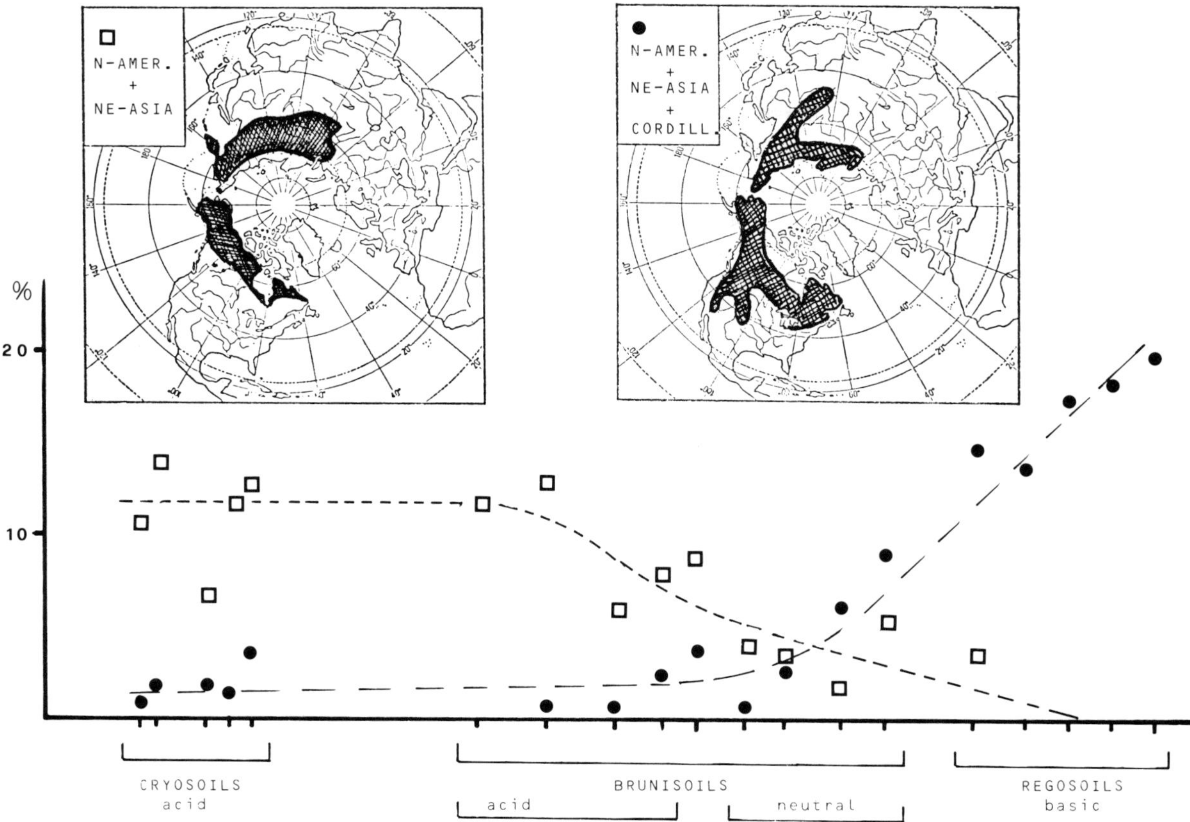

Fig. 46. Ecological behavior of two species groups present in boreal North America and in northeastern Asia. For explanations see caption to Figure 43 and the main text.

along the ecological gradient (Fig. 46). The species of the former subgroup are most frequent on acid Cryosols, and on Brunisols with impeded to moderate drainage, whereas those of the second subgroup are most frequent on neutral Brunisols and Regosols with moderate to excessive drainage. The curves representing the ecological behavior of the two subgroups cross each other at the point where the soil reaction turns from acid to neutral, and Regosols become prevalent. This peculiar behavior can be explained considering that only those species that were able to survive in a relatively arid environment had the possibility to expand southward along the deglaciated corridor during the time of ice retreat.

Figure 47 shows the ecological behavior of Cordilleran species (group 2, Fig. 37): they are rare both on Cryo- and on Brunisols, but become very frequent within the steppe grassland vegetation on Regosols. An analogous trend is shared by the species which are endemic to Alaska–Yukon (Fig. 48), which are particularly frequent in the community type G2, found only in the Kluane region. According to Lausi and Nimis (1985b), Cordilleran species are a common element also within the weed vegetation on Regosols along roadsides throughout the Yukon, and especially in the Kluane region, where natural steppes are particularly well developed.

The driest phases of the glacial period in western Siberia were characterized by the spreading of a tundra-steppe, dominated by *Artemisia* species, *Chenopodiaceae*, *Ephedra* species and *Selaginella sibirica*. According to Giteman and Golubeva (1967) this vegetation reached its maximum development during the

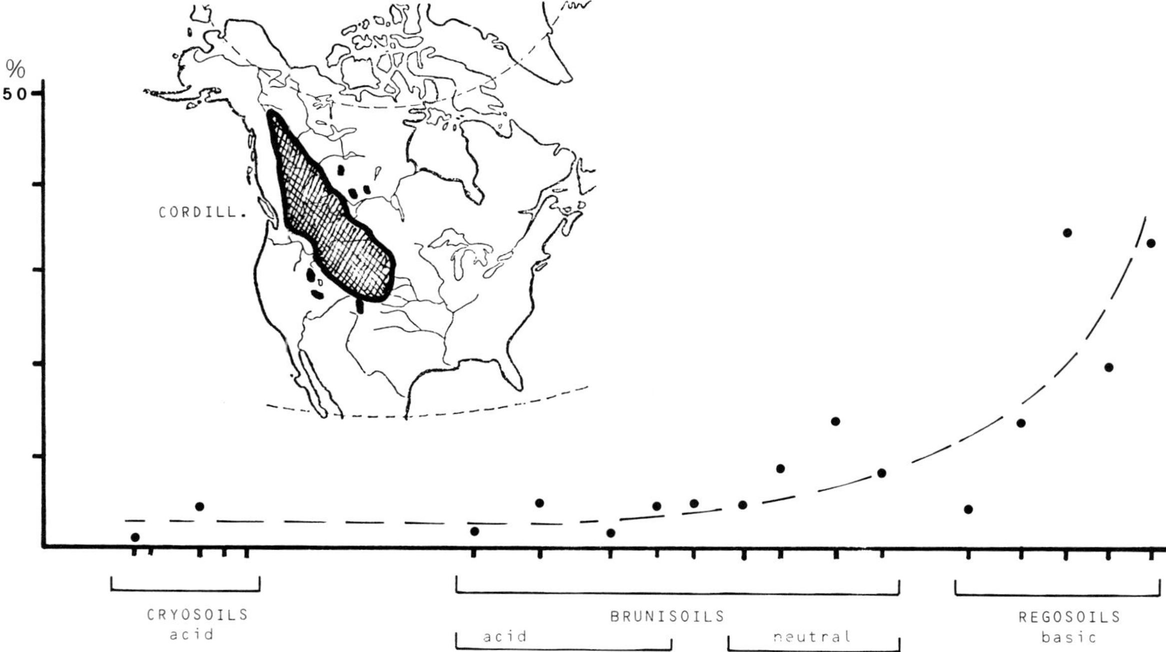

Fig. 47. Ecological behavior of Cordilleran species. For explanations see the caption to Figure 43.

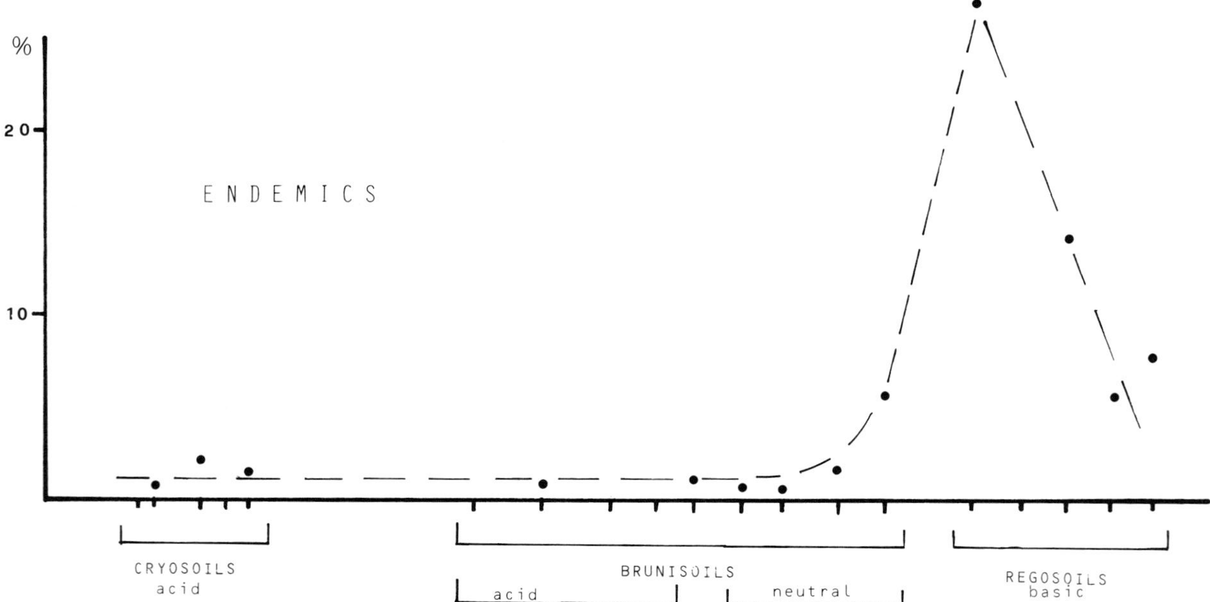

Fig. 48. Ecological behavior of endemic species. For explanations see the caption to Figure 43.

Samarov glaciation, when the climate became drier and the role of xerophytes increased, as they shifted northward to regions previously occupied by tundra and forest-tundra. According to palynological data (Colinvaux 1967: Cwynar and Ritchie 1980), *Artemisia* maxima are a common feature of the ancient herb spectra in Alaska. Such maxima have not yet been matched by surface spectra from anywhere in Arctic Alaska, so that Colinvaux (1967) concludes that an explanation for the ancient *Artemisia* maxima would be of great importance for our understanding of the vegetation in Beringia during the Pleistocene. Since studies of modern pollen have failed to identify a modern analog for the ancient tundra-steppes, some authors have assumed that both the plant and animal communities were reduced to their extinction in the Holocene (Matthews 1976; Hopkins et al. 1982). However, Young (1976a, b) and Yurtsev (1963, 1972, 1981) suggested that late-glacial steppe remnants persist in areas of northeastern Asia and northwestern North America. Murray et al. (1983), on the basis of a provisional study in Alaska and the Yukon, state that the steppes of Alaska–Yukon are comparable to those of the middle Yana and Indigirka drainages, and of the upper Kolyma River in eastern Siberia. At the start of our work, we also considered as most likely the hypothesis of a close connection between the steppes of Asia and those of Alaska–Yukon through the Bering Land Bridge. According to such a hypothesis, the periglacial steppes of the Yukon should be considered as the easternmost outposts of a formation type whose maximum development was reached in western Siberia. However, our results do not support such a hypothesis. From the chorograms of Fig. 37, and from Fig. 47 it is evident that the phytogeographic affinities between the periglacial steppes of Siberia and those of the Yukon are quite low. Most of the species present in the latter have a distribution restricted to the North American continent. Most of the true xerophytes are either endemic, or Cordilleran, or central North American prairie species with a disjunction in the Yukon. The only species providing a significant connection with the Asiatic steppes are those of group 4 (Fig. 39), plus several lichens and mosses (Nimis 1981a, b; Murray et al. 1983); however, they have extremely broad ranges in xeric regions of the world. The plants of group 6 (Fig. 41) have strongly disjunct ranges, indicative of a very old age of the species. Their appearance in the Yukon flora may date back to preglacial, Tertiary times (see also Murray et al. 1983). The high incidence of Cordilleran species in the steppes of the Yukon is most probably a consequence of the opening of the Cordilleran corridor, which was an effective migration way for xerophytic plants both from the refugia in southern Yukon southward and from the central North American plains northward. The scarcity of floristic connections at the species level with the Asiatic steppes may be due to two reasons. First, the periglacial steppes of Siberia were widespread mainly in the western portion of the region. In eastern Siberia maritime influences were still sufficiently strong to hinder the development of such a vegetation. As a consequence, the periglacial steppes of southwestern Yukon were separated from their Asiatic counterparts by a large area, extending up to the Lena River, with the main vegetation having consisted of tundra and open *Larix* woodland (Giteman and Golubeva 1967). Second, the Bering Land Bridge was not an effective migration route for xerophytic plants. According to Johnson and Packer (1967), throughout the Quaternary the Bering Bridge was covered by tundra, whose primary constituents were lowland species and moisture-tolerant herbs and shrubs. This is indirectly confirmed by our data which show a rapid decrease from hygrophytic to xerophytic conditions, in the incidence of species present at both sides of the Bering Strait.

ACKNOWLEDGMENTS

This paper is part of a broader cooperative effort between the Universities of Trieste (Italy) and London, Ontario (Canada), financed by the Italian C.N.R., the Canadian N.S.E.R.C. and the University of Trieste (Funds for International cooperation). We are indebted to Dr W. Stanek, Canadian Forestry Service, for coordination of the survey and for soil data, to Dr L. Orloci (University of Western Ontario) for organization. We would like to thank Dr P. Fewster (London, Ont.) for field assistance and Mrs G. Bolognini (Trieste) for her precious help in data elaboration. We are indebted to late Dr W. C. Steere (New York) and Dr D. H. Vitt (Edmonton) for the identification of critical briophytes, and to Dr J. Poelt (Graz) for the identification of critical lichens.

REFERENCES

Ahti, T. 1957. Poronjakalikoista peurojen asuma-alueina. Luonnon Tutk., 61, 76–79.

Ahti, T. 1959. Studies on the caribou lichen stands of Newfoundland. Anns. Bot. Soc. Zool. Bot. Fenn. Vanamo, 30 (4), 1–44.

Ahti, T. 1977. Lichens of the Boreal Coniferous Zone. In: Seaward M.R.D. (ed.), Lichen Ecology, pp. 145–181. Academic Press, London.

Anderberg, M. R. 1973. Cluster Analysis for Applications. Academic Press, New York.

Argus, G. W. 1984. The genus *Salix* in Alaska and the Yukon. Nat. Mus. Can. Publ. Bot., 2, 1–279.

Baxter, D. V. and Wadsworth, F. H. 1939. Forest and fungus succession in the lower Yukon Valley. Univ. Mich. School For. and Cons. Bull., 9, 1–52.

Benninghoff, W. S. 1952. Interaction of vegetation and soil frost phenomena. Arctic, 5, 34–44.

Black, R. A. and Bliss, L. C. 1978. Recovery sequence of *Picea mariana-Vaccinium uliginosum* forests after burning near Inuvik, Northwest Territories, Canada. Can. J. Bot., 56, 2020–2030.

Black, R. A. and Bliss, L. C. 1980. Reproductive ecology of *Picea mariana* at treeline near Inuvik, Northwest Territories, Canada. Ecol. Monogr. 50, 331–354

Bliss, R. A. 1975. Tundra grasslands, herblands and shrublands and the role of herbivores. Geosciences and Man, 10, 51–79.

Bostock, H. S. 1965. Physiography of the Canadian Cordillera, with special reference to the area north of the fifty-fifth parallel. Geol. Surv. Can., Memo 274, Ottawa.

Braun Blanquet, J., Sissingh, J. and Vlieger, J. 1939. Prodromus der Pflanzengesellschaften, 6. Klasse der *Vaccinio-Piceetea*. Comm. SIGMA, Montpellier, 1–123.

Brown, R. J. E. 1967. Permafrost in Canada. Geol. Surv. Can., Map 1246A. Ottawa.

Brown, R. J. E. 1973. Influence of vegetation on permafrost. Proc. Int. Permafrost Conf., pp. 20–25.

Burns, B. M. 1973. The climate of the Mackenzie Valley-Beaufort Sea, Vol. II. Environ Can., Atm. Environ. Climatol. Studies, 24, Ottawa.

Canada Soil Survey Committee 1978. The Canadian System of Soil Classification. Can. Dept. Agricult., Publ. 1646, Ottawa.

Carleton, T. J. and Maycock, P. F. 1978. Dynamics of the boreal forest south of James Bay. Can. J. Bot., 56, 1157–1173

Carleton, T. J. and Maycock, P. F. 1980. Understorey-canopy affinities in boreal forest vegetation. Can. J. Bot., 59, 1709–1716

Clements, F. E. 1916. Plant succession: an analysis of the development of vegetation. Carnegie Inst., Washington, Publ. 242.

Colinvaux, P. A. 1967. Quaternary vegetational history of Arctic Alaska. In: Hopkins, D. M. (ed.), The Bering Land Bridge, pp. 207–231. Stanford Univ. Press, Stanford, Cal.

Crovello, T. J. 1981. Quantitative Biogeography: An overview. Taxon, 30, 563–575.

Cwynar, L. C. and Ritchie, J. C. 1980. Arctic steppe-tundra: A Yukon perspective. Science, 208, 1375–1377.

Dillon, L. S. 1956. Wisconsin climate and life-zones in North America. Science, 123, 167–176.

Douglas, R. J. W. (ed.), 1970. Geology and Economic Minerals of Canada. Dept. En. Min. Res., Ottawa.

Douglas, G. W. 1974. Montane zone vegetation of the Alsek River Region, southwestern Yukon. Can. J. Bot., 52, 2505–2532.

Drury, W. H. Jr. 1956. Bog flats and physiographic processes in the upper Kuskokwim River region, Alaska. Contrib. Gray. Herb., 178, 1–130.

Dugle, J. R. and Bols, N. 1971. Variation in *Picea glauca* and *P. mariana* in Manitoba and adjacent areas. Atom. Energ. Comm. Can., Publ. AECL 3681.

Dyrness, C. T. and Grigal, D. F. 1979. Vegetation-soil relationships along a spruce forest transect in interior Alaska. Can. J. Bot., 57, 2644–2656.

Feoli, E. and Feoli Chiapella, L. 1980. Evaluation of ordination methods through simulated coenoclines: some comments. Vegetatio, 42, 35–41.

Foothills 1978. Terrain evaluation data maps (YK-04-0200-D series). Explanation of data categories, 7 pp.

Fremlin, G. 1973. The National Atlas of Canada (4th edn.) Can. Dept. En. Min. Res., Ottawa.

Giteman, R. E. and Golubeva, L. V. 1967. Vegetation of eastern Siberia during the anthropogene period. In: Hopkins, D. M. (ed.), The Bering Land Bridge, pp. 207–231. Stanford Univ. Press, Stanford, Cal.

Gjaerevoll, O. 1980. A comparison between the alpine plant communities of Alaska and Scandinavia. Acta Phytogeogr. Suec., 68, 83–88.

Guthrie, R. D. 1968. Palaeoecology of the large mammal community in interior Alaska during the late Pleistocene. Amer. Midl. Nat., 79, 346–363.

Haeupler, H. 1974. Statistische Auswertung von Punktrasterkarten der Gefässpflanzenflora Sud-Niedersachsen. Scripta Geobot., 8, 1–141.

Halliday, W. E. D. and Brown, A. W. 1943. Distribution of some important forest trees in Canada. Ecology, 24, 353–373.

Hansen, H. P. 1943. Palaeoecology of the sand dune bogs on the southern Oregon coast. Amer. J. Bot., 30, 335–340.

Hansen, H. P. 1947. Postglacial vegetation of the Northern Great Basin. Amer. J. Bot., 34, 164–171.

Hanson, H. C. 1950. Characteristics of some grasslands, marsh, and other plant communities in western Alaska. Ecol. Monogr., 21, 317–378.

Hanson, H. C. 1953. Vegetation types in northwestern Alaska and comparison with communities in other Arctic regions. Ecology, 34, 111–140.

Harington, C. R. 1978. Quaternary vertebrate faunas of Canada and Alaska and their suggested chronological sequence. Syllogeus, 15.

Hart, A. G. 1959. Silvical characteristics of balsam fir (*Abies balsamea*). US For. Serv. Pap. 122.

Hoefs, M., McCowan, I. and Krajina, V. J. 1975. Phytosociological analysis and synthesis of Sheep Mountain, Southwest Yukon Territory, Canada. Syesis, 8, 125–228.

Hoffmann, R. S. 1974. Terrestrial Vertebrates. In: Ives, D. J. and R. G. Barry (eds), Arctic and alpine Environments, pp. 475–568. Methuen, London.

Hopkins, D. M. (ed.) 1967. The Bering Land Bridge. Stanford Univ. Press, Stanford Cal.

Hopkins, D. M., Matthews, J. V., Schweger, C. E. and Young S. B. (eds). 1982. Palaeoecology of Beringia. Academic Press, New York.

Hughes, O. L., Campbell, R. B., Mueller, J. E. and Wheeler, J. O. 1969. Glacial limits and flow patterns, Yukon Territory, south of 65 degrees north latitude. Geol. Surv. Can., 68–34, Ottawa.

Hughes, O. L., Rampton, V. L. and Rutter, N. W. 1972. Quaternary geology and geomorphology, southern and central Yukon (N Canada). All. Int. Geol. Congr. 24th Sess.

Hultén, E. 1937. Outline of the history of Arctic and Boreal Biota during the Quaternary Period. Stockholm.

Hultén, E. 1968. Flora of Alaska and neighbouring Territories. Stanford Univ. Press, Stanford Cal.

Hultén, E. and Fries, M. 1980. Atlas of North European vascular plants north of the Tropic of Cancer. Koeltz, Koenigsstein.

Jaeger, E. 1968. Die pflanzengeographische Ozeanizitätsgliederung der Olarktis und die Ozeanizitätsbindung der Pflanzenareale. Feddes Rep., 79, 157–335.

Jaeger, E. 1970. Charakteristische Typen mediterran-mitteleuropäischer Pflanzenareale. Feddes Rep., 79, 157–335.

Jaeger, E. 1972. Comments on the history and ecology of continental European plants. In: Valentine, D. H. (ed.), Taxonomy, Phytogeography and Evolution, pp. 349–362. Academic Press, London.

Jalas, J. 1953. Rokua, suunnitellun kansallipuiston kasvillisuus ja kasvisto. Silva Fennica, 81, 1–98.

Jeffers, J. N. R. and Black, T. M. 1963. An analysis of variability in *Pinus contorta*. Forestry, 36, 199–218.

Jeffrey, W. W. 1959. Notes on plant occurrence along Lower Liard River, N. W. T. Nat. Mus. Can. Bull., 171, 32–115.

Johnson, A. W. and Packer, J. G. 1967. Distribution, ecology and cytology of the Ogotoruk Creek flora and the history of Beringia. In: Hopkins D. M. (ed.), The Bering Land Bridge, pp. 245–265. Stanford Univ. Press, Stanford, Cal.

Klement, O. 1955. Prodromus der mitteleuropäschen Flechtengesellschaften. Feddes Rep. Beih., 135, 5–194.

Knapp, R. 1965. Die Vegetation von Nord- und Mittelamerika und der Hawaii-Inseln. Fischer, Stuttgart.

Kornas, J. 1972. Corresponding taxa and their ecological background in the forests of temperate Eurasia and North America. In: Valentine, D. H. (ed.), Taxonomy, Phytogeography and Evolution, pp. 37–59. Academic Press, London.

Krajina, V. J. 1975. Some observations on the three subalpine biogeoclimatic zones in British Columbia, Yukon and Mackenzie District. Phytocoenologia, 2, 396–400.

Kubiena, W. L. 1958. Bestimmungsbuch und Systematik der Böden Europas. Enke, Stuttgart.

La Roi, G. H. 1967. Ecological studies in the Boreal spruce-fir forests of the North American taiga. I. Analysis of the vascular flora. Ecol. Monogr., 37, 229–253.

La Roi, G. H. and Stringer, M. H. L. 1976. Ecological studies in the Boreal spruce-fir forests of the North American taiga. II. Analysis of the bryophyte flora. Can. J. Bot., 54, 619–643.

Larsen, J. A. 1965. The vegetation of Ennadai Lake Area, N.W.T.: studies in subarctic and Arctic bioclimatology. Ecol. Monogr., 35, 37–59.

Larsen, J. A. 1970. Vegetation of Fort Reliance, Northwest Territories. Can. Field Nat., 85, 147–178.

Larsen, J. A. 1980. The Boreal Ecosystem. Academic Press, New York.

Laughlin, W. S. 1967. Human migration and permanent occupation in the Bering Sea area. In: Hopkins, D. M. (ed.), The Bering Land Bridge. Stanford Univ. Press, Stanford, Cal.

Lausi, D. and Nimis, P. L. 1985a. Quantitative phytogeography of the Yukon Territory (NW Canada) on a chorological-phytosociological basis. Vegetatio, 59, 9–20.

Lausi, D. and Nimis, P. L. 1985b. Roadside vegetation in Boreal south Yukon and adjacent Alaska. Phytocoenologia, 13, 103–138.

Liddicoet, A. R. and Righter, F. I. 1947. Trees of the Eddy Arboretum. Pac. Southwest For. Sta. Pap. 43, Placerville, Cal.

Loeve, D. and Freedman, N. J. 1956. A plant collection from southwest Yukon. Bot. Not., 109, 153–211.

Looman, J. 1964. The distribution of some lichen communities in the Prairie Provinces and adjacent parts of the Great Plains. Bryologist, 67, 209–224.

Maikawa, E. and Kershaw, K. A. 1976. Studies on lichen-dominated systems. XIX. The postfire recovery of black spruce-lichen woodland in the Abitan Lake Region, N.W.T. Can. J. Bot., 54, 2679–2689.

Matthews, J. V. Jr. 1976. Arctic steppe: an extinct biome. Abstr. IV Conf. Amer. Quatern. Ass., pp. 73–77. Tempe, Arizona.

Meusel, H. 1943. Vergleichende Arealkunde, 2vv. Berlin-Zehlendorf

Meusel, H., Jaeger, E. and Weinert, E. 1965. Vergleichende Chorologie der Zentraleuropaeische Flora. Fischer, Jena.

Mirov, N. T. 1967. The genus *Pinus*. Ronald Press, New York.

Morlan, R. E. 1980. Taphonomy and Archaeology in the

upper Pleistocene of the northern Yukon Territory: A glimpse of the peopling of the new world. Ach. Surv. Can., 64, Ottawa.

Moss, E. H. 1949. Natural pine hybrids in Alberta. Can. J. Res., Sect. C, 27, 218–229.

Mueller-Beck, H. 1967. On migrations of hunters across the Bering Land Bridge in upper Pleistocene. In: Hopkins, D. M. (ed.), The Bering Land Bridge, pp. 373–408. Stanford Univ. Press, Stanford, Cal.

Murray, D. F., Murray, B. M., Yurtsev, B. A. and Howenstein, R. 1983. Biogeographic significance of steppe vegetation in subarctic Alaska. Proc. IV Int. Conf. Permafrost, pp. 883–888, Nat Acad Press, Washington, D. C.

Nienstaedt, H. 1957. Silvical characteristics of white spruce (*Picea glauca*) U.S. For. Serv. Lake States For. Exp. St., Pap. 55.

Nimis, P. L. 1981a. Epigaeic lichen synusiae in the Yukon Territory. Cryptogamie Bryol. Lichenol, 2, 127–151.

Nimis, P. L. 1981b. *Caloplaca tominii* Savicz new to North America. Bryologist, 84, 222–225.

Nimis, P. L. 1982. Phytogeography of periglacial steppes in the Yukon Territory. Coll. Phytosociol., 11, 2–13.

Nimis, P. L. 1989. Phytogeographical analysis of a treeline community in northern Yukon (NW Canada). Vegetatio, 81, 209–215.

Orlóci, L. and Stanek, W. 1979. Vegetation survey of the Alaska Highway, Yukon Territory: types and gradients. Vegetatio, 41, 1–56.

Oswald, E. T. and Senyk, J. P. 1977. Ecoregions of the Yukon Territory. Can. Dept. Environ. Can. For. Serv., pp. 1–115, Victoria.

Parker, W. H. and McLachlan, D. G. 1978. Morphological variation in white and black spruce: investigation of natural hybridization between *Picea glauca* and *Picea mariana*. Can. J. Bot., 56, 2512–2520.

Porsild, A. E. 1945. The alpine flora of the east slope of Mackenzie Mountains, Northwest Territories. Nat. Mus. Can. Bull., 101, 1–35.

Porsild, A. E. 1951. Botany of the southeastern Yukon adjacent to the Canol Road. Nat. Mus. Can. Bull., 121, 1–400.

Porsild, A. E. 1966. Contributions to the flora of southwestern Yukon Territory. Nat. Mus. Can. Bull., 216, 1–86.

Porsild, A. E. 1974. Materials for a flora of central Yukon Territory. Nat. Mus. Publ. Bot., 4, 1–78.

Porsild, A. E. and Cody, W. J. 1980. Vascular plants of Continental Northwest Territories, Canada. Publ. Div. Nat. Mus. Canada, Ottawa.

Rabotnov, T. A. 1936. Ekologicheskie nablyudeniya nad lishaynikami v yuzhnoy Yakutii. Sov. Bot., 1936, 149–153.

Raup, H. M. 1947. Some natural floristic areas in Boreal America. Ecol. Monogr., 17, 221–234.

Ritchie, J. C. 1984. Past and Present Vegetation of the Far Northwest of Canada. Univ. Toronto Press, Toronto.

Roche, L. 1969. A genecological study of the genus *Picea* in British Columbia. New Phytol., 68, 505–554.

Rouse, W. R. and Kershaw, K. A. 1971. The effects of burning on the heat and water regimes of lichen-dominated subarctic surfaces. Arct. Alp. Res., 3, 291–304.

Rowe, J. S. 1977. Forest Regions of Canada. Can. Dept. Environ. Can. For. Serv. Publ. 1300, Ottawa.

Rudolf, E., Oswald, E. T. and Nyland, E. 1981. Chemosystematic studies in the genus *Picea*. V. Leaf oil terpene composition of white spruce from the Yukon Territory. Can. For. Serv. Res., Notes 1, 32–34.

Scoggan, H. J. 1978–79. The Flora of Canada. Nat. Mus. Can. Publ. Bot. 7 (1–4), Ottawa.

Scotter, G. W. 1964. Effect of forest fires on the winter range of barren-ground caribou in northern Saskatchewan. Wildl. Mgmt. Bull., Ottawa Ser.1, 18, 1–111.

Scotter, G. W. 1970. Wildfires in relation to the habitat of barren-ground caribou in the taiga of northern Canada. Proc. Ann. Tall. Timbers Fire Ecol. Congr., 85–105.

Sjoers, H. 1963. Bogs and fens of Attawapiskat River, northern Ontario. Bull. Mus. Can., 186, 45–133.

Spetzman, L. A. 1959. Vegetation of the Arctic slope of Alaska. US Geol. Surv. Prof. Pap., 302, 1–58.

Thomson, J. W. 1982. A further note on *Caloplaca tominii* Savicz in the Americas. Bryologist, 85, 251.

Trass, H. 1968. Analiz likhenoflory Estonii. Bot Inst AN SSSR, Tartu.

Uggla, E. 1958. Skogsbrandfalt i Muddus Nationalpark. Acta Phytogeogr. Suec., 41, 1–116.

Viereck, L. A. 1970. Forest succession and soil development adjacent to the Chena River in Interior Alaska. Arct. Alp. Res., 2, 1–26.

Vincent, A. B. 1965. Black spruce: A review of its silvics, ecology and silviculture. Can. Dept. For., Publ. 1100, Ottawa.

Vitt, D. H., Horton, D. and Pickard, J. 1987. An annotated list and the phytogeography of bryophytes of Keele Peak, Yukon, an isolated granitic mountain. Mem. NY Bot. Garden, 45, 198–210.

Walter, H. 1954. Grundlagen der Pflanzenverbreitung, 2 Teil: Arealkunde. Ludwigsburg.

Walter, H. 1979. Vegetation of the Earth (2nd edn). Springer, Berlin.

Walter, H. and Straka, H. 1970. Arealkunde, floristisch-historische Geobotanik (2 Aufl.). Stuttgart.

Welsh, H. 1974. Anderson's Flora of Alaska and adjacent parts of Canada. Provo, Utah.

Westhoff, V. and van der Maarel, E. 1973. The Braun-Blanquet Approach. In: Whittaker R. H. (ed.), Handbook of Vegetation Science, 5, 619–726.

Wildi, O. and Orlóci, L. 1980. Management and multivariate analysis of vegetation data. Swiss Fed. Inst. For. Res., Rep. 215, 1–68.

Wolfe, J. A. and Leopold, E. B. 1967. Neogene and early Quaternary vegetation of northwestern North America and northeastern Asia. In: Hopkins D. M. (ed.), The Bering Land Bridge, pp. 193–206. Stanford Univ. Press, Stanford, Cal.

Young, S. B. 1976a. Is steppe-tundra alive and well in Alaska? Amer. Quat. Ass. Abstr. 4, 84–88.

Young, S. B. 1976b. The environment of the Yukon-Charley rivers area, Alaska. Contr. Cent. Northern St., 9. Wolcott, Vermont.

Youngman, P. M. 1975. Mammals of the Yukon Territory. Nat. Mus. Can. Publ. Zool., 10, 1–192.

Yurtsev, B. A. 1963. On the floristic relations between steppes and prairies. Bot. Not., 116, 396–408.

Yurtsev, B. A. 1972. Phytogeography of northeastern Asia

and the problem of trans-Beringian floristic interrelations. In: Graham A. (ed.), Floristics and Palaeofloristics of Asia and Eastern North America, pp. 19–54. Elsevier, Amsterdam.

Yurtsev, B. A. 1974. Phytogeographical problems in north-eastern Asia. Nauka, Leningrad.

Yurtsev, B. A. 1981. The relict steppe complexes of north-eastern Asia (in Russian). Siberian Publishing House of 'Science', Novosibirsk.

Zoltai, S. C. and Pettapiece, W. W. 1973. Terrain, Vegetation and Permafrostt Relationships in the northern Part of the Mackenzie Valley and Northern Yukon. Environ Soc. Comm. Northern Pipelines Rep. 73–74.

Zoltai, S. C. and Tarnocai, C. 1974. Soils and Vegetation of hummocky Terrain. Environ Soc. Comm. Northern Pipelines, Rep. 74–75.

APPENDIX 1: LISTS OF SPORADIC SPECIES

Numbers in brackets following the species names refer to relevé numbers.

Sporadic species in Table 3

Vascular plants: *Abies lasiocarpa* (15), *Amerorchis rotundifolia* (7, 27), *Anemone richardsonii* (14), *Astragalus umbellatus* (27, 32, 47), *Betula occidentalis* (7, 11, 31), *Betula papyrifera* (18, 53), *Carex atratiformis* (33, 43), *Equisetum palustre* (21, 28, 29, 50), *Gentiana propinqua* (34, 50), *Habenaria obtusata* (19, 22, 27, 45), *Juniperus communis* (57), *Larix laricina* (14, 23), *Lupinus arcticus* (25, 47, 51, 54, 55), *Petasites frigidus* (9, 39), *Petasites palmatus* (28), *Polygonum viviparum* (30, 34), *Populus tremuloides* (18, 57), *Pyrola grandiflora* (40, 47, 49), *Rumex arcticus* (14, 36), *Salix arbusculoides* (5, 8, 38), *Salix lanata* (8), *Selaginella selaginoides* (15, 23, 28, 46, 57), *Shepherdia canadensis* (33), *Spiraea beauverdiana* (14), *Valeriana capitata* (23), *Viburnum edule* (51).

Lichens: *Cladonia cornuta* (14, 15, 23), *Cladonia deformis* (23, 34, 56, 57), *Cladonia pyxidata* (32, 48), *Nephroma arcticum* (57, 19).

Bryophytes: *Blepharostoma trichophyllum* (29), *Bryum bimum* (50), *Bryum caespiticium* (32), *Calliergon richardsonii* (29), *Catoscopium nigritum* (25), *Cephalotia* sp. (4), *Ceratodon purpureus* (30), *Cynclidium stygium* (25), *Distichium capillaceum* (32, 46), *Fissidens osmundioides* (30, 37), *Hedwigia ciliata* (50), *Leptobryum pyriforme* (50), *Lophotia* sp. (1), *Meesia uliginosa* (30, 37), *Odontoschisma sphagni* (14, 11), *Polytrichum commune* (21, 22), *Sphagnum fallax* (1), *Sphagnum fimbriatum* (29, 8), *Sphagnum girgensohnii* (8, 29), *Sphagnum molle* (16), *Sphagnum nemoreum* (19, 32, 48, 52), *Sphagnum obtusum* (14), *Sphagnum quinquefarium* (14), *Sphagnum russowii* (14), *Sphagnum subnitens* (19, 27, 34).

Sporadic species in Table 5

Vascular plants: *Abies lasiocarpa* (33, 43), *Agropyron yukonense* (63), *Anemone parviflora* (44), *Aquilegia brevistyla* (64), *Boschniakia rossica* (31), *Calamagrostis purpurascens* (63), *Carex aquatilis* (12), *Carex dioica* subsp. *gynocrates* (2), *Carex vaginata* (10, 13), *Delphinium glaucum* (44, 50, 51), *Equisetum arvense* (2, 4, 6, 19), *Equisetum sylvaticum* (44), *Equisetum variegatum* (13), *Fragaria virginiana* subsp. *glauca* (44), *Galium boreale* (32), *Listera borealis* (11, 25, 26), *Lycopodium complanatum* (44), *Pedicularis sudetica* subsp. *interior* (27, 34, 43, 66), *Petasites palmatus* (38), *Saussurrea angustifolia* (20), *Solidago decumbens* var. *oreophila* (56, 66), *Viola renifolia* (35, 51), *Zygadenus elegans* (28, 72).

Lichens: *Cetraria nivalis* (6, 28), *Cladina mitis* (49), *Cladina rangiferina* (16, 7, 38), *Cladonia deformis* (49, 58, 60), *Cladonia turgida* (28, 43), *Dactylina arctica* (5, 24, 54), *Stereocaulon tomentosum* (24, 48), *Thamnolia vermicularis* (20, 62, 73).

Bryophytes: *Bryum caespiticium* (13), *Catoscopium nigritum* (6), *Drepanocladus revolvens* (7, 14), *Grimmia ovalis* (53), *Hedwigia ciliata* (53), *Pohlia cruda* (53), *Polytrichum commune* (11, 55, 58), *Polytrichum juniperinum* (33, 49), *Ptilidium ciliare* (1, 2, 6), *Rhytidiadelphus triquetrus* (12), *Tortella fragillis* (13).

Sporadic species in Table 7

Vascular plants: *Antennaria rosea* var. *nitida* (32), *Arctostaphylos rubra* (11, 12), *Arnica cordifolia* (6), *Betula papyrifera* (10), *Calypso bulbosa* (4), *Delphinium glaucum* (34), *Juniperus communis* (34), *Pyrola grandiflora* (38), *Ribes triste* (9), *Rubus idaeus* subsp. *melanolasius* (33), *Salix bebbiana* (35), *Salix planifolia* (39), *Viola renifolia* var. *brainerdii* (10).

Lichens: *Cladonia cenotea* (21, 26), *Cladonia pyxidata* (21), *Dactylina arctica* (20).

Sporadic species in Table 9

Vascular plants: *Adoxa moschatellina* (14), *Agropyron yukonense* (55), *Androsace septentrionalis* (59, 61), *Anemone parviflora* (14), *Arabis hoelbelii* (59), *Artemisia arctica* (6), *Artemisia borealis* (19, 26), *Astragalus alpinus* subsp. *americanus* (28, 62), *Bromus pumpellianus* (24), *Calypso bulbosa* (40), *Carex aquatilis* (1), *Corallorhiza trifida* (8, 18), *Deschampsia caespitosa* (5), *Dryas drummondii* (33), *Empetrum nigrum* (14, 28), *Equisetum sylvaticum* (13, 27), *Erigeron* sp. (2), *Festuca brachyphylla* (29, 59), *Linum perenne* subsp. *lewisii* (22), *Lycopodium complanatum* (14), *Moneses uniflora* (5), *Polygonum viviparum* (2), *Potentilla diversifolia* (28), *Potentilla hookeriana* (26), *Rubus idaeus* subsp. *melanolasius* (16, 47), *Salix interior* (1), *Salix pulchra* (6), *Senecio lugens* (4), *Stellaria crassifolia* (28, 29).

Lichens: *Cetraria nivalis* (27, 30), *Cetraria tillesii* (49), *Cladina stellaris* (47), *Cladonia cornuta* (48), *Cladonia fimbriata* (48), *Cladonia gracilis* subsp. *turbinata* (17, 46), *Cladonia pocillum* (49), *Squamarina lentigera* (49), *Stereocaulon alpinum* (45, 47).

Bryophytes: *Mnium* sp. (2), *Pleurozium schreberi* (8), *Polytrichum commune* (46), *Polytrichum piliferum* (44, 45, 46, 47).

Sporadic species in Table 13

Vascular plants: *Aconitum delphinifolium* (6, 7), *Agropyron violaceum* (24, 25), *Alnus crispa* (31), *Andromeda polifolia* (17, 20), *Anemone richardsonii* (20), *Antennaria pulcherrima* (20, 21), *Arnica cordifolia* (27), *Astragalus aboriginus* (34), *Astragalus hudsonianum* (4, 18), *Betula papyrifera* (18), *Bromus pumpellianus* (32), *Carex aquatilis* (18), *Carex capillaris* (17), *Carex dioica* subsp. *gynocrates* (17, 20, 21), *Carex vaginata* (17, 20), *Chamaedaphne calyculata* (34), *Cornus canadensis* (8), *Crepis elegans* (32), *Crepis nana* (33), *Dryas integrifolia* (14, 20), *Eleocharis quinqueflora* (14, 20), *Equisetum pratense* (18, 21), *Erigeron elatus* (2, 19), *Erigeron purpuratus* (32), *Galium boreale* (5, 6, 25), *Habenaria obtusata* (20), *Hedysarum mackenzii* (5, 33, 34), *Juncus castaneus* (18), *Linnaea borealis* subsp. *americana* (23, 24), *Lloydia serotina* (5, 7), *Myosotis alpestris* (5, 7), *Oxytropis campestris* subsp. *gracilis* (37, 35), *Oxytropis deflexa* (23, 18), *Parnassia kotzebuei* (5, 7), *Pedicularis verticillata* (64), *Pinguicula vulgaris* (17), *Potentilla diversifolia* (12, 25), *Ranunculus hyperboreus* (15), *Rhodiola rosea* subsp. *integrifolia* (18, 6), *Rumex arcticus* (5, 7), *Senecio cymbalarioides* (20, 18), *Solidago decumbens* var. *oreophila* (8, 9, 25, 27), *Solidago multiradiata* (7, 18), *Spiranthes romanzoffiana* (13), *Viburnum edule* (20).

Lichens: *Alectoria nigricans* (2), *Alectoria ochroleuca* (1), *Cladina stellaris* (1, 2), *Cladonia deformis* (2, 3, 5), *Cladonia gracilis* subsp. *turbinata* (1, 2, 4), *Cladonia uncialis* (2), *Dactylina madreporiformis* (5), *Nephroma arcticum* (2).

Sporadic species in Table 15

Vascular plants: *Agropyron caninum* (5), *Amelanchier alnifolia* (31), *Androsace septentrionalis* (14), *Arnica cordifolia* (2, 3), *Aster sibiricus* (5), *Astragalus sibiricus* (5), *Barbarea orthoceras* (6), *Crepis elegans* (5), *Deschampsia caespitosa* (6), *Erigeron elatus* (2, 6), *Festuca altaica* (11, 14), *Hierochloe alpina* (26), *Lepidium multiflorum* (6), *Luzula rufescens* (2), *Matricaria matricarioides* (6, 7), *Melilotus alba* (7), *Mertensia paniculata* (3), *Minuartia stricta* (2), *Orobanche fasciculata* (18), *Oxytropis deflexa* (17), *Oxytropis sericea* (12), *Picea glauca* (34), *Populus balsamifera* (4, 5), *Ribes triste* (33), *Rumex fenestratus* (6), *Salix alaxensis* (6), *Salix glauca* (2), *Senecio* sp. (7), *Solidago multiradiata* (11), *Stipa comata* (26).

Lichens: *Cetraria laevigata* (14), *Cetraria tillesii*

(13), *Cladonia pyxidata* (13), *Peltigera canina* (24), *Peltigera rufescens* (21).

Bryophytes: *Polytrichum juniperinum* (24), *Ptilium crista-castrensis* (33).

APPENDIX 2

Species whose ranges have been utilized to construct the chorograms of Figures 36–42.

Fig. 36 – Incompletely circumboreal–temperate species

C3: *Arctostaphylos uva-ursi, Calypso bulbosa, Gymnocarpion dryopteris, Goodyera repens, Hierochloe odorata, Lycopodium complanatum, Mitella nuda, Moneses uniflora, Pyrola secunda* subsp. *secunda, Selaginella selaginoides*.

D3: *Arctostaphylos uva-ursi, Deschampsia caespitosa, Plantago major, Potentilla norvegica, Taraxacum officinale, Trifolium repens*.

Fig. 37 – Cordilleran species and endemics to Alaska–Yukon

A1: Anemone parviflora, Carex microchaeta, Delphinium glaucum, Mertensia paniculata, Pedicularis sudetica subsp. *interior, Senecio lugens, Veronica wormskjoldii*.

C2: *Abies lasiocarpa, Achillea borealis, Anemone multifida, Arabis hoelbelii, Arnica cordifolia, Betula papyrifera* subsp. *humilis, Delphinium glaucum, Dryas drummondii, Fragaria virginiana* subsp. *glauca, Gentiana propinqua, Geum macrophyllum* subsp. *perincisum, Hedysarum mackenzii, Lupinus arcticus, Mertensia paniculata, Oxytropis campestris* subsp. *gracilis, Oxytropis deflexa, Pinus contorta, Potentilla diversifolia, Salix arbusculoides, Salix scouleriana, Solidago decumbens* var. *oreophila, Zygadenus elegans*.

D4: *Achillea borealis, Agropyron violaceum, Agropyron yukonense, Aira cariophyllea, Amelanchier alnifolia, Anemone multifida, Antennaria rosea* var. *nitida, Arabis hoelbelii, Artemisia alaskana, Carex filifolia, Castilleja yukonis, Draba aurea, Erigeron caespitosum, Erigeron compositus, Erigeron elatus, Fragaria virginiana* subsp. *glauca, Gentiana propinqua, Geum macrophyllum* subsp. *perincisum, Hedysarum americanum, Juniperus horizontalis, Linum perenne* subsp. *lewisii, Lupinus arcticus, Oxytropis campestris* subsp. *gracilis, Penstemon gormanii, Penstemon procerus, Polemonium pulcherrimum, Potentilla pennsylvanica, Salix setchelliana, Saxifraga tricuspidata, Sedum lanceolatum, Shepherdia canadensis, Solidago decumbens* var. *oreophila, Solidago multiradiata* var. *scopulorum, Stipa comata*.

Fig. 38 – Boreal North American species

B2: *Betula glandulosa, Carex concinna, Carex membranacea, Carex scirpoidea, Cornus canadensis, Cypripedium passerinum, Dryas integrifolia, Geocaulon lividum, Hedysarum americanum, Juniperus horizontalis, Ledum groenlandicum, Mertensia paniculata, Pedicularis sudetica* subsp. *interior, Petasites sagittatus, Picea glauca, Picea mariana, Salix myrtillifolia, Salix planifolia, Salix scouleriana, Senecio lugens, Viola renifolia* subsp. *brainerdii*.

C1: *Alnus incana* subsp. *tenuifolia, Amerorchis rotundifolia, Aquilegia brevistyla, Betula glandulosa, Carex concinna, Cornus canadensis, Geocaulon lividum, Habenaria obtusata, Larix laricina, Ledum groenlandicum, Linnaea borealis* subsp. *americana, Picea glauca, Picea mariana, Populus balsamifera, Populus tremuloides, Ranunculus macounii, Ribes hudsonianum, Ribes oxyacanthoides, Salix myrtillifolia, Salix planifolia, Shepherdia canadensis, Viburnum edule*.

Fig. 39 – Arctic montane species

B1: *Calamagrostis canadensis, Carex aquatilis, Carex capillaris, Carex dioica* subsp. *gynocrates, Chamaedaphne calyculata, Equisetum arvense, Equisetum palustre, Equisetum scirpoides,*

Equisetum sylvaticum, Equisetum variegatum, Juniperus nana, Ledum palustre, Oxycoccus microcarpus, Pedicularis labradorica, Polygonum viviparum, Potentilla fruticosa, Pyrola asarifolia s. lat., *Rhododendron lapponicum, Rosa acicularis, Salix bebbiana, Salix glauca, Stellaria longipes.*

C6: *Alnus crispa* s. str., *Androsace septentrionalis, Calamagrostis canadensis, Corallorhiza trifida, Epilobium angustifolium* s. str., *Equisetum scirpoides, Equisetum sylvaticum, Galium boreale, Juniperus nana, Lycopodium annotinum, Pedicularis labradorica, Pyrola asarifolia* s. lat., *Rosa acicularis, Stellaria longipes.*

D1: *Dryopteris fragrans, Epilobium angustifolium* s. str., *Equisetum arvense, Galium boreale, Juniperus nana, Rubus ideaus* subsp. *melanolasius.*

Fig. 40 – Circumpolar (–boreal) species

A4: *Andromeda polifolia* s. str., *Empetrum nigrum* subsp. *hermaphroditum, Rubus arcticus, Rubus chamaemorus, Salix lanata, Vaccinium uliginosum* s. lat.

C5: *Artemisia borealis, Astragalus alpinus* s. str., *Carex vaginata, Empetrum nigrum* subsp. *hermaphroditum, Lycopodium clavatum* subsp. *monostachyon, Salix glauca, Stellaria crassifolia, Vaccinium uliginosum* s. lat., *Vaccinium vitis-idaea* subsp. *minus*,

Fig. 41 – Old preglacial relict species

A3: *Artemisia arctica* s. str., *Festuca altaica, Gentiana algida, Gentiana glauca, Luzula parviflora* s. str., *Spiraea beauverdiana.*

D5: *Artemisia frigida, Carex maritima, Carex obtusata, Festuca altaica, Pulsatilla patens* subsp. *multifida, Salix alaxensis.*

Fig. 42 – Amphiberingian species

A2: *Petasites hyperboreus, Poa arctica, Polemonium acutiflorum, Polygonum bistorta* subsp. *plumosum, Salix pulchra, Salix reticulata, Trisetum spicatum* s. lat., *Valeriana capitata.*

B3: Anemone richardsonii, Arctostaphylos rubra, Astragalus umbellatus, Carex lugens, Cassiope tetragona, Eriophorum vaginatum s. str., *Rumex arcticus, Salix alaxensis, Salix reticulata, Saussurea angustifolia, Tofieldia pussilla, Valeriana capitata.*

C4: *Aconitum delphinifolium, Anemone richardsonii, Arctostaphylos rubra, Boschniakia rossica, Calamagrostis purpurascens, Pyrola grandiflora, Ribes triste, Senecio atropurpureus, Salix alaxensis, Valeriana capitata.*

D2: Astragalus alpinus s. str., *Bromus pumpellianus, Calamagrostis purpurascens, Epilobium latifolium, Festuca brachyphylla, Hordeum jubatum, Juncus arcticus, Melandrium apetalum, Poa glauca, Potentilla hookeriana, Ribes triste.*

4. THE VASCULAR FLORA OF GROS MORNE NATIONAL PARK, NEWFOUNDLAND: A HABITAT CLASSIFICATION APPROACH BASED ON FLORISTIC, BIOGEOGRAPHICAL AND LIFE-FORM DATA

ANDRE BOUCHARD[1,2], STUART HAY[2], YVES BERGERON[3] AND ALAIN LEDUC[4]

[1] *Jardin botanique de la Ville de Montréal 4101 est, rue Sherbrooke, Montréal (Québec), Canada H1X 2B2*
[2] *Institut botanique de l'Université de Montréal, 4101 est, rue Sherbrooke, Montréal (Québec), Canada H1X 2B2*
[3] *Département de Sciences biologiques, Université du Québec à Montréal, C.P. 8888, succursale "A", Montréal (Québec), Canada H3C 3P8*
[4] *Centre de recherches écologiques de Montréal, Université de Montréal, 5858 chemin Côte-des-Neiges, bureau 400, Case postale 6128, succursale "A", Montréal (Québec), Canada H3C 3J7*

1. INTRODUCTION

Newfoundland's west coast has long been a center of attention for taxonomists and botanists interested in floristic and phytogeographical studies because of its peculiar flora. The importance of this area was first recognized by Fernald (1911, 1926, 1933) in the early part of this century. His pioneering explorations led to many important discoveries and resulted in two landmark papers concerning the phytogeography of the region of the Gulf of St Lawrence (Fernald 1924, 1925).

Many of the elements of the area's unique flora are plants with restricted or disrupted ranges, making them rare in Atlantic Canada. The phytogeographic origins of this flora have been discussed by various authors (Damman 1965, 1976; Drury 1969; Morisset 1971; Rousseau 1974; Scoggan 1950). In an earlier paper on the rare plants of Gros Morne National Park, Bouchard et al. (1986) discussed the situation for 43 of these taxa which are largely localized in restricted habitats ranging from tidal flats to limestone escarpments and alpine snowbeds.

Generally, the flora of Newfoundland remains under-studied in comparison with most parts of eastern North America. The first explorations include those of Joseph Banks in 1766, Bachelot de la Pylaie in 1816 and 1820, and A. C. Waghorne (1893, 1895, 1898). Fernald led a series of expeditions (1911, 1926–27, 1933) that produced the most documented work and stimulated much of the subsequent study of the Island's flora. Rouleau (1949, 1956, 1978) compiled three checklists, distribution maps and an index of botanists and their collections (unpublished files, MT). Bouchard et al. (1978, 1986) have outlined the history of floristic studies in the region prior to the creation of Gros Morne National Park. Subsequent botanical studies have been carried on by our research group (Bouchard et al. 1977, 1978; Bouchard and Hay 1974, 1976a, b).

In an earlier study of the Park flora, Bouchard et al. (1987) used canonical correlation techniques to analyse the general interrelationships between habitat types based on composition, phytogeographical affinities and biological life forms of the flora. A spectrum of 35 habitats (or vegetation types) were examined in function of the 711 taxa comprising the flora. In this paper, a floristic analysis, mainly based on indicator species, is performed by applying a classification approach based on a TWINSPAN analysis. To further refine the results obtained by canonical analysis (Bouchard et al. 1987), this classification approach was also used to analyse these habitats in respect to composition, phytogeographical and

P. L. Nimis and T. J. Crovello (eds): Quantitative Approaches to Phytogeography, pp. 123–157.

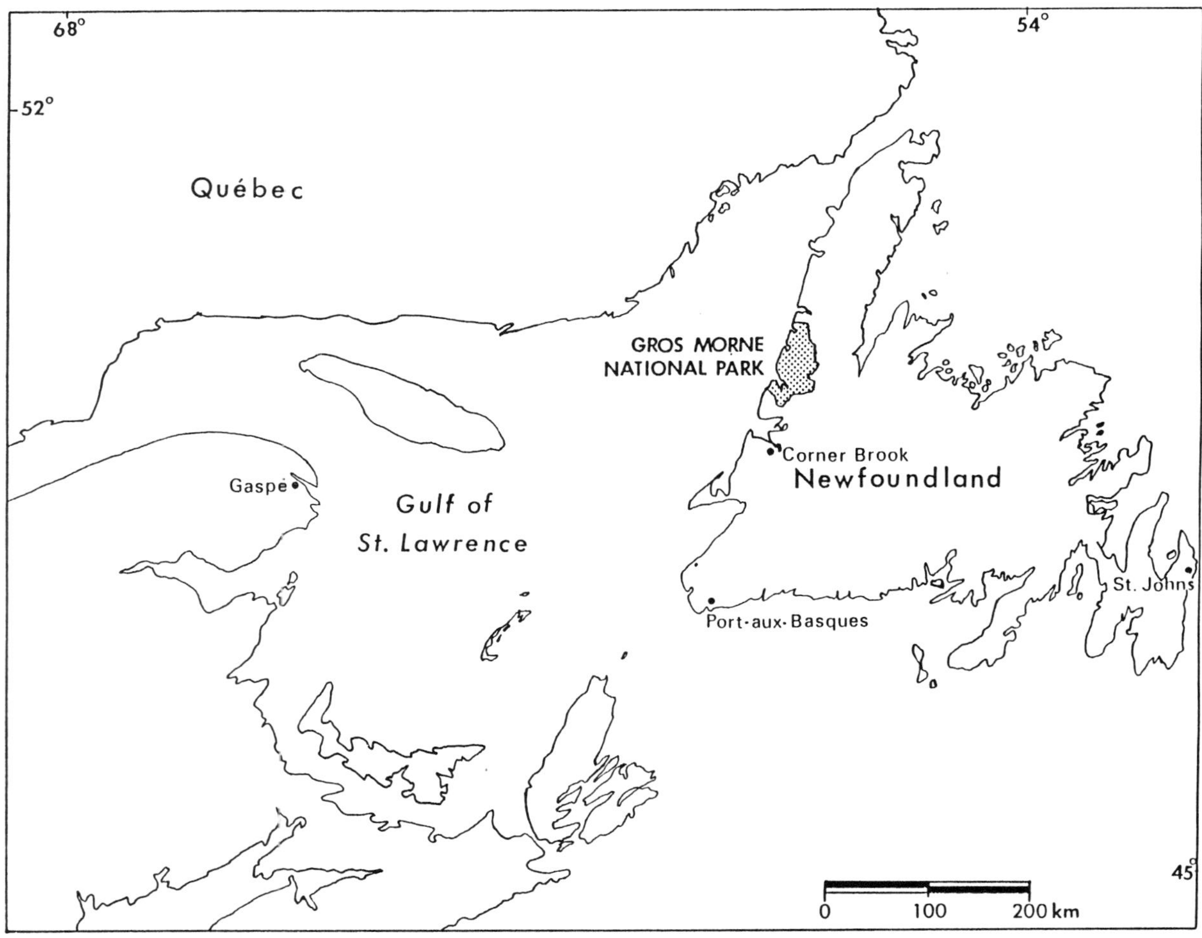

Fig. 1. Location of Gros Morne National Park, Newfoundland, in the Gulf of St. Lawrence.

life-form data. The three classifications are compared.

2. DESCRIPTION OF THE STUDY AREA

Gros Morne National Park is centered on Bonne Bay fjord which opens into the Gulf of St Lawrence (Fig. 1). It is situated in a particularly critical area of the west coast region of Newfoundland. The remarkable diversity of the flora is due in large part to the complex biophysiography which brings to bear a steep gradient of abiotic variables within the study area. Spanning almost 2000 km^2 (Fig. 2), the landscape is composed of distinct physiographic regions including the alpine plateau breached by fjords (Long Range Mountains), the serpentine tableland (Bay of Islands Serpentine Range), numerous limestone escarpments and the flat coastal plain mantled with extensive bogs (West Coast Lowland) (Bouchard et al. 1985b).

The Park is part of the western continental platform or Humber zone, one of four tectonic zones in Newfoundland (Rogerson 1983). The coastal plain (50–150 m elevation) is composed mainly of interbedded Cambrian and Ordovician sedimentary formations (ranging from limestone to sandstone) which lie in alternating bands parallel to the coast. Some limestone and dolomite exposures form high stratified cliffs in Bonne Bay. The alpine plateau (reaching elevations greater than 800 m) forms part of the Long Range Appalachian Mountains of the

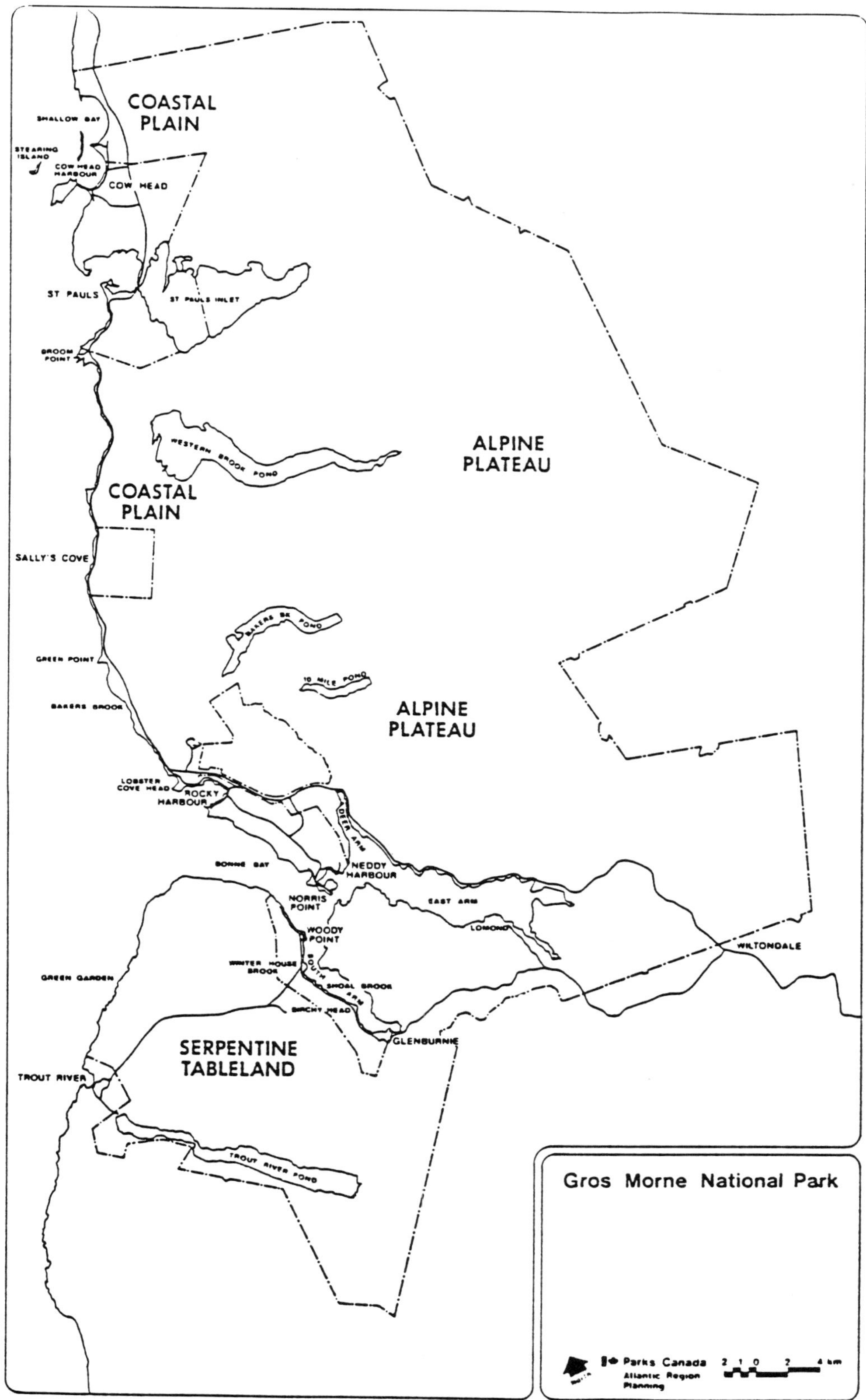

Fig. 2. Map of Gros Morne National Park, Newfoundland.

Northern Peninsula. It is an uplifted block of Precambrian Grenville Basement rock consisting principally of matamorphic and igneous granite or granite-gneiss (Cumming 1973). Several land-locked fjords such as Bakers Brook Pond and Western Brook Pond breach the western scarp of the highlands which marks its contact with the coastal plain. Important exposures of Cambrian sedimentary limestone and quartzites flank the abrupt escarpment such as at Gros Morne Mountain and Killdevil Mountain (Baird 1958). The serpentine tableland is an intrusive complex of iron and magnesium-rich rocks (Bay of Islands Complex, Cumming 1973). It is mainly comprised ultramafic serpentine, altered gabbroic rocks and quartz diorite that originated from the suboceanic crust (Smith, 1958).

Ancient episodes of block-faulting and klippen transport have shaped the Gros Morne region, however, since 50,000 B. P., glacial phases of the late Cenozoic have profoundly modified the landscape and the dominant geomorphological features of the Park reflect Pleistocene glaciation (Cumming 1973). These landforms include ice-scoured uplands, glacier-carved valleys, moraine-mantled lowlands and coastal rock terraces. There is recent glacial geological evidence by Grant (1977a, 1977b) for ice-free Wisconsin-age land surfaces in the Bonne Bay region.

Since the Pleistocene, the Gros Morne landscape has been modified to create numerous marine, fluvial, and eolian landforms. These include tidal flats, beaches, sand dunes and sea cliffs, as well as river and creekbeds, deltas and floodplains. Colluvial rockfalls and talus-screes occur throughout the Park, while extensive bogs and fens mantle the poorly drained coastal plain.

The climate of the Bonne Bay region of western Newfoundland has been described as "modified continental" based on a temperature regime of less than four summer months with mean temperatures above 100° C (Banfield 1983). The more important climatic indices are the cool temperatures with a short growing season, the moderating oceanic influence, a continual moisture excess, and strong prevailing winds.

Two main climatic zones may be recognized in the Park, one along the coast and coastal lowland, the other in the alpine region (Banfield 1983; Hare, 1952). A third climatic zone has been proposed for the Inner Bonne Bay Pond or Lomond Hills area (Nicholson and Bryant 1972; Watson 1974). These zones conform in part with the major physiographic land regions of the Park.

The coastal lowland climatic zone is characterized by moderate annual precipitation (900–1000 mm), cold and snow-abundant (300–350 mm) winters, relatively early spring and by moderately warm summers (maxima up to 30° C in sheltered valleys and inlets). Mean air temperature in July ranges from 14 to 16° C and −6 to −8° C in February. The short growing season (based on a 6.1° C threshold) ranges from 150 days in the southern sector to 140 days in the north. It begins about May 20 (Hare 1952).

The highland sector of the park, particularly the alpine plateau region extending north of Bonne Bay, experiences more rigorous climatic conditions. According to Watson (1974), average precipitation and snowfall are more than twice that of the coastal plain and snow cover is likely two months longer in duration. Some, extensive, almost snow-free tundra areas occur along the western edge of the plateau due to wind exposure. Mean air temperatures in July range from 12 to 14° C and from −8 to −10° C in February. The growing season may be as short as 130 days (Hare 1952) and even shorter in the late-lying snowbank or zabois habitats. The ocean has no moderating influence on winter temperatures and the highest elevations have been mapped by Damman (1976) within the coldest areas of Newfoundland during the growing season.

The Lomond–Inner Bonne Bay Pond area in the southern part of the Park affords significantly higher mean temperatures during the growing season (Damman 1976) which may be prolonged up to 160 days (Hare 1952). Mean temperatures in July are 16 to 17° C, and in February are about −6 to −7° C (Watson 1974).

Several other climatic factors may have an important effect on the vegetation and flora. Sustained high winds influence the structure of several plant communities, especially the coastal krummholz, upland tuckamoor, and exposed alpine barrens (Bouchard et al. 1978). The conditions of relatively high precipitation, low potential evapotranspiration (Hare 1952; 46 cm/yr) and poor drainage, due to the general flatness of both the coastal lowland and the alpine plateau, have resulted in the formation of extensive peatland (Bouchard et al. 1978) punctuated with numerous flashets or ponds. The late-lying snowbeds in the alpine areas of the Park create special ecological conditions which harbour some of the remarkable Arctic–Alpine elements of the flora.

The vegetation of the study area falls within the Boreal Forest region of Canada. The most common and characteristic tree species are *Abies balsamea, Betula papyrifera, Picea glauca* and *P. mariana*. Because of the marked regional differences in physiography and climate, Gros Morne embraces three boreal forest subsections (Rowe, 1972). The Northern Peninsula subsection encompasses the lowlying coastal plain as well as the forested flanks and piedmont area of the Long Range. The Newfoundland–Labrador Barrens subdivision includes the tundra vegetation of the high-altitude alpine plateau of the Long Range. The Corner Brook subsection, lying to the south of Bonne Bay, harbours more demanding southern tree species such as *Betula alleghaniensis* and *Fraxinus nigra*.

The main vegetation cover types of Gros Morne have been described in terms of the major biophysiographic land regions comprising the Park by Bouchard et al. (1978), Airphoto Analysis Associates (1975) and particularly for the coastal plain area (Bouchard 1974, 1975; Bouchard and Hay 1976b).

According to the ecoregions classification recently proposed for Newfoundland (Damman 1983), the Park overlaps three of the nine subdivisions defined. (a) the Northern Peninsula Forest ecoregion, comprising the coastal plain; (b) the Long Range Barrens ecoregion, comprising the alpine plateau; and (c) the Western Newfoundland ecoregion, corresponding to the sector south of Bonne Bay. These regions are characterized by distinctive, recurring patterns of vegetation and soil development which are controlled by regional climate. Within these ecoregions, Damman identifies four different subregions occurring in the Park.

(a) Within the Northern Peninsula Forest ecoregion. Damman describes the Coastal Plain subregion: "This includes the flat coastal plain and the western lower slope of the Long Range Mountains. Most of the coastal plain is occupied by ombrotrophic bogs. Forests are restricted mainly to the slopes of the Long Range." The vegetation of the coastal plain is relatively complex, it comprises many habitats, ranging from tidal flats, seashore cliffs, and peat bogs, to sedge meadows, dwarf black spruce scrub communities and American larch scrubs (Bouchard 1974; Bouchard and Hay 1976b).

(b) The Long Range Barrens ecoregion consists of alpine tundra barrens mostly dominated by dwarf shrub heaths, shallow patterned peatlands, "tuckamoor" or alpine krummholz, and some forest vegetation in sheltered valleys. This ecoregion is represented by the Northern Long Range subregion with the best developed zones of snowbank or zabois vegetation.

(c) The Western Newfoundland ecoregion includes two subregions within Gros Morne National Park. The Corner Brook subregion is "a heavily forested area with rugged topography; slates and limestones underly most of the area. The soils are generally nutrient-rich and productive. The peatlands are mainly soligenic and topogenic fens" (Damman 1983). Within this subregion the exposed limestone escarpments in the southern part of the Park are particularly interesting botanically. The Serpentine Range subregion, composed of serpentine and related ultramafic rock, does "not support forest cover even at sea level. The vegetation is very sparse with rocks and soil exposed on most of the surface, except in a few extensive fen areas on the plateaus" (Damman 1983). The limestone and serpentine areas harbour many of the Gulf of St Lawrence endemics and Cordilleran or Arctic-Alpine disjuncts for which this region is well known (Bouchard et al. 1985a).

3. METHODOLOGY

The inventory of herbarium collections, from the study area, from 1910 to 1970, was compiled by Professor E. Rouleau of the Université de Montréal. They are the fruit of his research on Newfoundland's flora in various North American and European herbaria as well as his own field studies. The more recent collections, dating from 1972 to 1984, were made mainly in the course of this study and related work on the vegetation of Gros Morne National Park. Our specimens were deposited in the following herbaria; National Herbarium of Canada, National Museums of Canada, Ottawa (CAN), Herbier Marie-Victorin, Institut botanique de l'Université de Montréal (MT) and Gros Morne National Park Herbarium, Rocky Harbour, Newfoundland.

All the information on herbarium labels and from field observations was compiled and stored in a computerized data bank. This information was previously used to define 35 habitats regrouped in 13 biophysiographic units and 4 land regions (Bouchard et al. 1985b). In addition, life form, taken from Scoggan (1978–79), and phytogeographical affinity for each taxon were included in the data bank. The distribution type was determined largely from sources such as Hultén (1958, 1962, 1968, 1971), Lavoie (1984), Payette and Lepage (1977) and Rousseau (1974). The nomenclature followed is based on the "Vascular Flora of Gros Morne National Park" (Bouchard et al. 1985b) and is a compromise between floras covering Newfoundland (Boivin 1966–67; Fernald 1950; Rouleau 1978; Scoggan 1978–79).

The distribution of species (presence/absence) in the 35 habitats was analysed using the TWINSPAN program (two-way indicator species analysis; Hill 1979). The algorithm is a polythetic divisive cluster analysis based on successive reciprocal averaging ordination (Hill 1973), that allows for a hierarchical division of habitats. Concurrently, a corresponding classification of species based on their fidelity to habitats is also produced. It becomes possible from this double classification, in the same way as with a phytosociological table, to identify indicator species associated to clustering of habitats (Gauch and Whittaker 1981).

The analysis was repeated for phytogeographical groups and life forms on the basis of the number of species in each class for each habitat. Previously, a double standardization (species standardization followed by habitats standardization) was performed in order to limit the importance of a few dominant descriptors (phytogeographical or life form) while preserving quantitative differences between habitats (Noy-Meir and Whittaker 1978). Since TWINSPAN requires binary or ordered discrete descriptors the double standardized values were partitioned into discrete states in such a way as to maximize equitability of the frequency in each class.

The topological distance (Podani and Dickinson 1984), defined by the number of the dendrogram nodes separating two objects, was computed for each pair of habitats. This was done for the three dendrograms. To evaluate their structure similarity, the topological distance matrices of each of the three dendrograms were compared using the Spearman rank correlation coefficient.

4. RESULTS AND DISCUSSION

4.1. Floristic analysis by indicator species

The dendrogram, based on a TWINSPAN analysis of the 711 species, presents the floristic structure of the 35 habitats (Fig. 3). The first division (I) splits the 35 habitats in two groups: (−) 26 habitats characterized by natural communities ranging from the zabois vegetation and the calcareous inland cliffs to the balsam fir forests and associated secondary successional communities and (+) nine coastal habitats that are either man-made, such as the ruderal areas and the trails, or characterized by open, unstable, natural, mostly herbaceous, communities often invaded by opportunistic weedy species. The first group has indicator species such as *Coptis trifolia* ssp. *groenlandica, Cornus canadensis, Gaultheria hispidula, Kalmia polifolia,*

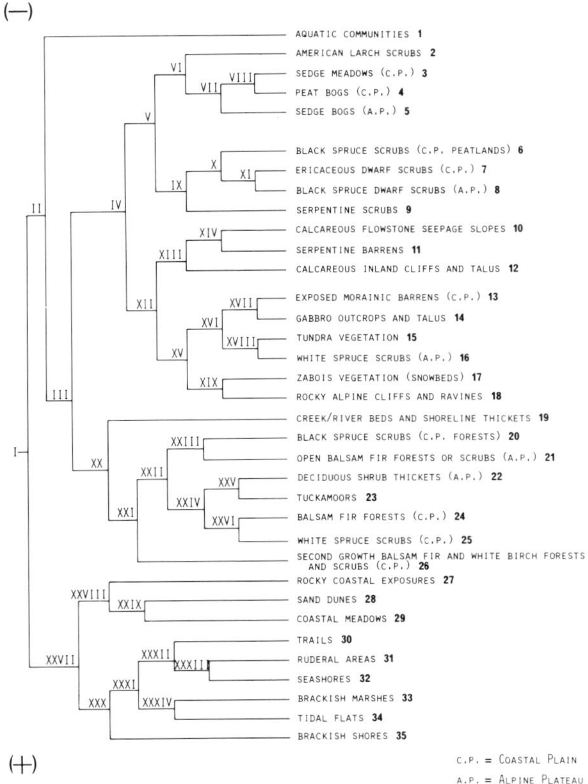

Fig. 3. Dendrogram showing the structure of 35 habitats, based on a TWINSPAN analysis of the flora.

Linnaea borealis, Picea mariana, Scripus cespitosus, and *Vaccinium angustifolium* while the second has indicator species such as *Agrostis stolonifera, Poa palustris, P. pratensis*, and *Potentilla anserina*.

The first group (habitats 1 to 26) includes a variety of habitats, ranging from aquatic communities to calcareous flowstone seepage slopes. The aquatic communities (1) constitute an isolated group (division II) because of the exclusive presence of several hydrophytes such as *Glyceria fluitans, Littorella uniflora* var. *americana, Lobelia dortmanna, Lysimachia terrestris, Myriophyllum tenellum, Potamogeton amplifolius, P. confervoides, Ranunculus hyperboreus, Sagittaria graminea, Scirpus subterminalis, Sparganium minimum* and *Utricularia vulgaris* and the exclusion of about one hundred species, mainly boreal forest and bog elements.

Division III splits a wide range of habitats, including the exposed alpine, calcareous and serpentine habitats and the more hydric and often eutrophic communities such as the sedge meadows, the American larch scrubs and the calcareous flowstone seepage slopes (habitats 2 to 18) from the more mesic and forested communities such as the second growth balsam fir and white birch forests and scrubs (C.P.) or the balsam fir forests (C.P.), together with the creek/river beds and shoreline thickets (habitats 19 to 26). Examples of indicator species for the first group of habitats are *Andromeda glaucophylla, Betula pumila, Empetrum nigrum, Juniperus communis, Kalmia polifolia, Ledum groenlandicum, Scirpus cespitosus*, and *Vaccinium uliginosum*, while *Equisetum sylvaticum, Gymnocarpium dryopteris, Habenaria obtusata, Lycopodium lucidulum, Moneses uniflora, Monotropa uniflora, Rubus idaeus*, and *Viola renifolia* are examples of the latter ones.

A large group, containing a wide range of habitats (2 to 18), including the exposed alpine, calcareous and serpentine habitats and the more hydric and often eutrophic communities, such as the sedge meadows, the American larch scrubs and the calcareous flowstone seepage slopes, is divided (IV) into the more hydric and often eutrophic communities together with the coniferous and the ericaceous scrubs on one hand (2 to 9), and the often open, wind-exposed habitats, either Arctic-Alpine, calcareous or serpentine on the other hand (10 to 18). For the first ones, *Amelanchier bartramiana, Chamaedaphne calyculata, Eriophorum angustifolium, Kalmia angustifolia*, and *Vaccinium oxycoccos* are good examples of indicator species, while *Alnus crispa, Arctostaphylos alpina, Carex scirpoidea, Empetrum nigrum* var. *eamesii, Juncus trifidus, Loiseleuria procumbens, Salix glauca, Solidago macrophylla, Streptopus roseus*, and *Vaccinium caespitosum* are good ones for the latter habitats.

The more hydric and often eutrophic communities together with the coniferous and the ericaceous scrubs (2 to 9) are divided (V) into two groups: (a) the more hydric and often eutrophic communities (2 to 5) and (b) the more mesic coniferous or ericaceous scrubs (6 to 9). The more hydric and often eutrophic communities have indicator species such as: *Carex exilis, C. rostrata, Lonicera villosa, Scirpus hudsonianus, Solidago uliginosa, Tofieldia glutinosa*, and *Triglochin maritima*, in opposition to *Alnus crispa, Arctostaphylos alpina, Carex trisperma, Epigaea repens, Geocaulon lividum, Nemopanthus mucronatus, Potentilla tridentata, Schizachne purpurascens, Sorbus decora, Vaccinium vitis-ideas*, and *Viburnum cassinoides*, for the more mesic coniferous or ericaceous scrubs. The American larch scrubs (2) are divided (VI) from the three other hydric communities (3, 4, 5) by several indicator species such as *Caltha palustris, Carex disperma, Cornus stolonifera, Cypripedium reginae, Geum rivale, Listera convallarioides, Mitella nuda, Osmunda cinnamomea, Salix serrissima*, and *Senecio aureus*. Because of their floristic similarities, the alpine plateau sedge bogs are classified with the coastal plain sedge meadows and peat bogs. The serpentine scrubs (9) are divided (IX) from the three other more oligotrophic scrub communities (6, 7, 8) by several indicator species such as *Acer rubrum, Adiantum pedatum* ssp. *calderi, Anaphalis margaritacea, Campanula rotundifolia, Carex scirpoidea, Glyceria canadensis, Oryzopsis asperifolia, Osmunda regalis, Pinus strobus*, and *Senecio pauperculus*.

The often open, wind-exposed habitats (10 to 18) are divided (XII) into two groups: (a) the calcareous and serpentine habitats (10, 11, 12), and (b) the alpine plateau habitats including the coastal plain exposed morainic barrens (13 to 18). Good examples of indicator species for the calcareous or serpentine habitats are *Anemone parviflora, Cypripedium calceolus, Erigeron hyssopifolius, Pinguicula vulgaris, Rhododendron lapponicum, Saxifraga aizoides*, and *Thalictrum alpinum*, in contrast to *Carex stylosa, Coptis trifolia* ssp. *groenlandica, Diapensia lapponica, Kalmia polifolia, Phyllodoce caerulea*, and *Vaccinium angustifolium*, for the alpine plateau habitats.

Within the six alpine habitats (13 to 18), including the exposed morainic barrens (C.P.), two habitats, the rocky alpine cliffs and ravines together with the zabois or snowbeds (17, 18) are split (XV) from the four other ones by indicator species such as *Cornus suecica, Epilobium hornemannii, Lycopodium alpinum, Salix glauca, S. herbacea, Stellaria calycantha, Vahlodea atropurpurea, Viola incognita, Viola pallens*, and *Viola palustris*. This mixture of Arctic–Alpine and hydric elements emphasizes the moist microenvironment of these restricted Arctic–Alpine habitats. On the other hand, preferential species such as *Abies balsamea, Linnaea borealis Lycopodium annotinum, Potentilla tridentata*, and *Vaccinium vitis-idaea*, indicate the boreal affinities of the four other habitats (13, 14, 15, 16).

The creek/river beds and shoreline thickets (19) are classified with the forest and scrub communities (20 to 26), because they share numerous species with these adjacent habitats, especially on the coastal plain. Nevertheless, they are characterized (XX) by numerous indicator species such as *Eleocharis palustris, Equisetum fluviatile, Fraxinus nigra, Galium asprellum, Impatiens capensis, Listera auriculata, Lycopus americanus, Onoclea sensibilis, Ranunculus reptans*, and *Thelypteris palustris*. Within the forest and scrub communities (20 to 26), three alpine plateau habitats are classified within the larger group of coastal plain coniferous forests or scrubs. These are, the tuckamoors, the deciduous shrub thickets and the open balsam fir forests or scrubs, habitats of boreal affinities rather than Arctic–Alpine. Within the group of mostly coniferous forests and scrubs (habitats 20 to 26), the second growth balsam fir and white birch forests and scrubs (26) are segregated (XXI) from the six other mostly natural communities (20 to 25) by several species, including *Agropyron repens, Agrostis tenuis, Corylus cornuta, Geranium robertianum, Mentha arvensis, Prunus pensylvanica, P. virginianus, Ranunculus abortivus, Stellaria graminea*, and *Urtica dioica*, species often favored by human or natural disturbances.

The nine coastal habitats, that are either manmade, such as the ruderal areas and the trails, or characterized by open, unstable, natural, mostly herbaceous, communities often invaded by opportunistic weedy species, are divided (XXVII) in two groups: (a) three coastal plain habitats, including the coastal meadows, sand dunes, and maritime cliffs, ridges, ledges, crests and talus (27, 28, 29) and (b) six habitats ranging from the trails to the brackish shores (30 to 35). Boreal forest elements such as *Abies balsamea, Actaea rubra, Botrychium virginianum, Juniperus communis, J. horizontalis, Picea glauca*, and *Vaccinium oxycoccos* on one hand, and more restricted species such as *Draba incana, Montia fontana* ssp. *fontana* and *Saxifraga caespitosa*, indicator species for the first group of habitats (27, 28, 29), underline their affinities to the boreal forest with the inclusion of elements particularly adapted to these coastal habitats, while *Agropyron repens, Anemone canadensis, Carex mackenziei, Galeopsis tetrahit, Glaux maritima, Hordeum jubatum, Linum catharticum, Spartina pectinata, Trifolium repens*, and *Veronica serpyllifolia*, indicator species of the second group (30 to 35), emphasize both the weedy and saline

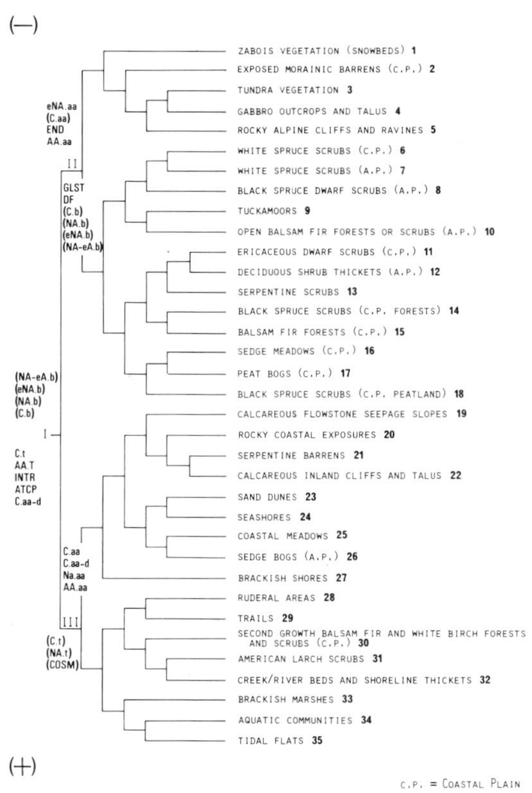

Fig. 4. Dendrogram showing the structure of 35 habitats, based on a TWINSPAN analysis of phytogeographical groups.

plant composition of the six other coastal plain habitats.

From the six habitats, ranging from the trails to the brackish shores (30 to 35), the last one (35) is isolated (XXX) from the other five by several indicator species such as *Astragalus alpinus, Elymus virginicus* and *Solidago sempervirens* while the latter (30 to 34) have common weedy species such as *Agropyron repens, Agrostis stolonifera, Festuca rubra*, and *Poa pratensis*, for indicator species. It appears that, within Gros Morne National Park, the open brackish sand or gravel shores, which are of very limited extent, are perhaps less disturbed by human activities. The trails, the ruderal areas, the seashores, the brackish marshes and the tidal flats (30 to 34) are divided (XXXI) in two groups. The first three habitats (30, 31, 32) have, for indicator species, weedy plants such as *Anemone canadensis, Fragaria virginiana, Galeopsis tetrahit, Linum catharticum*, and *Veronica seryplllifolia*, while the two latter ones (33, 34) have, for indicator species, halophytes such as *Carex mackenziei, Eleocharis halophila, Juncus gerardii, Ruppia maritima, Spergularia canadensis* and *Triglochin gaspense*.

4.2. Phytogeographical analysis

The dendrogram, based on a TWINSPAN analysis of the 28 phytogeographical groups (Table 1), presents the phytogeographical structure of the 35 habitats (Fig. 4). The first division (I) splits the 35 habitats in two groups: (–) 18 habitats characterized by natural communities (habitats 1 to 18), usually oligotrophic and often dominated by conferous species, ranging from the zabois vegetation to the black spruce scrubs, and (+) 17 habitats (19 to 35) characterized by the open, unstable, natural, mostly herbaceous communities such as the calcareous inland cliffs and ridges or the serpentine barrens, the man-made habitats such as the ruderal areas and the trails, and finally the saline habitats such as the brackish marshes and the tidal flats. Within these mostly coastal habitats (19 to 35), are also included the aquatic communities.

Table 1. Phytogeographical spectrum of the total flora.

Phytogeographical groups			Percent of total species (711)
Arctic-Alpine	Circumpolar	C.aa	4.9
	Circumpolar with widely disrupted range	C.aa-d	0.9
	Amphi-Atlantic	AA.aa	1.6
	North American and east-Asian	NA-eA.aa	0.1
	North American	NA.aa	0.3
	Eastern North American	eNA.aa	0.2
		aa-subtotal	8.0
Boreal	Circumboreal	C.b	24.8
	Circumboreal with widely disrupted range (including coastal species)	C.b-d	3.0
	Amphi-Atlantic	AA.b	2.2
	Amphi-Beringian, disjunct in eastern North America	AB.b-d	0.6
	North American and east-Asian	NA-eA.b	4.7
	North American	NA.b	24.0
	North American with widely disrupted range	NA.b-d	0.3
	Eastern North American	eNA.b	9.6
	Eastern North American and east-Asian	eNA-eA.b	0.2
	Cordilleran, disjunct in eastern North America	CORD.b-d	0.8
		b-subtotal	70.2
Temperate	Circumboreal and temperate with widely disrupted range (including coastal species)	C.b/t-d	0.2
	Circumtemperate	C.t	0.7
	Amphi-Atlantic	AA.t	0.6
	North American	NA.t	1.8
	Deciduous forest region of eastern North America and eastern Asia	DF-eA	0.1
	Deciduous forest region of eastern North America	DF	1.6
	Appalachian and/or Great Lakes-St. Lawrence Region and eastern Asia	GLST-eA	0.2
	Appalachian and/or Great Lakes-St. Lawrence Region	GLST	6.8
		t-subtotal	11.8
	Cosmopolitan in the northern hemisphere	COSM	0.9
	Atlantic coastal plain	ATCP	1.1
	Endemic, centered mainly on the Gulf of St. Lawrence	END	1.2
	Introduced	INTR	6.0
	Undetermined (hybrid taxa)	?	0.6

The first group (habitats 1 to 18) has for indicator phytogeographical groups the circumboreal, the boreal North American, the boreal eastern North American, and the boreal North American and east Asian. This wide range of habitats, from the Arctic-Alpine zabois (1) or the rocky alpine cliffs and ravines (5) to the coastal plain balsam fir forests (15) or the peat bogs (17) is grouped together by the strong component of boreal elements. On the other hand, the second group (habitats 19 to 35) is characterized by more exclusive phytogeographical groups such as circumtemperate, temperate amphi-Atlantic, Atlantic coastal plain, circumpolar Arctic–Alpine with widely disrupted range and introduced elements. This wide range of coastal habitats, from calcareous flowstone seepage slopes (19) to brackish marshes (33) is tied together by temperate elements in contrast to the first boreal group of habitats (1 to 18), in addition to the presence of other phytogeographical elements such as those of the Atlantic coastal plain. The introduced elements are important in this wide range of either naturally disturbed or man-made habitats.

Division II splits the 18 habitats characterized

by natural communities (habitats 1 to 18), which are usually oligotrophic and often dominated by coniferous species, in two groups: (a) the alpine plateau habitats, including the coastal plain morainic barrens (habitats 1 to 5) and (b) the several coniferous scrubs and forests, together with the sedge meadows and the peat bogs (habitats 6 to 18). The first group (habitats 1 to 5) is characterized by Arctic–Alpine (circumpolar, amphi-Atlantic, eastern North American) elements together with the endemics, while the second group (habitats 6 to 18) is characterized by boreal (circum, North American, eastern North American, North American and east Asian) together with the more temperate elements (deciduous forest region of eastern North America and Appalachian and/or Great Lakes–St Lawrence region). The Arctic–Alpine habitats (1 to 5) are only isolated from the other boreal ones (6 to 18) at the second level of division (II), emphasizing the overall boreal structure of Gros Morne National Park's flora, with alpine inclusions.

The second group of habitats (19 to 35), tied together by temperate elements in contrast to the first boreal group of habitats (1 to 18), is divided (III) in two groups: (a) the four open rocky calcareous, serpentine or coastal habitats (19, 20, 21 and 22), four coastal plain seashore habitats (23, 24, 25 and 27) together with the alpine plateau sedge bogs, and (b) a wide spectrum of habitats ranging from second growth balsam fir and white birch forests and scrubs to aquatic communities, in addition to man-made habitats such as ruderal areas and trails (habitats 28 to 35). The first group (habitats 19 to 27) is characterized by Arctic–Alpine elements (circumpolar, circumpolar with widely disrupted range, North American and amphi–Atlantic), especially well represented in the four open rocky calcareous, serpentine or coastal habitats (19, 20, 21 and 22). The second group (habitats 28 to 35) is characterized by temperate (circum, North American) and cosmopolitan elements. This latter group has climatically protected habitats such as the creek/river beds and shoreline thickets in addition to habitats with proportionately more cosmopolitan species such as the aquatic communities, the tidal flats and the ruderal areas, habitats that are rather similar across the Northern Hemisphere.

Similarly to the Arctic–Alpine habitats (of the alpine plateau (1 to 5)), isolated from the boreal ones at the second level of division, the four open rocky calcareous, serpentine or coastal habitats (19, 20, 21 and 22) are isolated by Arctic–Alpine elements but within the temperate part of the diagram.

4.3. Life-form analysis

The dendrogram, based on a TWINSPAN analysis of the 16 life-form groups (Table 2), presents the life-form structure of the 35 habitats (Fig. 5). The first division (I) splits the 35 habitats in two groups: (−) 15 habitats characterized by mostly open, herbaceous, communities, often hydric, saline or eutrophic, together with the

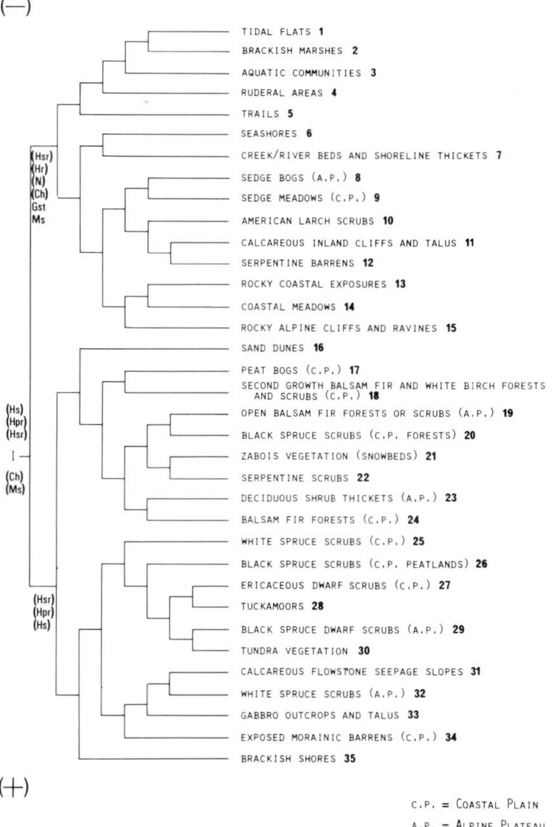

Fig. 5. Dendrogram showing the structure of 35 habitats, based on a TWINSPAN analysis of life-form groups.

Table 2. Life-form spectrum of the total flora.

Life forms		Percent of total species (711)
Mesophanerophytes	Ms	2.1
Microphanerophytes	Mc	2.9
Nanophanerophytes	N	6.2
	phanerophytes subtotal:	11.2
Chamaephytes	Ch	10.9
Protohemicroptophytes without runners	Hp	3.2
Protohemicrophytes with runners	Hpr	6.8
Hemicryptophytes, semi-rosette, without runners	Hs	18.5
Hemicryptophytes, semi-rosette, with runners	Hsr	7.5
Hemicryptophytes, rosette, without runners	Hr	7.1
Hemicryptophytes, rosette, with runners	Hrr	2.3
	hemicryptophytes subtotal:	45.4
Rhizome geophytes	Grh	19.0
Stem-tuber geophytes	Gst	1.3
Root-tuber geophytes	Grt	1.8
Helophytes	Hel	2.1
Hydrophytes	HH	4.2
	cryptophytes subtotal:	28.4
Therophytes	T	4.1

man-made habitats such as the ruderal areas and trails (habitats 1 to 15), and (+) 20 habitats (habitats 16 to 35) regrouping most of the oligotrophic coniferous scrubs and forests, such as the balsam fir forests and the black spruce dwarf scrubs, together with several of the Arctic–Alpine habitats such as zabois or tundra vegetation. The first group (habitats 1 to 15) has for indicator life-form groups the hemicryptophytes, semi-rosette, without runners (Hs), the protochemicryptophytes with runners (Hpr), the hemicryptophytes, semi-rosette, with runners (Hsr), and the therophytes (T). The hemicryptophytes, characteristic of temperate climates (Scoggan 1978–79), are important elements in the formation of the first group (habitats 1 to 15), thus emphasizing the temperate nature of this group. Actually, 13 out of 15 of these habitats (habitats 1 to 15, Fig. 5) are also found in the positive side of the previous dendrogram (habitats 19 to 35, Fig. 4), with a wide range of coastal habitats, tied together by temperate elements. The therophytes are also important in this wide range of coastal, naturally disturbed or man-made habitats, such as the seashores (6) or the ruderal areas (4). On the other hand, the second group (habitats 16 to 35) is characterized by the importance of chamaephytes (Ch), most abundant in regions that have a long winter (Scoggan 1978–79), and the mesophanerophytes (Ms), represented by coniferous trees such as *Abies balsamea* and *Picea mariana*, emphasizing the boreal forest nature of these habitats (16 to 35). Similarly to the first group, 16 out of 20 of these habitats (habitats 16 to 35, Fig. 5) are also found in the negative side of the previous dendrogram (habitats 1 to 18, Fig. 4), where habitats are grouped together by the strong component of boreal elements.

The other divisions do produce, at lower levels, some groups that are similar on both dendrograms (Figs 4 and 5), such as the tidal flats, the brackish marshes and the aquatic communities (1, 2, 3 on Fig. 5; 33, 34, 35 on Fig. 4), or the calcareous inland cliffs, ridges... and the serpentine barrens (11, 12 on Fig. 5; 21, 22 on

Fig. 4). These few groups are therefore similarly structured by either a phytogeographical or a life-form analysis. On the other hand the well-defined Arctic–Alpine group obtained in the phytogeographical analysis (habitats 1 to 5, Fig. 4) is dispersed in the life-form analysis (habitats 15, 21, 30, 33 and 34).

Although the two latter dendrograms, analysing the phytogeographical and the life-form structures, are more similar (0.3152; $P < 0.001$), the phytogeographical one is closer to the habitats dendrogram (0.2140; $P < 0.001$).

5. CONCLUSIONS

Gros Morne National Park has a rich flora of more than 700 species within an area of almost 2000 km^2. Although essentially a boreal flora, it has a high β diversity due to the wide range of habitats ranging from the seashores to the tundra. Therefore, a habitat classification approach based on floristic, biogeographical and life-form data is found to be a useful tool to present and analyse this diversified flora.

The first one, the floristic analysis by indicator species, emphasizes natural or man-made disturbances as the first stratificator while climatic and edaphic variables are more influential at secondary divisions. Nevertheless, the overall classification, based on 711 species distributed in 35 habitats, presents easily interpretable groupings such as zabois vegetation (snowbeds) and rocky alpine cliffs on one hand or brackish marshes and tidal flats on the other hand. The two other dendrograms, based on phytogeographical and life-form groups, although less discriminant at the finer structure level (groupings of two or three habitats), provide essentially a climatic classification for the first division of both classifications. At lower levels of divisions, the phytogeographical groups are more discriminant than the life forms.

The three dendrograms, based on TWINSPAN analyses, provide different but complementary classifications, most useful to understand the flora of Gros Morne National Park, in the Gulf of St. Lawrence.

ACKNOWLEDGMENTS

We are grateful to E. Rouleau, Université de Montréal, for information from unpublished files concerning pre-1970 herbarium collections from the study area. We thank C. Gauvin and I. Saucier, Université de Montréal, for field assistance and data processing. Typing of the manuscript was done by Y. Bourget. Financial assistance came from Parks Canada (research contract GM-83-20) and Natural Sciences and Engineering Research Council of Canada.

REFERENCES

Airphoto Analysis Associates. 1975. Biophysical resource inventory: Gros Morne National Park, Newfoundland. Parks Canada, Atlantic Region, Vol. 1, 83 pp.; Vol. 2, 126 pp.; Vol. 3, 114 pp.

Baird, D. M. 1958. Geology; Sandy Lake (west half) Newfoundland. 1/253,440. Geol. Surv. Can., map 47–1959.

Banfield, C. E. 1983. Climate. Biogeography and Ecology of the Island of Newfoundland (ed. by G. R. South), pp. 37–106. Junk, Boston.

Boivin, B. 1966–67. Enumération des plantes du Canada. Naturaliste Can., 93, 253–274, 371–437, 583–646, 989–1063; 94, 131–157, 471–528, 625–655.

Bouchard, A. 1974. The coastal plain vegetation of the Gros Morne National Park. Parks Canada, Ottawa, contract no. 71–186. mimeogr., 171 pp.

Bouchard, A. 1975. Natural resources analysis of a section of Gros Morne National Park, in Newfoundland, Canada. Contract 74–70, Dept. of Indian Affairs and Northern Development, Parks Canada, Ottawa. 125 pp.

Bouchard, A. and Hay, S. 1974. Addition à la flore de Terre-Neuve: *Lycopodium alpinum* L. Naturaliste Can., 101, 803.

Bouchard, A. and Hay, S. 1976a. *Thelypteris limbosperma* in eastern North America. Rhodora, 78, 552–553.

Bouchard, A. and Hay, S. 1976b. The vascular flora of the Gros Morne National Park coastal plain, in Newfoundland. Rhodora, 78, 207–260.

Bouchard, A., Barabe, D. and Hay, S. 1977. An isolated colony of *Oreopteris limbosperma* (All.) Holub in Gros Morne National Park, Newfoundland, Canada. Naturaliste Can. 104, 239–244.

Bouchard, A., Hay, S. and Rouleau, E. 1978. The vascular flora of St. Barbe South District, Newfoundland; an interpretation based on biophysiographic areas. Rhodora, 80, 228–308.

Bouchard, A., Barabe, D., Bergeron, Y., Dumais, M. and Hay, S. 1985a. La phytogéographie des plantes vasculaires rares du Québec. Naturaliste Can. (Rev. Ecol. Syst.), 112, 283–300.

Bouchard, A., Hay, S., Gauvin, C. and Bergeron, Y. 1985b. The vascular flora of Gros Morne National Park, Newfoundland, Canada. Contract GM83-20 with Parks Canada, Gros Morne National Park, Rocky Harbour, Newfoundland, 104 pp. + appendices.

Bouchard, A., Hay, S., Gauvin, C. and Bergeron, Y. 1986. Rare vascular plants of Gros Morne National Park, Newfoundland, Canada. Rhodora, 88 (856), 481–502.

Bouchard, A., Hay, S., Bergeron, Y. and Leduc, A. 1987. Phytogeographical and life-form analysis of the vascular flora of Gros Morne National Park, Newfoundland Canada. J. Biogeog. 14, 343–358.

Cumming, L. M. 1973. Geology of Gros Morne National Park, western Newfoundland. Parks Canada, Ottawa, mimeogr. 1, 88 pp.; 2, 223 pp.

Damman, A. W. H. 1965. The distribution patterns of northern and southern elements in the flora of Newfoundland. Rhodora, 67, 363–392.

Damman, A. W. H. 1976. Plant distribution in Newfoundland especially in relation to summer temperatures measured with the sucrose inversion method. Can. J. Bot. 54, 1561–1585.

Damman, A. W. H. 1983. An ecological subdivision of the Island of Newfoundland. Biogeography and Ecology of the Island of Newfoundland (ed. by G. R. South), pp. 163–206. Junk, Boston.

Drury, W. H. Jr, 1969. Plant persistence in the Gulf of St. Lawrence. Essays in Plant Geography and Ecology (ed. by K. N. H. Greenidge), pp. 105–148. Nova Scotia Museum, Halifax.

Fernald, M. L. 1911. A botanical expedition to Newfoundland and southern Labrador. Rhodora, 13, 109–162.

Fernald, M. L. 1924. Isolation and endemism in northeastern America and their relation to the Age-and-Area hypothesis. Amer. J. Bot. 11, 558–582.

Fernald, M. L. 1925. The persistence of plants in unglaciated areas of boreal America. Mem. Amer. Acad. Arts. Sci. 15, 239–342.

Fernald, M. L. 1926. Two summers of botanizing in Newfoundland. Rhodora, 28, 49–63, 74–87, 89–111, 115–129, 145–155, 161–178, 181–204, 210–225, 234–241. Reprinted in: Contr. Gray Herb., 76, 49–241.

Fernald, M. L. 1933. Recent discoveries in the Newfoundland flora. Rhodora, 35, 1–16, 47–63, 80–107, 120–140, 161–185, 203–223, 231–247, 265–283, 298–315, 327–346, 364–386, 395–403. Reprinted in: Contr. Gray Herb., 101, 1–403.

Fernald, M. L. 1950. Gray's Manual of Botany (8th edn.), American Book Co., New York. 1632 pp.

Gauch, H. G. Jr. and Whittaker, R. H. 1981. Hierarchical classification of community data. J. Ecol., 69, 537–557.

Grant, D. R. 1977a. Glacial style and ice limits, the Quaternary stratigraphic record, and changes of land and ocean level in the Atlantic provinces, Canada. Géogr. Phys. Quat., 31, (3/4), 247–260.

Grant, D. R. 1977b. Altitudinal weathering zones and glacial limits in western Newfoundland, with particular reference to Gros Morne National Park. Geol. Surv. Can. Paper 77–1A, 455–463.

Hare, F. K. 1952. The climate of the Island of Newfoundland: a geographical analysis. Geogr. Bull, 2, 36–88.

Hill, M. O. 1973. Reciprocal averaging: an eigen vector method for ordination. J. Ecol., 61, 237–249.

Hill, M. O. 1979. TWINSPAN – a FORTRAN program for arranging multivariate data in ordered two way tables by classification of individuals and attributes. Cornell Univ., Ithaca, New York. 90 pp.

Hultén, E. 1958. The Amphi-Atlantic Plants and their Phytogeographical Connections. Almquist & Wiksell, Stockholm. 340 pp.

Hultén, E. 1962. The circumpolar plants. I. Vascular cryptogams, conifers, monocotyledons. Kungl. Svenska Vetenskapsakademiens Handl. IV, 8, 1–275.

Hultén, E. 1968. Flora of Alaska and Neighboring Territories. Stanford Univ. Press, Stanford, Cal. 1008 pp.

Hultén, E. 1971. The Circumpolar Plants. II. Dicotyledons. Almquist and Wiksell, Stockholm. 463 pp.

Lavoie, G. 1984. Contributions à la connaissance de la flore vasculaire et invasculaire de la Moyenne-et-Basse-Côte-Nord, Québec/Labrador. Provancheria (Université Laval, Québec) no. 17, 1–149.

Morisset, P. 1971. Endemism in the vascular plants of the Gulf of St. Lawrence region. Naturaliste Can., 98, 167–177.

Nicholson, J. and Bryant, D. G. 1972. Climatic zones of insular Newfoundland: a principal component analysis. Can. Forestry Serv. Publ. no. 1299. 13 pp.

Noy-Meir, I. and Whittaker, R. H. 1978. Recent developments in continuous multivariate techniques. In: Whittaker, R. H. (ed.). Ordination of Plant Communities, pp. 337–378. Junk, The Hague.

Payette, S. and Lepage, E. 1977. La flore vasculaire du golfe de Richmond, baie d'Hudson, Nouveau-Québec. Provancheria (Université Laval, Québec), no. 7, 1–68.

Podani, J. and Dickinson, T. A. 1984. Comparison of dendrograms: a multivariate approach. Can. J. Bot., 62, 2765–2778.

Rogerson, R. J. 1983. Geological Evolution. Biogeography and ecology of the Island of Newfoundland (ed. by G. R. South), pp. 5–106. Junk, Boston.

Rouleau, E. 1949. *Enumeratio plantarum vascularum Terrae-Novae.* Contrib. Inst. Bot. Univ. Montréal, 64, 61–83.

Rouleau, E. 1956. A checklist of the vascular plants of the province of Newfoundland. Contrib. Inst. Bot. Univ. Montréal, 69, 41–103.

Rouleau, E. 1978. List of the vascular plants of the province of Newfoundland (Canada). Oxen Pond Botanic Park, St. John's, Newfoundland. 132 pp.

Rousseau, C. 1974. Géographie floristique du Québec-Labrador. Distribution des principales espèces vasculaires. Les Presses de l'Université Laval, Québec. 799 pp.

Rowe, J. S. 1972. Forest Regions of Canada. Can. Forestry Serv. Ottawa, Publ. no. 1300. 172 pp.

Scoggan, H. J. 1950. The flora of Bic and the Gaspé Peninsula, Québec. Natl. Mus. Can. Bull., 115. 399 pp.

Scoggan, H. J. 1978–79. The flora of Canada. Natl. Mus. Can. Publ. Bot. no. 7. 4 parts. 1711 pp.

Smith, C. H. 1958. Bay of Islands igneous complex, western Newfoundland. Geol. Surv. Can. Memoir 290.

Waghorne, A. C. 1893. The flora of Newfoundland, Labrador and St-Pierre et Miquelon. Trans. Nova Scotian Inst. Sci., Ser. 2, Vol. 1, 359–373.

Waghorne, A. C. 1895. The flora of Newfoundland, Labrador and St-Pierre et Miquelon. Trans. Nova Scotian Inst. Sci., Ser. 2, Vol. 2, 83–100.

Waghorne, A. C. 1898. The flora of Newfoundland, Labrador and St-Pierre et Miquelon. Trans. Nova Scotian Inst. Sci., Ser. 2, Vol. 2, 361–401.

Watson, W. B. 1974. The climate of Gros Morne National Park, Newfoundland. Atmospheric Environment Serv., Dept. of Environment, Canada.

APPENDIX 1

Vascular flora, alphabetical list with classification by habitats. The nomenclature followed is a compromise between floras covering Newfoundland (Boivin 1966–67; Fernald 1950; Rouleau 1978; Scoggan 1978–79

1, aquatic communities; 2, second growth balsam fir and white birch forests and scrubs (C.P.); 3, balsam fir forests (C.P.); 4, white spruce scrubs (C.P.); 5, deciduous shrub thickets (A.P.); 6, exposed morainic barrens (C.P.); 7, rocky alpine cliffs and ravines; 8, zabois vegetation (snowbeds); 9, open balsam fir forests or scrubs (A.P.); 10,white spruce scrubs (A.P.); 11, tuckamoors; 12, tundra vegetation; 13, gabbro outcrops and talus; 14, ericaceous dwarf scrubs (C.P.); 15, black spruce dwarf scurbs (A.P.);16, sedge bogs (A.P.); 17, peat bogs (C.P.); 18, sedge meadows (C.P.); 19, black spruce scrubs (C.P. peatlands); 20, serpentine scrubs; 21, American larch scrubs; 22, black spruce scrubs (C.P. forests); 23, creek/river beds and shoreline thickets; 24, calcareous inland cliffs, and talus; 25, serpentine barrens; 26, calcareous flowstone seepage slopes; 27, trails; 28, ruderal areas; 29, coastal meadows; 30, seashores; 31, sand dunes; 32, brackish marshes; 33, tidal flats; 34, brackish shores; 35, rocky coastal exposures.

C.P. = Coastal Plain; A.P. = Alpine Plateau.

Flora	Habitats																																		
	1	2	3	4	5	6	7	8	9	10	11	12	13	14	15	16	17	18	19	20	21	22	23	24	25	26	27	28	29	30	31	32	33	34	35
Abies balsamea	–	x	x	x	x	–	–	–	x	x	x	x	x	x	x	x	x	x	x	x	x	x	x	x	x	–	–	–	x	–	x	–	–	–	–
Acer rubrum	–	–	x	–	–	–	–	–	–	–	–	–	–	–	–	–	–	–	–	x	–	x	–	–	–	–	–	–	–	–	–	–	–	–	–
Acer spicatum	–	x	x	–	x	–	–	–	x	–	–	–	–	–	–	–	–	–	–	–	x	x	x	x	–	–	–	–	–	–	–	–	–	–	–
Achillea millefolium var. borealis	–	–	–	–	–	–	–	–	–	–	–	–	–	–	–	–	–	–	–	–	–	–	–	–	–	–	–	–	–	–	–	–	–	–	x
Achillea millefolium	–	–	–	x	–	–	–	–	–	–	–	–	–	–	–	–	–	–	–	–	–	–	–	x	x	–	–	x	x	–	–	x	–	–	x
Achillea ptarmica	–	–	–	–	–	–	–	–	–	–	–	–	–	–	–	–	–	–	–	–	–	–	–	–	–	–	–	x	–	–	–	–	–	–	–
Aconitum bicolor	–	–	–	–	–	–	–	–	–	–	–	–	–	–	–	–	–	–	–	–	–	–	x	–	–	–	–	x	–	–	–	–	–	–	–
Actaea rubra	–	–	–	–	–	–	–	–	x	–	–	–	–	–	–	–	–	–	–	–	–	–	x	x	x	–	–	–	–	–	x	–	–	–	x
Adiantum pedatum var.aleuticum	–	–	–	–	–	–	–	–	–	–	–	–	–	–	–	–	–	–	–	x	–	–	–	x	x	–	–	–	–	–	–	–	–	–	–
Agrimonia striata	–	–	–	–	–	–	–	–	–	–	–	–	–	–	–	–	–	–	–	–	–	–	–	–	–	–	x	x	–	x	x	–	–	–	–
Agropyron repens	–	x	–	–	–	–	–	–	–	–	–	–	–	–	–	–	–	–	–	x	–	–	x	x	–	–	x	x	–	x	–	x	–	–	–
Agropyron trachycaulum	–	–	–	–	–	–	–	–	–	–	–	–	–	–	–	–	–	–	–	–	x	–	–	x	x	–	x	–	x	x	–	–	–	–	x
Agrostis borealis	–	–	x	–	–	–	x	–	–	–	–	–	–	–	–	–	–	–	–	–	–	–	–	x	x	–	x	–	–	–	–	–	–	–	–
Agrostis hyemalis var. geminata	–	–	–	–	–	–	–	–	–	–	–	–	–	–	–	–	–	–	–	–	–	–	–	–	–	–	–	–	–	x	–	–	–	–	x
Agrostis hyemalis var. tenuis	–	–	–	–	–	–	x	–	–	–	–	–	–	–	–	–	–	x	–	–	x	–	x	x	x	–	x	–	–	x	–	–	–	–	–
Agrostis stolonifera	–	–	–	–	–	–	–	–	–	–	–	–	–	–	–	–	–	–	–	–	x	–	x	x	–	–	x	x	x	x	x	x	x	–	x
Agrostis tenuis	–	x	–	–	–	–	–	–	–	x	–	–	–	–	–	–	–	–	–	–	–	–	–	–	–	–	–	–	x	–	–	–	–	–	–
Alchemilla filicaulis	–	–	–	–	–	–	x	–	–	–	–	–	–	–	–	–	–	–	–	–	–	–	x	x	–	–	–	–	–	x	–	–	–	–	x
Allium schoenoprasum	–	–	–	–	–	–	–	–	–	–	–	–	–	–	–	–	–	–	–	–	–	–	–	x	–	–	–	–	–	x	–	–	–	–	–
Alnus crispa	–	x	x	–	x	–	–	x	x	x	x	x	x	–	x	–	–	–	–	x	–	x	x	x	x	–	–	x	–	x	x	–	–	–	–
Alnus rugosa	–	x	x	–	x	–	–	–	–	–	–	–	–	–	–	–	–	x	x	x	x	x	x	–	–	–	–	–	x	–	–	–	–	–	–
Alopecurus aequalis	–	–	–	–	–	–	–	x	–	–	–	–	–	–	–	–	–	–	–	–	–	–	x	–	–	–	–	–	–	–	–	–	x	–	–
Alopecurus pratensis	–	–	–	–	–	–	–	–	–	–	–	–	–	–	–	–	–	–	–	–	–	–	–	–	–	–	–	–	–	–	–	–	–	–	x
Amelanchier bartramiana	–	–	x	–	x	–	–	–	x	–	x	–	–	x	x	–	x	x	x	x	x	x	–	x	–	–	–	x	–	–	–	–	–	–	–
Amelanchier laevis	–	–	–	–	–	–	–	–	–	–	–	–	–	x	–	–	–	–	x	x	x	–	–	–	–	–	–	–	–	–	–	–	–	–	x

Amelanchier spicata	– x x – – – – – – – – – – x –
Ammophila breviligulata	– x x x – – – –
Anaphalis margaritacea	– – – – – – – – – – – – – – – – – – – x – – – x x – – x x – – – – – x
Andromeda glaucophylla	– – – – – – – – x – – – x x x x x x x x x – – x x – – x – – – – – – –
Androsace septentrionalis	– x – – – – – – – – – – –
Anemone canadensis	– x x – x – – – – –
Anemone multifida	– – – – – – – – – – – – x – – – – – – – – – – x – – – – – – – – – – x
Anemone parviflora	– x x x – – – – – – – – x
Angelica atropurpurea	– – – – – – – – – – – – – – – – – x – – x x x x – – – – – – – – – – –
Antennaria bayardii	– x – – – – – – – – – – –
Antennaria brunnescens	– x – – – – – – – – – – –
Antennaria cana	– x – – – – – – – – – x
Antennaria confusa	– x – – – – – – – – – – –
Antennaria eucosma	– x – – – – – – – – – –
Antennaria gaspensis	– x – – – – – – – – – – –
Antennaria neodioica	– x – – – – – – – – – – –
Antennaria rupicola	– – – – – x –
Antennaria spathulata	– x – – – – – – – – – – x
Antennaria straminea	– x – – – – – – – – – – –
Anthoxanthum odoratum	– x – x – – – – – –
Arabis alpina	– x
Arabis drummondii	– x – – – – – – – – – – –
Aralia nudicaulis	– – x – – – – – x – – – – – – – – – – – – x x x – – x – – – – – – – –
Arceuthobium pusillum	– – – – – – – – – – – – – – – – – – – x – – – – – – – – – – – – – – –
Arctostaphylos alpina	– – – – – x x – – – – x x x x – – – – – – – – x x – – – – – – – – – –
Arctostaphylos uva–ursi	– – – – – – – – – – – x x – – – – – – x – – – x x – – – – – – – – – –
Arenaria dawsonensis	– x – – – – – – – – – –
Arenaria humifusa	– x – – – – – – – – – –
Arenaria lateriflora	– x – – – x x – – – – – –
Arenaria marcescens	– x – – – – – – – – – –
Arenaria peploides	– x x x – – – x
Arenaria rubella	– x x – – – – – – – – – x
Arethusa bulbosa	– – – – – – – – – – – – – – – – x x – – – – – – – – – – – – – – – – –
Armeria maritima	– x – – – – – – – – – –
Arnica lonchophylla ssp. chionopappa	– x – – – – – – – – – – –
Arnica tomentosa	– x – – – – – – – – – – –
Aronia melanocarpa	– – – – – – – – – – – – – x – x x x – – – – x – – – – – – – – – – – –
Artemisia borealis	– x – – – – – – – – – – x
Artemisia canadensis	– x x – – – – – – – – – –
Asplenium viride	– x – – – – – – – – – – –
Aster foliaceus	x – – – – – – – – – – – x – – x – – – – x – x x x – – – x – – x x – x

Flora	Habitats 1	2	3	4	5	6	7	8	9	10	11	12	13	14	15	16	17	18	19	20	21	22	23	24	25	26	27	28	29	30	31	32	33	34	35
Aster nemoralis	x	—	—	—	—	—	—	—	—	—	—	—	—	—	—	x	x	x	—	—	—	—	x	—	—	—	—	—	—	—	—	—	—	—	—
Aster novi–belgii	—	—	—	—	—	—	—	—	—	—	—	—	—	—	—	—	—	—	—	—	—	—	—	x	x	—	—	—	—	x	—	—	—	—	—
Aster puniceus	—	x	—	—	—	—	—	x	—	—	—	—	—	—	—	—	—	—	—	x	—	x	x	—	x	—	x	—	x	—	—	—	x	—	—
Aster radula	x	—	—	—	—	—	—	x	x	—	—	x	—	—	x	x	—	x	—	x	x	x	x	—	x	—	—	—	—	—	—	—	x	—	—
Aster umbellatus	—	—	—	—	—	—	—	—	—	—	—	—	—	—	—	—	—	x	—	—	—	—	x	—	x	—	x	x	—	—	—	—	—	—	—
Astragalus alpinus	—	—	—	—	—	—	—	—	—	—	—	—	—	—	—	—	—	—	—	—	—	—	—	—	—	—	—	—	—	—	—	—	—	x	x
Athyrium distentifolium var. americanum	—	—	—	—	—	—	—	x	—	—	—	—	—	—	—	—	—	—	—	—	—	—	—	—	—	—	—	—	—	—	—	—	—	—	—
Athyrium filix–femina	—	x	x	—	—	—	—	—	x	—	—	—	—	—	—	—	—	—	—	x	—	x	x	—	x	—	—	x	x	—	—	—	—	—	—
Atriplex patula	—	—	—	—	—	—	—	—	—	—	—	—	—	—	—	—	—	—	—	—	—	—	—	—	—	—	—	—	—	x	—	—	x	—	x
Avena fatua	—	—	—	—	—	—	—	—	—	—	—	—	—	—	—	—	—	—	—	—	—	—	—	—	—	—	—	x	—	—	—	—	—	—	—
Avena sativa	—	—	—	—	—	—	—	—	—	—	—	—	—	—	—	—	—	—	—	—	—	—	—	—	—	—	—	x	—	—	—	—	—	—	—
Barbarea vulgaris	—	—	—	—	—	—	—	—	—	—	—	—	—	—	—	—	—	—	—	—	—	—	—	—	—	—	—	x	—	—	—	—	—	—	—
Bartonia paniculata	—	—	—	—	—	—	—	—	—	—	—	—	—	—	—	x	—	—	—	—	—	—	—	—	—	—	—	—	—	—	—	—	—	—	—
Bellis perennis	—	—	—	—	—	—	—	—	—	—	—	—	—	—	—	—	—	—	—	—	—	—	—	—	—	—	—	x	x	—	—	—	—	—	—
Betula alleghaniensis	—	—	x	—	—	—	—	—	—	—	—	—	—	—	—	—	—	—	—	—	—	—	—	—	—	—	—	—	—	—	—	—	—	—	—
Betula glandulosa	—	—	—	—	—	—	—	x	x	x	—	x	—	—	—	x	—	—	—	—	—	—	—	—	—	—	—	—	—	—	—	—	—	—	—
Betula michauxii	x	—	—	—	—	—	—	—	—	—	—	—	—	—	—	x	x	x	—	—	—	—	—	—	—	—	—	—	—	—	—	—	—	—	—
Betula minor	—	—	—	—	—	—	x	—	—	—	—	—	—	—	—	—	—	—	—	—	—	—	—	x	—	—	—	—	—	—	—	—	—	—	—
Betula papyrifera	—	x	x	x	x	—	—	—	x	x	x	x	—	x	x	—	x	x	x	x	—	x	x	x	—	—	—	—	—	—	—	—	—	—	—
Betula pumila	—	—	—	—	—	—	—	x	x	—	—	x	x	x	x	x	x	x	x	x	x	x	x	x	x	x	—	—	—	—	—	—	—	—	x
Betula x dutillyi	—	—	—	—	—	—	—	—	—	—	—	x	x	—	—	—	—	—	x	—	—	—	—	x	x	—	—	—	—	—	—	—	—	—	—
Botrychium lunaria	—	—	—	—	—	—	—	—	—	—	—	—	—	—	—	—	—	—	—	—	—	—	—	—	—	—	—	—	x	x	—	—	—	—	x
Botrychium matricariifolium	—	—	—	—	—	—	—	—	—	—	—	—	—	—	—	—	—	—	—	—	—	—	—	x	—	—	—	—	—	—	x	—	—	—	—
Botrychium multifidum	—	—	—	—	—	—	—	—	—	—	—	—	—	—	—	—	—	—	—	—	—	—	x	—	—	—	—	—	x	—	—	—	—	—	x
Botrychium virginianum	—	x	—	—	—	—	—	—	—	—	—	—	—	—	—	—	—	x	—	—	x	x	—	—	—	—	—	—	x	—	—	—	—	—	x
Brachyelytrum erectum	—	—	—	—	—	—	—	—	—	—	—	—	—	—	—	—	—	—	—	x	—	—	x	—	—	—	—	—	—	x	—	—	—	—	—
Bromus ciliatus	—	—	—	—	—	—	—	—	—	—	—	—	—	—	—	—	—	—	—	—	x	—	x	—	—	—	x	—	—	—	—	—	—	x	—
Cakile edentula	—	—	—	—	—	—	—	—	—	—	—	—	—	—	—	—	—	—	—	—	—	—	—	—	—	—	—	x	—	x	x	—	—	—	—
Calamagrostis canadensis	x	—	—	—	x	—	—	x	x	—	—	x	—	x	—	x	x	x	—	x	—	x	x	x	—	—	—	x	x	—	—	x	—	—	—
Calamagrostis inexpansa	—	—	—	—	—	—	—	—	—	x	—	—	x	—	—	x	—	x	—	—	—	—	—	x	x	—	—	—	—	x	—	—	—	—	x
Calamagrostis neglecta	—	—	—	—	—	—	—	—	—	—	—	—	—	—	—	—	—	—	—	—	—	—	—	x	x	—	—	—	x	—	—	—	—	—	—
Calamagrostis pickeringii	—	—	—	—	—	—	—	x	—	x	—	—	—	—	x	x	—	—	—	x	—	—	—	—	x	—	—	—	—	—	—	—	—	—	—
Callitriche anceps	x	—	—	—	—	—	—	—	—	—	—	—	—	—	—	—	—	—	—	—	—	—	x	—	—	—	—	—	—	—	—	—	—	—	—
Callitriche hermaphroditica	—	—	—	—	—	—	—	—	—	—	—	—	—	—	—	—	—	—	—	—	—	—	x	—	—	—	—	—	—	—	—	x	—	—	—
Callitriche verna	x	—	—	—	—	—	—	—	—	—	—	—	—	—	—	—	—	—	—	—	—	—	x	—	—	—	—	—	—	—	—	—	—	—	—
Calopogon tuberosus	—	—	—	—	—	—	—	—	—	—	—	—	—	—	—	x	x	—	—	—	—	—	—	—	—	—	—	—	—	—	—	—	—	—	—
Caltha palustris	—	—	—	—	—	—	—	—	—	—	—	—	—	—	—	—	—	—	x	—	x	x	x	—	—	—	—	—	x	—	—	x	—	—	—

Campanula rotundifolia	- - - - - - - - - - - - x - - x - x - x - - - x x x - - x - - - - - x
Capsella bursa-pastoris	- x - - - - - - -
Cardamine pensylvanica	x - - - - - x - x x - - - - - -
Carex aquatilis	x - - - - - - - - - - - - - - - x x - - x - - - - - - - - - - - - - -
Carex arctata	- x - - - - - - - - - - - - - - - - - x - - - x - - x - - - - - - - -
Carex atratiformis	- x - - - - x x - - - - x
Carex aurea	- x - x - - - x x x - - - - x
Carex bebbii	- - - - - - - - - - - - - - - - - x - - - - - - - - - - - - - - - - -
Carex bigelowii	- - - - - x x x - - - x - - x x - - - - - - - - x - - - - - - - - - -
Carex brunnescens	- - - - x - x x x x - - - - - - - - - - - - x - - - x - x - - - - - -
Carex buxbaumii	- - - - - - - - - - - - - - - - - x - - - - x - - - - - x - - - - - -
Carex canescens	x - - - x - - x x - - - - - - - - - - - x - x - - - x - x - - - x - -
Carex capillaris	- x - - x - x x - - - - x
Carex castanea	- x -
Carex chordorrhiza	- - - - - - - - - - - - - - - - - x - - - - - - - - - - - - - x - - -
Carex concinna	- - - - x - x
Carex crawfordii	- x - x - - x - - - - -
Carex debilis	- x - x - - - - - -
Carex deflexa	- - - - - - x - x - - - - -
Carex deweyana	- x - x - - - x - - x - - - -
Carex diandra	- x - - - - - - - - - - - - - - x - - - x - x - - - x - - - - - - - -
Carex disperma	- x x x - - - - - - - - - - - -
Carex eburnea	- x x - - - - - - - - - -
Carex exilis	- - - - - - - - - - - - - - - x x x - - x - x - x - - - - - - - - - -
Carex flava	x - - - - - - - - - - - - - - - - x - - x x - - - - x - - - - x - - -
Carex glacialis	- x - - - - - - - - - - x
Carex gracillima	- x - x - - - - - - - -
Carex gynocrates	- - - - - - - - - - - - - x - - x x - - x - - - - - - - - - - - - - -
Carex hormathodes	- x - - - - - - - - x - - -
Carex hostiana	- x - - - - -
Carex interior	- - - - - - - - - - - - - - - - - x - - x x x - - - x - x - - x x - -
Carex intumescens	- - - - - - - - - - - - - - - - - - - x - - x - - - x - - - - - - - -
Carex lasiocarpa	x - - - - - - - - - - - - - - - x x - - x x x - - - x - - - - x - - -
Carex lenticularis	- x - - - - - - x - - - - x
Carex lepidocarpa	- - - - - - - - - - - - - - - - x x - - x - - - - - x - - x - - - - -
Carex leporina	- x - - - - - - - - - - - -
Carex leptalea	- - - - - - - - - - - - - - - - - x - - x x x - - - - - - - - - - - -
Carex leptonervia	- x - - - x - x - - - - - -
Carex limosa	x - - - - - - x - - - - - - - x x x x - x - - - - - - - x - - - x - -
Carex livida	x - - - - - - - - - - - - - - - x x - - x x x - - - - - - - - - - - -
Carex mackenziei	- x x x -

Flora	Habitats 1	2	3	4	5	6	7	8	9	10	11	12	13	14	15	16	17	18	19	20	21	22	23	24	25	26	27	28	29	30	31	32	33	34	35
Carex maritima	—	—	—	—	—	—	—	—	—	—	—	—	—	—	—	—	—	—	—	—	—	—	—	—	—	—	—	—	x	—	x	—	—	—	x
Carex michauxiana	x	—	—	—	—	—	—	—	—	—	—	—	—	—	—	x	—	—	—	—	—	—	x	—	—	—	—	—	—	—	—	—	—	—	—
Carex muricata var. angustata	x	—	—	—	—	—	—	x	—	—	—	—	—	—	—	x	—	x	—	—	—	—	x	—	—	—	x	—	—	—	—	—	—	—	—
Carex muricata var. cephalantha	x	—	—	—	—	—	—	x	—	—	—	—	—	—	—	—	—	x	—	—	—	—	—	—	—	—	x	—	—	—	—	—	—	—	—
Carex muricata	—	—	—	—	—	—	—	—	x	—	—	—	—	—	—	—	—	—	—	—	—	—	—	—	x	—	—	—	—	—	—	—	—	—	—
Carex nigra	—	—	—	—	—	—	—	—	—	—	—	—	—	—	—	—	—	x	—	—	—	—	—	—	—	—	x	x	x	—	—	x	—	—	x
Carex oligosperma	—	—	—	—	—	—	—	x	x	—	—	—	—	—	—	x	—	x	—	—	—	—	—	—	x	—	x	—	—	—	—	—	—	—	—
Carex paleacea	—	—	—	—	—	—	—	—	—	—	—	—	—	—	—	—	—	—	—	—	—	—	x	—	—	—	x	—	—	—	—	x	x	—	—
Carex pallescens	—	—	—	—	—	—	—	—	—	—	—	—	—	—	—	—	—	—	—	—	—	—	—	—	—	—	x	x	—	—	—	—	—	—	—
Carex pauciflora	—	—	—	—	—	—	—	—	x	—	—	—	—	x	x	x	x	x	x	—	—	—	—	—	—	—	—	—	—	—	—	—	—	—	—
Carex paupercula	—	—	x	—	—	—	—	x	x	x	—	—	—	—	—	x	—	x	x	—	x	x	x	—	—	—	—	—	—	—	—	—	—	—	—
Carex pedunculata	—	x	—	—	—	—	—	—	—	—	—	—	—	—	—	—	—	—	—	—	—	—	—	—	—	—	—	—	—	—	—	—	—	—	x
Carex petricosa var. misandroides	—	—	—	—	—	—	—	—	—	—	—	—	—	—	—	—	—	—	—	—	—	—	—	x	—	—	—	—	—	—	—	—	—	—	—
Carex projecta	—	x	—	—	—	—	—	—	—	—	—	—	—	—	—	—	—	—	—	—	—	—	—	—	—	—	x	—	—	—	—	—	—	—	—
Carex rariflora	—	—	—	—	—	—	—	—	—	—	—	—	—	—	—	—	x	x	—	—	—	—	—	—	—	—	—	—	—	—	x	—	—	—	—
Carex rostrata	x	—	—	—	—	—	—	—	—	—	—	—	—	—	—	x	x	x	—	—	x	x	x	x	—	—	x	—	—	—	—	—	—	—	—
Carex rupestris	—	—	—	—	—	—	—	—	—	—	—	—	—	—	—	—	—	—	—	—	—	—	—	x	—	—	—	—	—	—	—	—	—	—	—
Carex salina	—	—	—	—	—	—	—	—	—	—	—	—	—	—	—	—	—	—	—	—	—	—	x	—	—	—	—	—	x	x	—	x	x	—	—
Carex saxatilis var. miliaris	—	—	—	—	—	—	—	x	—	—	—	—	—	—	—	x	—	—	—	—	—	—	—	—	x	—	x	—	—	x	x	—	—	—	—
Carex scirpoidea	—	—	—	—	—	—	x	—	—	—	—	x	x	—	—	x	—	—	—	x	—	—	—	x	x	x	—	—	—	—	—	—	—	—	x
Carex scoparia	—	x	—	—	—	—	—	—	—	—	—	—	—	—	—	—	—	—	—	—	—	—	—	—	—	—	x	—	x	—	—	—	x	—	—
Carex sterilis	—	—	—	—	—	—	—	—	—	—	—	—	—	—	—	—	—	—	—	—	x	—	—	x	—	—	—	—	—	—	—	—	—	—	—
Carex stipata	—	x	—	—	—	—	—	—	—	—	—	—	—	—	—	—	—	x	—	—	—	x	x	—	—	—	x	—	x	—	—	—	—	—	—
Carex stylosa	—	—	—	—	—	—	x	x	x	x	x	x	—	—	—	x	—	—	—	—	—	—	—	—	—	—	—	—	—	—	—	—	—	—	—
Carex tenuiflora	—	—	—	—	—	—	—	—	—	—	—	—	—	—	—	—	—	x	—	—	x	—	—	—	—	—	—	—	—	—	—	—	—	—	—
Carex trisperma	—	—	x	x	x	—	—	—	x	—	—	—	—	x	x	—	—	—	x	x	x	x	x	—	—	—	x	x	—	—	—	—	—	—	—
Carex vaginata	—	—	—	—	—	—	—	—	—	—	—	—	—	—	—	—	—	—	—	—	x	x	—	x	—	—	—	—	—	—	—	—	—	—	—
Carex vesicaria	—	—	—	—	—	—	—	—	—	—	—	—	—	—	—	x	—	x	—	—	—	—	—	—	—	—	x	—	—	—	—	—	—	—	—
Carex viridula	—	—	—	—	—	—	—	—	—	—	—	—	—	—	—	—	x	—	—	—	—	—	x	—	x	—	x	x	—	x	x	x	—	—	x
Carex wiegandii	—	x	—	—	—	—	—	—	—	—	—	—	—	—	—	—	—	—	—	—	—	x	—	—	—	—	—	—	—	—	—	—	—	—	—
Carex x pieperiana	—	—	—	—	—	—	—	—	—	—	—	—	—	—	—	—	—	—	—	—	x	—	—	—	—	—	x	—	—	—	—	—	—	—	—
Carex x subviridula	—	x	—	—	—	—	—	—	—	—	—	—	—	—	—	—	—	—	—	—	—	—	—	—	—	—	—	—	—	—	—	—	—	—	—
Carex x xanthina	—	—	—	—	—	—	—	—	—	—	—	—	—	—	—	—	—	—	—	—	—	—	—	—	—	—	—	—	—	x	—	—	—	—	—
Carum carvi	—	—	—	—	—	—	—	—	—	—	—	—	—	—	—	—	—	—	—	—	—	—	—	—	—	—	—	x	—	—	—	—	—	—	—
Cassiope hypnoides	—	—	—	—	—	—	x	—	—	—	—	x	x	—	—	—	—	—	—	—	—	—	—	—	—	—	—	—	—	—	—	—	—	—	—
Castilleja pallida var. septentrionalis	—	—	x	x	—	—	—	—	—	—	—	—	—	—	—	—	—	—	—	—	—	—	—	x	x	—	—	—	x	—	x	—	—	—	x
Catabrosa aquatica	—	—	—	—	—	—	—	—	—	—	—	—	—	—	—	—	—	—	—	—	—	—	—	—	—	—	—	—	x	—	x	—	—	—	x

Centaurea nigra	-	-	-	-	-	-	-	-	-	-	-	-	-	-	-	-	-	-	-	-	-	-	-	-	-	-	x	x	-	-	-	-	-	-	-
Cerastium arvense	-	-	-	-	-	-	-	-	-	-	-	-	-	-	-	-	-	-	-	-	-	-	-	-	x	-	-	-	x	-	-	-	-	-	x
Cerastium beeringianum	-	-	-	-	-	-	-	-	-	-	-	-	-	-	-	-	-	-	-	-	-	-	-	-	x	-	-	-	-	-	-	-	-	-	-
Cerastium vulgatum	-	-	-	-	-	-	-	-	-	-	-	-	-	-	-	-	-	-	-	-	-	-	-	x	-	-	x	x	x	-	-	-	-	-	x
Chamaedaphne calyculata	-	-	-	-	-	-	-	-	-	-	-	-	-	x	x	x	x	x	x	x	x	x	x	-	-	-	-	-	-	-	-	-	-	-	-
Chenopodium album	-	-	-	-	-	-	-	-	-	-	-	-	-	-	-	-	-	-	-	-	-	-	-	-	-	-	-	x	-	-	-	-	-	-	-
Chrysanthemum leucanthemum	-	-	-	-	-	-	-	-	-	-	-	-	-	-	-	-	-	-	-	-	-	-	-	-	-	-	-	x	-	-	-	x	-	-	-
Cichorium intybus	-	-	-	-	-	-	-	-	-	-	-	-	-	-	-	-	-	-	-	-	-	-	-	-	-	-	-	x	-	-	-	-	-	-	-
Cinna latifolia	-	-	-	-	-	-	-	-	-	-	-	-	-	-	-	-	-	x	-	x	-	-	x	-	x	-	-	-	-	-	-	-	-	-	x
Circaea alpina	-	x	x	x	-	-	-	-	-	-	-	-	-	-	-	-	-	-	-	-	-	-	x	-	-	-	x	-	-	x	x	-	-	-	-
Cirsium arvense	-	-	-	-	-	-	-	-	-	-	-	-	-	-	-	-	-	-	-	-	-	x	x	-	-	-	-	x	x	x	-	-	-	-	-
Cirsium muticum	-	-	-	-	-	-	-	-	-	-	-	-	-	-	-	-	-	-	-	x	-	x	x	x	x	-	x	x	x	-	-	-	-	-	-
Cirsium vulgare	-	-	-	-	-	-	-	-	-	-	-	-	-	-	-	-	-	-	-	-	-	-	-	-	-	-	-	x	-	-	-	-	-	-	-
Clintonia borealis	-	x	x	-	x	-	-	x	x	x	x	x	-	-	-	-	-	-	-	x	-	x	x	x	-	-	x	-	-	-	-	-	-	-	-
Cochlearia cyclocarpa	-	-	-	-	-	-	-	-	-	-	-	-	-	-	-	-	-	-	-	-	-	-	-	-	-	-	-	x	-	x	-	x	-	-	x
Comandra umbellata var. umbellata	-	-	-	-	-	-	-	-	-	-	-	-	-	-	-	-	x	-	-	-	-	-	-	-	-	-	-	-	-	-	-	-	-	-	-
Conioselinum chinense	-	-	-	-	-	-	-	x	-	-	-	-	-	-	-	-	-	-	-	x	x	x	x	x	-	-	-	-	x	x	-	-	-	-	x
Convolvulus sepium	-	-	-	-	-	-	-	-	-	-	-	-	-	-	-	-	-	-	-	-	-	-	-	-	-	-	-	-	x	x	-	-	-	-	-
Coptis trifolia ssp. groenlandica	-	-	x	-	x	-	x	x	x	x	x	x	-	x	-	x	x	x	x	x	x	x	-	-	-	-	-	-	-	-	-	-	-	-	-
Corallorhiza maculata	-	x	x	-	-	-	-	-	-	-	-	-	-	-	-	-	-	-	-	-	-	-	-	-	-	-	-	-	-	-	-	-	-	-	-
Corallorhiza trifida	-	x	-	-	-	-	-	-	-	-	-	-	-	-	-	-	-	-	-	-	-	-	-	x	-	-	-	-	-	-	-	-	-	-	-
Cornus canadensis	-	x	x	x	x	-	x	x	x	x	x	x	x	x	x	x	x	-	x	x	x	x	x	x	-	-	-	-	x	-	-	-	-	-	-
Cornus stolonifera	-	x	-	-	-	-	-	-	-	-	-	-	-	-	-	-	-	-	-	-	x	x	x	x	-	-	-	x	x	x	-	-	-	-	-
Cornus suecica	-	-	-	x	-	-	x	x	x	-	-	-	-	-	-	-	-	-	-	-	-	x	-	-	-	-	-	-	-	x	-	-	-	-	-
Cornus x unalaschkensis	-	-	-	-	-	-	-	-	-	-	-	-	-	-	-	-	-	-	-	-	-	-	-	-	-	-	-	-	x	-	-	-	-	-	-
Corylus cornuta	-	x	-	-	-	-	-	-	-	-	-	-	-	-	-	-	-	-	-	-	-	-	x	-	-	-	-	-	-	x	-	-	-	-	x
Cryptogramma stelleri	-	-	-	-	-	-	-	-	-	-	-	-	-	-	-	-	-	-	-	-	-	-	-	x	-	-	-	-	-	-	-	-	-	-	x
Cypripedium acaule	-	-	-	-	-	-	-	-	-	-	-	-	-	x	-	-	-	-	-	-	-	x	x	x	-	-	-	-	-	-	-	-	-	-	x
Cypripedium calceolus	-	-	-	-	-	-	-	-	-	-	-	-	-	-	-	-	-	-	-	-	-	x	-	x	x	x	-	-	-	-	-	-	-	-	-
Cypripedium reginae	-	-	-	-	-	-	-	-	-	-	-	-	-	-	-	-	-	-	-	-	x	-	-	-	-	-	-	-	-	-	-	-	-	-	-
Cystopteris bulbifera	-	-	-	-	-	-	-	-	-	-	-	-	-	-	-	-	-	-	-	-	-	-	-	x	-	-	-	-	-	-	-	-	-	-	-
Cystopteris fragilis	-	-	-	-	-	-	-	-	-	-	-	-	-	-	-	-	-	-	-	-	-	-	-	x	-	-	-	-	-	-	-	-	-	-	x
Dactylis glomerata	-	-	-	-	-	-	-	-	-	-	-	-	-	-	-	-	-	-	-	-	-	-	-	-	-	-	-	x	-	-	-	-	-	-	-
Danthonia intermedia	-	-	-	-	-	-	-	-	-	-	-	-	x	-	-	-	-	-	-	-	-	-	-	-	x	-	-	-	-	-	-	-	-	-	-
Danthonia spicata	-	-	-	-	-	-	-	-	-	-	-	-	x	-	-	-	-	-	-	x	-	-	x	x	x	-	-	x	x	-	-	-	-	-	-
Deschampsia cespitosa	-	-	-	-	-	-	-	-	-	-	-	-	-	-	-	x	-	-	-	-	-	-	-	-	x	-	-	-	-	-	-	-	-	-	-
Deschampsia flexuosa	-	x	-	x	x	-	x	x	x	x	x	x	x	x	x	-	-	x	-	x	x	-	-	x	x	-	x	-	x	-	-	-	-	-	x
Diapensia lapponica	-	-	-	-	-	x	x	x	-	-	-	x	x	-	-	-	-	-	-	-	-	-	-	-	-	-	-	-	-	-	-	-	-	-	-
Digitalis purpurea	-	-	-	-	-	-	-	-	-	-	-	-	-	-	-	-	-	-	-	-	-	-	-	-	-	-	-	x	-	-	-	-	-	-	-
Draba arabisans	-	-	-	-	-	-	-	-	-	-	-	-	-	-	-	-	-	-	-	-	-	-	-	x	-	-	-	-	-	-	-	-	-	-	x
Draba glabella	-	-	-	-	-	-	-	-	-	-	-	-	-	-	-	-	-	-	-	-	-	-	-	x	-	-	-	-	-	-	-	-	-	-	x

Flora	Habitats																																		
	1	2	3	4	5	6	7	8	9	10	11	12	13	14	15	16	17	18	19	20	21	22	23	24	25	26	27	28	29	30	31	32	33	34	35
Draba incana	–	–	–	–	–	–	–	–	–	–	–	–	–	–	–	–	–	–	–	–	–	–	–	–	–	–	–	–	x	–	x	–	–	–	x
Draba lactea	–	–	–	–	–	–	–	–	–	–	–	–	–	–	–	–	–	–	–	–	–	–	–	–	x	–	–	–	–	–	–	–	–	–	–
Drosera anglica	x	–	–	–	–	–	–	–	–	–	–	–	–	–	–	x	x	x	–	–	–	–	–	–	–	–	–	–	–	–	–	–	–	–	–
Drosera intermedia	x	–	–	–	–	–	–	–	–	–	–	–	–	–	–	x	x	x	–	–	–	–	–	–	–	–	–	–	–	–	–	–	–	–	–
Drosera linearis	–	–	–	–	–	–	–	–	–	–	–	–	–	–	–	–	x	x	–	–	–	–	–	–	–	–	–	–	–	–	–	–	–	–	–
Drosera rotundifolia	–	–	–	–	–	–	–	–	x	–	–	–	–	x	–	x	x	x	x	–	x	x	x	–	x	–	–	–	–	–	–	x	–	–	–
Drosera x obovata	–	–	–	–	–	–	–	–	–	–	–	–	–	–	–	x	–	–	–	–	–	–	–	–	–	–	–	–	–	–	–	–	–	–	–
Dryas integrifolia	–	–	–	–	–	–	–	–	–	–	–	–	–	–	–	–	–	–	–	–	–	–	–	x	–	x	–	–	–	–	–	–	–	–	–
Dryopteris cristata	–	–	–	–	–	–	–	–	–	–	–	–	–	–	–	–	–	–	–	–	x	–	x	–	–	–	x	–	–	–	–	–	–	–	–
Dryopteris filix-mas	–	–	–	–	–	–	–	–	–	–	–	–	–	–	–	–	–	–	–	–	–	–	–	x	–	–	x	–	–	–	–	–	–	–	–
Dryopteris fragrans	–	–	–	–	–	–	–	–	–	–	–	–	x	–	–	–	–	–	–	–	–	–	–	–	–	–	–	–	–	–	–	–	–	–	–
Dryopteris spinulosa	–	x	x	x	x	–	–	x	x	x	x	x	–	–	–	–	–	–	–	–	–	x	x	x	–	–	–	–	–	–	–	–	–	–	–
Dryopteris x boottii	–	–	–	–	–	–	–	–	–	–	–	–	–	–	–	–	–	–	–	–	–	–	–	–	–	–	x	–	–	–	–	–	–	–	–
Echium vulgare	–	–	–	–	–	–	–	–	–	–	–	–	–	–	–	–	–	–	–	–	–	–	–	–	–	–	–	x	–	–	–	–	–	–	–
Eleocharis acicularis	–	–	–	–	–	–	–	–	–	–	–	–	–	–	–	–	–	–	–	–	–	–	–	–	–	–	–	–	–	–	–	–	x	–	–
Eleocharis halophila	–	–	–	–	–	–	–	–	–	–	–	–	–	–	–	–	–	–	–	–	–	–	x	–	–	–	–	–	–	–	–	x	x	–	–
Eleocharis nitida	–	–	–	–	–	–	–	–	–	–	–	–	–	–	–	x	–	–	–	–	–	–	–	–	x	–	–	–	–	–	–	–	–	–	–
Eleocharis palustris	x	–	–	–	–	–	–	–	–	–	–	–	–	–	–	–	–	–	–	–	–	–	x	–	–	–	–	–	–	–	–	–	–	–	–
Eleocharis parvula	–	–	–	–	–	–	–	–	–	–	–	–	–	–	–	–	–	–	–	–	–	–	–	–	–	–	–	–	–	–	–	–	x	–	–
Eleocharis quinqueflora	–	–	–	–	–	–	–	–	–	–	–	–	–	–	–	x	–	–	–	–	–	–	–	–	–	–	–	–	–	–	–	–	–	–	–
Eleocharis smallii	x	x	–	–	–	–	–	–	–	–	–	–	–	–	–	–	–	–	–	–	–	–	x	–	–	–	x	–	x	–	–	–	x	–	–
Eleocharis tenuis var. borealis	–	–	–	–	–	–	–	–	–	–	–	–	–	–	–	–	–	–	–	–	–	–	–	–	–	–	–	–	–	x	–	–	–	–	–
Elymus arenarius	–	–	–	–	–	–	–	–	–	–	–	–	–	–	–	–	–	–	–	–	–	–	–	–	–	–	–	–	–	x	x	–	–	–	x
Elymus virginicus	–	–	–	–	–	–	–	–	–	–	–	–	–	–	–	–	–	–	–	–	–	–	–	–	–	–	–	–	–	–	–	–	–	x	–
Empetrum nigrum var. eamesii	–	–	–	–	–	x	x	–	–	–	–	x	x	–	–	x	–	–	–	–	–	–	–	x	–	–	–	–	–	–	–	–	–	–	–
Empetrum nigrum	–	–	–	–	–	x	x	x	x	x	x	x	x	x	x	x	x	x	x	x	–	x	–	x	x	–	–	–	x	–	–	–	–	–	x
Epigaea repens	–	–	x	–	–	–	–	–	–	–	x	–	–	x	–	–	–	–	–	x	–	x	–	x	–	–	–	–	–	–	–	–	–	–	–
Epilobium alpinum	x	–	–	–	–	–	x	–	–	–	–	–	–	–	–	–	–	–	–	–	–	–	–	x	–	–	–	–	–	x	–	–	–	–	–
Epilobium anagallidifolium	–	–	–	–	–	–	–	x	–	–	–	–	–	–	–	–	–	–	–	–	–	–	–	–	–	–	–	–	–	–	–	–	–	–	–
Epilobium angustifolium	–	–	–	–	–	–	–	–	–	–	–	–	–	–	–	–	–	–	–	–	–	x	–	x	–	–	–	x	x	x	–	–	–	–	–
Epilobium ciliatum	–	x	–	–	–	–	–	–	–	–	–	–	–	–	–	–	–	–	–	–	–	–	–	–	–	–	–	–	–	–	–	–	–	–	–
Epilobium glandulosum	x	x	–	–	–	–	–	–	–	–	–	–	–	–	–	–	x	x	–	–	x	–	x	–	–	–	x	x	–	–	–	x	–	–	x
Epilobium hornemannii	x	–	–	–	x	–	x	x	–	–	–	–	–	–	–	–	–	–	–	–	–	–	x	–	–	–	–	–	–	–	–	–	–	–	x
Epilobium latifolium	–	–	–	–	–	–	–	–	–	–	–	–	–	–	–	–	–	–	–	–	–	–	x	–	–	–	–	–	–	x	–	–	–	–	–
Epilobium palustre	–	–	–	–	–	–	x	–	–	–	–	–	–	–	–	–	x	x	–	–	x	–	x	–	–	–	–	x	x	–	–	–	x	–	x
Equisetum arvense	–	x	x	–	–	–	–	–	–	–	–	–	–	–	–	–	–	–	x	–	x	x	x	–	–	–	–	x	x	x	x	–	–	–	–
Equisetum fluviatile	x	–	–	–	–	–	–	–	–	–	–	–	–	–	–	–	x	x	x	–	x	–	x	x	–	–	–	x	–	–	–	–	–	–	–
Equisetum palustre	x	–	–	–	–	–	–	–	–	–	–	–	–	–	–	–	x	x	–	–	x	x	x	x	–	–	x	–	–	x	–	x	–	–	–

Equisetum scirpoides	-	-	-	-	-	-	-	-	-	-	-	-	-	-	-	-	-	-	-	-	-	-	-	-	-	-	-	-	x	x	-	-	-	-	-
Equisetum sylvaticum	-	x	x	-	-	-	-	x	x	-	-	-	-	-	-	-	-	-	-	-	-	x	x	-	-	-	x	-	-	-	-	-	-	-	-
Equisetum variegatum	-	-	-	-	-	-	-	-	-	-	-	-	-	-	-	-	-	x	-	-	x	x	-	-	-	-	-	-	-	-	-	-	-	-	-
Erigeron hyssopifolius	-	-	-	-	-	-	-	-	-	-	-	-	-	-	-	-	-	-	-	-	-	-	-	x	x	x	-	-	-	-	-	-	-	-	-
Erigeron philadelphicus	-	-	-	-	-	-	-	-	-	-	-	-	-	-	-	-	-	-	-	-	-	-	-	-	-	-	x	-	-	-	-	-	-	-	-
Erigeron strigosus	-	-	-	-	-	-	-	-	-	-	-	-	-	-	-	-	-	-	-	-	-	-	-	-	-	-	x	-	-	-	-	-	-	-	-
Eriocaulon septangulare	x	-	-	-	-	-	-	-	-	-	-	-	-	-	-	-	-	x	-	-	-	-	-	-	-	-	-	-	-	-	-	-	-	-	-
Eriophorum angustifolium	-	-	-	-	-	-	-	-	-	-	-	-	-	x	-	x	x	x	x	x	x	-	x	-	x	-	-	-	-	-	-	-	-	-	-
Eriophorum chamissonis	x	-	-	-	-	-	-	-	-	-	-	-	-	-	-	x	x	-	-	-	-	-	-	-	x	-	-	-	-	-	-	-	-	-	-
Eriophorum gracile	-	-	-	-	-	-	-	-	-	-	-	-	-	-	-	-	x	-	-	-	-	-	-	-	-	-	-	-	-	-	-	-	-	-	-
Eriophorum tenellum	x	-	-	-	-	-	-	-	-	-	-	-	-	-	-	-	x	-	-	-	-	-	-	-	-	-	-	-	-	-	-	-	-	-	-
Eriophorum vaginatum ssp. spissum	-	-	-	-	-	-	-	x	-	-	-	x	-	x	x	x	x	x	x	-	-	x	-	-	-	-	-	-	-	-	-	-	-	-	-
Eriophorum virginicum	x	-	-	-	-	-	-	-	-	-	-	-	-	-	-	x	x	x	-	-	-	-	-	-	-	-	x	-	-	-	-	-	-	-	-
Eriophorum viridicarinatum	-	-	-	-	-	-	-	-	-	-	-	-	-	-	-	x	-	x	-	-	-	x	x	-	-	-	x	-	-	-	-	-	-	-	-
Eriophorum x pylaieanum	-	-	-	-	-	-	-	-	-	-	-	-	-	-	-	-	x	-	x	-	-	-	-	-	-	-	-	-	-	-	-	-	-	-	-
Eupatorium purpureum var. maculatum	-	-	-	-	-	-	-	-	-	-	-	-	-	-	-	-	-	-	-	-	-	x	x	x	-	-	-	-	-	-	-	x	-	-	-
Euphrasia americana	-	-	-	-	-	-	-	-	-	-	-	-	-	-	-	-	-	-	-	-	-	-	-	-	-	-	-	x	x	-	x	x	-	-	-
Euphrasia arctica	-	-	-	-	-	-	-	-	-	-	-	-	-	-	-	x	-	-	-	-	-	-	x	x	x	-	-	-	x	-	-	-	-	-	x
Euphrasia oakesii	-	-	-	-	-	-	x	-	-	-	-	-	-	-	-	-	x	-	-	-	-	-	x	x	-	-	-	-	x	x	x	x	-	-	x
Euphrasia rigidula	-	-	-	-	-	-	-	-	-	-	-	-	-	-	-	-	-	-	-	-	-	-	-	-	-	-	-	x	-	-	-	-	-	-	-
Festuca altaica	-	-	-	-	-	-	-	-	-	-	-	-	-	-	-	-	-	-	-	-	-	-	-	x	x	-	-	-	-	-	-	-	-	-	-
Festuca brachyphylla	-	-	-	-	-	-	-	-	-	-	-	-	-	-	-	-	-	-	-	-	-	-	-	x	-	-	-	-	-	-	-	-	-	-	-
Festuca elatior	-	x	-	-	-	-	-	-	-	-	-	-	-	-	-	-	-	-	-	-	-	-	-	-	-	-	-	-	-	-	-	-	-	-	x
Festuca ovina var. vivipara	-	-	-	-	-	-	-	-	-	-	-	-	-	-	-	-	-	-	-	-	-	-	-	x	-	-	-	-	-	-	-	-	-	-	-
Festuca ovina	-	-	-	-	-	-	-	-	-	-	-	-	-	-	-	-	-	-	-	-	-	-	-	x	-	-	-	-	-	-	-	-	-	-	-
Festuca rubra var. prolifera	-	-	-	-	-	-	-	-	-	-	-	-	-	-	-	-	-	-	-	-	-	-	-	x	-	-	-	-	-	-	-	-	-	-	-
Festuca rubra	-	-	-	-	-	-	-	-	-	-	-	-	x	-	-	x	-	-	-	x	x	-	-	x	x	-	-	x	x	x	x	x	x	-	x
Festuca saximontana	-	-	-	-	-	-	-	-	-	-	-	-	-	-	-	-	-	-	-	-	-	-	-	x	-	-	-	-	-	-	-	-	-	-	-
Fragaria virginiana	-	-	-	-	-	-	-	-	-	-	-	-	-	-	-	-	-	-	-	-	-	-	-	x	-	-	x	x	x	x	-	-	-	-	-
Fraxinus nigra	-	-	-	-	-	-	-	-	-	-	-	-	-	-	-	-	-	-	-	-	-	-	x	x	-	-	x	-	-	-	-	-	-	-	-
Galeopsis tetrahit	-	-	-	-	-	-	-	-	-	-	-	-	-	-	-	-	-	-	-	-	-	-	-	-	-	-	x	x	-	x	-	-	-	-	-
Galium asprellum	-	-	-	-	-	-	-	-	-	-	-	-	-	-	-	-	-	-	-	-	-	-	x	-	-	-	-	-	-	-	-	-	-	-	-
Galium kamtschaticum	-	-	-	-	-	-	-	-	x	-	-	-	-	-	-	-	-	-	-	-	-	x	-	-	-	-	x	-	-	-	-	-	-	-	-
Galium labradoricum	-	-	-	-	-	-	-	-	-	-	-	-	-	-	-	-	x	x	-	-	x	x	x	-	-	-	-	-	x	-	-	x	-	-	-
Galium palustre	-	-	-	-	-	-	-	-	-	-	-	-	-	-	-	-	-	-	-	-	-	-	-	-	-	-	-	-	x	x	x	-	x	-	-
Galium trifidum	-	-	-	-	-	-	-	-	-	-	-	-	-	-	-	-	-	-	-	-	-	-	x	-	-	-	-	-	x	x	-	-	-	x	-
Galium triflorum	-	x	-	-	-	-	-	-	-	-	-	-	-	-	-	-	-	-	-	-	-	x	x	x	-	-	x	-	-	-	-	-	-	-	x
Gaultheria hispidula	-	x	x	x	x	-	x	-	x	x	x	x	-	x	x	x	x	-	x	x	-	x	-	x	-	-	-	-	-	-	-	-	-	-	-
Gaylussacia baccata	x	-	-	-	-	-	-	-	-	-	-	-	-	x	-	-	x	-	-	-	-	-	x	-	-	-	-	-	-	-	-	-	-	-	-
Gaylussacia dumosa	-	-	-	-	-	-	-	-	-	-	-	-	-	-	-	-	x	-	-	-	-	-	-	-	-	-	-	-	-	-	-	-	-	-	-
Gentianella detonsa ssp. nesophila	-	-	-	-	-	-	-	-	-	-	-	-	-	-	-	-	x	-	-	-	-	-	-	-	-	-	-	-	x	x	-	-	-	-	x

Flora	Habitats 1	2	3	4	5	6	7	8	9	10	11	12	13	14	15	16	17	18	19	20	21	22	23	24	25	26	27	28	29	30	31	32	33	34	35
Geocaulon lividum	–	–	–	–	–	–	x	–	–	–	–	–	–	x	–	–	–	–	x	–	–	–	–	x	–	–	–	–	–	–	–	–	–	–	x
Geranium pratense	–	–	–	–	–	–	–	–	–	–	–	–	–	–	–	–	–	–	–	–	–	–	x	–	–	–	–	–	–	–	–	–	–	–	–
Geranium robertianum	–	x	–	–	–	–	–	–	–	–	–	–	–	–	–	–	–	–	–	–	–	–	–	x	–	–	–	–	–	x	–	–	–	–	–
Geum macrophyllum	–	x	–	–	–	–	–	–	–	–	–	–	–	–	–	–	–	–	–	–	x	x	x	x	–	–	–	x	–	–	–	–	–	–	–
Geum rivale	–	x	–	–	–	–	–	–	–	–	–	–	–	–	–	–	–	–	x	–	x	x	x	–	–	–	–	–	x	–	–	–	–	–	–
Glaux maritima	–	–	–	–	–	–	–	–	–	–	–	–	–	–	–	–	x	–	–	–	–	–	–	–	–	–	–	–	–	x	–	x	x	x	–
Glechoma hederacea	–	–	–	–	–	–	–	–	–	–	–	–	–	–	–	–	–	–	–	–	–	–	–	x	–	–	–	–	–	–	–	–	–	–	–
Glyceria borealis	x	–	–	–	–	–	–	–	–	–	–	–	–	–	–	–	–	–	–	–	–	–	x	–	–	–	x	–	x	x	–	–	–	–	–
Glyceria canadensis	x	–	–	–	–	–	–	–	–	–	–	–	–	–	–	–	–	–	–	x	–	–	–	–	–	–	–	–	–	–	–	–	–	–	–
Glyceria fluitans	x	–	–	–	–	–	–	–	–	–	–	–	–	–	–	–	–	–	–	–	–	–	–	–	–	–	–	–	–	–	–	–	–	–	–
Glyceria grandis	–	–	–	–	–	–	–	–	–	–	–	–	–	–	–	–	–	–	–	–	–	–	–	–	–	–	–	–	–	x	–	–	–	–	–
Glyceria striata	–	x	–	–	–	–	–	–	–	–	–	–	–	–	–	–	–	x	–	x	–	x	x	–	–	–	–	–	–	–	–	–	–	x	–
Gnaphalium norvegicum	–	–	–	–	–	–	–	x	–	–	–	–	–	–	–	–	–	–	–	–	–	–	–	–	–	–	–	–	–	–	–	–	–	–	–
Gnaphalium supinum	–	–	–	–	–	–	x	x	–	–	–	x	–	–	–	–	–	–	–	–	–	–	–	–	–	–	–	–	–	–	–	–	–	–	–
Gnaphalium sylvaticum	–	–	–	–	–	–	–	–	–	–	–	–	–	–	–	–	–	–	–	–	–	–	–	–	–	–	–	–	–	–	–	–	–	–	–
Gnaphalium uliginosum	–	–	–	–	–	–	–	–	–	–	–	–	–	–	–	–	–	–	–	–	–	–	–	–	–	–	–	x	–	–	–	–	–	–	–
Goodyera repens	–	–	x	–	–	–	–	–	–	–	–	–	–	–	–	–	–	–	–	–	–	–	–	x	–	–	–	–	–	–	–	–	–	–	–
Goodyera tesselata	–	x	x	–	–	–	–	–	–	–	–	–	–	–	–	–	–	–	–	–	–	–	–	–	–	–	x	–	–	–	–	–	–	–	–
Gymnocarpium dryopteris	–	x	x	–	–	–	–	–	x	–	–	–	–	–	–	–	–	–	–	–	–	x	x	x	–	–	–	–	–	–	–	–	–	–	–
Gymnocarpium robertianum	–	–	–	–	–	–	–	–	–	–	–	–	–	–	–	–	–	–	–	–	–	–	–	x	–	–	–	–	–	–	–	–	–	–	–
Habenaria blephariglottis	–	–	–	–	–	–	–	–	–	–	–	–	–	–	–	x	–	–	–	–	–	–	–	–	–	–	–	–	–	–	–	–	–	–	–
Habenaria clavellata	x	–	x	–	–	–	–	–	–	–	–	–	–	–	–	x	x	x	–	–	–	–	–	–	–	–	–	–	–	–	–	–	–	–	–
Habenaria dilatata	–	–	–	–	–	–	–	x	x	–	–	–	–	–	–	x	–	x	–	–	x	x	x	–	x	–	x	x	x	–	–	x	–	–	–
Habenaria hyperborea	–	–	–	–	–	–	–	–	–	–	–	–	–	–	–	–	–	x	–	–	x	–	–	x	–	–	–	–	x	x	–	x	–	–	–
Habenaria obtusata	–	x	x	x	–	–	–	–	x	–	–	–	–	–	–	–	–	–	–	–	–	x	–	x	–	–	–	–	–	–	–	–	–	–	x
Habenaria orbiculata	–	x	x	–	–	–	–	–	–	–	–	–	–	–	–	x	–	–	–	–	–	–	–	–	–	–	x	–	–	–	–	–	–	–	–
Habenaria psycodes	–	–	–	–	–	–	–	–	–	–	–	–	–	–	–	x	x	x	–	–	–	–	x	–	x	–	x	x	x	–	–	–	–	–	–
Habenaria viridis	–	–	–	–	–	–	x	–	–	–	–	–	–	–	–	–	–	–	–	–	–	–	–	x	–	–	–	–	–	–	–	–	–	–	–
Halenia deflexa	–	–	–	–	–	–	–	–	–	–	–	–	–	–	–	–	–	–	–	–	–	–	–	x	–	–	–	–	x	–	x	–	–	–	x
Hedysarum alpinum	–	–	–	–	–	–	–	–	–	–	–	–	–	–	–	–	–	–	–	–	–	–	–	x	–	–	–	–	–	–	–	–	–	–	x
Heracleum lanatum	–	x	–	–	–	–	–	–	–	–	–	–	–	–	–	–	–	–	–	–	–	x	x	x	–	–	–	x	x	–	–	–	–	–	–
Hieracium aurantiacum	–	–	–	–	–	–	–	–	–	–	–	–	–	–	–	–	–	–	–	–	–	–	–	x	–	–	x	x	–	–	–	–	–	–	–
Hieracium florentinum	–	–	–	–	–	–	–	–	–	–	–	–	–	–	–	–	–	–	–	–	–	–	–	x	–	–	–	–	–	–	–	–	–	–	x
Hieracium floribundum	–	–	–	–	–	–	–	–	–	–	–	–	–	–	–	–	–	–	–	–	–	–	–	–	–	–	–	x	–	x	–	–	–	–	–
Hieracium kalmii	–	–	–	–	–	–	–	–	–	–	–	–	–	–	–	–	–	–	–	–	–	–	–	x	–	–	x	–	–	–	–	–	–	x	x
Hierochloe alpina	–	–	–	–	–	–	x	–	–	–	–	x	x	–	–	–	–	–	–	–	–	–	–	–	x	–	–	–	–	–	–	–	–	–	–
Hierochloe odorata	–	–	–	–	–	–	–	–	–	–	–	–	–	–	–	–	–	–	–	–	–	–	–	–	–	–	–	–	–	–	x	x	–	–	–
Hippuris vulgaris	x	–	–	–	–	–	–	–	–	–	–	–	–	–	–	–	x	–	–	–	–	–	x	–	–	–	–	–	–	–	–	–	–	–	–

Hordeum jubatum	-	-	-	-	-	-	-	-	-	-	-	-	-	-	-	-	-	-	-	-	-	-	-	-	-	-	-	-	-	x	-	x	x	x	-
Hypericum boreale	x	-	-	-	-	-	-	-	-	-	-	-	-	-	-	-	-	-	-	-	-	-	-	-	-	-	-	-	-	-	-	-	-	-	-
Hypericum virginicum	x	-	-	-	-	-	-	-	-	-	-	-	-	-	-	-	-	x	-	-	-	-	-	-	-	-	-	-	-	-	-	x	x	-	-
Impatiens capensis	-	-	-	-	-	-	-	-	-	-	-	-	-	-	-	-	-	-	-	-	-	-	x	-	-	-	-	x	x	x	-	-	-	-	-
Iris setosa var. canadensis	-	-	-	-	-	-	-	-	-	-	-	-	-	-	-	x	-	-	-	-	-	-	-	-	-	-	-	-	x	x	-	x	x	-	x
Iris versicolor	-	-	-	-	-	-	-	-	-	-	-	-	-	-	-	-	-	x	-	-	x	-	x	-	-	-	-	-	x	-	-	x	x	-	-
Isoetes echinospora var. muricata	x	-	-	-	-	-	-	-	-	-	-	-	-	-	-	x	-	-	-	-	-	-	-	-	-	-	-	-	-	-	-	-	-	x	-
Isoetes macrospora	x	-	-	-	-	-	-	-	-	-	-	-	-	-	-	-	-	-	-	-	-	-	-	-	-	-	-	-	-	-	-	-	-	-	-
Isoetes tuckermanii	x	-	-	-	-	-	-	-	-	-	-	-	-	-	-	-	-	-	-	-	-	-	-	-	-	-	-	-	-	-	-	-	-	-	-
Juncus albescens	-	-	-	-	-	-	-	-	-	-	-	-	-	-	-	-	-	-	-	-	-	-	-	-	x	-	-	-	-	-	-	-	-	-	-
Juncus alpinus	-	-	-	-	-	-	-	-	-	-	-	-	-	-	-	-	-	-	-	-	-	-	x	-	x	-	-	-	-	x	x	x	-	-	-
Juncus articulatus	-	-	-	-	-	-	-	-	-	-	-	-	-	-	-	x	-	x	-	-	x	-	x	-	-	-	-	x	-	x	-	x	-	-	-
Juncus balticus	-	-	-	-	-	-	-	-	-	-	-	-	-	-	-	x	x	-	x	x	x	-	x	-	x	-	-	-	-	-	-	x	x	-	-
Juncus brevicaudatus	-	-	-	-	-	-	-	-	-	-	-	-	-	-	-	x	-	x	-	-	-	-	-	-	x	-	x	x	x	x	-	-	x	-	x
Juncus bufonius	-	x	-	-	-	-	-	-	-	-	-	-	-	-	-	-	-	-	-	-	-	-	-	-	-	-	x	x	-	-	x	x	x	-	-
Juncus canadensis	-	-	-	-	-	-	-	-	-	-	-	-	-	-	-	-	-	x	-	-	x	-	x	-	-	-	-	-	-	-	-	-	-	-	-
Juncus dudleyi	-	-	-	-	-	-	-	-	-	-	-	-	-	-	-	-	-	-	-	-	-	-	x	-	-	-	-	-	-	-	-	-	-	-	-
Juncus effusus	-	x	-	-	-	-	-	-	-	-	-	-	-	-	-	-	-	-	-	-	-	-	-	-	x	-	x	x	-	-	-	-	-	-	-
Juncus filiformis	-	-	-	-	-	-	-	x	x	x	-	-	-	-	-	x	-	-	-	-	x	-	x	-	-	-	x	-	x	-	-	x	-	-	-
Juncus gerardii	-	-	-	-	-	-	-	-	-	-	-	-	-	-	-	-	-	-	-	-	-	-	-	-	-	-	-	-	-	-	-	x	x	-	-
Juncus nodosus	-	-	-	-	-	-	-	-	-	-	-	-	-	-	-	-	-	-	-	-	-	-	x	-	-	-	-	-	-	-	-	-	-	-	-
Juncus pelocarpus	x	-	-	-	-	-	-	-	-	-	-	-	-	-	-	-	-	-	-	-	-	-	x	-	-	-	-	-	-	-	-	-	-	-	-
Juncus stygius	x	-	-	-	-	-	-	-	-	-	-	-	-	-	-	x	x	x	-	-	-	-	-	-	-	-	-	-	-	-	-	-	-	-	-
Juncus tenuis	-	x	-	-	-	-	-	-	-	-	-	-	-	-	-	-	-	-	-	-	-	-	x	-	-	-	x	x	-	x	-	-	-	-	-
Juncus trifidus	-	-	-	-	-	x	x	x	-	-	-	x	x	-	-	x	-	-	-	-	-	-	-	x	x	-	-	-	-	-	-	-	-	-	-
Juncus x alpiniformis	-	-	-	-	-	-	-	-	-	-	-	-	-	-	-	-	-	-	-	-	-	-	x	-	-	-	-	-	-	-	-	-	-	-	-
Juncus x nodosiformis	-	-	-	-	-	-	-	-	-	-	-	-	-	-	-	-	-	-	-	-	-	-	x	-	-	-	-	-	-	-	-	-	-	-	-
Juniperus communis	-	x	-	x	-	-	-	-	-	-	-	x	x	x	x	x	x	x	x	x	x	-	-	x	x	x	-	-	x	-	-	-	-	-	x
Juniperus horizontalis	-	-	-	-	-	-	-	-	x	-	-	-	x	-	-	x	x	x	x	x	-	-	-	x	x	x	-	-	-	-	x	-	-	-	x
Kalmia angustifolia	-	-	x	-	-	-	-	-	-	-	x	x	-	x	x	x	x	x	x	x	x	x	-	-	-	-	-	-	-	-	-	-	-	-	-
Kalmia polifolia	-	-	-	-	-	-	x	x	x	x	x	x	x	x	x	x	x	x	x	x	x	x	-	-	-	-	-	-	-	-	-	-	-	-	-
Kobresia simpliciuscula	-	-	-	-	-	-	-	-	-	-	-	-	-	-	-	-	-	-	-	-	-	-	-	x	-	-	-	-	-	-	-	-	-	-	-
Lactuca biennis	-	-	-	-	-	-	-	-	-	-	-	-	-	-	-	-	-	-	-	-	-	-	-	-	-	-	x	x	-	-	-	-	-	-	-
Lamium purpureum	-	-	-	-	-	-	-	-	-	-	-	-	-	-	-	-	-	-	-	-	-	-	-	-	-	-	-	x	-	-	-	-	-	-	-
Larix laricina	-	x	-	-	-	-	-	-	-	-	-	-	x	x	x	x	x	x	x	x	x	x	x	x	x	x	-	-	-	-	-	-	-	-	-
Lathyrus japonicus	-	-	-	-	-	-	-	-	-	-	-	-	-	-	-	-	-	-	-	-	-	-	-	-	-	-	-	x	x	x		-	-	-	x
Lathyrus palustris	-	-	-	-	-	-	-	-	-	-	-	-	-	-	-	-	-	-	-	-	-	-	x	-	-	-	-	x	x	-	x	x	x	-	-
Ledum groenlandicum	-	-	-	-	-	-	-	-	-	-	x	x	x	x	x	x	x	x	x	x	x	x	-	x	x	-	-	-	-	-	-	-	-	-	-
Leontodon autumnalis	-	-	-	-	-	-	-	-	-	-	-	-	-	-	-	-	-	-	-	-	-	-	x	-	-	-	-	x	x	x	-	x	-	-	-
Lesquerella arctica var. purshii	-	-	-	-	-	-	-	-	-	-	-	-	-	-	-	-	-	-	-	-	-	-	-	x	-	-	-	-	-	-	-	-	-	-	-
Ligusticum scothicum	-	-	-	-	-	-	-	-	-	-	-	-	-	-	-	-	-	-	-	-	-	-	-	x	-	-	-	-	x	x	-	-	-	-	x

Flora	1	2	3	4	5	6	7	8	9	10	11	12	13	14	15	16	17	18	19	20	21	22	23	24	25	26	27	28	29	30	31	32	33	34	35
	Habitats																																		
Linnaea borealis	–	x	x	x	x	–	–	–	x	x	x	x	x	–	–	–	–	x	x	x	x	x	x	x	–	–	–	–	–	–	–	–	–	–	–
Linum catharticum	–	–	–	–	–	–	–	–	–	–	–	–	–	–	–	–	–	–	–	–	–	–	–	–	–	–	x	x	–	x	–	–	–	–	–
Listera auriculata	–	–	–	–	–	–	–	–	–	–	–	–	–	–	–	–	–	–	–	–	–	–	x	–	–	–	–	–	–	–	–	–	–	–	–
Listera convallarioides	–	–	–	–	–	–	–	–	–	–	–	–	–	–	–	–	–	–	–	–	x	x	x	–	–	–	x	–	–	–	–	–	–	–	–
Listera cordata	–	–	x	–	x	–	–	–	–	–	x	–	–	–	–	–	–	–	–	–	x	x	–	–	–	–	–	–	–	–	–	–	–	–	–
Littorella uniflora var. americana	x	–	–	–	–	–	–	–	–	–	–	–	–	–	–	–	–	–	–	–	–	–	–	–	–	–	–	–	–	–	–	–	–	–	–
Lobelia dortmanna	x	–	–	–	–	–	–	–	–	–	–	–	–	–	–	–	–	–	–	–	–	–	–	–	–	–	–	–	–	–	–	–	–	–	–
Lobelia kalmii	x	–	x	–	–	–	–	–	–	–	–	–	–	–	–	–	–	–	–	–	x	–	–	x	–	–	–	x	–	–	–	x	–	–	–
Loiseleuria procumbens	–	–	–	–	–	x	x	x	–	–	–	x	x	x	–	x	–	–	–	–	–	–	–	x	x	–	–	–	–	–	–	–	–	–	–
Lolium multiflorum	–	–	–	–	–	–	–	–	–	–	–	–	–	–	–	–	–	–	–	–	–	–	–	–	–	–	–	x	–	–	–	–	–	–	–
Lomatogonium rotatum	–	–	–	–	–	–	–	–	–	–	–	–	–	–	–	–	–	–	–	–	–	–	–	–	–	–	x	–	x	–	x	x	–	–	–
Lonicera villosa	–	x	–	–	–	–	–	–	–	–	–	–	–	–	–	x	x	x	–	–	x	x	x	x	–	–	–	–	x	–	–	–	–	–	x
Lotus corniculatus	–	–	–	–	–	–	–	–	–	–	–	–	–	–	–	–	–	–	–	–	–	–	–	–	–	–	–	x	–	–	–	–	–	–	–
Luzula multiflora	–	x	–	–	–	–	–	–	–	–	–	–	–	–	–	–	–	–	–	–	–	–	–	–	–	–	x	x	–	x	–	–	–	–	–
Luzula parviflora	x	–	–	–	x	–	–	x	x	x	–	x	–	–	–	–	–	–	–	–	–	–	x	–	x	–	–	–	x	x	–	–	–	–	–
Luzula spicata	–	–	–	–	–	–	x	–	–	–	–	–	–	–	–	–	–	–	–	–	–	–	–	x	–	–	–	–	–	–	–	–	–	–	x
Luzula sudetica	–	–	–	–	–	–	–	–	–	–	–	–	–	–	–	–	–	–	–	–	–	–	–	–	–	–	–	–	–	–	–	–	–	–	x
Lychnis alpina	–	–	–	–	–	–	–	–	–	–	–	–	–	–	–	x	–	–	–	–	–	–	–	–	x	–	–	–	–	–	–	–	–	–	–
Lycopodium alpinum	–	–	–	–	–	–	x	x	–	–	–	–	–	–	–	x	–	–	–	–	–	–	–	–	–	–	–	–	–	–	–	–	–	–	–
Lycopodium annotinum	–	–	–	–	–	–	–	–	x	x	–	x	x	x	–	x	x	–	–	x	–	–	–	–	–	–	x	–	–	–	–	–	–	–	–
Lycopodium clavatum	–	x	–	–	–	–	–	–	–	–	–	–	–	–	–	–	–	–	–	–	–	–	–	–	–	–	–	–	–	–	–	–	–	–	–
Lycopodium complanatum	–	–	–	–	–	–	–	–	–	–	–	–	–	x	–	–	–	–	–	–	–	–	–	–	–	–	–	–	–	–	–	–	–	–	–
Lycopodium inundatum	–	–	–	–	–	–	–	–	–	–	–	–	–	–	–	x	–	–	–	–	–	–	–	–	–	–	–	–	–	–	–	–	–	–	–
Lycopodium lucidulum	–	x	x	–	–	–	–	–	x	–	–	–	–	–	–	–	–	–	x	–	–	x	x	–	–	–	–	–	–	–	–	–	–	–	–
Lycopodium obscurum	–	x	x	–	–	–	–	–	–	–	–	–	–	–	–	–	–	–	–	–	–	–	–	–	–	–	–	–	–	–	–	–	–	–	–
Lycopodium sabinifolium	–	–	–	–	–	–	x	x	–	–	–	x	–	x	–	–	–	–	–	–	–	–	–	–	–	–	–	–	–	–	–	–	–	–	–
Lycopodium selago	–	–	–	–	–	–	–	x	–	–	–	x	x	–	–	–	–	–	–	–	–	–	–	–	x	–	–	–	–	–	–	–	–	–	x
Lycopus americanus	–	–	–	–	–	–	–	–	–	–	–	–	–	–	–	–	–	–	–	–	–	–	x	–	–	–	–	–	–	–	–	–	–	–	–
Lycopus uniflorus	x	–	–	–	–	–	–	–	–	–	–	–	–	–	–	–	–	x	–	–	–	–	x	–	–	–	–	–	x	–	–	x	–	–	–
Lysimachia terrestris	x	–	–	–	–	–	–	–	–	–	–	–	–	–	–	–	–	–	–	–	–	–	–	–	–	–	–	–	–	–	–	–	–	–	–
Lythrum salicaria	–	–	–	–	–	–	–	–	–	–	–	–	–	–	–	–	–	–	–	–	–	–	–	–	–	–	–	x	–	–	–	x	–	–	–
Maianthemum canadense	–	x	x	x	x	–	–	x	x	x	x	x	x	x	x	x	x	–	–	x	–	x	x	x	–	–	–	–	x	–	–	–	–	–	–
Malaxis monophyllos var. brachypoda	–	–	–	–	–	–	–	–	–	–	–	–	–	–	–	–	x	–	–	–	–	–	–	x	–	–	–	–	–	x	–	–	–	–	x
Malaxis unifolia	–	–	–	–	–	–	–	–	–	–	–	–	–	–	–	x	–	x	–	–	x	–	x	–	–	–	–	–	–	–	–	–	–	–	–
Matricaria maritima	–	–	–	–	–	–	–	–	–	–	–	–	–	–	–	–	–	–	–	–	–	–	–	–	–	–	–	x	–	–	–	–	–	–	–
Matricaria matricarioides	–	–	–	–	–	–	–	–	–	–	–	–	–	–	–	–	–	–	–	–	–	–	–	–	–	–	–	x	–	–	–	–	–	–	–
Matteucia struthiopteris	–	x	–	–	–	–	–	–	–	–	–	–	–	–	–	–	–	–	–	–	–	–	x	–	–	–	–	–	–	–	–	–	–	–	–
Mentha arvensis	x	x	–	–	–	–	–	–	–	–	–	–	–	–	–	–	–	–	–	–	–	–	x	–	–	–	–	–	–	x	–	x	–	–	–

Menyanthes trifoliata	x	-	-	-	-	-	-	-	-	-	-	-	-	-	-	x	-	x	-	-	-	-	x	-	-	-	-	-	x	-	x	-	-	-	-
Mertensia maritima	-	-	-	-	-	-	-	-	-	-	-	-	-	-	-	-	-	-	-	-	-	-	-	-	-	-	-	-	-	x	x	-	-	-	-
Milium effusum	-	-	-	-	-	-	x	-	-	-	-	-	-	-	-	-	-	-	-	-	-	-	x	-	-	-	-	-	x	-	-	-	-	-	-
Mimulus moschatus	x	-	-	-	-	-	-	-	-	-	-	-	-	-	-	-	-	-	-	-	-	-	-	-	-	-	x	x	-	-	-	-	-	-	-
Mitella nuda	-	x	x	-	-	-	-	-	-	-	-	-	-	-	-	-	-	-	-	-	x	x	x	x	-	-	-	-	-	-	-	-	-	-	x
Moneses uniflora	-	x	x	x	x	-	-	-	x	-	x	-	-	-	-	-	-	-	-	-	x	-	-	-	-	-	-	-	-	-	-	-	-	-	x
Monotropa hypopithys	-	-	x	-	-	-	-	-	-	-	-	-	-	-	-	-	-	-	-	-	-	-	-	x	-	-	-	-	-	-	-	-	-	-	-
Monotropa uniflora	-	x	x	x	x	-	-	-	x	-	x	-	-	-	-	-	-	-	-	-	-	x	x	x	-	-	-	-	-	-	-	-	-	-	-
Montia fontana ssp. fontana	-	-	-	-	-	-	x	-	-	-	-	-	-	-	-	-	-	-	-	-	-	-	-	-	-	-	-	-	x	-	-	-	-	-	x
Muhlenbergia glomerata	-	-	-	-	-	-	-	-	-	-	-	-	-	-	-	-	-	-	-	-	x	-	-	-	-	-	-	-	-	-	-	x	-	-	-
Myosotis laxa	-	-	-	-	-	-	-	-	-	-	-	-	-	-	-	-	-	-	-	-	-	-	x	-	-	-	-	-	x	-	-	-	-	-	-
Myosotis scorpioides	-	-	-	-	-	-	-	-	-	-	-	-	-	-	-	-	-	-	-	-	-	-	-	-	-	-	-	-	x	-	-	-	-	-	-
Myrica gale	-	-	-	-	-	-	-	x	x	-	-	-	x	x	x	x	x	x	x	x	x	x	x	-	x	-	-	-	x	x	-	x	x	-	-
Myriophyllum alterniflorum	x	-	-	-	-	-	-	-	-	-	-	-	-	-	-	-	-	-	-	-	-	-	x	-	-	-	-	-	-	-	-	-	-	-	-
Myriophyllum tenellum	x	-	-	-	-	-	-	-	-	-	-	-	-	-	-	-	-	-	-	-	-	-	-	-	-	-	-	-	-	-	-	-	-	-	-
Nemopanthus mucronatus	-	x	x	-	-	-	-	-	x	-	x	-	-	x	x	-	x	-	x	x	-	x	x	x	-	-	-	-	-	x	-	-	-	-	-
Nuphar variegatum	x	-	-	-	-	-	-	-	-	-	-	-	-	-	-	x	-	-	-	-	-	-	-	-	-	-	-	-	-	-	-	-	-	-	-
Odontites serotina	-	-	-	-	-	-	-	-	-	-	-	-	-	-	-	-	-	-	-	-	-	-	-	-	-	-	-	x	-	-	-	-	-	-	-
Oenothera biennis	-	-	-	-	-	-	-	-	-	-	-	-	-	-	-	-	-	-	-	-	-	-	-	-	-	-	-	x	-	-	-	-	-	-	-
Oenothera parviflora	-	-	-	-	-	-	-	-	-	-	-	-	-	-	-	-	-	-	-	-	-	-	-	-	-	-	-	x	-	-	-	-	-	-	x
Oenothera perennis	-	-	-	-	-	-	-	-	-	-	-	-	-	-	-	-	-	-	-	-	-	-	-	-	-	-	-	-	-	x	-	-	-	-	-
Onoclea sensibilis	-	-	-	-	-	-	-	-	-	-	-	-	-	-	-	-	-	-	-	-	-	-	x	-	-	-	-	-	x	-	-	-	-	-	-
Orobanche uniflora	-	-	-	-	-	-	-	-	-	-	-	-	-	-	-	-	-	-	-	-	-	-	-	x	x	-	-	-	-	-	-	-	-	-	x
Oryzopsis asperifolia	-	-	-	-	-	-	-	-	-	-	-	-	-	-	-	-	-	-	-	x	-	x	-	-	x	-	-	-	-	-	-	-	-	-	-
Oryzopsis canadensis	-	-	-	-	-	-	-	-	-	-	-	-	-	-	-	-	-	-	-	-	-	-	-	-	x	-	-	-	-	-	-	-	-	-	-
Osmorhiza depauperata	-	x	-	-	-	-	-	-	-	-	-	-	-	-	-	-	-	-	-	-	-	-	-	-	-	-	-	-	-	-	-	-	-	-	-
Osmunda cinnamomea	-	-	x	-	x	-	-	-	x	-	-	-	-	-	-	-	-	-	x	x	x	x	x	-	-	-	-	-	-	-	-	-	-	-	-
Osmunda claytoniana	-	-	x	-	x	-	-	x	x	-	-	-	-	-	-	-	-	-	-	-	-	-	-	-	-	-	-	x	-	-	-	-	-	-	-
Osmunda regalis	-	x	-	-	-	-	-	-	-	-	-	-	-	-	-	-	-	-	-	x	-	-	-	-	x	-	-	-	-	-	-	-	-	-	-
Oxyria digyna	-	-	-	-	-	-	-	x	-	-	-	-	-	-	-	-	-	-	-	-	-	-	-	-	-	-	-	-	-	-	-	-	-	-	x
Oxytropis campestris var. johannensis	-	-	-	-	-	-	-	-	-	-	-	-	-	-	-	-	-	-	-	-	-	-	-	x	-	-	-	-	x	-	-	-	-	-	x
Parnassia glauca	-	-	-	-	-	-	-	-	-	-	-	-	-	-	-	-	-	x	-	-	-	-	-	-	-	-	-	-	-	-	-	-	-	-	-
Parnassia parviflora	-	-	-	-	-	-	-	-	-	-	-	-	-	-	-	-	-	-	-	-	-	-	-	x	-	-	-	x	x	-	-	-	-	-	x
Pastinaca sativa	-	-	-	-	-	-	-	-	-	-	-	-	-	-	-	-	-	-	-	-	-	-	-	-	-	-	-	x	-	-	-	-	-	-	-
Pedicularis palustris	-	-	-	-	-	-	-	-	-	-	-	-	-	-	-	-	x	x	-	-	-	-	-	-	-	-	-	x	x	-	-	-	-	-	-
Phalaris arundinacea	-	-	-	-	-	-	-	-	-	-	-	-	-	-	-	-	-	-	-	-	-	-	-	-	-	-	-	x	-	-	-	-	-	-	-
Phleum alpinum	-	-	-	-	-	-	-	x	-	-	-	-	-	-	-	-	-	-	-	-	-	-	-	x	-	-	-	-	-	-	-	-	-	-	-
Phleum pratense	-	-	-	-	-	-	-	-	-	-	-	-	-	-	-	-	-	-	-	-	-	-	-	-	-	-	-	x	x	-	-	-	-	-	-
Phyllodoce caerulea	-	-	-	-	-	-	x	x	-	-	-	x	x	-	-	-	-	-	-	-	-	-	-	-	-	-	-	-	-	-	-	-	-	-	-
Picea glauca	-	x	x	x	-	-	-	-	x	x	x	x	-	-	-	-	-	-	-	-	-	-	x	x	-	x	-	-	x	-	x	-	-	-	-

Flora	Habitats 1	2	3	4	5	6	7	8	9	10	11	12	13	14	15	16	17	18	19	20	21	22	23	24	25	26	27	28	29	30	31	32	33	34	35
Picea mariana	—	—	x	x	x	—	—	—	x	—	x	x	x	x	x	x	x	x	x	x	x	x	x	x	x	—	—	—	—	—	—	—	—	—	—
Pinguicula vulgaris	—	—	—	—	—	—	—	—	—	—	—	—	—	—	—	x	—	x	—	x	—	—	x	x	x	x	—	—	x	—	—	—	—	—	—
Pinus strobus	—	—	x	—	—	—	—	—	—	—	—	—	—	—	—	—	—	—	—	x	—	—	—	x	x	—	—	—	—	—	—	—	—	—	—
Plantago lanceolata	—	—	—	—	—	—	—	—	—	—	—	—	—	—	—	—	—	—	—	—	—	—	x	—	—	—	—	—	—	—	—	—	—	—	—
Plantago major	—	—	—	—	—	—	—	—	—	—	—	—	—	—	—	—	—	—	—	—	—	—	—	—	—	—	—	—	x	x	—	x	—	—	—
Plantago maritima	—	—	—	—	—	—	—	—	—	—	—	—	—	—	—	—	—	—	—	—	—	—	x	x	x	—	—	x	x	x	—	—	x	—	x
Poa alpina	—	—	—	—	—	—	x	—	—	—	—	—	—	—	—	—	—	—	—	—	—	—	—	x	—	—	—	—	—	—	—	—	—	—	x
Poa annua	—	—	—	—	—	—	—	—	—	—	—	—	—	—	—	—	—	—	—	—	—	—	—	—	—	—	—	x	—	—	—	—	x	—	x
Poa compressa	—	—	—	—	—	—	—	—	—	—	—	—	—	—	—	—	—	—	—	—	—	—	—	—	—	—	—	x	—	x	—	—	—	—	—
Poa eminens	—	—	—	—	—	—	—	—	—	—	—	—	—	—	—	—	—	—	—	—	—	—	—	—	—	—	—	—	—	x	—	x	—	—	x
Poa fernaldiana	—	—	—	—	—	—	x	—	—	—	—	—	—	—	—	—	—	—	—	—	—	—	—	—	—	—	—	—	—	—	—	—	—	—	—
Poa glauca	—	—	—	—	—	—	—	—	—	—	—	—	—	—	—	—	—	—	—	—	—	—	—	x	—	—	—	—	—	—	—	—	—	—	x
Poa nemoralis	—	—	—	—	—	—	—	—	—	—	—	—	—	—	—	—	—	—	—	—	—	—	—	x	—	—	—	—	—	—	—	—	—	—	x
Poa palustris	—	—	—	—	—	—	—	—	—	—	—	—	—	—	—	—	—	—	—	—	—	—	x	x	—	—	—	x	x	x	—	—	x	x	x
Poa pratensis var. alpigena	—	—	—	—	—	—	x	—	—	—	—	—	—	—	—	—	—	—	—	—	—	—	—	x	—	—	—	—	—	x	—	—	—	—	—
Poa pratensis	—	—	—	—	—	—	x	—	—	—	—	—	—	—	—	—	—	—	—	—	—	—	x	x	—	—	—	x	x	x	x	x	x	—	x
Poa saltuensis	—	x	—	—	—	—	—	—	—	—	—	—	—	—	—	—	—	—	—	—	—	—	—	—	—	—	—	—	—	—	—	—	—	—	—
Poa trivialis	—	x	—	—	—	—	—	—	—	—	—	—	—	—	—	—	—	—	—	—	—	—	—	x	—	—	x	—	—	—	—	—	—	—	—
Pogonia ophioglossoides	—	—	—	—	—	—	—	—	—	—	—	—	—	—	—	x	—	x	—	—	—	—	—	—	—	—	—	—	—	—	—	—	—	—	—
Polygonum amphibium	—	—	—	—	—	—	—	—	—	—	—	—	—	—	—	—	—	—	—	—	—	—	—	—	—	—	—	—	x	—	—	—	—	—	—
Polygonum aviculare	—	—	—	—	—	—	—	—	—	—	—	—	—	—	—	—	—	—	—	—	—	—	—	—	—	—	—	x	x	—	x	—	—	x	—
Polygonum convolvulus	—	—	—	—	—	—	—	—	—	—	—	—	—	—	—	—	—	—	—	—	—	—	—	—	—	—	—	—	—	x	—	—	—	—	—
Polygonum fowleri	—	—	—	—	—	—	—	—	—	—	—	—	—	—	—	—	—	—	—	—	—	—	—	—	—	—	—	—	x	x	—	—	—	—	—
Polygonum hydropiper	—	—	—	—	—	—	—	—	—	—	—	—	—	—	—	—	—	—	—	—	—	—	—	—	—	—	—	—	x	—	—	—	—	—	—
Polygonum oxyspermum	—	—	—	—	—	—	—	—	—	—	—	—	—	—	—	—	—	—	—	—	—	—	—	—	—	—	—	—	—	x	—	—	—	—	—
Polygonum persicaria	—	—	—	—	—	—	—	—	—	—	—	—	—	—	—	—	—	—	—	—	—	—	x	—	—	—	—	x	—	—	—	x	—	—	—
Polygonum viviparum	—	—	—	—	—	—	—	x	—	—	—	—	—	—	—	—	—	—	—	—	—	—	—	x	—	—	—	—	x	x	—	—	—	—	x
Polypodium virginianum	—	—	x	—	—	—	—	—	—	—	—	—	—	—	—	—	—	—	—	—	—	—	—	x	—	—	x	—	—	—	—	—	—	—	x
Polystichum braunii	—	—	—	—	—	—	—	—	x	—	—	—	—	—	—	—	—	—	—	—	—	—	x	x	—	—	—	—	—	—	—	—	—	—	x
Polystichum lonchitis	—	—	—	—	—	—	—	—	—	—	—	—	—	—	—	—	—	—	—	—	—	—	—	x	—	—	—	—	—	—	—	—	—	—	—
Populus balsamifera	—	x	x	—	—	—	—	—	—	—	—	—	—	—	—	—	—	—	—	—	—	—	—	—	—	—	—	—	—	x	—	—	—	—	—
Populus tremuloides	—	—	x	—	—	—	—	—	—	—	—	—	—	—	—	—	—	—	—	—	—	—	—	—	—	—	—	x	—	—	—	—	—	—	—
Populus x gileadensis	—	—	—	—	—	—	—	—	—	—	—	—	—	—	—	—	—	—	—	—	—	—	—	—	—	—	—	x	—	—	—	—	—	—	—
Potamogeton alpinus	x	—	—	—	—	—	—	—	—	—	—	—	—	—	—	—	—	—	—	—	—	—	x	—	—	—	—	—	—	—	—	—	—	—	—
Potamogeton amplifolius	x	—	—	—	—	—	—	—	—	—	—	—	—	—	—	—	—	—	—	—	—	—	—	—	—	—	—	—	—	—	—	—	—	—	—
Potamogeton confervoides	x	—	—	—	—	—	—	—	—	—	—	—	—	—	—	—	—	—	—	—	—	—	—	—	—	—	—	—	—	—	—	—	—	—	—
Potamogeton epihydrus	x	—	—	—	—	—	—	—	—	—	—	—	—	—	—	—	—	—	—	—	—	—	x	—	—	—	—	—	—	—	—	—	—	—	—
Potamogeton filiformis	—	—	—	—	—	—	—	—	—	—	—	—	—	—	—	—	—	—	—	—	—	—	x	—	—	—	—	—	—	—	—	x	x	—	—

Potamogeton gramineus	x	-	-	-	-	-	-	-	-	-	-	-	-	-	-	-	-	-	-	-	-	-	x	-	-	-	-	-	-	-	-	-	-	-	-
Potamogeton natans	x	-	-	-	-	-	-	-	-	-	-	-	-	-	-	-	-	-	-	-	-	-	x	-	-	-	-	-	-	-	-	-	-	-	-
Potamogeton oakesianus	x	-	-	-	-	-	-	-	-	-	-	-	-	-	-	x	-	-	-	-	-	-	-	-	-	-	-	-	-	-	-	-	-	-	-
Potamogeton perfoliatus	-	-	-	-	-	-	-	-	-	-	-	-	-	-	-	-	-	-	-	-	-	-	x	-	-	-	-	-	-	-	-	-	x	-	-
Potentilla anserina	-	-	-	-	-	-	-	-	-	-	-	-	-	-	-	-	-	-	-	-	-	-	x	-	-	-	-	x	x	x	x	x	x	-	-
Potentilla egedii	-	-	-	-	-	-	-	-	-	-	-	-	-	-	-	-	-	-	-	-	-	-	-	-	-	-	-	-	-	x	-	-	-	-	x
Potentilla fruticosa	-	-	-	-	-	-	-	-	-	-	-	x	x	-	-	x	-	x	x	x	x	x	x	x	x	x	-	-	-	x	-	-	-	-	-
Potentilla nivea	-	-	-	-	-	-	-	-	-	-	-	-	-	-	-	-	-	-	-	-	-	-	-	x	-	-	-	-	-	-	-	-	-	-	-
Potentilla norvegica	-	-	-	-	-	-	x	-	-	-	-	-	-	-	-	-	-	-	-	-	-	-	x	x	-	-	-	x	-	x	-	-	-	-	-
Potentilla palustris	-	-	-	-	-	-	-	-	-	-	-	-	-	-	-	-	-	-	-	-	-	-	x	-	-	-	x	-	x	-	-	x	-	-	-
Potentilla tridentata	-	-	-	-	-	x	-	-	-	-	-	x	x	-	x	-	-	-	-	x	-	-	-	x	x	-	-	-	x	-	-	-	-	-	-
Prenanthes trifoliolata	-	-	-	-	x	-	-	-	x	-	-	-	x	-	-	x	-	x	-	x	-	-	x	x	x	x	x	-	x	-	-	-	-	-	x
Primula laurentiana	-	-	-	-	-	-	x	-	-	-	-	-	-	-	-	-	-	-	-	-	-	-	-	x	-	-	-	-	x	-	-	x	-	-	x
Primula mistassinica	-	-	-	-	-	-	-	-	-	-	-	-	-	-	-	-	-	x	-	-	x	-	x	x	x	-	-	-	-	-	-	-	-	-	x
Prunella vulgaris	-	-	-	-	-	-	-	-	-	-	-	-	-	-	-	-	-	x	-	-	x	-	x	x	-	-	-	x	-	-	-	-	-	-	-
Prunus pensylvanica	-	x	-	-	-	-	-	-	-	-	-	-	-	-	-	-	-	-	-	-	-	-	-	-	-	-	-	-	-	-	-	-	-	-	-
Prunus virginiana	-	x	-	-	-	-	-	-	-	-	-	-	-	-	-	-	-	-	-	-	-	-	x	x	-	-	-	-	-	-	-	-	-	-	x
Pteridium aquilinum	-	-	-	-	-	-	-	-	-	-	-	-	-	-	-	-	-	-	-	-	-	x	-	-	-	-	-	-	-	-	-	-	-	-	-
Puccinellia coarctata	-	-	-	-	-	-	-	-	-	-	-	-	-	-	-	-	-	-	-	-	-	-	-	-	-	-	-	-	-	-	-	-	-	-	x
Puccinellia paupercula	-	-	-	-	-	-	-	-	-	-	-	-	-	-	-	-	-	-	-	-	-	-	-	-	-	-	-	-	x	-	-	-	x	x	x
Pyrola asarifolia	-	-	-	-	-	-	-	-	-	-	-	-	-	-	-	-	-	x	-	-	x	-	x	-	-	-	-	-	-	-	-	-	-	-	x
Pyrola minor	-	x	x	-	-	-	-	-	-	-	-	-	-	-	-	-	-	-	-	-	-	-	-	-	x	-	x	-	-	-	-	-	-	-	x
Pyrola rotundifolia	-	x	-	-	-	-	-	-	-	-	-	-	-	-	-	-	-	-	-	-	-	-	-	-	x	-	-	-	-	-	-	-	-	-	-
Pyrola secunda	-	-	x	-	-	-	-	-	-	-	-	-	-	-	-	-	-	x	-	-	x	x	-	-	-	-	x	-	-	-	-	-	-	-	-
Pyrola virens var. virens	-	-	-	x	-	-	-	-	-	-	-	-	-	-	-	-	-	-	-	-	-	-	-	-	-	-	-	-	-	-	-	-	-	-	-
Ranunculus abortivus	-	x	-	-	-	-	x	-	-	-	-	-	-	-	-	-	-	-	-	-	-	-	x	x	-	-	-	-	x	-	-	-	-	-	-
Ranunculus acris	-	-	-	-	-	-	-	-	-	-	-	-	-	-	-	-	-	-	-	-	-	-	x	-	-	-	-	x	x	-	-	-	-	-	-
Ranunculus cymbalaria	-	-	-	-	-	-	-	-	-	-	-	-	-	-	-	-	-	-	-	-	-	-	-	-	-	-	-	-	-	x	x	x	x	-	x
Ranunculus hyperboreus	x	-	-	-	-	-	-	-	-	-	-	-	-	-	-	-	-	-	-	-	-	-	-	-	-	-	-	-	-	-	-	-	-	-	-
Ranunculus pedatifidus	-	-	-	-	-	-	x	-	-	-	-	-	-	-	-	-	-	-	-	-	-	-	-	-	-	-	-	-	-	-	-	-	-	-	-
Ranunculus pensylvanicus	-	-	-	-	-	-	-	-	-	-	-	-	-	-	-	-	-	-	-	-	-	-	x	-	-	-	-	x	-	x	-	-	-	-	-
Ranunculus repens	-	-	-	x	-	-	-	-	-	-	-	-	-	-	-	-	-	-	-	-	-	-	x	-	-	-	-	x	x	-	x	x	-	-	-
Ranunculus reptans	x	-	-	-	-	-	-	-	-	-	-	-	-	-	-	-	-	-	-	-	-	-	x	-	-	-	-	-	-	-	-	-	-	-	-
Rhamnus alnifolius	-	-	-	-	-	-	-	-	-	-	-	-	-	-	-	-	-	x	-	-	x	x	x	x	-	-	x	-	-	-	-	-	-	-	x
Rhinanthus borealis	-	-	-	-	-	-	-	-	-	-	-	-	-	-	-	x	-	-	-	-	-	-	-	x	-	-	-	-	x	-	-	-	-	-	x
Rhinanthus crista-galli	-	-	-	-	-	-	-	-	-	-	-	-	x	-	-	-	-	x	-	-	x	-	-	x	-	-	-	x	x	-	-	-	-	-	x
Rhododendron canadense	-	-	-	-	-	-	-	-	x	-	-	-	-	x	x	-	x	x	x	x	-	x	x	-	-	-	-	-	-	-	-	-	-	-	-
Rhododendron lapponicum	-	-	-	-	-	-	-	-	-	-	-	-	-	-	-	-	-	-	-	-	-	-	-	x	x	x	-	-	-	-	-	-	-	-	-
Rhynchospora alba	x	-	-	-	-	-	-	-	-	-	-	-	-	-	-	x	x	x	-	-	-	-	-	-	-	-	-	-	-	-	-	-	-	-	-
Rhynchospora capillacea	-	-	-	-	-	-	-	-	-	-	-	-	-	-	-	-	-	x	-	-	-	-	-	-	-	-	-	-	-	-	-	-	-	-	-
Rhynchospora fusca	-	-	-	-	-	-	-	-	-	-	-	-	-	-	-	x	-	-	x	-	-	-	-	-	-	-	-	-	-	-	-	-	-	-	-

Flora	Habitats 1	2	3	4	5	6	7	8	9	10	11	12	13	14	15	16	17	18	19	20	21	22	23	24	25	26	27	28	29	30	31	32	33	34	35
Ribes glandulosum	–	x	x	–	x	–	x	–	x	x	–	–	–	–	–	–	–	–	–	–	–	–	x	–	–	–	–	–	–	–	–	–	–	–	–
Ribes hirtellum	–	–	–	–	–	–	–	–	–	–	–	–	–	–	–	–	–	–	–	–	–	–	–	x	–	–	–	–	–	–	–	–	–	–	x
Ribes lacustre	–	x	x	–	–	–	x	–	–	–	–	–	–	–	–	–	–	–	x	–	–	x	x	x	–	–	–	–	–	–	–	–	–	–	–
Ribes triste	–	x	–	–	–	–	–	–	–	–	–	–	–	–	–	–	–	–	x	–	x	x	x	–	–	–	–	–	–	–	–	–	–	–	–
Rorippa islandica	–	–	–	–	–	–	–	–	–	–	–	–	–	–	–	–	–	–	–	–	–	–	x	–	–	–	–	–	–	x	–	–	–	–	–
Rosa nitida	–	–	–	–	–	–	–	–	–	–	–	–	–	–	–	–	x	x	x	x	x	x	x	–	x	–	–	–	–	–	–	–	–	–	–
Rosa virginiana	–	–	–	–	–	–	–	–	–	–	–	–	–	–	–	–	–	–	–	–	–	–	x	–	–	–	–	–	–	x	–	–	–	–	–
Rubus acaulis	–	–	–	–	–	–	–	–	–	–	–	–	–	–	–	–	–	–	–	–	–	–	x	–	–	–	–	–	x	x	–	–	–	–	x
Rubus chamaemorus	–	–	–	–	–	–	x	x	x	–	–	x	–	x	x	x	x	x	x	–	–	x	–	–	x	–	–	–	–	–	x	–	–	–	–
Rubus idaeus	–	x	x	–	x	–	–	–	–	–	–	–	–	–	–	–	–	–	–	–	–	x	x	x	–	–	x	x	x	–	–	–	–	–	–
Rubus pubescens	–	x	x	x	–	–	x	–	–	–	–	–	–	–	–	–	–	–	–	x	x	x	x	x	–	–	x	x	x	–	–	–	–	–	–
Rumex acetosa	–	–	–	–	–	–	–	–	–	–	–	–	–	–	–	–	–	–	–	–	–	–	–	–	–	–	–	–	x	–	–	–	–	–	–
Rumex acetosella	–	–	–	–	–	–	–	–	–	–	–	–	–	–	–	–	–	–	–	–	–	–	–	–	–	–	–	x	x	–	–	–	–	–	–
Rumex crispus	–	–	–	–	–	–	–	–	–	–	–	–	–	–	–	–	–	–	–	–	–	–	–	–	–	–	–	x	–	–	–	–	–	–	–
Rumex longifolius	–	–	–	–	–	–	–	–	–	–	–	–	–	–	–	–	–	–	–	–	–	–	–	–	–	–	–	–	x	–	–	–	x	–	–
Rumex mexicanus	–	–	–	–	–	–	–	–	–	–	–	–	–	–	–	–	–	–	–	–	–	–	–	–	–	–	–	–	x	x	–	–	–	–	–
Rumex obtusifolius	–	–	–	–	–	–	–	–	–	–	–	–	–	–	–	–	–	–	–	–	–	–	–	–	–	–	–	–	–	x	–	–	–	–	–
Rumex orbiculatus	–	–	–	–	–	–	–	–	–	–	–	–	–	–	–	–	–	x	–	–	x	–	–	x	–	–	–	–	x	x	–	x	–	–	x
Rumex pallidus	–	–	–	–	–	–	–	–	–	–	–	–	–	–	–	–	–	–	–	–	–	–	–	–	–	–	–	–	–	x	x	–	–	–	–
Ruppia maritima	–	–	–	–	–	–	–	–	–	–	–	–	–	–	–	–	–	–	–	–	–	–	–	–	–	–	–	–	–	–	–	x	x	–	–
Sagina nodosa	–	–	–	–	–	–	–	–	–	–	–	–	–	–	–	–	–	–	–	–	–	–	–	x	x	–	–	–	x	x	x	–	–	–	x
Sagina procumbens	–	–	–	–	–	–	–	–	–	–	–	–	–	–	–	–	–	–	–	–	–	–	–	–	–	–	–	–	x	x	–	–	–	–	x
Sagittaria graminea	x	–	–	–	–	–	–	–	–	–	–	–	–	–	–	–	–	–	–	–	–	–	–	–	–	–	–	–	–	–	–	–	–	–	–
Salicornia europaea	–	–	–	–	–	–	–	–	–	–	–	–	–	–	–	–	–	–	–	–	–	–	–	–	–	–	–	–	–	–	–	–	x	–	–
Salix arctica	–	–	–	–	–	–	–	–	–	–	–	–	–	–	–	–	–	–	–	–	–	–	–	–	x	–	–	–	–	–	–	–	–	–	–
Salix bebbiana	–	–	–	–	–	–	–	–	–	–	–	–	–	–	–	–	–	–	–	–	–	–	x	x	–	–	–	x	–	x	–	–	–	–	x
Salix candida	–	–	–	–	–	–	–	–	–	–	–	–	–	–	–	–	–	x	–	–	x	–	x	x	–	–	–	–	x	x	–	–	–	–	x
Salix cordata	–	–	–	–	–	–	–	–	–	–	–	–	–	–	–	–	–	–	–	–	–	–	x	–	x	–	–	–	–	x	–	–	–	–	–
Salix cordata var. rigida	x	–	–	–	–	–	–	–	–	–	–	–	–	–	–	–	–	–	–	–	–	–	–	–	–	–	–	–	–	x	–	–	–	–	–
Salix discolor	–	x	x	–	–	–	–	–	–	–	–	–	–	–	–	–	–	–	–	–	x	x	x	–	–	–	–	x	–	x	–	–	–	–	–
Salix glauca	–	–	–	–	–	–	x	x	–	–	–	–	–	–	–	–	–	–	–	–	–	–	–	x	x	x	–	–	x	–	–	–	–	–	–
Salix herbacea	–	–	–	–	–	–	x	x	–	–	–	–	–	–	–	–	–	–	–	–	–	–	–	–	–	–	–	–	–	–	–	–	–	–	–
Salix laurentiana f. glaucophylla	–	x	–	–	–	–	–	–	–	–	–	–	–	–	–	–	–	–	–	–	–	–	–	–	–	–	–	x	–	x	–	–	–	–	–
Salix lucida	x	–	–	–	–	–	–	–	–	–	–	–	–	–	–	–	–	–	–	–	–	–	x	–	–	–	–	–	–	x	–	–	–	–	–
Salix myrtillifolia	–	–	–	–	–	–	–	–	–	–	–	–	–	–	–	–	–	–	–	–	–	–	–	x	–	–	–	–	x	–	–	–	–	–	x
Salix pellita	–	–	–	–	–	–	–	–	–	–	–	–	–	–	–	–	–	–	–	–	x	–	x	–	–	–	–	–	–	x	–	–	–	–	–
Salix planifolia	–	–	–	–	–	–	x	–	–	–	–	–	–	–	–	–	–	–	–	–	–	–	x	–	–	–	–	–	–	x	–	–	–	–	–
Salix serrissima	–	–	–	–	–	–	–	–	–	–	–	–	–	–	–	–	–	–	–	–	x	–	–	–	–	–	–	–	–	–	–	–	–	–	–

Salix uva-ursi	–	–	–	–	–	–	x	–	–	–	–	x	x	–	–	–	–	–	–	–	–	–	–	x	–	–	–	–	–	–	–	–	–	–	–
Salix vestita	–	–	–	–	–	–	–	–	–	–	–	–	–	–	–	–	–	–	–	–	–	–	–	x	–	–	–	–	–	–	–	–	–	–	–
Salsola kali	–	–	–	–	–	–	–	–	–	–	–	–	–	–	–	–	–	–	–	–	–	–	–	–	–	–	–	–	–	x	x	–	–	–	–
Sambucus racemosa var. pubens	–	x	x	–	–	–	–	–	–	–	–	–	–	–	–	–	–	–	–	–	–	–	x	–	–	–	–	x	–	–	–	–	–	–	–
Sanguisorba canadensis	–	x	–	–	x	–	–	x	x	x	–	x	x	–	–	x	–	x	x	x	x	x	x	–	x	–	–	x	x	x	–	x	x	–	–
Sanicula marilandica	–	x	–	–	–	–	–	–	–	–	–	–	–	–	–	–	–	–	–	–	x	–	–	x	–	–	–	–	–	–	–	–	–	–	x
Sarracenia purpurea	–	–	–	–	–	–	–	–	–	–	–	x	–	x	–	x	x	x	x	x	x	x	x	–	x	–	–	–	–	–	–	–	–	–	–
Satureja vulgaris	–	–	–	–	–	–	–	–	–	–	–	–	–	–	–	–	–	–	–	–	–	–	–	x	–	–	–	x	–	–	–	–	–	–	–
Saxifraga aizoides	–	–	–	–	–	–	–	–	–	–	–	–	–	–	–	–	–	–	–	–	–	–	–	x	x	x	–	–	–	–	–	–	–	–	x
Saxifraga aizoon	–	–	–	–	–	–	–	–	–	–	–	–	–	–	–	–	–	–	–	–	–	–	–	x	–	–	–	–	–	–	–	–	–	–	x
Saxifraga caespitosa	–	–	–	–	–	–	–	–	–	–	–	–	–	–	–	–	–	–	–	–	–	–	–	–	–	–	–	–	x	–	–	–	–	–	x
Saxifraga oppositifolia	–	–	–	–	–	–	–	–	–	–	–	–	–	–	–	–	–	–	–	–	–	–	–	x	x	–	–	–	–	–	–	–	–	–	–
Saxifraga rivularis	–	–	–	–	–	–	x	–	–	–	–	–	–	–	–	–	–	–	–	–	–	–	–	–	–	–	–	–	–	–	–	–	–	–	–
Saxifraga stellaris var. comosa	–	–	–	–	–	–	x	–	–	–	–	–	–	–	–	–	–	–	–	–	–	–	–	–	–	–	–	–	–	–	–	–	–	–	–
Scheuchzeria palustris	–	–	–	–	–	–	–	–	–	–	–	–	–	–	–	x	–	x	–	–	–	–	–	–	–	–	–	–	–	–	–	–	–	–	–
Schizachne purpurascens	–	–	–	–	–	–	–	–	–	–	–	–	–	–	x	–	–	–	–	x	–	x	–	x	x	–	–	–	–	–	–	–	–	–	–
Schizaea pusilla	x	–	–	–	–	–	–	–	–	–	–	–	–	–	–	x	–	–	–	–	–	–	–	–	–	–	–	–	–	–	–	–	–	–	–
Scirpus americanus	–	–	–	–	–	–	–	–	–	–	–	–	–	–	–	–	–	–	–	–	–	–	–	–	–	–	–	–	–	–	–	x	x	–	–
Scirpus atrocinctus	–	x	–	–	–	–	–	–	–	–	–	–	–	–	–	–	–	–	–	–	–	–	x	–	–	–	x	–	–	–	–	–	–	–	–
Scirpus cespitosus	–	–	–	–	–	–	–	x	x	x	–	x	x	x	x	x	x	x	x	x	x	x	x	–	x	x	–	–	–	–	–	–	–	–	–
Scirpus hudsonianus	–	–	–	–	–	–	–	–	–	–	–	–	–	–	–	x	x	x	–	–	x	–	x	–	–	–	–	–	–	–	–	x	–	–	–
Scirpus lacustris ssp. glaucus	x	–	–	–	–	–	–	–	–	–	–	–	–	–	–	–	–	–	–	–	–	–	–	–	–	–	–	–	–	–	–	x	x	–	–
Scirpus microcarpus var. rubrotinctus	–	x	–	–	–	–	–	–	–	–	–	–	–	–	–	–	–	–	–	–	–	–	–	–	–	–	–	x	–	x	–	–	–	–	–
Scirpus rufus	–	–	–	–	–	–	–	–	–	–	–	–	–	–	–	–	x	–	–	–	–	–	–	–	–	–	x	–	–	–	–	x	x	x	–
Scirpus subterminalis	x	–	–	–	–	–	–	–	–	–	–	–	–	–	–	–	–	–	–	–	–	–	–	–	–	–	–	–	–	–	–	–	–	–	–
Scrophularia nodosa	–	–	–	–	–	–	–	–	–	–	–	–	–	–	–	–	–	–	–	–	–	–	x	–	–	–	–	–	–	–	–	–	–	–	–
Scutellaria epilobiifolia	–	–	–	–	–	–	–	–	–	–	–	–	–	–	–	–	–	–	–	–	–	–	x	x	–	–	–	–	x	x	–	–	–	–	x
Scutellaria lateriflora	–	–	–	–	–	–	–	–	–	–	–	–	–	–	–	–	–	–	–	–	–	–	x	x	–	–	–	–	–	–	–	–	–	–	–
Sedum rosea	–	–	–	–	–	–	–	–	–	–	–	–	–	–	–	–	–	–	–	–	–	–	–	–	–	–	–	–	–	–	–	–	–	–	x
Selaginella selaginoides	–	–	–	–	–	–	–	–	–	–	–	–	x	–	–	x	x	x	–	x	x	–	x	x	x	x	–	–	–	–	–	–	–	–	x
Senecio aureus	–	–	–	–	–	–	–	–	–	–	–	–	–	–	–	–	–	–	–	–	x	x	x	x	–	–	–	x	–	–	–	–	–	–	–
Senecio gaspensis	–	–	–	–	–	–	–	–	–	–	–	–	–	–	–	–	–	–	–	–	–	–	–	–	x	–	–	–	–	–	–	–	–	–	–
Senecio pauperculus	–	–	–	–	–	–	–	–	–	–	–	x	x	–	–	x	–	x	–	x	–	–	x	x	x	x	–	–	–	–	–	–	–	–	x
Senecio pseudo-arnica	–	–	–	–	–	–	–	–	–	–	–	–	–	–	–	–	–	–	–	–	–	–	–	–	–	–	–	–	x	x	–	–	–	–	–
Senecio vulgaris	–	–	–	–	–	–	–	–	–	–	–	–	–	–	–	–	–	–	–	–	–	–	–	–	–	–	–	–	x	–	–	–	–	–	x
Shepherdia canadensis	–	–	–	x	–	–	–	–	–	–	–	–	–	–	–	–	–	–	–	–	–	–	–	x	–	–	–	–	–	x	x	–	–	–	x
Sibbaldia procumbens	–	–	–	–	–	–	x	x	–	–	–	x	–	–	–	–	–	–	–	–	–	–	–	–	–	–	–	–	–	–	–	–	–	–	–
Silene acaulis	–	–	–	–	–	–	–	–	–	–	–	–	–	–	–	–	–	–	–	–	–	–	–	x	x	–	–	–	–	–	–	–	–	–	x
Sisyrinchium montanum	–	–	–	–	–	–	–	–	–	–	–	–	–	–	–	–	–	–	–	–	–	–	x	–	–	–	–	–	x	–	–	x	–	–	x
Smilacina stellata	–	–	–	–	–	–	–	–	–	–	–	–	–	–	–	–	–	–	–	–	–	–	x	x	–	–	–	–	x	x	x	–	–	–	x
Smilacina trifolia	–	–	–	–	–	–	–	x	x	–	–	–	x	x	–	x	x	x	x	–	x	x	x	–	–	–	–	–	x	–	x	–	–	–	–

Flora	Habitats 1	2	3	4	5	6	7	8	9	10	11	12	13	14	15	16	17	18	19	20	21	22	23	24	25	26	27	28	29	30	31	32	33	34	35
Solidago canadensis	–	–	–	–	–	–	–	–	–	–	–	–	–	–	–	–	–	–	–	–	–	–	–	–	–	–	x	–	–	–	–	–	–	–	–
Solidago hispida	–	–	–	–	–	–	–	–	–	–	–	–	x	–	–	–	–	–	–	x	–	x	–	x	x	x	–	–	–	–	–	–	–	–	x
Solidago lepida	–	–	–	–	–	–	–	–	–	–	–	–	–	–	–	–	–	–	–	–	–	–	–	–	–	–	–	–	–	x	–	–	–	–	–
Solidago macrophylla	–	x	x	–	x	–	x	x	x	x	x	x	x	–	–	–	–	–	–	–	–	x	x	x	–	–	x	–	–	–	–	–	–	–	–
Solidago multiradiata	–	–	–	–	–	–	–	–	–	–	–	–	–	–	–	–	–	–	–	–	–	–	–	x	x	–	–	–	–	–	–	–	–	–	x
Solidago purshii	–	–	–	–	–	–	–	–	–	–	–	–	–	–	–	x	–	–	–	–	–	–	–	x	–	–	–	–	–	–	–	–	–	–	–
Solidago rugosa	–	x	–	–	–	–	–	–	–	–	–	–	–	–	–	–	–	x	–	–	–	x	x	–	x	–	–	x	–	–	–	–	–	–	–
Solidago sempervirens	–	–	–	–	–	–	–	–	–	–	–	–	–	–	–	–	–	–	–	–	–	–	–	–	–	–	–	–	–	–	–	–	–	x	–
Solidago uliginosa	–	–	–	–	–	–	–	–	–	–	–	–	–	–	–	x	x	x	–	–	x	x	x	x	–	–	–	–	x	–	–	–	–	–	–
Solidago x calcicola	–	–	–	–	–	–	–	–	–	–	–	–	–	–	–	–	–	–	–	–	–	–	x	–	–	–	–	–	–	–	–	–	–	–	x
Sonchus arvensis	–	–	–	–	–	–	–	–	–	–	–	–	–	–	–	–	–	–	–	–	–	–	–	–	–	–	–	x	x	–	–	–	–	–	–
Sonchus oleraceus	–	–	–	–	–	–	–	–	–	–	–	–	–	–	–	–	–	–	–	–	–	–	–	–	–	–	–	x	–	–	–	–	–	–	–
Sorbus americana	–	–	x	–	–	–	–	–	x	–	–	–	–	–	–	–	–	–	–	–	–	x	–	–	–	–	–	–	–	–	x	–	–	–	x
Sorbus decora	–	x	x	x	x	–	–	x	x	–	x	–	–	x	–	–	–	–	x	x	–	x	x	x	–	–	–	–	–	–	–	–	–	–	–
Sparganium angustifolium	x	–	–	–	–	–	–	–	–	–	–	–	–	–	–	–	–	–	–	–	–	–	x	–	–	–	–	–	x	–	–	–	–	–	–
Sparganium chlorocarpum	x	–	–	–	–	–	–	–	–	–	–	–	–	–	–	–	–	–	–	–	–	–	x	–	–	–	–	–	–	–	–	–	–	–	–
Sparganium eurycarpum	–	–	–	–	–	–	–	–	–	–	–	–	–	–	–	–	–	–	–	–	–	–	x	–	–	–	–	x	–	–	–	–	–	–	–
Sparganium hyperboreum	x	–	–	–	–	–	–	–	–	–	–	–	–	–	–	x	–	–	–	–	–	–	–	–	–	–	–	–	–	–	–	–	–	–	–
Sparganium minimum	x	–	–	–	–	–	–	–	–	–	–	–	–	–	–	–	–	–	–	–	–	–	–	–	–	–	–	–	–	–	–	–	–	–	–
Spartina pectinata	–	–	–	–	–	–	–	–	–	–	–	–	–	–	–	–	–	–	–	–	–	–	x	–	–	–	–	–	–	x	–	x	x	–	–
Spergularia canadensis	–	–	–	–	–	–	–	–	–	–	–	–	–	–	–	–	–	–	–	–	–	–	–	–	–	–	–	–	–	–	–	x	x	–	–
Sphenopholis intermedia	–	–	–	–	–	–	–	–	–	–	–	–	–	–	–	–	–	–	–	–	–	–	–	x	–	–	–	–	–	–	–	–	–	–	–
Spiraea alba var. latifolia	–	–	–	–	–	–	–	–	–	–	–	–	–	–	–	–	–	x	–	–	–	x	x	–	–	–	–	–	x	–	–	–	x	–	–
Spiranthes romanzoffiana	–	–	–	–	–	–	–	–	–	–	–	–	x	–	–	–	–	x	–	–	x	–	–	–	–	–	x	–	–	x	–	–	–	–	–
Stachys palustris	–	–	–	–	–	–	–	–	–	–	–	–	–	–	–	–	–	–	–	–	–	–	–	–	–	–	x	–	–	–	–	–	–	–	–
Stellaria calycantha	–	–	–	–	–	–	x	x	–	–	–	–	–	–	–	–	x	–	–	–	–	–	–	–	–	–	–	–	–	–	–	–	–	–	x
Stellaria graminea	–	x	–	–	–	–	–	–	–	–	–	–	–	–	–	–	–	–	–	–	–	–	–	–	–	–	–	x	–	x	–	–	–	–	–
Stellaria humifusa	–	–	–	–	–	–	–	–	–	–	–	–	–	–	–	–	–	–	–	–	–	–	–	–	–	–	–	–	–	x	–	x	–	–	–
Stellaria longipes	–	–	–	–	–	–	–	–	–	–	–	–	–	–	–	–	–	–	–	–	–	–	–	–	–	–	–	x	x	–	–	–	–	–	x
Stellaria media	–	–	–	–	–	–	–	–	–	–	–	–	–	–	–	–	–	–	–	–	–	–	–	–	–	–	x	–	x	–	–	–	–	–	x
Streptopus amplexifolius	–	x	x	–	x	–	–	x	x	x	–	–	–	–	–	–	–	–	–	–	–	–	x	–	–	–	–	–	–	–	–	–	–	–	–
Streptopus roseus	–	x	x	–	x	–	–	x	x	x	–	x	–	–	–	–	–	–	–	x	–	x	x	x	x	–	–	–	–	–	–	–	–	–	–
Tanacetum huronense	–	–	–	–	–	–	–	–	–	–	–	–	–	–	–	–	–	–	–	–	–	–	–	–	x	–	–	–	–	–	–	–	–	–	–
Tanacetum vulgare	–	–	–	–	–	–	–	–	–	–	–	–	–	–	–	–	–	–	–	–	–	–	–	–	–	–	–	x	–	–	–	–	–	–	–
Taraxacum officinale	–	–	x	x	–	–	–	–	–	–	–	–	–	–	–	–	–	–	–	–	–	–	x	x	–	–	–	x	x	x	–	–	–	–	x
Taxus canadensis	–	x	x	–	x	–	–	–	x	–	x	–	–	–	–	–	–	–	–	x	–	x	x	x	–	–	–	–	–	–	–	–	–	–	–
Thalictrum alpinum	–	–	–	–	–	–	–	–	–	–	–	–	–	–	–	x	–	x	–	–	x	–	x	x	x	x	–	x	–	–	–	–	–	–	x
Thalictrum pubescens	–	x	–	–	–	–	–	–	–	–	–	–	–	–	–	–	–	x	–	x	x	x	x	x	x	–	x	–	x	x	–	x	x	–	–

Thelypteris limbosperma	- - - - - - - - x -
Thelypteris noveboracensis	- - x - - - - - - - - - - - - - - - - - - x - - - - - - - - - - - - -
Thelypteris palustris	- x - - - - - - - - - - - -
Thelypteris phegopteris	- x x - x - - x x - - - - - - - - - - - - x x - x - - - - - - - - - -
Tofieldia glutinosa	x - - - - - - - - - - - - - - x x x - - x - x - x x - - x - x - - - -
Tofieldia pusilla	- x - x - - - - - - - - -
Trientalis borealis	- x x x x - - x x - x x - x x x x x x x x x x x - - x - x - - - - - -
Trifolium agrarium	- x - - - - - - -
Trifolium hybridum	- x - - - - -
Trifolium pratense	- x x - - - - - -
Trifolium repens	- x - x - x - - -
Triglochin gaspense	- x x - -
Triglochin maritima	x - - - - - - - - - - - - - - x x x - - x - x - x - - - - - - x - - -
Triglochin palustris	x - - - - - - - - - - - - - - - x x - - - - - - x - - x x x x - - - -
Trillium cernuum	- x - - - - - - - - - - - -
Trisetum melicoides	- x - - - - -
Trisetum spicatum	- x x - - - x - - - - x x
Tussilago farfara	- x - - - - - - -
Typha latifolia	- x - - - - x - - - - - - -
Urtica dioica	- x - x - - - - x - x - - - - -
Utricularia cornuta	x - - - - - - - - - - - - - - x x x - - - - - - - - - - - - - - - - -
Utricularia intermedia	x - - - - - - - - - - - - - - - - x - - - - - - - - - - - - - - - - -
Utricularia minor	x - - - - - - - - - - - - - - x - x - - - - - - - - - - - - - - - - -
Utricularia vulgaris	x -
Vaccinium angustifolium	- x x - - x x x x - x x x x x - x x x x - x x - - - - - - - - - - - -
Vaccinium caespitosum	- - - - x - x x - x - x - - - - x - - - - - - x - - - - - - - - - - -
Vaccinium macrocarpon	x - - - - - - - - - - - - - - x x x - - - x x - - - - - - - - - x - -
Vaccinium ovalifolium	- - x - - - - x x - x - - - - - - - - - - x - - - - - - - - - - - - -
Vaccinium oxycoccos	- - - - - - - x x - - - - x x x x x x x x x x x - - - - x - x - - - x
Vaccinium uliginosum	- - - - x x x x x x - x x x x x x x x - - x - x x - - - x - - - - - x
Vaccinium vitis-idaea	- - x - - x - - x - x x x x x - x - x x - x - x x - - - x - - - - - -
Vahlodea atropurpurea	- - - - - - x x -
Verbascum thapsus	- x - - - - x - - - - - - -
Veronica americana	- x - - - x x - - - x - - x
Veronica arvensis	- x - - - - - - -
Veronica officinalis	- x - - - - -
Veronica scutellata	- x - - - - - -
Veronica serpyllifolia	- x x - x - - - - -
Veronica tenella	x - - - - - x -
Viburnum cassinoides	- x x - x - - - x - x - - x x - - - x x - x x - - - - - - - - - - - -
Viburnum edule	- x - - x - - - x - - - - - - - - - x - - x x x - - x - - x - - - - -

Flora	Habitats 1	2	3	4	5	6	7	8	9	10	11	12	13	14	15	16	17	18	19	20	21	22	23	24	25	26	27	28	29	30	31	32	33	34	35
Viburnum trilobum	—	—	x	—	x	—	—	—	—	—	—	—	—	—	—	—	—	—	—	—	—	—	x	—	—	—	—	—	—	x	—	—	—	—	—
Vicia cracca	—	—	—	—	—	—	—	—	—	—	—	—	—	—	—	—	—	—	—	—	—	—	—	—	—	—	—	x	—	x	x	—	—	—	—
Viola adunca	x	—	—	x	—	—	—	—	—	—	—	—	—	—	—	—	—	—	—	—	—	—	—	x	x	—	—	—	—	—	—	—	—	—	—
Viola cucullata	—	—	—	—	—	—	—	—	—	—	—	—	—	—	—	—	—	x	—	—	x	—	x	—	—	—	x	—	x	x	—	—	—	—	—
Viola incognita	—	x	x	x	x	—	x	x	x	—	—	—	—	—	—	—	x	—	—	x	x	—	x	—	—	—	x	—	x	x	—	—	—	—	x
Viola nephrophylla	—	—	—	—	—	—	—	—	—	—	—	—	—	—	—	—	—	—	—	—	—	—	x	—	—	—	—	—	—	—	—	—	—	—	x
Viola pallens	—	—	—	—	x	—	x	x	x	—	—	—	—	—	—	x	—	—	—	—	x	—	x	—	—	—	x	—	x	—	—	x	—	—	—
Viola palustris	—	—	—	—	—	—	x	x	—	—	—	—	—	—	—	—	—	—	—	—	—	—	—	—	—	—	—	—	—	—	—	—	—	—	—
Viola renifolia	—	x	x	x	—	—	—	—	—	—	—	—	—	—	—	—	—	—	—	—	—	x	x	x	—	—	x	—	x	—	—	—	—	—	x
Viola selkerkii	—	—	—	—	—	—	—	—	—	—	—	—	—	—	—	—	—	—	—	—	—	—	—	x	—	—	—	—	—	—	—	—	—	—	—
Viola septentrionalis	—	—	—	—	—	—	—	—	—	—	—	—	—	—	—	—	—	—	—	—	—	—	x	—	—	—	—	—	—	—	—	—	—	—	—
Woodsia alpina	—	—	—	—	—	—	—	—	—	—	—	—	—	—	—	—	—	—	—	—	—	—	—	x	—	—	—	—	—	—	—	—	—	—	—
Woodsia glabella	—	—	—	—	—	—	—	—	—	—	—	—	—	—	—	—	—	—	—	—	—	—	—	x	—	—	—	—	—	—	—	—	—	—	—
Woodsia ilvensis	—	—	—	—	—	—	—	—	—	—	—	—	—	—	—	—	—	—	—	—	—	—	—	x	—	—	—	—	—	—	—	—	—	—	x
Zannichellia palustris	—	—	—	—	—	—	—	—	—	—	—	—	—	—	—	—	—	—	—	—	—	—	—	—	—	—	—	—	—	—	—	x	x	—	—
Zostera marina	—	—	—	—	—	—	—	—	—	—	—	—	—	—	—	—	—	—	—	—	—	—	—	—	—	—	—	—	—	—	—	—	x	—	—

APPENDIX 2

Repartition of the flora by habitats and phytogeographical affinites

1, aquatic communities; 2, second growth balsam fir and white birch forests and scrubs (C.P.); 3, balsam fir forests (C.P); 4, white spruce scrubs (C.P.); 5, deciduous shrub thickets (A.P.); 6, exposed morainic barrens (C.P.); 7, rocky alpine cliffs and ravines; 8, zabois vegetation (snowbeds); 9, open balsam fir forests or scrubs (A.P.); 10, white spruce scrubs (A.P.); 11, tuckamoors; 12, tundra vegetation; 13, gabbro outcrops and talus; 14, ericaceous dwarf scrubs (C.P.); 15, black spruce dwarf scurbs (A.P.); 16, sedge bogs (A.P.); 17, peat bogs (C.P.); 18, sedge meadows (C.P.); 19, black spruce scrubs (C.P. peatlands); 20, serpentine scrubs; 21, American larch scrubs; 22, black spruce scrubs (C.P. forests); 23, creek/river beds and shoreline thickets; 24, calcareous inland cliffs, and talus; 25, serpentine barrens; 26, calcareous flowstone seepage slopes; 27, trails; 28, ruderal areas; 29, coastal meadows; 30, seashores; 31, sand dunes; 32, brackish marshes; 33, tidal flats; 34, brackish shores; 35, rocky coastal exposures.

C.P. = Coastal Plain: A.P. = Alpine Plateau.

Phytogeographical	Habitats																																		
groups	1	2	3	4	5	6	7	8	9	10	11	12	13	14	15	16	17	18	19	20	21	22	23	24	25	26	27	28	29	30	31	32	33	34	35	Total
C.aa	3.7	0	1.4	0	4.4	41.7	22.4	15.3	2.6	3.1	3.1	14.0	13.2	7.8	10.5	3.6	2.2	0.8	3.4	1.2	0	1.7	0.8	12.0	9.6	4.0	0	0.7	4.3	3.5	4.0	0	0	12.5	10.7	4.9
C.aa–d	0	0	0	0	0	0	3.0	0	1.3	0	0	0	0	0	0	0.9	0	0.8	0	0	0.9	0	0.4	1.3	1.6	8.0	0	0.7	0.6	2.8	4.0	1.2	0	0	3.3	0.9
AA.aa	0	0	0	0	0	16.7	9.0	6.9	0	0	0	7.0	5.7	0	0	2.7	0	0	0	0	0	0	0.8	2.2	4.8	4.0	0	0	1.2	0.7	2.0	0	0	0	5.3	1.6
NA–eA.aa	0	0	0	0	0	0	0	0	0	0	0	0	0	0	0	0	0	0	0	0	0	0	0	0	1.6	0	0	0	0	0	0	0	0	0	0	0.1
NA.aa	0	0	0	0	0	0	0	1.4	0	0	0	0	0	0	0	0.9	0	0	0	0	0	0	0	0.4	0.8	4.0	0.9	0	0	0.7	2.0	0	0	0	0	0.3
eNA.aa	0	0	0	0	0	0	1.5	0	0	0	0	1.8	1.9	0	0	0	0	0	0	0	0	0	0	0.9	0	0	0	0	0	0	0	0	0	0	0	0.2
																																			Arctic sub-Total:	8.0
C.b	37.8	19.6	20.5	24.1	24.4	8.3	14.9	26.4	28.6	31.3	15.6	21.1	24.5	27.5	21.1	33.6	35.5	35.1	29.3	22.0	33.0	25.2	26.2	17.3	23.2	28.0	17.6	16.2	27.4	20.3	28.0	28.8	23.6	18.8	21.3	24.8
C.b–d	2.4	0.9	1.4	6.9	2.2	0	4.5	2.8	2.6	3.1	3.1	1.8	1.9	2.0	5.3	0.9	3.2	2.3	0	3.7	1.7	2.5	2.0	3.6	4.0	4.0	4.6	0.7	3.0	4.2	4.0	7.5	10.9	18.8	2.7	3.0
AA.b	4.9	0.9	0	0	0	0	4.5	2.8	0	0	0	0	1.9	0	0	0.9	2.2	4.6	3.4	0	3.5	1.7	2.5	1.8	0.8	0	4.6	1.4	3.7	4.2	6.0	3.7	1.8	0	2.0	2.2
AB.b–d	0	0	1.4	0	0	0	0	1.4	2.6	0	3.1	0	0	0	0	0	0	0	0	1.2	0	1.7	0	0.9	1.6	0	0.9	0	0.6	1.4	0	1.2	0	0	0.7	0.6
NA–eA.b	2.4	7.5	8.2	6.9	6.7	0	3.0	6.9	5.2	6.3	9.4	5.3	5.7	5.9	5.3	1.8	4.3	3.1	5.2	4.9	7.0	7.6	5.7	3.6	3.2	0	4.6	4.2	3.7	2.1	6.0	5.0	0	0	5.3	4.7
NA.b	6.1	32.7	27.4	48.3	35.6	8.3	22.4	19.4	28.6	37.5	34.4	31.6	26.4	31.4	31.6	20.9	20.4	22.1	31.0	29.3	25.2	31.1	24.6	27.1	23.2	40.0	25.0	16.9	20.7	19.6	14.0	11.2	5.5	18.8	23.3	24.0
NA.b–d	1.2	0.9	0	0	0	0	0	0	0	0	0	0	1.9	0	0	0	0	0	0	0	0	0	0	0.4	2.4	0	0	0	0	0	0	1.2	1.8	0	0.7	0.3
eNA.b	12.2	10.3	11.0	6.9	11.1	8.3	7.5	12.5	13.0	9.4	18.8	12.3	9.4	11.8	15.8	11.8	12.9	13.7	10.3	13.4	12.2	12.6	9.8	7.1	8.8	4.0	8.3	4.2	7.3	7.0	4.0	6.3	12.7	0	5.3	9.6
eNA–eA.b	1.2	0	0	0	0	0	0	0	0	0	0	0	0	0	0	0.9	0	0	0	0	0	0	0.4	0	0	0	0	0	0.6	0	0	0	1.8	6.3	0.7	0.2
CORD.b–d	1.2	0	0	0	0	0	1.5	1.4	1.3	3.1	3.1	1.8	1.9	0	0	1.8	0	0	0	0	0.9	0	0.4	1.3	0.8	0	0	0	1.8	0.7	0	1.2	1.8	0	0.7	0.8
																																			Boreal sub-total:	70.2
C.b/t–d	0	0	0	0	0	0	0	0	0	0	0	0	0	0	0	0	0	0	0	0	0	0	0	0	0	0	0	0	0	0	0	2.5	9.1	0	0	0.2
C.t	0	0.9	1.4	0	0	0	0	0	0	0	0	0	0	0	0	0	0	0	0	0	0.9	0	0.8	1.3	0.8	0	1.9	2.1	1.8	0.7	2.0	1.2	1.8	0	0.7	0.7
AA.t	1.2	0.9	0	0	0	0	0	0	0	0	0	0	0	0	0	1.8	1.1	0.8	0	0	0.9	0	0.4	0.4	0	0	1.9	1.4	0.6	1.4	2.0	0	1.8	0	0	0.6
NA.t	2.4	4.7	5.5	0	2.2	0	1.5	0	0	0	0	0	0	0	0	0.9	0	0.8	0	1.2	1.7	1.7	2.9	0.9	1.6	0	3.7	2.1	1.8	4.9	2.0	2.5	5.5	0	0.7	1.8
DF–eA	0	0	0	0	0	0	0	0	0	0	0	0	0	0	0	0	0	0	0	0	0	0	0.8	0	0	0	0	0	0.6	0	0	0	0	0	0	0.1
DF	4.9	1.9	4.1	0	2.2	0	0	0	1.3	0	0	0	0	2.0	0	4.5	4.3	3.1	1.7	6.1	1.7	2.5	2.0	0.4	0.8	0	3.7	0	0.6	0.7	0	0	0	0	0	1.6
GLST–eA	0	0	1.4	0	2.2	0	0	1.4	1.3	0	0	0	0	0	0	0	0	0	0	0	0.9	0	0	0	0	0	0	0.7	0	0	0	0	0	0	0	0.2
GLST	11.0	11.2	13.7	0	8.9	0	0	0	11.7	0	9.4	0	1.9	11.8	7.9	5.5	9.7	11.5	10.3	14.6	6.1	9.2	10.2	6.2	6.4	4.0	9.3	3.5	3.0	4.9	2.0	1.2	3.6	0	4.7	6.8
																																			Temperate sub-total:	11.8
COSM	1.2	1.9	1.4	0	0	0	0	0	0	0	0	0	0	0	0	0	1.1	0.8	1.7	0	1.7	1.7	1.2	0.9	0	0	0.9	2.1	0.6	0.7	4.0	1.2	3.6	0	0.7	0.9
ATCP	3.7	0	0	0	0	0	0	0	0	0	0	0	0	0	0	3.6	1.1	0	0	0	0	0	0.8	0	0	0	0	0.7	1.8	3.5	6.0	7.5	5.5	12.5	0.7	1.1
END	0	0	0	0	0	16.7	3.0	1.4	0	3.1	0	1.8	1.9	0	2.6	1.8	1.1	0	0	1.2	0	0	0	4.4	3.2	0	0	0.7	0.6	1.4	0	2.5	1.8	0	2.7	1.2
INTR	2.4	4.7	1.4	6.9	0	0	1.5	0	0	3.1	0	0	0	0	0	0	0	0.8	0	1.2	0.9	0.8	5.7	4.9	0	0	10.2	40.8	13.4	14.0	8.0	13.7	7.3	12.5	7.3	6.0
?	0	0.9	0	0	0	0	0	0	0	0	0	1.8	1.9	0	0	0.9	1.1	0	3.4	0	0.9	0	1.2	0.4	0.8	0	1.9	0.7	0	0.7	0	0	0	0	0.7	0.6
Total species	84	107	76	28	45	12	66	73	77	32	32	57	53	51	38	110	93	131	58	82	114	119	239	225	125	24	107	139	164	140	51	79	55	16	151	711

5. FLORISTIC DATABANKS AND THE PHYTOGEOGRAPHIC ANALYSIS OF A TERRITORY

An example concerning northeastern Italy

LIVIO POLDINI, FABRIZIO MARTINI, PAOLA GANIS AND MARISA VIDALI

Department of Biology, The University of Trieste, I–34127 Trieste, Italy

INTRODUCTION

In recent years there have been rapid developments in the field of floristic databanks (Brenan et al. 1975; Allkin and Bisby 1984). Such banks allow a quick and efficient retrieval of a great number of phytogeographic data otherwise scattered in the literature or in herbaria. They can produce computerized distribution maps, or complex matrices of floristic, ecological and geographic data. The distribution maps may be processed by classification programs, to obtain groups of species with similar distribution (chorological types, geoelements); the matrices may be analyzed by multivariate methods to reveal and causally explain trends of phytogeographic variation. The computerization of floristic data and their numerical analysis led to a radical change in the methodological bases of phytogeography: Qualitative Phytogeography, where the formulation of hypotheses is mostly based on intuitive thinking, is moving towards Quantitative Phytogeography (Crovello 1981), where data analysis and inference must be based on a strict logical consistency.

In Italy, after the creation of a unified coding system (Cristofolini et al. 1969; Pignatti 1973, 1976, 1978, 1981), and of a central databank for the Italian Flora (Anzaldi and Mirri 1979 and 1980; Nimis 1981a, b, 1984c; Nimis et al. 1984), some local databanks have been also created, containing floristic-phytogeographic data at regional level. Some examples of such databanks are offered by Feoli Chiapella and Feoli (1979), Napoleone (1982), Nimis (1984a, b, d), Scimone et al. (1985, 1987).

The databank for the Floristic Cartography of Friuli–Venezia Giulia (Poldini et al. 1985; Poldini and Vidali 1985, 1986), based on a software by Lagonegro et al. (1982), contains ca. 100,000 field data, organized on a reference grid of 71 basic areas (Fig. 1) delimited as in the Project for the Floristic Mapping of Central Europe (Ehrendorfer and Hamann 1965; Niklfeld 1971; Karrer 1986); it contains also literature and herbarium data. Similar databanks exist also in neighboring regions outside Italy, e.g. Slovenia

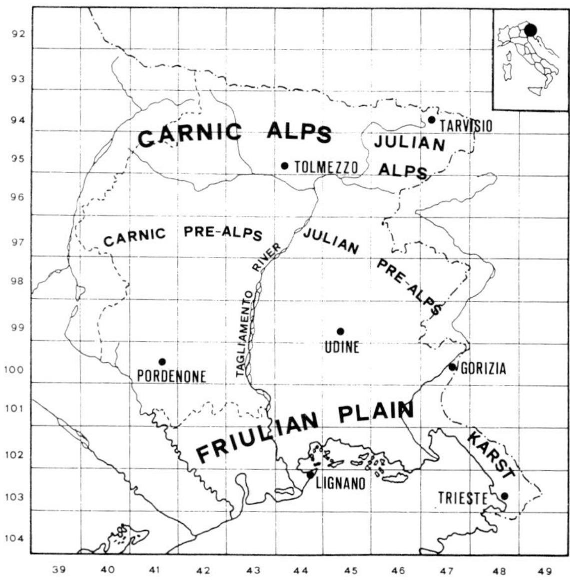

Fig. 1. Map of the Friuli–Venezia Giulia region, with the subdivision in basic areas (OGUs) according to the Project of Floristic Cartography of Central Europe.

P. L. Nimis and T. J. Crovello (eds): Quantitative Approaches to Phytogeography, pp. 159–181.

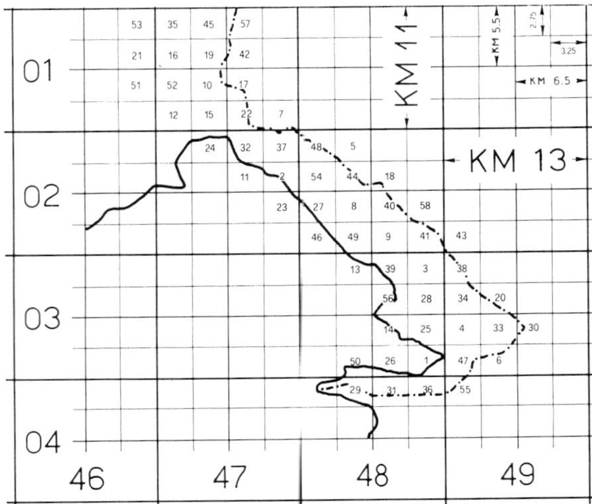

Fig. 2. Map of the north Adriatic Karst area, with the subdivision in basic areas, quadrants, and sections (see text). The sections, numbered from 1 to 58, were used as OGUs in the phytogeographic analysis of this area.

(Wraber 1968; Wraber and Skoberne 1989) and Carinthia (Hartl and Radic 1988; Otto and Kuhn 1988). The program package, initially addressed to the production of distribution maps on the model of Cadbury et al. (1971), Jalas and Suominen (1972–76), Billensteiner (1978), Voeth and Loeschl (1978) and Welten and Sutter (1982), was complemented by integration softwares (Ganis 1985) which have higher efficiency standards in data retrieval and organization.

In a preliminary paper on the northern Adriatic Karst region (Fig. 2) Poldini and Vidali (1985) used several outputs of the Databank to discuss the main methodologies of data retrieval and organization for phytogeographic purposes. Other results were published by Poldini and Vidali (1986) and Poldini et al. (1988).

Our approach attempts to clarify, on the basis of quantitative data, the correlations between qualitative, quantitative and structural elements of the flora (taxa, geoelements, life forms), and the grid units which organize the informations on the main abiotic factors (temperature, rainfall, potential solar radiation, lithology, etc.).

This paper presents examples of the use of the databank for the phytogeographic analysis of an area, and for the solution of phytogeographic problems, based on a quantitative approach.

2. DATA AND METHODS

2.1. Structure of the databank

Three types of Operational Geographic Units (OGUs, Crovello 1981) have been selected: (1) basic areas: 1/4th of a sheet of the IGM map of Italy (1:50,000); (2) Quadrants: 1/4th of a basic area; (3) Sections, 1/4th of a quadrant.

Floristic relevés, i.e. complete species lists, are available of each OGU. They are the basic data of the bank.

The bank includes two Operational Geographic Sets (OGS): (1) the entire Friuli–Venezia Giulia region; (2) the northern Adriatic Karst. In the first case the OGUs are 71 basic areas, in the second, being the area smaller and better explored, the OGUs are 58 sections.

The data are implemented by the ARCBASE program (Lagonegro et al. 1982) which creates archives formed by two types of records (Poldini and Vidali 1985):

(a) Locality record (leader), with the following information:

1. Number of the section
2. Number of the quadrant
3. Number of the basic area
4. Number of species records following the leader
5. Name of locality.

(b) Species record (follower), with the following information:

6. Pignatti's code (see Pignatti 1981)
7. Indication of the critical status of a species
8. Code of the subspecies
9. Free field available for data, chosen by the user
10. Name of the species.

The FASTDIZ program constructs dictionaries of species and localities. The species dictionary contains, for each species:

1. Chromosome number
2. Life form
3. Altitudinal distribution (minimum and maximum height)

4. Geoelement (chorological type).
5. Lithotype
6. Environment types
7. Landolt's ecological indices (Landolt 1977)
8. Phytosociological rank
9. Data source: Literature (A), Herbarium (E), Field (C).

The locality dictionary gives an alphabetical list of localities with the informations stored in the locality records.

2.2. Methods of data elaboration

The FUSAF package (Ganis 1985) gives absolute and percent values of species characters (life and growth forms, chorological types, etc.) in the OGUs. An example of output is shown in Figure 3, reporting the numbers of Arctic–Alpine taxa in the OGUs of the entire Friuli–Venezia Giulia region. A program of automatic mapping (SURFER), when applied to such data, transforms the discrete data structure into a continuous distributional model, producing isoporic maps (sensu Rothmaler 1955) such as that shown in Figure 4.

Evenness values (Pielou 1975) of the OGUs relative to genera and families have been calculated with programs described by Scimone et al. (1987).

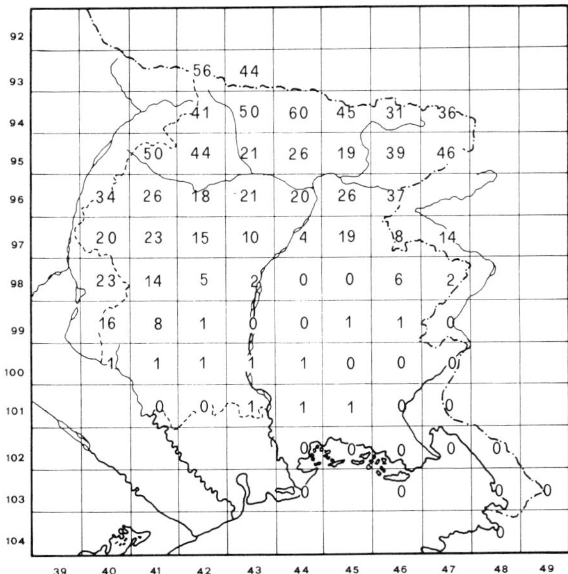

Fig. 3. Number of Arctic–Alpine species in each OGU.

Fig. 4. Isoporic map of the Arctic–Alpine geoelement.

The matrices of OGUs and characters have been analyzed by PCA (see Orlóci 1978) with the BIPLOT program (Lagonegro et al. 1985), which allows a reciprocal ordination of OGUs and characters.

In order to attain an optimal biogeographic subdivision of the Karst region on the basis of floristic, structural, chorological and ecological data, a matrix of species and OGUs has been constructed (1300 species and 58 OGUs). This matrix has been multiplied by the matrices of OGUs and, respectively, life forms, chorological types and ecological indices, obtaining three reduced matrices. In the case of the ecological indices, being these expressed on a 5-class ordinal scale, the data in the reduced matrix are weighted averages.

Hierarchic classifications of OGUs (Anderberg 1973) have been obtained from the three matrices through the EIVAVE program (Orlóci 1978). The discriminating power of the variables (species, life forms, chorological types and ecological indices) at different hierarchical levels of the classifications has been calculated with the NESTOFL program (Feoli et al. 1984).

The main factors affecting the distribution of the most important tree species in the Karst region have been studied by joint mapping of selected species and of some life forms, chorological types and ecological indices; this allows to relate the distribution of a species to factors which can be deduced by characters of the entire flora. These indirect data have been used to relate the Mediterranean dendroflora to aridity and to the percentage of Stenomediterranean species in each OGU.

Finally, the subdivision of the Karst based on the joint variable life-form–chorological type allowed to evaluate the degree of anthropization of the area. The analysis was made using the CROSSTAB program of the SPSS package (Nie et al. 1975), based on a contingency table with double entry (life-form–chorological type) as proposed by Feoli and Ganis (1984).

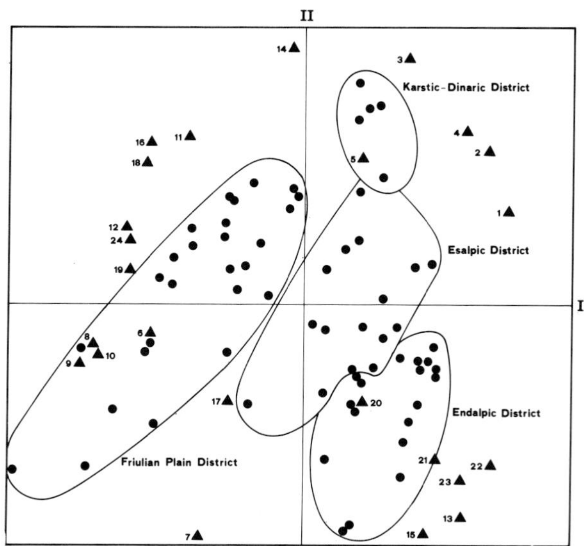

Fig. 5. Reciprocal ordering by PCA of OGUs (dots) and different variables. 1, no. of species; 2, no. of genera; 3, no. of families; 4, Shannon's index (SI) of genera; 5, SI of families; 6, evenness of genera; 7 evenness of families; 8, ratio families/genera; 9, ratio families/species; 10, ratio genera/species; 11, % of macrothermic species; 12, % of mesothermic species; 13, % of microthermic species; 14, % of anthropochore species; 15, % of endemic species; 16, temperature index; 17, humidity index; 18, eutrophication index; 19, soil texture index; 20, humus type index; 21, precipitation index; 22, elevation; 23, refugial stations; 24, % of artificial channels (see text).

3. RESULTS

3.1. Analysis of floristic diversity

The analysis of floristic diversity in the entire Friuli–Venezia Giulia region was based on a matrix of the 71 OGUs and the following variables: absolute number (richness), Shannon's indices (Shannon 1949) and evenness (Pielou 1975) of species, genera and families; ratios of families/genera, families/species, and genera/species, expressing the degree of taxonomic poorness of an area (Feoli and Scimone 1984); percentages of macro-, meso- and microthermic species (see later); percentages of anthropochore and of endemic species; weighted averages of Landolt's indices relative to: temperature, humidity, eutrophication, soil texture, humus type; mean yearly precipitation; elevation (expressed as difference between the highest and the lowest point in each OGU); refugial areas (unglaciated surfaces with high number of endemics); percent of surface occupied by artificial channels.

This matrix has been subjected to reciprocal ordering with PCA, and the results are shown in Figure 5. The main character points with positive scores on the first Principal Component refer to: number of species, genera and families, precipitation, elevation, refugial areas, percent of microthermic and endemic species; the character points with negative scores are: percents of meso-, macrothermic and anthropochore species, temperature, eutrophication, soil texture and humidity indices, surfaces occupied by artificial channels, evenness values, taxonomic poorness. The arrangement of the OGUs points is as follows: the OGUs of the Karst region, the Pre-Alps (Esalpic district) and the Alps (Endalpic district) are grouped into distinct point clusters and have positive scores on the first Principal Component; the lowland areas of Friuli have negative scores on the first Principal Component.

The results can be interpreted as follows: the mesooligotrophic areas (Karst, Pre-Alps, Alps) have high richness values and low evenness, while the eutrophicated areas of the Friulian Plain have low richness and high evenness values.

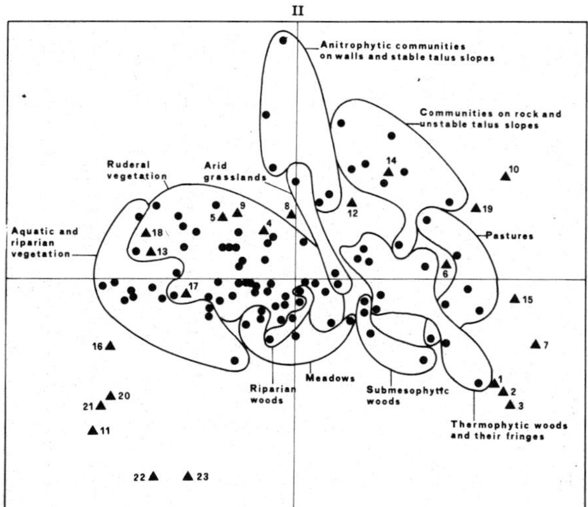

Fig. 6. Reciprocal ordering of 92 plant communities (dots) and some main variables, numbered as follows: 1, no. of species; 2, no. of genera; 3, no. of families; 4, evenness of genera; 5, evenness of families; 6, no. of life forms; 7, no. of geoelements; 8, evenness of life forms; 9, evenness of geoelements; 10, % of macrothermic species; 11, % of mesothermic species; 12, % of microthermic species; 13, % of antropochore species; 14, % of endemic species; 15, % of entomophilous species; 16, % of anemophilous species; 17, % of hydrophilous species; 18, % of autogamous species; 19, pH index; 20, eutrophication index; 21, humidity index; 22, soil texture index; 23, humus type index.

Floristic diversity seems to be related to the following main factors: elevation (range of ecological variation), precipitation, presence of refugial areas, and degree of oligotrophism. Noteworthy is that eutrophication seems to be related to a general decrease of floristic richness; this however could be due to the fact that in the region eutrophicated areas are those with lowest ecological diversity (lowland areas). In order to test this point, a similar analysis has been carried out on a matrix of 92 plant communities from a single, geomorphologically and climatically homogeneous area, the Karst region (Poldini 1989), and the set of characters reported in the legend to Figure 6. This matrix was submitted to Reciprocal Ordering with PCA, and the results are shown in Figure 6. In this figure the community types are grouped in 10 clusters obtained by numerical classification of the matrix of species and communities. The first Principal Component reveals a gradient of eutrophication, from seminatural woods and grasslands (positive scores) to ruderal and riparian vegetation (negative scores). Also in this case, the arrangement of the points relative to diversity values indicates a negative correlation, along the gradient, between eutrophication and diversity. Noteworthy is the arrangement of the character points relative to impollination types. Entomophilous impollination is related to low eutrophication values; anemophilous and hydrophilous impollination and autogamy are related to high eutrophication. Poldini and Vidali (1987) suggest that in oligotrophic systems plants tend to develop energy-saving adaptations, so that the energy-expensive anemophilous impollination is negatively selected. Autogamy is related to extreme environments (chiefly trampled areas) where plants can ensure reproduction also in absence of pollinating insects.

The results of the elaboration of community-data are in agreement with those deriving from geographic data; both indicate a strong influence of eutrophication on diversity values.

3.2. Interpretation of isoporic maps

This section presents examples of the interpretation of isoporic maps obtained from the data-bank in terms of climatical factors.

Fig. 7. Isoporic map of the genus *Gentiana* (see text).

The first example shows the correspondence of the isoporic maps of a taxon (genus *Gentiana*, Fig. 7) and of a geoelement (Arctic–Alpine, Fig. 4). Both have a maximum in the endalpic system (Poldini 1974) and sharply decrease south of the external Pre-Alps. Furthermore, both mark, with the isopory line of 1, a part of the Friulian Plain with extensive emergence of cold ground water, characterized by the occurrence of several microthermic glacial relicts.

The isoporic map of the Stenomediterranean element (Fig. 8) shows a very different distribution pattern, with a maximum on the Trieste Karst, and minor irradiations towards the eastern Pre-Alps. The isopory line of 7 marks a macrothermic area in the Friuli region, corresponding to relict stations of *Quercus ilex* on the southern slopes of the first mountain chains.

Those isoporic maps that can easily interpreted in ecological terms can be used for explaining the distribution patterns of given taxa. For example, they can substitute isotherms and isoietes, as is shown in Figure 9, where the distribution of the genus *Euphorbia*, rich in thermophytic species, corresponds well with the isoporic map of the Stenomediterranean element. Rather similar is also the distribution of the family *Orchidaceae* (Fig. 10), a group which is highly dependent on the thermic factor. In Figure 10 two main chorocenters are evident, one on the Karst region, the other in the mountains. They allow us to distinguish, within the family, two ecologically and phytogeographically distinct groups: (a) the microthermic group, rich in mono- or oligospecific genuses with circumboreal (*Ceologlossum, Corallorhiza, Goodyera, Listera, Malaxis*) or Arctic–Alpine (*Chamaeorchis, Nigritella, Pseudorchis*) distribution; (b) the macrothermic group, with

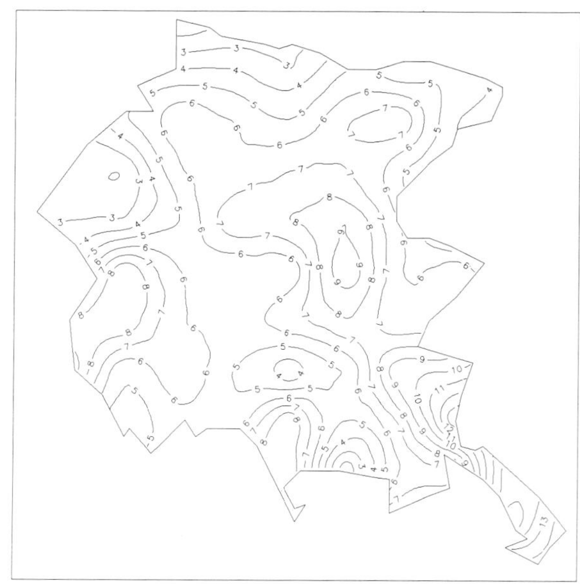

Fig. 9. Isoporic map of the genus *Euphorbia*.

Fig. 8. Isoporic map of the Stenomediterranean geoelement.

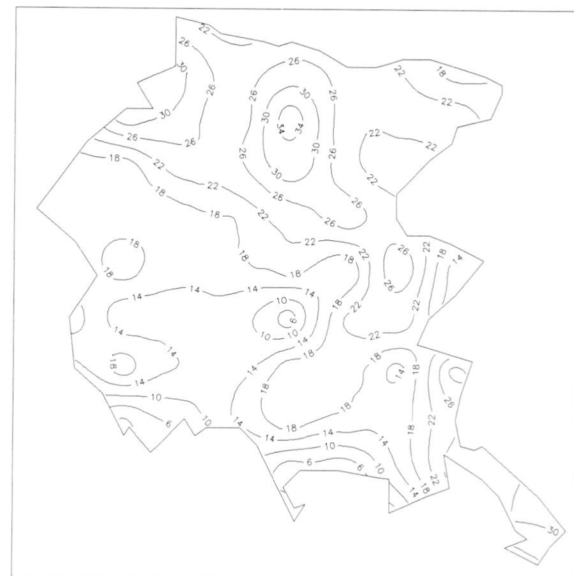

Fig. 10. Isoporic map of the *Orchidaceae*.

Fig. 11. Isoporic map of the *Genisteae.*

Fig. 13. Isoporic map of the Subatlantic geoelement.

genuses of mainly Mediterranean distribution (*Himanthoglossum, Ophrys, Serapias*).

The isoporic maps of the *Genisteae* (Fig. 11), and of the Endemic geoelement (Fig. 12) both have two chorocenters in the Pre-Alpine region. This phenomenon, according to Poldini (1974) and Martini (1987), is due to the existence of relict periglacial stations in the case of the endemics, and to present climatic conditions in the case of the *Genisteae*.

A last example concerns the distribution of the Subatlantic group in Friuli (Fig. 13), corresponding to areas with suboceanic climate in the north-eastern portion of the region, as is evident from the precipitation map (Fig. 14).

Following this kind of approach it is possible to test the existence of ecological–phytogeographic thresholds in a given area, using isoporic maps showing the distribution of taxa or of structural features of the flora.

Fig. 12. Isoporic map of the endemic geoelement.

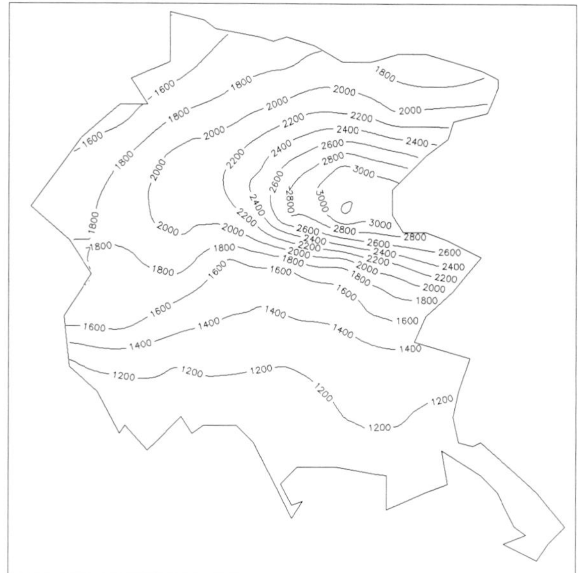

Fig. 14. Isoporic map of average yearly precipitations.

3.3. Phytogeographic subdivision of the entire region

A matrix reporting the percentages of 22 geoelements in the 71 OGUs has been submitted to numerical classification of the OGUs and to Reciprocal Ordering with PCA. The results are summarized in Figure 15, that shows the arrangement of the OGUs and of the geoelements according to the first two Canonical Variates, and the subdivision of the OGUs in four main groups, obtained by numerical classification. The four OGUs groups correspond to four main phytogeographic districts of the region, already recognized by Poldini (1974, 1987); according to the analysis of the discriminant power of the variables, the four districts are characterized by the following geoelements:

(1) Karstic–Dinaric district: southeast European, Pontic, Mediterranean–Pontic, Stenomediterranean, Eurimediterranean.

(2) Friulian Plain district: Palaeotemperate, Cosmopolitan, Mediterranean–Atlantic, Eurosiberian, Eurasiatic, Anthropochore.

(3) Esalpic district: European, Subatlantic.

(4) Endalpic district: Circumboreal, Alpine, east Alpine, Alpine–Carpathian, Arctic–Alpine, Mediterranean–montane, north Illyric, Endemic.

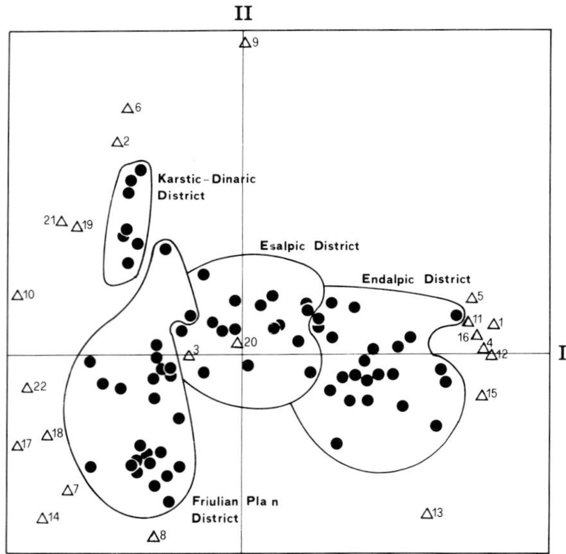

Fig. 15. Reciprocal Ordering by PCA of the OGUs (dots) and of the geoelements. The OGUs are grouped into four clusters (districts), obtained by numerical classification of the matrix of OGUs by geoelements. The geoelements are designed by a number, as follows: 1, Mediterranean–montane; 2, Pontic; 3, European; 4, Alpine, 5, N Illyric; 6, SE European; 7, Eurosiberian; 8, Eurasiatic; 9, S Illyric; 10, Eurimediterranean; 11, Endemic; 12, E Alpine, 13, Circumboreal, 14, Cosmopolitan, 15, Arctic–Alpine; 16, Alpine–Carpathic, 17, Palaeotemperate; 18, Mediterranean–Atlantic, 19, Stenomediterranean; 20, Subatlantic; 21, Mediterranean–Pontic; 22, Anthropochore.

The Eurimediterranean geoelement characterizes jointly the Karstic–Dinaric and the Friulian Plain districts. The south Illyric element joins the Karstic and the Esalpic districts.

The arrangement of the districts along the first Canonical Variate corresponds with a thermic gradient, from the Karst to the Endalpic district, where the highest elevations of the region are reached. The arrangement of the geoelements allows to distinguish three main groups, that appear to be conditioned mainly by temperature, i.e. the microthermic, macrothermic and mesothermic groups. The first two groups include geoelements showing clear trends of latitudinal and altitudinal variation. The mesothermic type is more or less constant throughout the study area, with fluctuations depending on local factors.

The existence of variation trends depending on elevation, or on latitude, is typical of "marginal" geoelements, i.e. of those that are at the margins of their main distributional centers. The microthermic marginal group includes the following geoelements: Arctic–Alpine, Circumboreal, Alpine, Alpine–Carpathian, eastern Alpine, north Illyric, Mediterranean–montane. The macrothermic marginal group includes the geoelements: Stenomediterranean, Eurimediterranean, south Illyric, southeastern European, Pontic and Mediterranean–Pontic. The Arctic–Alpine and Stenomediterranean elements appear as the most characteristic in the respective groups. The mesothermic group is "central" with respect to the study area, i.e. it includes geoelements whose range completely surrounds and includes the region, such as: European, Eurosiberian, Eurasiatic, Palaeotemperate, Cosmopolitan and Mediterranean–Atlantic geoelements.

3.4. Vectorial representation of the geoelements

In order to evaluate the main trends of phytogeographic variation within the region, variation diagrams of the geoelements have been constructed. From the geographic baricenter of the region, eight main axes have been drawn, corresponding to the main directions in the compass card. The OGUs (71 basic areas) have been grouped in eight subgroups of nine OGUs each, in such a way that each axis intersects a subgroup of OGUs. The eight subgroups roughly correspond to eight main phytogeographic subdivisions of the region (Alpine system, western and eastern high plain, Karst region, etc.). The average frequency of a given geoelement in each subgroup of OGUs has been reported on the respective axis in a variation diagram. The variation diagrams show the orientation of the main phytogeographic trends within the region.

Figure 16 reports the diagrams of the Arctic–Alpine and the Stenomediterranean geoelements, with opposed thermic requirements. Figures 17 and 18 show the diagram and the isoporic map of the European geoelement, without any significant geographic trend. Since the two geoelements have opposite thermic requirements, one would expect variation trends oriented in a N–S direction. However, in Figure 16 two directions of maximal anisotropism are evident, one oriented in a NNW–SSE (Arctic–Alpine geoelement), the other in a NW–SE direction (Stenomediterranean geoelement).

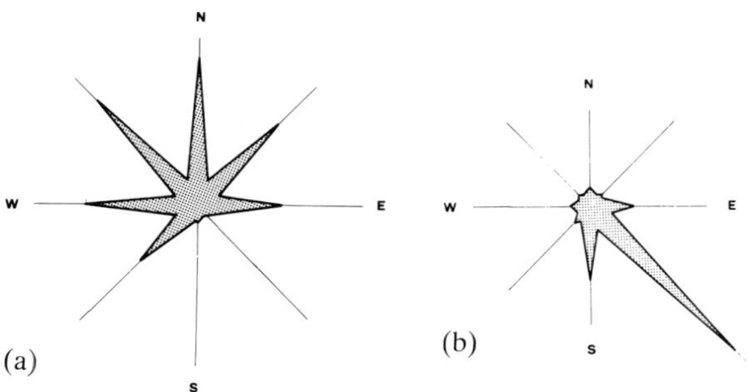

Fig. 16. Variation diagrams of the Arctic–Alpine (a) and Steno-mediterranean (b) elements (see text).

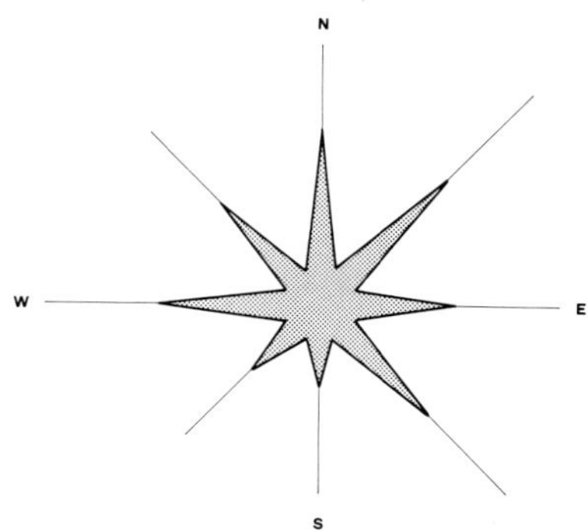

Fig. 17. Variation diagram of the European element (see text).

The former depends on the fact that the northwestern part of the region is characterized by the prevalence of acidic substrates and by a more continental climate, favoring the occurrence of Arctic–Alpine plants. The latter is due to the prevailing SE–NW orientation of the northern Dinaric Mountains and of the northern Adriatic Karst Plateau, which were preferential migration ways for southern plants during the postglacial period. This example shows that the direction of maximum floristic variation of a geoelement may be due both to present (ecological–climatical) and to historical factors.

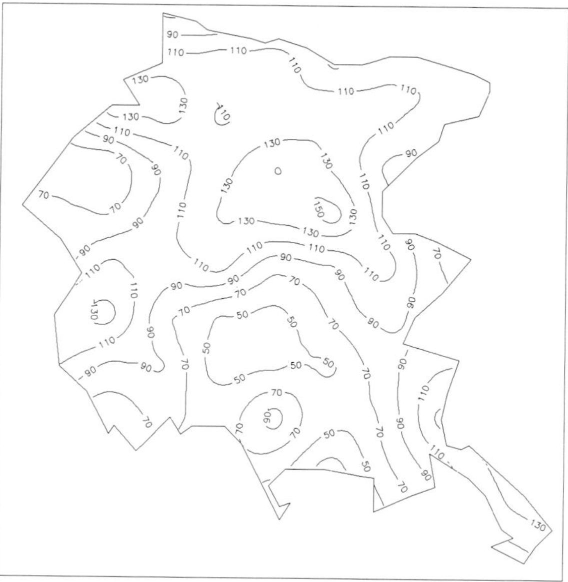

Fig. 18. Isoporic map of the European geoelement.

3.5. Phytogeographic subdivision of the northern Adriatic Karst

The phytogeographic subdivision of the northern Adriatic Karst has been attempted on the basis of four data sets: geoelements, life forms, floristic data, ecological indices.

The chorological spectrum of the flora of the northern Adriatic Karst (Fig. 19) shows a predominance of Eurasiatic taxa (33.1%), with a good incidence of the Pontic–Illyric and Eurimediterranean geoelements.

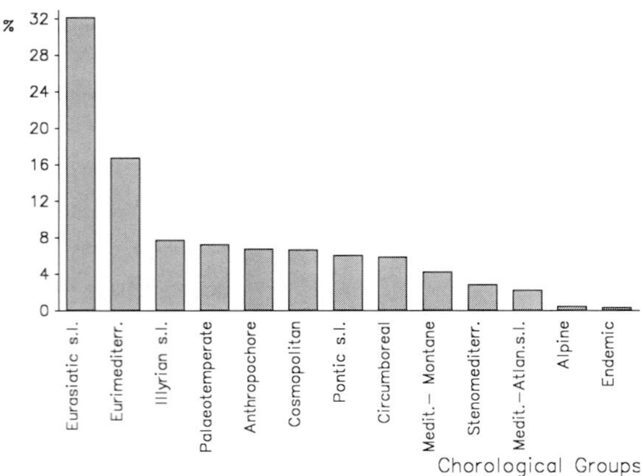

Fig. 19. Chorological spectrum of the flora of the Karst region.

Table 1. Chorological spectra of three different areas of the Friuli–Venezia Giulia region (see text).

	Trieste and Gorizia Karst	Osoppo Hills	Prescudin Reserve
Eurasiatic s.l	32.2	39.9	33.2
Eurimediterranean	16.8	10.9	2.6
Palaeotemperate	7.3	10.0	6.3
Illyric s.l.	7.8	5.1	6.3
Pontic s.l.	6.1	4.3	1.6
Cosmopolitan	6.7	4.9	2.4
Circumboreal	5.9	8.5	11.2
Mediterranean–montane	4.3	7.7	21.4
Anthropochore	6.8	4.0	0.6
Mediterr.–Atlantic s.l.	2.3	1.2	0.6
Stenomediterranean	2.9	0.8	–
Alpine s.l.	0.5	0.6	6.3
Endemic	0.4	0.8	3.8
Arctic–Alpine	–	0.6	3.2
Total of species	1300	466	490

Eurasiatic s.l. = Eurasiatic + European + Eurosiberian
Illyric s.l. = S-Illyric + N-Illyric + SE-European
Pontic s.l. = Pontic + Mediterranean-Pontic
Mediterr.–Atlantic s.l. = Mediterr.–Atlantic + Subatlantic
Alpine s.l. = Alpine + E-Alpine + W-Alpine + Alpine-Carpathic

This spectrum has been compared with the analogous spectra of two areas located along the Pre-Alpine arc of the Friuli region, to detect possible trends of phytogeographic variation from the coastal Karst to the Pre-Alpine sector. The two areas are: Colli di Osoppo, in the Province of Udine (Fornaciari 1961) and the Experimental Reserve of Prescudin, in the Province of Pordenone (Poldini 1986). The three spectra are shown in Table 1: the incidence of the large group of Eurasiatic species, including European and Eurosiberian species, is almost constant, in spite of the orographical and mesoclimatical differences among the three areas; this group defines the main phytoclimatical characters of the entire regional flora, indicating a relatively cool macroclimate. The phytogeographic differentiation between Coastal Karst and Pre-Alpine area is given by two geoelements: the Mediterranean–Pontic, which prevails in the Karst region, and the Mediterranean–montane, prevailing along the Pre-Alps.

The life form spectrum of the Karst flora is shown in Figure 20; hemicryptophytes are the most frequent life-form type, followed by therophytes, geophytes and phanerophytes. Since this area is transitional between the Mediterranean and the temperate vegetation zones, these data have been compared with analogous spectra of different parts of the world (Table 2) published by Walter (1960), in order to evaluate the at-

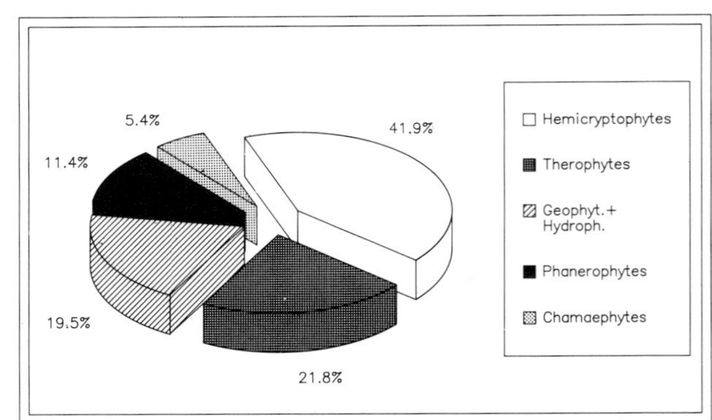

Fig. 20. Life form spectrum of the flora of the Karst region.

Table 2. Life form spectra of different climatic zones, compared with that of the north Adriatic Karst.

	P	Ch	H	G	T
Tropics:					
1. Seychelles	61.0	6.0	12.0	5.0	16.0
Desert zone:					
2. Libyan desert	12.0	21.0	20.0	5.0	42.0
3. Cyrenaika	9.0	14.0	19.0	8.0	50.0
4. Dead Sea area	4.0	7.5	1.5	5.0	82.0
Mediterranean zone:					
5. Italy	12.0	6.0	29.0	11.0	42.0
6. Palestine	5.5	7.5	23.0	13.0	51.0
Temperate zone:					
7. Parisian basin	8.0	6.5	51.5	25.0	9.0
8. Central Switzerland	10.0	5.0	50.0	15.0	20.0
9. Denmark	7.0	3.0	50.0	22.0	18.0
Artic zone:					
10. Spitzbergen	1.0	22.0	60.0	15.0	2.0
11. Greenland	–	27.5	52.5	18.0	2.0
Snow zone:					
12. Alps	–	24.5	68.0	4.0	3.5
13. Karst	11.5	5.5	41.9	19.3	21.8

G (Geophytes) = Helophytes + Idrophytes + Geophytes
P (Phanerophytes) = Phanerophytes + Nanophanerophytes

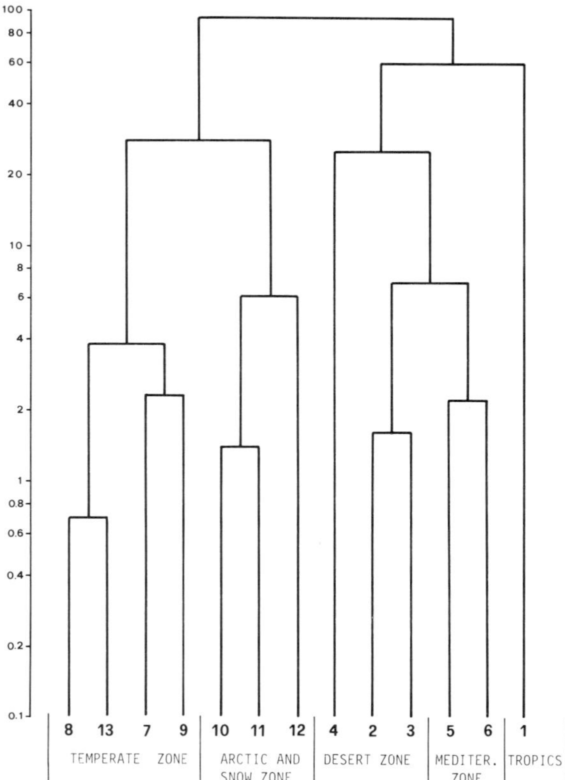

Fig. 21. Classification of different floras on the basis of life form spectra (see Table 2). The numbers refer to different regions, as in Table 2.

tribution of the area to either zone on the basis of structural features. The results of classification and ordination, based on the data in Table 2, are shown in Figures 21 and 22, respectively. The Karst clearly pertains to the temperate zone, characterized by a high incidence of hemicryptophytes, and not to the Mediterranean zone, due to the low incidence of therophytes.

3.5.1. Subdivision based on floristic data

The numerical classification of a matrix of 1300 species and 58 OGUs produced four main clusters of OGUs, whose location is shown in Figure 23. Program NESTOFL allowed to detect, for each cluster, a group of species with the highest discriminant value with respect to the other clusters. The main characteristics of the four groups of discriminant species are as follows:

– Cluster I: Eurimediterranean, Stenomediterranean and south Illyrian species with optimum on calcareous cliffs, on reefs and in the Mediterranean maquis; mostly phanerophytes and chamephytes.
– Cluster II: European, SE European, Circumboreal (and Alpine) species of submesophytic woods, scrub vegetation and meadows; mostly hemicryptophytes and geophytes.
– Cluster III: Pontic, Mediterranean–montane, Eurosiberian and Eurasiatic species with optimum in stony pastures, thermophytic woods and their margins; mostly chamaephytes.
– Cluster IV: Anthropochore, Palaeotemperate and Cosmopolitan species; mostly therophytes and hemicryptophytes of anthropized and disturbed environments.

The clusters I and II of OGUs correspond to the most inaccessible zone of the coast (I) and to the internal Karst (II), i.e. to the less disturbed, more 'natural' parts of the study area. Clusters

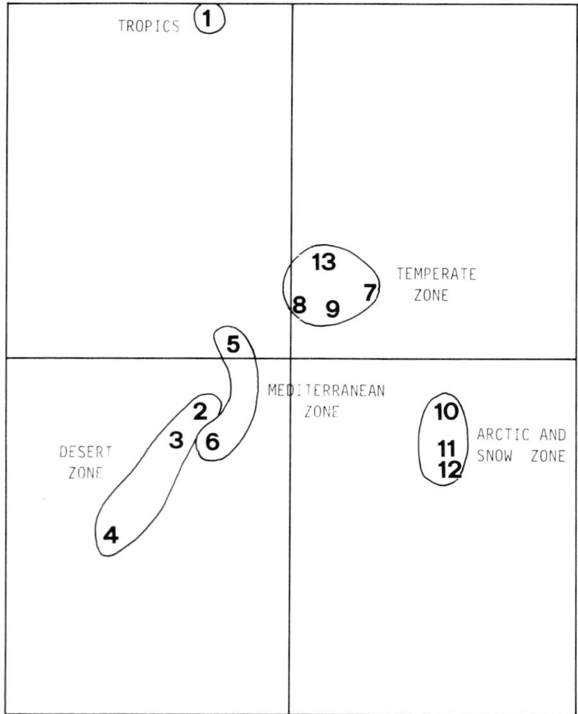

Fig. 22. Ordination of different regions on the basis of their life form spectra (see Table 2). The regions are numbered as in Table 2.

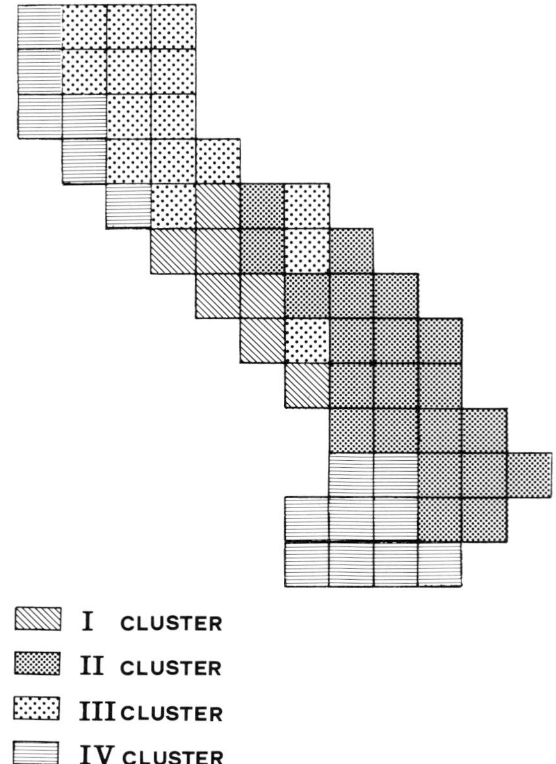

Fig. 23. Subdivision of the north Adriatic Karts into four clusters, obtained by the numerical classification of the matrix of OGUs by species.

III and IV include OGUs with highest anthropic disturbance.

Among the discriminant species of cluster III it is possible to distinguish two species sets, one including archeophytes such as *Ballota nigra* L. subsp. *foetida* Hayek, and *Papaver rhoeas* L., and apophytes such as *Ranunculus acris* L. and *Leontodon hispidus* L., occurring in manured meadows. The other set includes species whose occurrence is bound to phenomena of secondary succession in abandoned pastures, such as *Dictamnus albus* L., *Genista tinctoria* L., *Mercurialis ovata* Sternb. & Hoppe.

Cluster IV includes the OGUs with the highest concentration of settlements and industrial areas; it is characterized by species such as *Silene alba* (Mill.) Kranje, *Reseda lutea* L., *Conyza albida* Willd., *Sorghum halepense* (L.) Pers., *Verbascum blattaria* L., *Rumex obtusifolius* L., *Amaranthus deflexus* L., *Carduus acanthoides* L., *Eleusine indica* (L.) Gaertn.

Summarizing, the classification based on floristic data subdivides the territory into two main parts; the first, less influenced by man's action, includes two sectors, one with a Mediterranean (cluster I) the other with a submontane (cluster II) climate. The second part, more influenced by man's action, also includes two sectors, one with prevailing industrial activities (cluster IV), the other mainly exploited for agricultural purposes (cluster III).

These results show that the use of floristic data for the phytogeographic subdivision of a territory does not always allow to distinguish phytoclimatically homogeneous areas. In our case, the subdivision of the "natural" area in two sectors (clusters I and II) actually reflects climatical differences, whereas the distinction of clusters III and IV depends on the degree and nature of anthropization. In order to obtain a better phytoclimatical subdivision of the territory it is necessary to rely on other types of data, such as life forms, geoelements, etc., which, in any case, are derived from the basic floristic lists.

3.5.2. *Subdivision based on life forms and chorotypes*

The numerical classification of OGUs based on the percent of life forms produced two main groups (Fig. 24), one including the coastal area, the other the innermost Karst region.

Hemicryptophytes have a high incidence in both OGUs groups; they indicate a temperate macroclimate. The variables with the highest discrimination power are scapose therophytes, and suffruticose chamaephytes. The former sharply decrease from the coast (19.7%) towards the interior (11.4%), the latter have an opposite behavior (4.4% in the coast, 6.1% in the interior). The high percentage of therophytes in the coastal strip is due both to climatic factors and to anthropization (see later).

The classification of OGUs based on the percentages of geoelements produced three main clusters (Fig. 25). The European, Eurimediterranean and Stenomediterranean geoelements have the highest discriminant power (Table 3). Cluster I includes all OGUs whith fragments of evergreen Mediterranean vegetation and is characterized by Stenomediterranean species. Cluster II includes all other coastal areas, and is characterized by Eurimediterranean species. Cluster III includes the interior Karst, and is characterized by European species.

Table 3. Average percentage values of the most discriminant variables in the classification of OGUs based on chorological groups (see text).

	Cluster		
Chorological groups	I	II	III
Stenomediterranean	5.3	1.75	0.9
Eurimediterranean	21.65	17.6	14.4
European	10.95	11.0	16.6

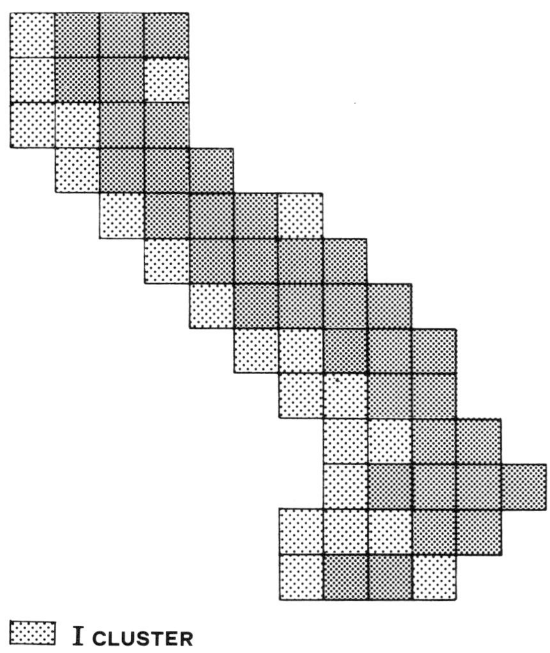

Fig. 24. Subdivision of the north Adriatic Karst into two clusters, on the basis of the numerical classification of the matrix of OGUs and life forms.

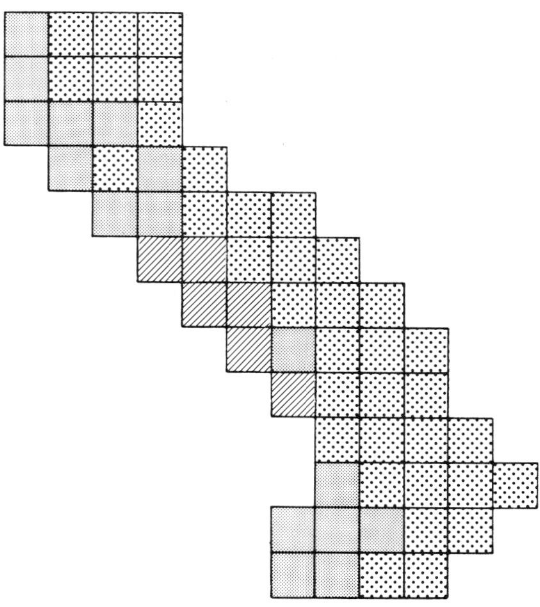

Fig. 25. Subdivision of the north Adriatic Karst into three clusters on the basis of the numerical classification of the matrix of OGUs and geoelements.

The partitions obtained on the basis of life forms (Fig. 24) and of chorological groups (Fig. 25) reflect better a phytoclimatical subdivision of the territory, since they correspond well to the distribution of the thermal energy.

3.5.3. *Subdivision based on ecological indices*

Three Landolt indices have been taken in consideration for this analysis, i.e. those concerning humidity of the soil, thermic range (altimetric belts) and continentality.

The classification of the matrix of OGUs and of the five classes of the humidity index produced four main clusters (Fig. 26): two clusters (II–III) include the innermost, mesic sections; cluster IV includes the OGUs with a more humid climate (presence of small lakes and ponds); cluster I includes the OGUs of the coast, with a relatively high aridity, as shown in Figure 27 and Table 4.

Table 4. Average percentage values of the most discriminant variables in the classification of the OGUs based on the humidity index (see text).

	Cluster			
	I	II	III	IV
Water-soaked soil	0.5	0.2	1.9	7.2
Humid to very humid soil	1.5	3.7	6.3	9.3
Medium dry to humid soil	19.7	31.9	31.9	26.9
Dry soil	39.9	42.1	37.8	30.5
Very dry soil	40.3	20.7	21.4	21.1

Four clusters have been also obtained by the classification based on the temperature index (Fig. 28). Cluster I includes the coastal Mesomediterranean sections with arid soils and a maritime climate; cluster II defines a more internal, lower Supramediterranean belt, with intermediate climatical characters. Cluster III includes the higher Supramediterranean belt and cluster IV the inner, higher Supramediterranean belt with Oromediterranean tendencies, where temperature is the most discriminant factor (Ozenda 1975). Figure 29 and Table 5 show the decreasing thermic gradient from the coastal belt to the interior.

In the case of the continentality index, due to the small area of the Karst (about 554 km^2) we took into consideration only two of the five index classes (those indicating maritime to weakly

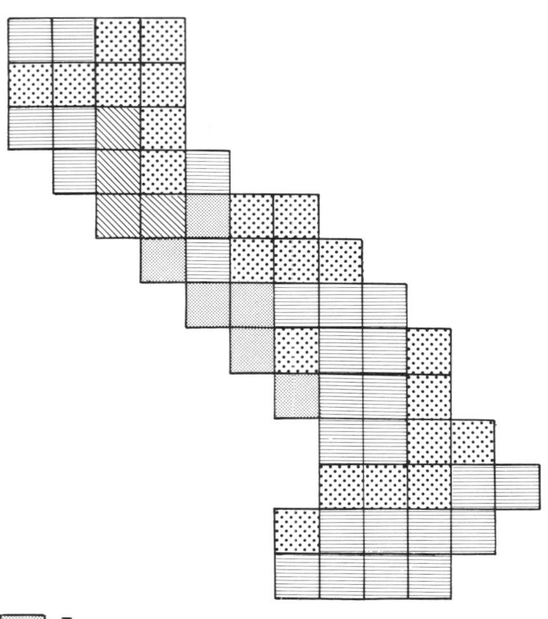

Fig. 26. Subdivision of the north Adriatic Karst into four clusters, on the basis of the numerical classification of the matrix of OGUs and five classes of the humidity index (see text).

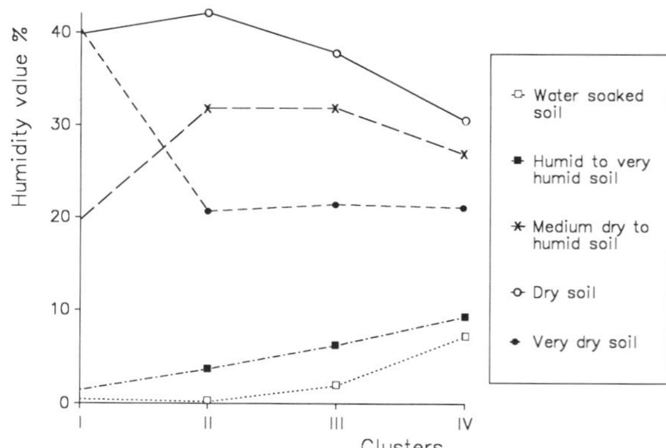

Fig. 27. Diagram showing the values of five different classes of the humidity index in the four clusters of OGUs shown in Figure 26.

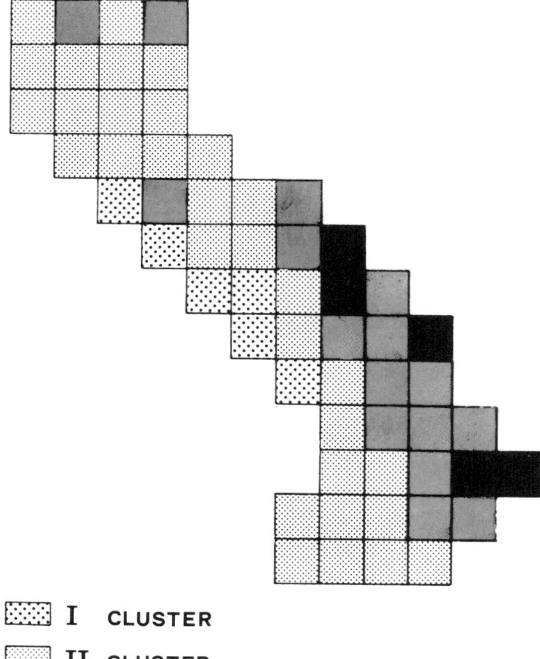

Fig. 28. Subdivision of the north Adriatic Karst into four clusters, on the basis of the classification of the matrix of OGUs and five classes of the temperature index (see text).

continental climates). The matrix has not been submitted to numerical analysis, and the subdivision of the territory has been obtained directly by the weighted averages. The results are reported in the map of Figure 30, showing a coastal with a maritime climate and a inner area with a weakly continental climate; the two areas presumably have different precipitation regimes.

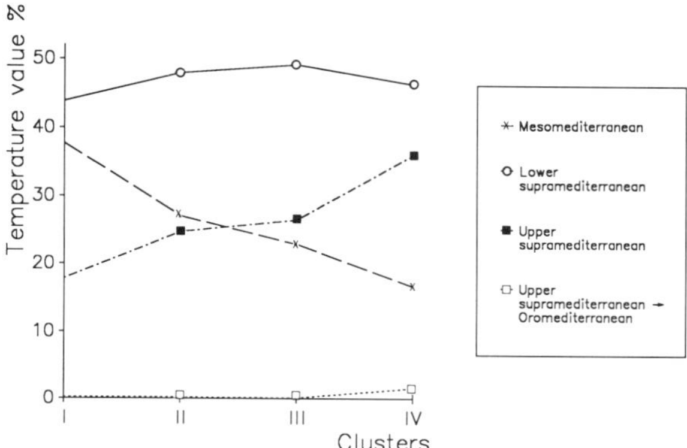

Fig. 29. Diagram showing the values of five different classes of the temperature index in the four clusters of OGUs shown in Figure 28.

Table 5. Average percentage values of the most discriminant variables in the classification of OGUs based on the temperature index.

	Cluster			
	I	II	III	IV
Mesomediterranean	37.7	27.1	22.9	16.7
Lower Supramediterranean	43.9	48.0	49.2	46.3
Upper Supramediterranean	17.9	24.8	26.5	36.0
Upper Supram.→Oromedit.	0.2	0.2	1.0	1.6
Oromediterranean	–	–	0.1	0.1

3.6. Indirect evaluation of the ecology of a species

In order to clarify the ecological behavior of a species within a territory one can rely on direct ecological data. This, however, is not always possible. The alternative is an indirect evaluation based on ecological parameters derived from the indicator value of the flora.

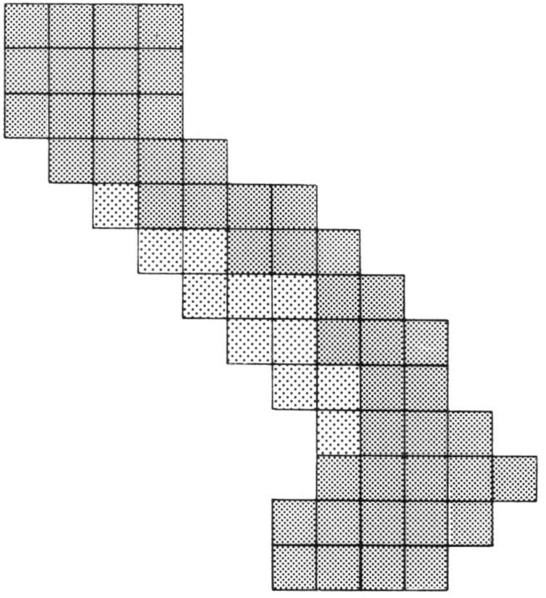

Fig. 30. Subdivision of the north Adriatic Karst into two clusters, based on the continentality index (see text).

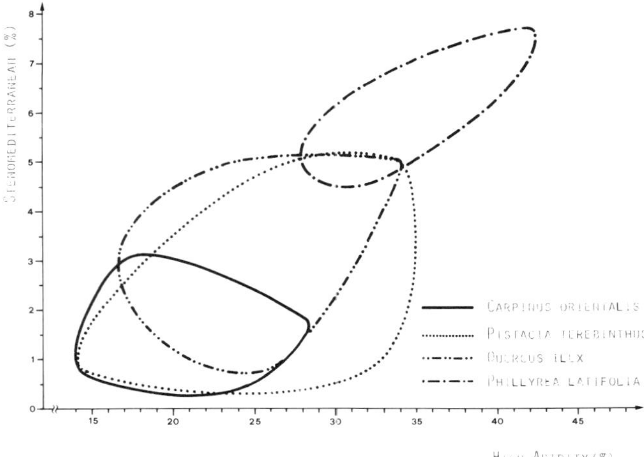

Fig. 31. Schematic representation of the ecological range of four tree species, with respect to the percentage of Stenomediterranean species and of the aridity index in each OGU.

The example reported here concerns the following tree species: *Quercus ilex* L., *Carpinus orientalis* Mill., *Pistacia terebinthus* L., *Phyllirea latifolia* L., *Carpinus betulus* L. Their local distribution has been compared with the distribution patterns of the following combined variables: life form (geophytes), geoelements (Stenomediterranean and European), abiotical factors derived from Landolt's indices (humidity/aridity of the soil).

Figure 31 shows the arrangement of the first 4 species, that have a Mediterranean-Submediterranean distribution, according to two related factors: index of aridity and percentage, in each OGU, of Stenomediterranean species; a clear gradient is evident, where: *Carpinus orientalis*, a Submediterranean tree, appears as the most mesophytic of the four species; its local behavior corresponds with its ecology in southeastern Europe. *Quercus ilex* and *Pistacia terebinthus* have a higher tolerance for aridity and temperature, the latter with a somewhat broader local distribution indicating a slightly wider ecological range. *Phyllirea latifolia* is restricted to conditions of high aridity and temperature.

To analyze the ecology of *Carpinus betulus*, a more northern species, locally restricted to humid woods rich in geophytes, we compared its distribution with that of the geophytes (Fig. 32) and

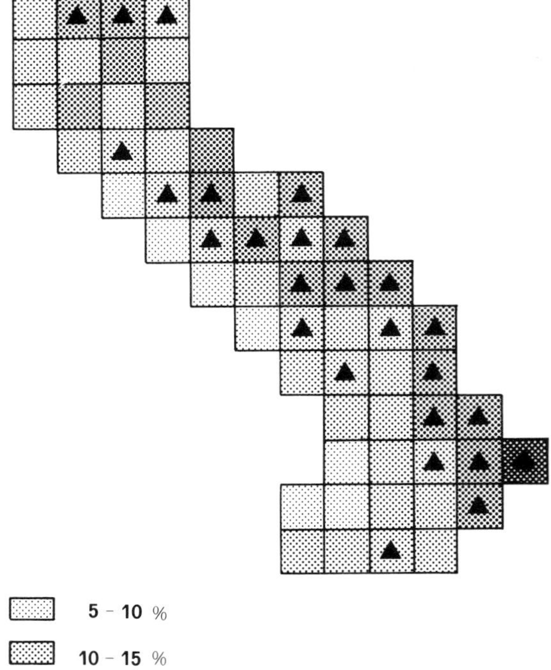

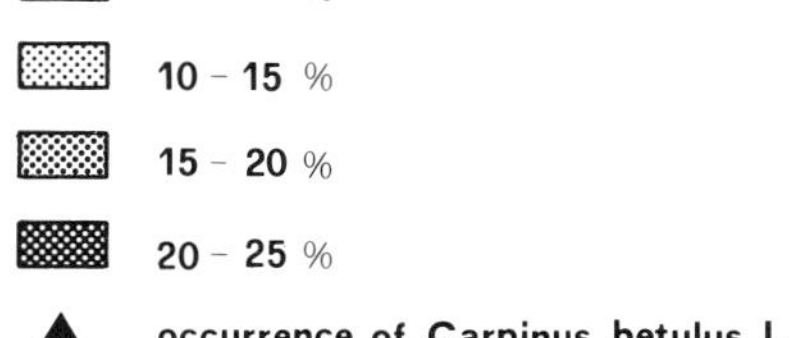

Fig. 32. Distribution of *Carpinus betulus* with respect to the percentage occurrence of geophytes in each OGU.

of the European species (Fig. 33). The combination : presence of *Carpinus betulus*, high frequency of geophytes and of European species allows to distinguish the parts of the territory with a cooler climate.

3.7. The 'floristic pollution' of a territory

Man's impact on natural ecosystems has led to the creation of environments where ecological conditions radically differ from those prevailing in the natural vegetation. In the temperate zone, anthropogenous environments generally have higher thermic ranges and a lower moisture regime than in the buffered microclimates of the climax vegetation. This causes the selection of a peculiar flora, with particular kinds of dispersal mechanisms, and of morphological and physiological adaptations to the new niches available

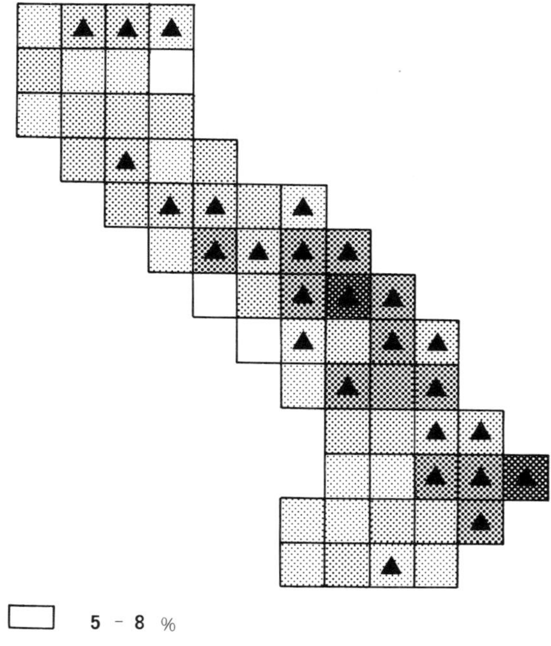

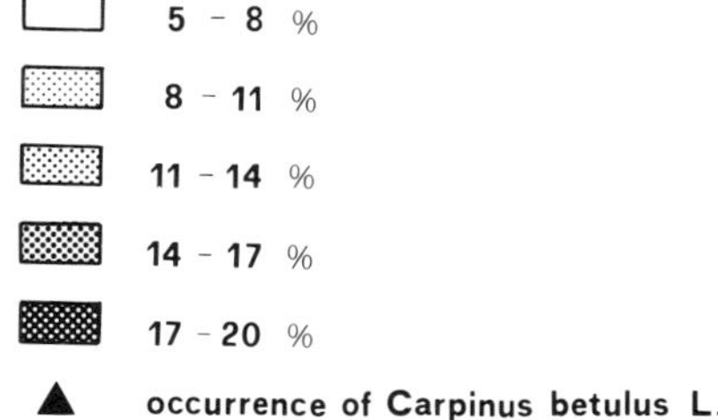

Fig. 33. Distribution of *Carpinus betulus* with respect to the percentage occurrence of European species.

Fig. 34. Isoporic map of anthropophytic species.

for plant life. These new niches were also the scene of impressive phenomena of naturalization, in particular regarding migrants from drier regions (Lausi and Nimis 1985).

In his floristic catalogue of Friuli-Venezia Giulia, Poldini (1980) traces a first balance of the alteration of the autochthonous flora by the introduction and naturalization of esotic species. The degree of "floristic pollution" of a territory can be estimated as the percentage of introduced species on the total flora. Compared with other European countries (Poland 57%, Czechoslovakia 32%) the floristic pollution in Friuli-Venezia Giulia appears rather low (6.5%). Figure 34 shows the distribution of the anthropophytes within the region. The lowlands, where most of the settlements are located, have more anthropophytes than the mountains. Within the lowlands, there is a good correlation between the maximum concentrations of anthropophytes and the location of the main urban and industrial settlements: Trieste, Monfalcone, Pordenone, Udine, Lignano and most of the coasts. An exception, marked by the isopory line 11 are some semi-natural areas (risorgive) under environmenatal protection. The progress of the isopory line 15 from the plain northward marks the penetration of anthropophytes into the Alpine territory along the Tagliamento–Fella valleys, crossed by important highways and railways.

3.8. The anthropization in the northern Adriatic Karst

Two main aspects of anthropization are considered in the following. They are: (a) variations in floristic composition, in the structural characters of the flora and in its phytogeographic affinities; (b) the role of the autochthonous flora in the recolonization of disturbed biotopes (apophytism).

Among the colonizers of disturbed biotopes there are both foreign anthropochore species and apophytes; the latter may be subdivided into autapophytes (shifting from primary to secondary biotopes) and deuterapophtes (autochthonous

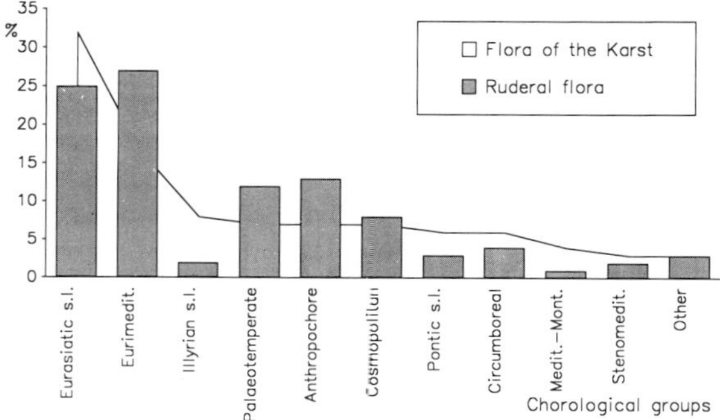

Fig. 35. Comparison between the phytogeographic spectra of the ruderal flora and of the total Karst flora.

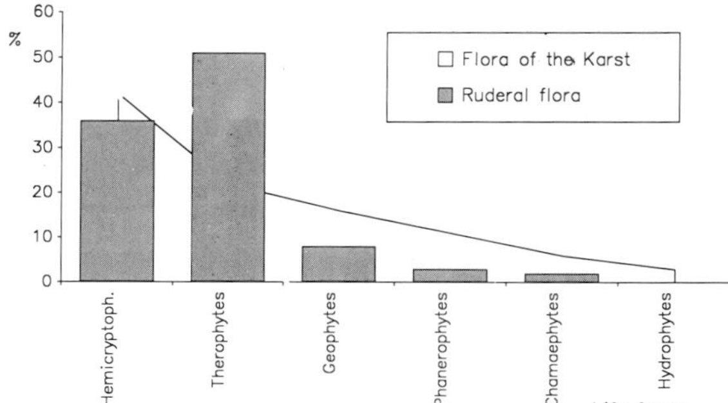

Fig. 36. Comparison between the life form spectra of the ruderal flora and of the total Karst flora.

species, occurring only in secondary biotopes) (Holub and Jirasek 1967).

We define as "synanthropic" both anthropochore and deuterapophyte species. In the entire region, 429 species occur in heavily disturbed sites, of which 244 are synanthropic and 185 are autapophytes.

Figure 35 shows the chorological spectrum of the total Karst flora and of the ruderal environments (synanthropic species and autapophytes). Table 6 reports also the variation in the percentages of the chorological groups between the two spectra. The Mediterranean–montane, Illyrian and Pontic geoelements are the most sensitive to disturbance, while the Anthropochore, Palaeotemperate, Eurimediterranean and Cosmopolitan

Table 6. Chorological spectra of the total flora of the northern Adriatic Karst, of its ruderal flora, and their variation.

Chorological groups	Total flora	Ruderal flora	Variation
Anthropochore	6.8	13.0	91.2
Palaeotemperate	7.3	12.5	71.2
Eurimediterranean	16.8	27.5	63.7
Cosmopolitan	6.7	8.0	19.4
Mediterr.–montane	4.3	0.7	−83.7
Illyric s.l.	7.8	1.7	−78.2
Pontic s.l.	6.1	2.6	−57.4
Stenomediterranean	2.9	1.8	−39.6
Circumboreal	5.9	4.0	−32.2
Eurasiatic s.l.	23.2	25.5	−20.8
Other	3.2	2.8	−14.1

Eurasiatic s.l. = Eurasiatic + European + Eurosiberian
Illyric s.l. = S-Illyric + N-Illyric + SE-European
Pontic s.l. = Pontic + Mediterranean–Pontic
Other = Atlantic + Endemic + Alpine

species appear to be favored by human action. The species with a broad distribution seem to be less sensitive to disturbance; this might be due to their broader ecological range.

The life-form spectra of the total Karst flora and of the ruderal environments are compared in Figure 36, and Table 7 shows the variation in the percentages of the life forms between the two floras. Phanerophytes, chamaephytes and geophytes prevail in natural environments, hemicryptophytes are more or less constant, while therophytes increase by 131.4% in the ruderal environments.

Among the several classification of OGUs, based on different characters, there is one which appears to be of particular interest for the study of anthropization; this is the classification based on the joined variable life form–chorological group. This classification produced five main clusters, whose distribution is shown in Figure 37. The most discriminant variables are reported in Figure 38 and in Table 8. The scapose–anthro-

Table 7. Life form spectra of the total flora of the northern Adriatic Karst, of its ruderal flora, and their variation.

Life forms	Total flora	Ruderal flora	Variation
Therophytes	21.8	50.5	131.4
Phanerophytes	11.4	3.6	− 68.4
Chamaephytes	5.4	2.0	− 64.1
Geophytes	19.5	8.0	− 59.0
Hemicryptophytes	41.9	35.9	− 14.9

Geophytes = Geophytes + Hydrophytes + Helophytes
Phanerophytes = Phanerophytes + Nanophanerophytes

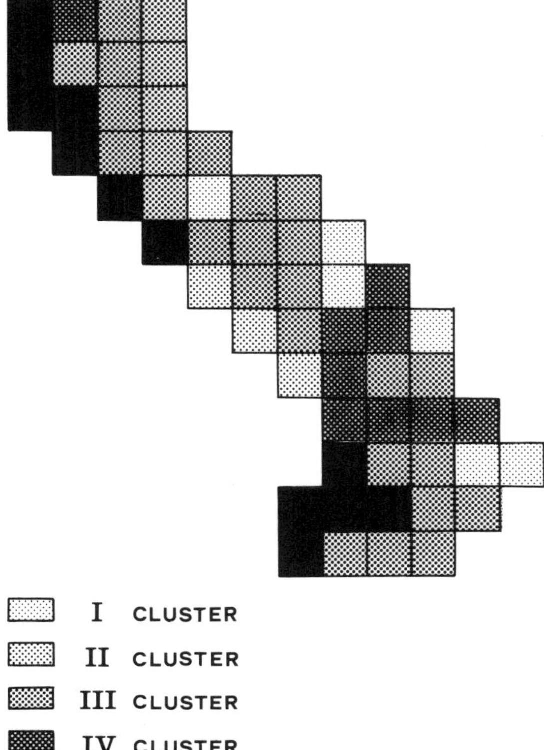

Fig. 37. Subdivision of the north Adriatic Karst into five clusters on the basis of the numerical classification of the matrix of OGUs by the joint variables life form geoelement.

pochore and the scapose–cosmopolitan therophytes reach their maximum in the parts of the coasts with highest urbanization, coinciding with cluster V of OGUs. The Eurasiatic–rhizomatose geophytes characterize the cooler areas, with dolines hosting *Carpinus betulus* L. (cluster I of

Table 8. Average percentage values of the joint variable life form geoelement in the classification of the OGUs based on this variable.

Life forms – Chorological groups	Cluster I	II	III	IV	V
T scap–Anthropochore	0.5	0.9	1.2	1.8	3.6
Ch suffr–S Illyric	0.8	1.6	0.7	0.9	0.2
P caesp–Eurimediterranean	–	1.3	0.5	0.5	0.4
T scap–Cosmopolitan	1.5	2.1	2.3	2.7	4.4
G rhiz–Eurasiatic	1.9	0.4	0.7	1.1	0.3

OGUs). The south Illyrian-suffruticose chamaephytes and the Eurimediterranean-cespitose phanerophytes characterize cluster II of OGUs, corresponding to rocky, impervious areas and to the Mediterranean belt. Clusters III and IV have no discriminant variables, since they include areas where heavy disturbance and semi-natural conditions coexist. In cluster IV there is a weak increase of Eurasiatic rhizomatous geophytes, due to the presence, in this area, of several dolines.

Figure 39 shows the primary habitats of the 185 autapophytic species. The recolonization of disturbed biotopes occurs mainly by means of species of the stony grasslands, followed by those of humid environments, meadows, etc. The sequence of the habitats in Figure 39 probably depends on three main factors: (a) the extension of a given habitat in the survey area; (b) the fact that disturbed biotopes are more favorable to species growing in early successional stages of

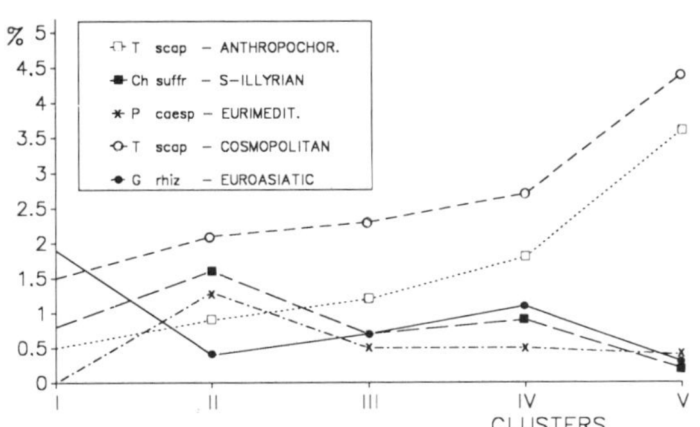

Fig. 38. Diagram showing the values of the joint variables life form geoelements in the five clusters shown in Figure 37 (see Table 8).

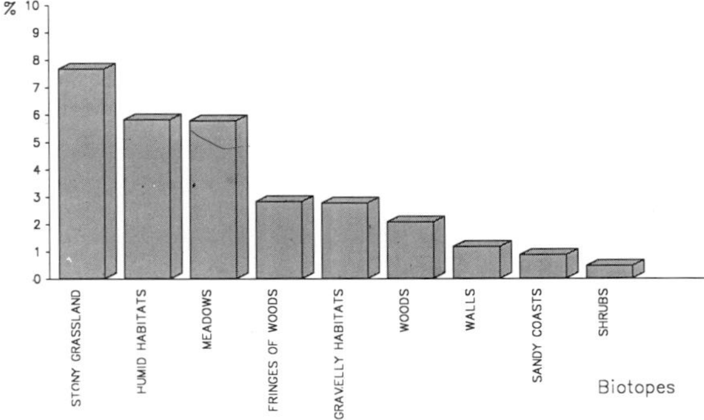

Fig. 39. Percentages of autapophytes colonizing ruderal environments within natural environments in the north Adriatic Karst.

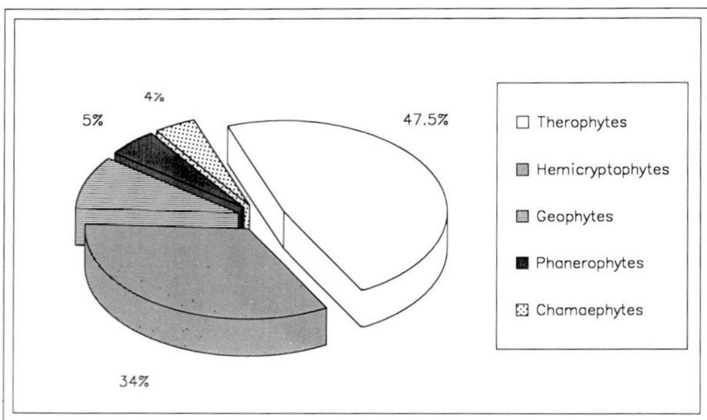

Fig. 40. Life-form spectrum of the autapophytes of the north Adriatic Karst.

	Family	Provenance	Life form	Chorol.group	Habitat
Anthropochore	Compositae Gramineae	N-America	Therophyte	Adventitious	Waste places
Deuterapophytes	Compositae Gramineae	Autochthon.	Therophyte	Eurimedit.	Waste places
			Hemicrypt.	Palaeotemp. Cosmopolit. Eurasiatic	
Autapophytes	Compositae Gramineae	Autochthon.	Therophyte	Eurimedit. Palaeotemp. Cosmopolit.	Stony grassl. Gravel.habit. Meadows
			Hemicrypt.	Palaeotemp. Eurasiatic Eurimedit.	Humid habit. Meadows Stony grassl.

Fig. 42. Schematic representation of the phytogeographic and ecological characteristics of synanthropic species of the north Adriatic Karst.

the xeroseries (stony grasslands); (c) the high degree of eutrophication in disturbed sites, selecting species from relatively eutrophized semi-natural or natural environments (meadows).

Figure 40 shows the life-form spectrum of the autapophytic species. The predominant life forms are therophytes and hemicryptophytes. Figure 41 gives a synthetic view of the phytogeographic and ecological characteristics of autapophytes with these two life forms. The most important role in the recolonization of disturbed biotopes is played by Eurimediterranean therophytes spreading from stony grasslands, gravelly habitats and meadows, followed by Palaeotemperate hemicryptophytes of humid biotopes, meadows and stony pastures.

In the classification of OGUs based on life forms (cf. Fig. 24) the scapose therophytes characterize the coastal strip. Since they prevail in ruderal environments also, the problem arises whether their high incidence along the coasts (Fig. 24) is due to climatic factors or to disturbance. Figure 41 shows that, in the whole of the survey area, Eurimediterranean therophytes play an important role in the colonization of disturbed sites. The Eurimediterranean geoelement is the main reservoir of therophytes in the survey area; their presence in disturbed environments is probably due to microclimatical factors (aridity, strong insolation). However, the prevalence, along the coastal strip, of anthropochore and cosmopolitan therophytes (Fig. 37), and not of Eurimediterranean ones, indicates that the high incidence of this life form along the coasts is not due solely to climatic reasons, but is probably influenced also by the high degree of anthropization of the coastal area.

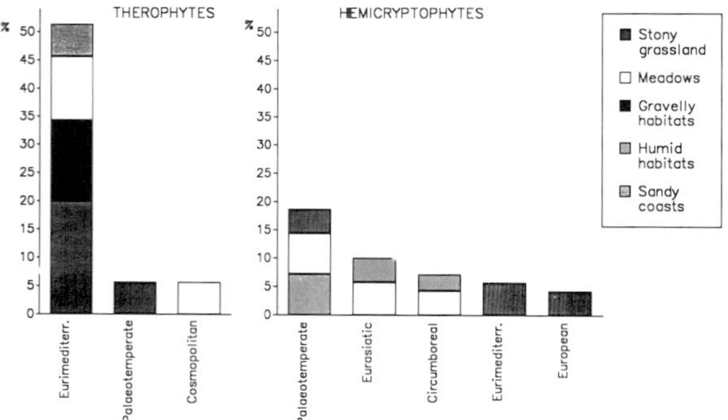

Fig. 41. Subdivision of the autapophytes of the north Adriatic Karst by geoelements, habitats and life forms (only the two most frequent life forms are considered).

Figure 42 gives a synthesis on the main taxonomic, phytogeographic and ecological features of the synanthropic flora in the survey area.

3.9. Relation between autapophytic flora and environment

In order to study the relations between the autapophytic flora and some main environmental factors, chiefly temperature, the OGUs (basic areas) surrounding the main town of the entire region have been analyzed in terms of: ratio therophytes/hemicryptophytes, ratio Eurimediterranean/mesothermic species, percentage of

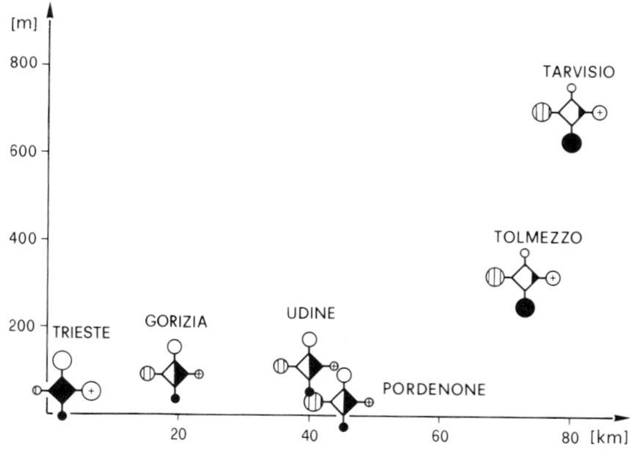

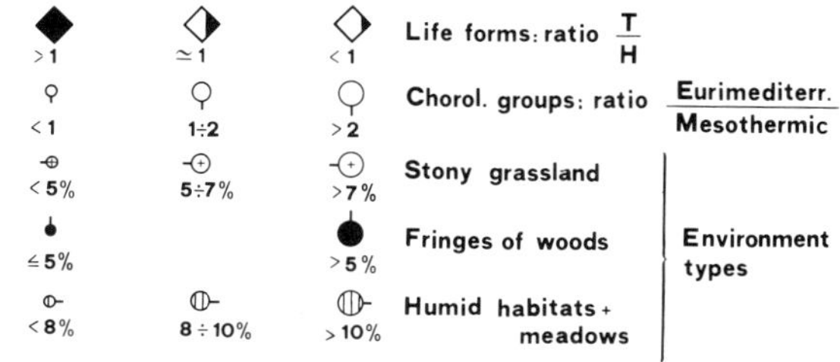

Fig. 43. Variation of some structural and ecological characters of the autapophytic flora surrounding the main settlements of the region Friuli–Venezia Giulia, in relation to elevation and distance from the sea (see text).

autapophytes growing in the three environment-types that are the main source of autapophytic species, i.e. stony grasslands, fringes of woods, and humid environments and meadows. Figure 43 summarizes the results. A clear gradient, related to elevation and distance from the sea is evident. Along this gradient there is a decrease of Eurimediterranean species and of therophytes, an increase of mesothermic species (Eurosiberian, Palaeotemperate, Cosmopolitan) and of hemicryptophytes. In the coastal Karst area and in the mountains most autapophytes derive froms stony grasslands, while humid environments and meadows are the principal source of autapophytes in the Friulian Plain. From the sea to the mountains there is an increase of the role of wood fringes as a main source of autapophytic species. In this case these are mostly nitrophytic plants naturally occurring in the fringes of humid woods (*Glechometalia hederaceae* Tx. in Tx. et Brun Hool 75).

These results underline the importance of natural biotopes for the recolonization of disturbed areas.

ACKNOWLEDGMENTS

This study has been financed by the Italian Ministry for Public Instruction, funds 40% and 60%.

REFERENCES

Allkin, R. and Bisby, F. A. 1984. Databases in Systematics. Syst. Ass. sp. val., 26, pp. 329, Academic Press.

Anderberg, M. R. 1973. Cluster Analysis for Applications. New York and London, Academic Press. 359 pp.

Anzaldi, C. and Mirri, L. 1979. Un esperimento di strutturazione di dati floristici e vegetazionali. Pubbl. IAC Roma, 195, 1–148.

Anzaldi, C. and Mirri, L. 1980. La banca dei dati floristici e vegetazionali: controllo e formalizzazione dei dati di input. Coll. Progr. Fin. Qualità Ambiente, CNR AQ/5/29. Roma.

Billensteiner. H. 1978. Beobachtungen an Orchideen im Oberen Gailtal. Carinthia II, 168/88, 279–320.

Brenan, J. P. M., Franks, J. W., Rajnal, J. and Cullen, J. 1975. Report of working party on electronic data processing in major European plant collections. Adansonia, 15, 7–24.

Cadbury, D. A., Hawkes, J. G. and Readett, R. C. 1971. A Computer-Mapped Flora. A study of the County of Warwickshire. Academic Press, London, New York.

Cristofolini, G., Lausi, D. and Pignatti, S. 1969. Survey of the system for coding plant sociological records used by the Trieste group. Trieste.

Crovello, T. J. 1981. Quantitative Biogeography: an Overview. Taxon, 30 (3), 563–575.

Ehrendorfer, F. and Hamann, U. 1965. Vorschläge zu einer floristischen Kartierung von Mitteleuropa. Ber. Deutsch Bot. Ges., 78, 35–50.

Feoli Chiapella, L. and Feoli, E. 1979. Predizione ambientale basata su flore locali. Un esempio di applicazione della banca dati TAXIR. In: Ferrari, C. et al. (eds), Le comunità vegetali come indicatori ambientali, pp. 111–131. Bologna.

Feoli, E. and Ganis, P. 1984. On the application of numerical and computer methods in plant taxonomy and plant geography: an integrated information system for data banking and numerical classifications and ordinations. Webbia, 38, 165–184.

Feoli, E. and Scimone, M. 1984. Hierarchical diversity: an application to broad-leaved woods of the Apennines. Gior. Bot. Ital., 118 (1–2), 1–15.

Feoli, E., Lagonegro, M. and Orloci, L. 1984. Information analysis of vegetation data. Task for Vegetation Science, 10. Junk, The Hague, Boston.

Fornaciari, G. 1961. Osservazioni sulla flora e sulla vegetazione dei Colli di Osoppo. Acad. Sci. Lett. Arti., 7 (2), 1–135.

Ganis, P. 1985. FUSAF: Manuale per l'uso di programmi a integrazione della banca dati SBAFT. Quaderni del Gruppo Elaborazione Automatica Dati Ecologia Quantitativa, 2. Trieste.

Hartl, H. and Radic, J. 1988. Florenkartierung mit Biodat. Flor. Rundbr., 22 (2), in press.

Holub, J. and Jirasek, V. 1967. Zur Vereinheitlichung der Terminologie in der Phytogeographie. Folia Geobot. Phytotaxon, 1 (2), 69–113.

Jalas, J. and Suominen, J. 1972–76. Atlas *Florae Europaeae*. Distribution of Vascular Plants in Europe, Vols, 1, 2, 3. Helsinki.

Karrer, G. 1986. Quantitative Analyse von Arealgrösse und Disjunctionsgrad an Artengarnituren von Pflanzengesellschaften des Alpinenostrandes. Sauteria, 1, 89–134.

Lagonegro, M. 1985. SBAFT: software per banche dati di flore territoriali. Quaderni del Gruppo Elaborazione Automatica Dati Ecologia Quantitative, 1. Trieste.

Lagonegro, M. and Feoli, E. 1985. Analisi multivariata di dati. Manuale d' uso di programmi BASIC per personal computers. Libreria Goliardica, Trieste.

Lagonegro, M., Ganis, P., Feoli, E., Poldini, L. and Canavese, T. 1982. Un software per banche dati di flore territoriali, estendibile alla vegetazione. Quaderni C.N.R., AQ/5/38, 1–160.

Landolt, E. 1977. Ökologische Zeigerwerte zur Schweizer Flora. Ber. Geobot. Inst. ETH Zürich, 64, 64–207.

Lausi, D. and Nimis, P. L. 1985. Roadside vegetation in boreal South Yukon and adjacent Alaska. Phytocoenologia, 13 (1), 103–138.

Martini, F. 1987. L'endemismo vegetale nel Friuli-Venezia Giulia. Biogeographia, 13, 339–399.

Napoleone, I. 1982. Prototipo di banca dati sulle specie foraggere sperimentate in Italia. Quaderni C.N.R., AQ/5/37, Roma.

Nie, N. H., Hull, C. H., Jenkins, J. G., Steinbrenner, K. and Bent, D. H. 1975. Statistical Package for the Social Science (SPSS). McGraw-Hill, New York.

Niklfeld, H. 1971. Bericht über die Kartierung der Flora Mitteleuropas. Taxon, 20 (4), 545–571.

Nimis, P. L. 1981a. La banca dati relativa alla flora e vegetazione d'Italia. Quaderni C.N.R., AC/1/105, 83–86, Roma.

Nimis, P. L. 1981b. La banca dati sulla flora e vegetazione d'Italia: utenza e gestione. Quaderni C.N.R., AC/5/14, 41–57, Roma.

Nimis, P. L. 1984a. Contributions to quantitative phytogeography of Sicily. I: Correlations between phytogeographical categories and environment-types. Webbia, 38, 123–137.

Nimis, P. L. 1984b. Contributions to quantitative phytogeography of Sicily. II: Correlations between phytogeographical categories and elevation. Studia Geobot., 3 (4), 49–62.

Nimis, P. L. 1984c. Lichenological studies in north east Italy. I. The computerization of the TSB lichen herbarium. Gortania, 6, 139–146.

Nimis, P. L. 1984d. Contributions to quantitative phytogeography of Sicily. III: Correlations between phytogeographical categories elevation, and environment types. Arch. Bot. Biogeogr. Ital., 60 (3–4), 11–40.

Nimis, P. L., Feoli, E. and Pignatti, S. 1984. The Network of Databanks for the Italian Flora and Vegetation. In: Allkin, R. and Bisby, F. A. (eds), Databases in Systematics, pp. 113–124. Academic Press, London and Orlando.

Orlóci, L. 1978. Multivariate analysis in vegetation research (2nd edn.). Junk, The Hague.

Otto, A. and Kuhn, C. 1988. Florkart – Ein Kartierungsprogramm mit Kartengrafik auf einem Mikrocomputer. Flor. Rundbr., 21 (2), 126–133.

Ozenda, P. 1975. Sur les étages de végétation dans les montagnes du bassin méditerranéen. Doc. Cart. Écol., 16, 1–32.

Pielou, E. C. 1975. Ecological Diversity. Wiley, New York.

Pignatti, S. 1973. Problemi di codifica dei dati floristici in fitosociologia. Not. Fitosociol., 7, 17–20.

Pignatti, S. 1976. A system for coding plant species for data-processing in phytosociology. Vegetatio, 33, 23–32.

Pignatti, S. 1978. Dieci anni di cartografia floristica nell'Italia di Nord-Est. Inform. Bot. Ital., 10, 212–219.

Pignatti, S. (1980), 1981. Check-list of the flora of Italy with codified plant names for computer use. Quaderni C.N.R., AQ/5/13, pp. 256. Roma.

Poldini, L. 1974. Primo tentativo di suddivisione fitogeografica delle Alpi Carniche. In Alto, 58, 258–279. Udine.

Poldini, L. 1980. Catalogo floristico del Friuli-Venezia Giulia e dei territori adiacenti. Studia Geobot., 1 (2), 313–474.

Poldini, L. 1986. Il Paesaggio vegetale. In: AA. VV. (eds), Suoli, vegetazione e foreste del Prescudin, 59–96. Udine.

Poldini, L. 1987. La suddivisione fitogeografica del Friuli-Venezia Giulia. Biogeographia, 13, 41–56.

Poldini, L. 1989. La vegetazione del Carso isontino e triestino. Lint, Trieste.

Poldini, L., Lagonegro, M., Gamis P. and Vidali, M. 1985. Flora computerizzata del Carso triestino e goriziano. In: Poldini, L. (ed.), Studio naturalistico del Carso triestino e goriziano, pp. 39–52. Region. Auton. Friuli-Venezia Giulia and Univer. Trieste.

Poldini, L. and Vidali, M. 1985. Utilizzazione di una banca dati per la suddivisione fitogeografica di un territorio. Biogeographia, 11, 247–259.

Poldini, L. and Vidali, M. 1986. Die Anwendung einer Datenbank für die pflanzengeographische Gliederung eines Gebietes. In: Reichl, E. R. (ed.), Computers in Biogeography. Schriftenreihe für Informatik, 1–12. Trauner Verlag, Linz.

Poldini, L. and Vidali, M. 1987. Lo stress ambientale e il risparmio energetico nei meccanismi di impollinazione nelle cenosi erbacee. Biogeographia, 13, 179–207.

Poldini, L., Martini, F. and Pertot, M. 1988. Variazioni strutturali ed ecologiche del geoelemento pontico nel passaggio dal Carso litoraneo alle Alpi sudorientali. Studia Geobot., in press.

Rothmaler, W. 1955. Allgemeine Taxonomie und Chorologie der Pflanzen (II ed.). Jena.

Scimone, M., Feoli, E. and Parente, G. 1985. La Banca Dati sulle Foraggere nella programmazione territoriale delle aree marginali. Atti del Convegno: Fattori di marginalità e sviluppo nell'economia montana. Barcis.

Scimone, M., Ganis, P. and Feoli, E. 1987. Programmi Basic per il calcolo di misure di diversità in comunità ecologiche. Quaderni del Gruppo Elaborazione Automatica Dati Ecologia Quantitativa, 5. Trieste.

Shannon, C. E. and Weaver, W. 1949. The Mathematical Theory of Communication. Univ. of Illinois Press, Urbana.

Vöth, W. and Löschl, E. 1978. Zur Verbreitung der Orchideen an der östlichen Adria. Linzer. Biol. Beitr., 10 (2), 369–430.

Walter, H. 1960. Grundlagen der Pflanzenverbreitung: Standortslehre (analytisch – ökologische Geobotanik), III, 1. Verlag E. Ulmer, Stuttgart.

Welten, M. and Sutter, R. 1982. Verbreitungsatlas der Farn- und Blütenpflanzen der Schweiz, Vol. 1, 2.

Wraber, T. 1968. Razsirjenost rastlinskih vrst v Sloveniji. Proteus (Ljubljana), 30, 252–253.

Wraber, T. and Skoberne, P. 1989. The Red Data List of Threatened Vascular Plants in Socialistic Republic of Slovenia. Varstvo nar., 14–15, pp. 428. Ljubljana.

6. THE USE OF SATELLITE IMAGERY IN QUANTITATIVE PHYTOGEOGRAPHY: A CASE STUDY OF PATAGONIA (ARGENTINA)

JOSÉ M. PARUELO, MARTÍN R. AGUIAR, ROLANDO J. C. LEÓN, RODOLFO A. GOLLUSCIO AND WILLIAM B. BATISTA

Departamento de Ecología, Facultad de Agronomia, Universidad de Buenos Aires, Av. San Martín 4453, 1417–Buenos Aires, Argentina

1. INTRODUCTION

Ecological surveys of arid zones, both with basic and applied phytogeographical objectives, are very difficult because of their magnitude and the low density of their renewable natural resources. The lack of adequate knowledge of the heterogeneity and the functioning of the vegetation is common in these regions. This leads to increasing damage of the environment as a consequence of the inbalance between supply and demand in the use of their natural resources (Soriano 1986).

Most processes associated with desertification (decrease in net primary production, increase of erosion) take place on a regional scale. Regional vegetation studies in temperate areas are limited, generally, to the growing season. This makes it difficult to have the proper perception of the desertification phenomenon, especially if classical methods of vegetation studies are used. Nevertheless, in processes that take place at a global scale, multiyear changes become very important. These constraints of the classical methods produce a disagreement between the scales of the phenomenon under study and that of the variables measured (Allen and Starr 1982). Remote sensing information provided by satellites permits a better adjustment between the scales of the processes and the observations (i.e. changes in the proportion of communities association with desertification, drop in the net primary production, etc.).

With the launching of the first meteorological satellite of the TIROS series (TIROS I), in 1960, the age of remote sensing of the Earth from space was started (Smith et al. 1986). ERTS I (after LANDSAT I), launched in 1972, was the first satellite designed specifically to make land surface observations. With the regular operation of the LANDSAT satellite the spectral information began to be used broadly in the survey of natural resources and specifically in vegetation studies.

The use of remote-sensing data provided by satellites allowed completion of previous descriptions of vegetation and inprovement of phytogeographical knowledge for several reasons:

– Information is obtained for entire regions, including places with difficult access; this also improves boundaries perception and area estimation.
– Sequential information is accumulated, allowing increased understanding of both seasonal dynamics and multiyear changes in plant cover.
– Numerical data become available that can be analyzed using multivariate methods.

In this chapter, the known phytogeographic information on the extra–Andean Patagonia is reviewed. It is compared with the results obtained using spectral data provided by the meteorological satellites of the NOAA/AVHRR series. This analysis on a regional or phytogeographical scale is similar to the one used by Soriano (1956a) and tries to answer the following questions:

(1) What is the structure of the satellite data?
(2) Are there any differences among phytogeographical districts which can be detected

P. L. Nimis and T. J. Crovello (eds): Quantitative Approaches to Phytogeography, pp. 183–204.

from the spectral data, and in that case, which are the directions of these differences?

(3) Are the differences among districts the main sources of the spectral data variation?

The first section of this chapter reviews the principal characteristics of the satellite information and its use for vegetation studies at a regional scale. The main phytogeographic studies of the Patagonian region are summarized in the second section. In the last part of the chapter the analysis and interpretation of the spectral data provided by the NOAA satellites are presented to answer the above questions.

2. REMOTE SENSING IN VEGETATION STUDIES

Green vegetation shows a characteristic pattern of spectral reflectance. This pattern results from high absorption of incoming radiation in the visible wavelengths (0.45–0.75 μm) and high reflectance in near infrared (0.75–1.35 μm) (Myers 1983; Swain and Davis 1978). Photosynthetic pigments (chlorophyll a and b, carotenes and xanthophyll) dominate spectral responses of vegetation in the visible wavelengths. Leaf scattering due to the presence of spongy mesophylls, explains reflectance rise in near infrared (Myers 1983; Tucker and Sellers 1986).

Remote-sensing studies of vegetation are based on the analysis of radiation reflected by plant cover in specific wavelength regions. They were selected because they provided strong signals from the vegetation and adequate contrast from the background, e.g. soil, rock, water, etc. (Tucker and Sellers 1986). Such background in contrast with vegetation, exhibit a low reflectance in the near infrared (Haralick and King Su Fu 1983).

Several indices have been proposed to improve the contrast between plant cover and other surfaces (Tucker 1979; Jackson et al. 1983; Curran 1981). They allow the spectral characterization of vegetation to be relatively independent of the underlying material or the topography (Curran 1981; Holben and Justice 1981). One of the most common indices is the Normalized Difference Vegetation Index (NDVI):

$$\text{NDVI} = (\text{NIR} - \text{R}) / (\text{NIR} + \text{R})$$

where R is the reflectance in the red band and NIR is the reflectance in the near infrared band.

The physical meaning of these indices is still under study (Goward et al. 1985). The NDVI shows a nonlinear correlation with the leaf area index and green biomass of the canopy (Tucker et al. 1981; Hatfield 1983). Canopy architecture, optical properties of green parts, density of biomass and soil cover modify the shape of this relationship (Horler and Barber 1981).

Kumar and Monteith (1981) and Asrar et al. (1984) found a close relationship between NDVI and intercepted photosynthetically active radiation in crops. Tucker and Sellers (1986) pointed out the fact that reflectance data provide an indication of the instantaneous rates associated with plant canopies (gross primary production, transpiration) rather than reliable estimates of any state associated with the vegetation such as leaf area index or biomass.

The first data used in vegetation studies, both natural and cultivated, came from LANDSAT satellites. These images are a potentially useful tool in sequential surveys of vegetation. However, this objective can not be fully accomplished in very large areas. With LANDSAT, an area is only monitored every 16 or 18 days. Due to problems of cloudiness, etc., this frequency rarely produces satisfactory temporal coverage of an area. It also makes it difficult to have contemporaneous images for large areas (Coldwell and Hicks 1985).

Spectral data provided by satellites of the NOAA series (National Oceanographic Atmospheric Agency) began to be used in vegetation studies at the start of the 80s. The Advanced Very High Resolution Radiometer (AVHRR) sensor on board NOAA satellites provides digital data in the visible (0.58–0.68 μm), near-infrared (0.725–1.1 μm) and thermal channels of the electromagnetic spectrum (Kidwell 1984).

Plate 1. Typical closed basin between hills and basaltic tablelands. Alluvial clay – plain with a temporary pool and a basaltic cone. A mosaic of shrub steppe and semidesert ("erial") vegetation. (Photo by Paruelo and León.)

Plate 2. A meadow ("mallín") with *Berberis* shrubs, in a semidesert. Grazing sheep. *Salix* sp. (right) and *Populus* sp. (back left) planted along water ways or irrigation channels. Basaltic plateaus on the horizon. (Photo by Piccinini.)

Plate 3(a). Verbena tridens shrubs steppe in the Central district. Plateau SW of Comandante Piedra Buena, Sta. Cruz.

Plate 3(b). Detail of a shrub steppe. In the foreground *Verbena tridens, Festuca pyrogea* and *Stipa ibari*. (Photo by Movia.)

Plate 4(a). Semidesert ("erial") in the Central district. In the background a typical basaltic tableland.

Plate 4(b). A detail of the semidesert ("erial") in the Central district. Soft cushion-like habit shrub: *Nassauvia glomerulosa*. (Photo by Paruelo and León.)

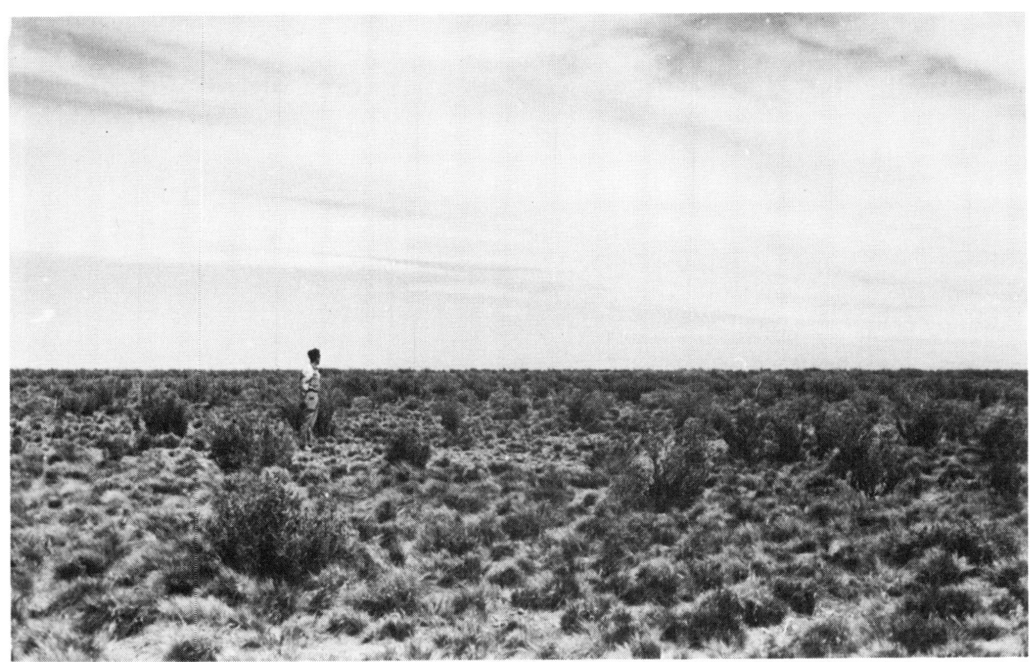

Plate 5. Shrub-herbaceous steppe in the Occidental district. Tussock grasses: *Stipa speciosa, S. humilis* and *Poa ligularis;* low shrubs: *Mulinum spinosum* and *Senecio filaginoides*; tall shrubs: *Adesmia campestris* and *Berberis heterophylla.* Flat plain in Paso Río Mayo, Chubut. (Photo by Piccinini.)

Plate 6(a). Semidesert ("erial") in the Central district. In the background a shrub steppe with *Chuquiraga avellanedae*. Badlands on the steep slopes.

Plate 6(b). A detail of the semidesert ("erial") in the Central district: erosion pavement, cushion plants of *Chuquirage aurea* and small tussocks of *Stipa ibari*. (Photo by Paruelo and León.)

Plate 7(a). Shrub steppe in the Gulf district. Rolling country, east of Pampa del Castillo, Cañadón Ferraris, Comodoro Rivadavia, Chubut.

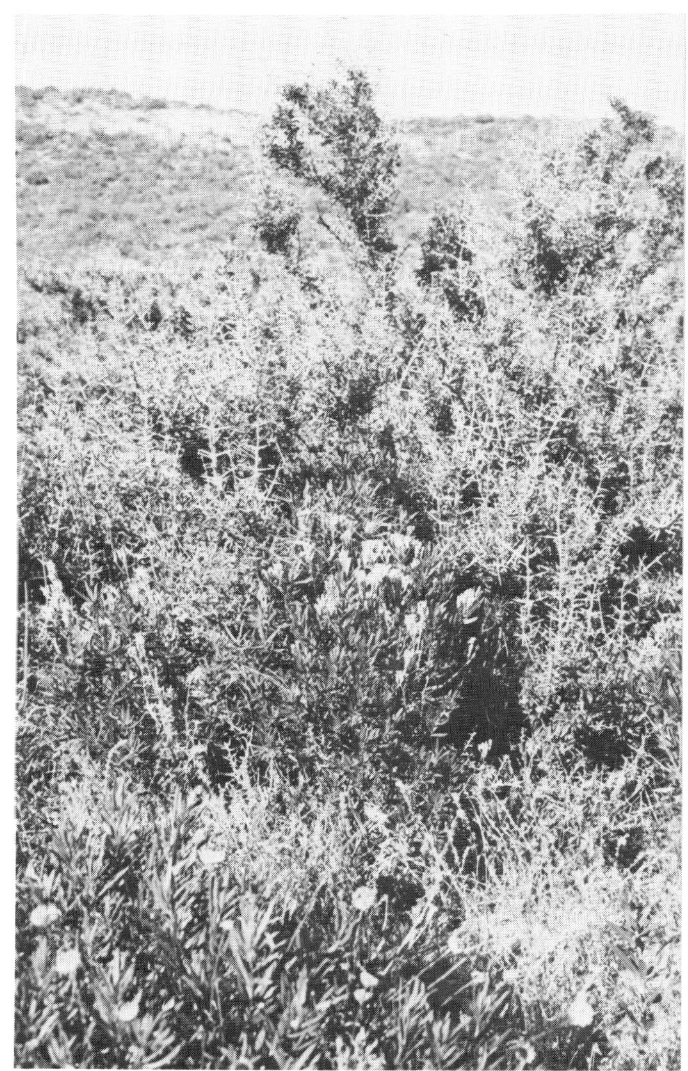

Plate 7(b). A detail of the shrub steppe in the Gulf district. *Colliguaya integerrima* in the foreground with a tall shrub of *Trevoa patagonica*. (Photo by Paruelo and León.)

Plate 8. Herbaceous steppe in the Gulf district. *Festuca pallescens* with flowering culms and low shrub of *Mulinum spinosum*, *Nardophyllum obtusifolium* and *Verbena* sp. Tall shrub species is *Adesmia campestris*. Two active oil pumps in the background. Pampa del Castillo, Chubut. (Photo by Paruelo and León.)

Plate 9. Herbaceous steppe in the Subandean district. *Festuca pallescens* tussocks with flowering culms. Fluvio-glacial landscape east of Río Pico, Chubut. (Photo by Soriano.)

Plate 10. Shrub steppe on the slope of a basaltic tableland. *Colliguaya integerrima* is the dominant species. The tall shrubs species are *Trevoa patagónica* and *Schinus polygamus*. Central district, El Canquel, Chubut. Upper left corner: basaltic scarp. (Photo by Paruelo and León.)

The spatial (1.1 km) and temporal (daily) resolution of the Large Area Coverage Data make it useful for vegetation studies at the regional or global scale. The swath width (2250 km) of a single orbit of the NOAA satellite permits easier reconnaissance of a broad pattern than the LANDSAT's. The high temporal frequency of coverage allows the detection of both intra- and interseasonal changes in vegetation.

Several authors have used NOAA/AVHRR spectral data in regional or global vegetation studies. Norwine and Greegor (1983) analyzed the vegetation distribution across a climatic gradient in Texas. Tucker et al. (1984a) used NOAA/AVHRR data to monitor the development of crops through the growing season in the Nile delta. They related the seasonal changes of NDVI with environmental conditions and cultural practices. In the Sahel, Tucker et al. (1984b) studied biomass production of grasslands. For large vegetation units of North America, Goward et al. (1985) found a clear association between net primary production and the integral of NDVI across the growing season. Using NOAA data, Tucker et al. (1985) classified land cover and studied temporal dynamics of vegetation in Africa. Their classification shows a good correspondence with previous vegetation maps. Using modified NOAA/AVHRR data, Justice et al. (1985) analyzed phenological changes of vegetation in South America, Africa and southeast Asia. Malingreau (1986) made a similar analysis of different vegetation units of Asia. Townshend et al. (1987) evaluated different methods of multivariate analysis to devise classification of the land cover in South America.

3. THE EXTRA-ANDEAN PATAGONIA

3.1. General features

Patagonia extends approximately from 39° to 50°S. It includes the Andean Cordillera and the plateaus (mesetas), plains and hills lying between the Andes and the South Atlantic Ocean (Fig. 1). Plateaus which begin at the Subandean foot-

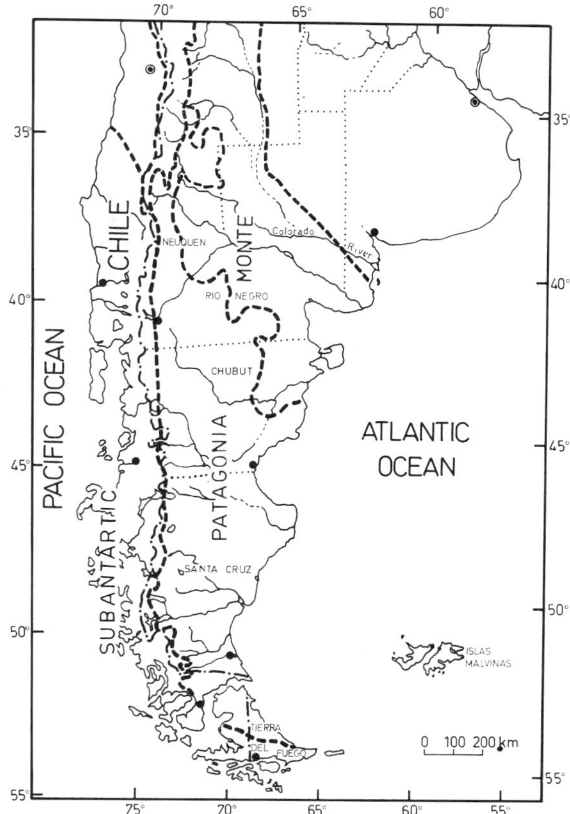

Fig. 1. Phytogeographical provinces in southern Argentina.

hills and descend to reach the Atlantic coast characterize Extra-andean Patagonia from 42° to 46°S (in Chubut Province). Several Extra-andean orographic units interrupt this landscape, the most important ones being the Sierra de Tecka and the Sierras Centrales de Patagonia (Volkheimer 1983).

This vast region has a cold temperate (Soriano 1983) climate. Maximum annual rainfall values (2000 mm) occur in the Andean zone (Fig. 2). In the central part of Patagonia mean annual rainfall oscillates between 125 and 500 mm, concentrated in the winter months (Barros et al. 1979). Decreasing precipitation from west to east determines a gradient of vegetation types: grass steppes, shrub-grass steppes, shrub steppes and semideserts (Soriano 1956a; Movia et al. 1987).

From a phytogeographic viewpoint, three provinces can be distinguished in Patagonia (Cabrera, 1971) (Fig. 1): (1) Subantartic Forest Province, dominated by *Nothofagus*, (2) Monte Province,

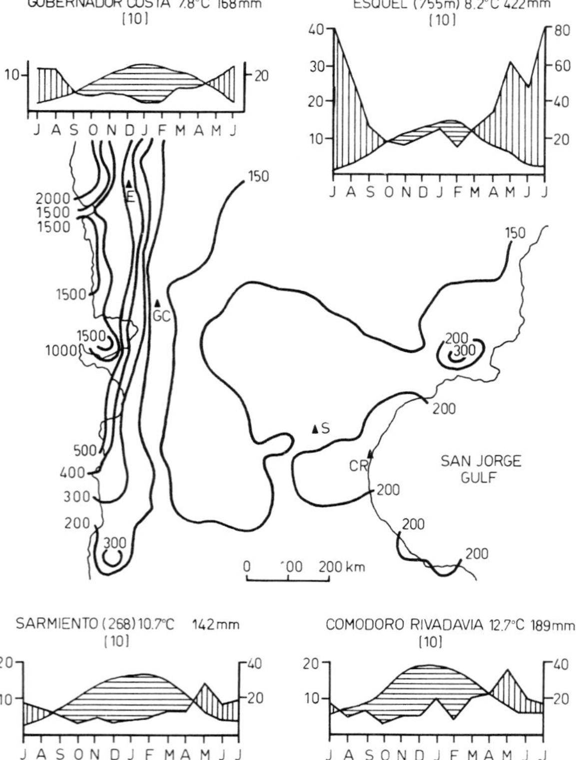

Fig. 2. Isohyets map of the study area and climatic diagrams for four extra-Andean districts: E, Esquel; GC, Gobernador Costa; S, Sarmiento; CR, Comodoro Rivadavia.

where scrub of *Larrea* spp. is the principal vegetation type, and (3) Patagonian Province, with scrubs, shrub steppes, grass steppes and semideserts as the main vegetation types. The Subantartic Forest covers a small area in Argentina. Within this unit, in places with more than 2000 mm of rainfall perennial forest prevails; with 1000 to 2000 mm deciduous forest is the dominant type, and mosaics of deciduous forests, scrubs and grass steppes are present in areas with 500 to 1000 mm rainfall (Hueck and Seibert 1981). The Monte Province, with a predominance of Chaqueño Domain species, occupies only the northeastern part of Patagonia.

The Patagonian Province is a southern floristic extension of the High Andean Domain. This Province is the most characteristic of the southern part of South America, because of its high endemism and the number of species with xeromorphic adaptation. In this chapter we deal only with that portion of the Patagonian region corresponding to the Patagonian Province.

Shrub steppe is the most conspicuous vegetation in the Patagonian Province. Two types of shrub steppe are particularly frequent: the medium height (20–80 cm) and medium density shrub steppe (1 shrub/6 m^2), with a relatively high tussock-grass cover, and the low shrub steppe or semidesert, with small cushion-like shrubs (5–20 cm height), sparse grasses and low total cover. These two physiognomies often appear associated forming mosaics.

Grass steppes are another important vegetation type. They show up to 70% of total cover. Tall scrubs 60–200 cm high, with more or less closed canopies and dominated by shrubs, are common over hilly slopes and near the Atlantic coast. Meadows (locally named "mallines"), generally located on valley bottoms but also frequently associated with springs over hilly slopes are dominated by mesophytic grasses and sedges. Although they only represent small areas, they are very important due to their high productivity. When these areas are overgrazed, salinization processes occur, and wind and water erosion transform these meadows into true deserts and starting points of "sand tongues" (Movia 1972). As in all the arid and semiarid regions of Argentina, halophytic steppes are frequent in Patagonia. In them, very sparse shrubs and ephemerals prevail, the latter being dominant in humid years.

In Patagonia, as in other arid zones of the world, vegetation has been modified through grazing by exotic animals. Sheep were introduced at the end of the nineteenth century (Morrison 1917), and nowadays it is impossible to find areas where sheep and European hare grazing has not taken place (Soriano 1983). At present, the sheep industry remains the principal rural activity.

Patagonia has serious problems of desertification and erosion (Soriano and Movia 1986). Aridity, persistent and strong winds, low plant cover and scanty soil material cohesion are factors that increase the systems fragility. Nevertheless, desertification is triggered by sheep grazing (Soriano and Movia 1986), that leads to broad erosion processes (Movia 1972) and vegetation changes (Anchorena 1985; León and Aguiar 1985; Borelli et al. 1985; Movia et al. 1988).

3.2. Phytogeographic characterization

The Andean Domain in the Argentine territory extends from the Puna de Atacama, in the north, to Tierra del Fuego, in the southern extreme of the continent. Several species, such as *Doniophyton anomalum, Nassauvia axilaris* and *Senecio filaginoides*, can be found all along this domain. Although the domain exhibits a high degree of floristic and physiognomic uniformity, Cabrera (1971) distinguished three provinces within it: Altoandina, Puneña and Patagonian. All of them have predominance of xerophytic grasses of the genera *Festuca, Stipa, Poa* and *Deyeuxia*. The Patagonian Province is distinguished from the others by several important endemics, such as the genera *Pantacantha, Benthamiella, Duseniella, Saccardophyton, Ameghinoa, Xerodraba, Lepidophyllum*, etc.

Many species growing in the Andean Domain show several types of morphophysiological adaptations to arid conditions:

- Aphyllous shrubs or shrubs with small coriaceous, or pubescent leaves or leaves reduced to spiny segments (Ruthsatz 1978; Ancibor 1980; Boelcke 1957).
- Small, hemispherical shrubs, with cushion-like growth or compact shrubs with heteroblastic growth (small branches with limited growth and a rossete of little leaves on the top or with very short internodes) (Hager 1986).
- Tussock grasses with dead standing material or with folding or convoluted linear, spiny leaves, and thick cuticle (Cabrera 1971). In some of them, the dead part forms the center of the tussock which is encircled by a ring of tillers, bearing green and recently dead leaves (Soriano et al. 1987).
- Low geophyte or hemicryptophyte perennial herbs growing under the protection of tussocks or spiny shrubs. Ephemeral herbs able to complete their life cycle quickly.

Description of the Patagonian flora began in the eighteenth century. Pioneer explorers and naturalists made their observations and floristic collections near the Atlantic coast, in insular areas, and along the main rivers and the Magellan Strait. During the first half of the twentieth century, continental areas received more attention, and the first phytogeographic works were published (Hauman 1920, 1926; Soriano 1949a; Auer 1951). Soriano (1956a) described the internal heterogeneity of the Patagonian Province and proposed its division into phytogeographic districts. This study solved many previous phytogeographic discrepancies, which were due to incomplete exploration.

Since 1970 numerous studies have helped to improve the floristic (INTA 1969, 1971, 1978, 1984a, b), plant geographic (Bertiller et al. 1981a, b; Golluscio et al. 1982; Roig et al. 1985; Movia et al. 1987) and ecological (Soriano et al. 1976; Soriano et al. 1980; Soriano and Sala 1983, 1986; Bertiller 1984; Soriano 1981; Sala et al. 1989) knowledge of the Patagonian vegetation.

Several authors also studied some particular phytogeographic features of the Patagonian Province. Ruiz Leal (1972) and Roig et al. (1980) studied the north and northeast boundaries of the Province in its ecotone with the Monte Province. Cabrera (1971) included Payunia as a new district in the northwest. León and Facelli (1981) analyzed floristic gradients of central Patagonia. Bertiller et al. (1981b) and Aguiar et al. (1988) studied the limits among districts.

Grazing effects on vegetation structure and function received increasing attention. Early studies on sheep grazing-induced changes in the region made by Soriano (1949b, 1956b, 1959), Soriano and Brun (1973) and Boelcke (1957) have been continued by León and Aguiar (1985), Borelli et al. (1985), Anchorena (1985), Bonvisuto et al. (1983), Facelli and León (1987), Schlichter et al. (1978) and Lores et al. (1983).

Taking into account all the existing phyto-

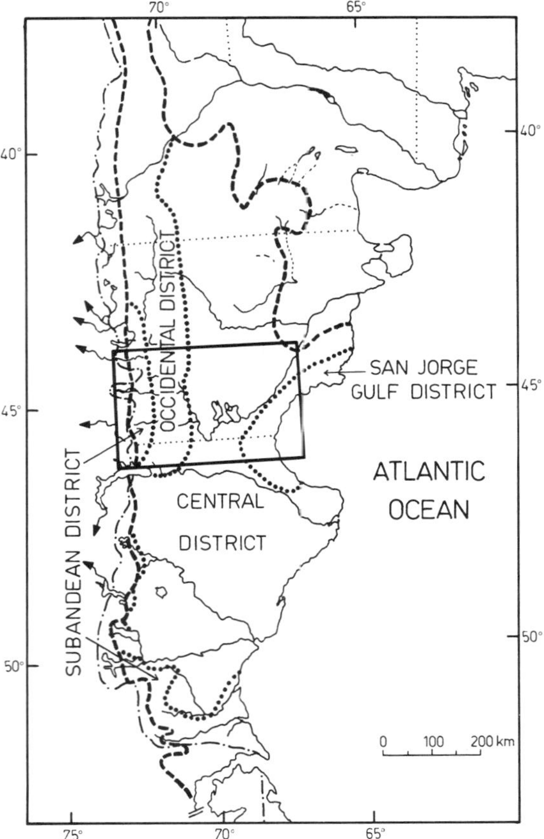

Fig. 3. Patagonian region showing phytogeographical district boundaries (•••••), provincial state limits (·····) and international boundaries (–·–·–). The boxed section corresponds to the study area.

geographic studies, six districts can be defined in the Patagonian Province (Cabrera 1976) (Fig. 3):

1. Payunia
2. Occidental
3. Central
4. San Jorge Gulf
5. Subandean
6. Fueguian

In this section the continental part of the Province is described giving special emphasis to the central portion (between 44° and 47°S) (Fig. 3). We chose this area because its heterogeneity represents most of the physiognomy types of the region and it is the richest in recent studies. The Payunia district has not been considered because of the lack of available information (Soriano 1983).

3.3. Occidental District

The Occidental district is a strip 100–200 km wide that extends west of the 70°W meridian from Picún Leufú River in Neuquén (39°S) to Lake Buenos Aires in Santa Cruz (46°30′S). The western limit of the Occidental district forms a broad ecotone with the Subandean district (León and Facelli 1981). The Occidental district is characterized by an open shrub steppe with a total plant cover of approximately 50% and height ranging from 60 to 180 cm. Grasses (locally named "coirones") and herbs account for 68% of mean total plant cover (Golluscio et al. 1982).

The most important community in the southern portion of this district has been described phytosociologically by Golluscio et al. (1982) and named the "*Stipa speciosa, Stipa humilis, Adesmia campestris, Berberis heterophylla*, and *Poa lanuginosa* community" (Table 1). Soriano et al. (1976) referred to this same community as "pastizal de coirón amargo". In addition to the above-mentioned species, there are also important because of their constancy and/or cover, (Table 1) the shrubs *Senecio filaginoides, Mulinun spinosum, Ephedra frustillata, Lycium chilense, Schinus polygamus*; the grasses *Bromus setifolius*, *Hordeum comosum, Poa liqularis* and the herbs *Adesmia lotoides, Perezia recurvata, Oenothera contorta, Verbena minutifolia, Doniophyton patagonicum, Microsteris gracilis*, and *Calceolaria polyrhiza*. The physiognomy of the northern part is not different from the one described above. However, in the south, some floristic elements such as *Stillingia patagonica, Corynabutilon bicolor, Tetraglochin ameghinoi* or *Nardophyllum parvifolium* are absent or have low frequencies (Speck 1982; Loras et al. 1983).

This shrub steppe sometimes alternates with a shorter shrub steppe or semidesert ("erial") of low total cover. Frequency and extension of semidesert patches increase towards the east (Golluscio et al. 1982). The "erial" is the typical physiognomy of the Central district, while in the Occidental district it only occurs as small patches associated with particular topographic positions such as terrace edges, or in places where erosion

Table 1. Phytosociological characterization of the SW Chubut vegetation. Modified from Golluscio, Leon and Perelman (1982)

Group	Sub-Group	Species	A Const.	A Cov. %	B Const.	B Cov. %	C Const.	C Cov. %
1	1	*Festuca pallescens*	V	45				
		Rhytidosperma picta	V	5				
		Nassauvia aculeata	V	1				
		Relbunium richardianum	V	1				
		Senecio sericeonitens	V	1				
		Erigeron andicola	IV	+				
		Lathyrus magellanicus	IV	1				
		Luzula chilensis	IV	+				
		Festuca magellanica	IV	+				
		Vicia biyuga	IV	1				
		Koeleria grisebachii	III	1				
		Acaena pinnatifida	III	1				
		Taraxacum offinale	III	1				
		Koeleria permollis	III	1				
		Thlaspi magellanica	III	+				
		Acaena splendens	III	5				
		Gamochaeta nivalis	III	+				
		Oxalis adenophylla	III	+				
			III	+				
	2	*Viola maculata*	V	1	I	1		
		Sisyrinchium macrocarpum	IV	+	I	+	I	+
		Acaena sp.	III	+			I	
		Adesmia corymbosa	III	r	I	1		
	3	*Sisyrinchium* sp.	IV	+	II	+	I	r
		Silene sp.	III	+	II	+		
	4	*Agoseris coronopifolia*	II	+				
		Armeria sp.	II	+				
		Poa sp.	II	1				
		Astragalus sp.	II	+				
		Chloraea sp.	II	+				
		Draba magellanica	II	r				
		Geranium molle	II	+				
		Myosotis stricta	II	+				
		Nassauvia darwinii	II	1				
		Aira caryophyllea	I	+				
		Anemone decapetala	I	r				
		Cruckshanksia glacialis	I	5				
		Quinchamalium chilense	I	+				
		Trifolium repens	I	+				
	5	*Leuceria candidissima*	II	1	I	r		
		Triptilion achilleae	II	+			I	+
		Agrostis sp.	II	+			I	+
		Azorella ameghinoii	II	1			I	+
		Colobanthus subulatus	II	+			I	1
2	1	*Hypochoeris* sp.	V	+	IV	+		
		Loasa sp.	III	1	II	+		
	2	*Carex* sp.	V	1	III	+	I	r
		Calceolaria sp.	V	+	II	+	I	r
	3	*Bromus setifolius*	V	1	V	1	II	+
		Mulinum spinosum	IV	5	V	5	II	r
	5	*Festuca argentina*	II	1	I	1	I	r

Table 1 (Continued)

Group	Sub-Group	Species	A Const.	A Cov. %	B Const.	B Cov. %	C Const.	C Cov. %
3	1	*Lycium chilense*			III	+		
	2	*Adesmia campestris*	I	r	V	5		
		Berberis heterophylla	I	r	V	+	I	r
		Poa Lanuginosa			IV	1	I	1
		Verbena minutifolia			III	+	I	r
		Schinus polygamus			III	+	I	r
	3	*Oenothera contorta*	II	+	IV	+	II	r
		Perezia recurvata	II	+	IV	+	II	+
	4	*Gilia laciniata*			II	r		
		Stipa psilantha			II	1		
		Plantago patagonica			II	r		
		Verbena ligustrina			II	+		
		Austrocactus sp.			I	r		
		Acaena platyacantha			I	r		
		Heliotropium paronychioides			I	+		
		Tweedia o'donelli			I	r		
	5	*Calandrinia patagonica*	I	r	II	r		
		Euphorbia collina			II	+	I	+
		Leuceria millefolium			II	+	I	r
		Acantholipia seriphioides			I	+	I	1
		Pantacantha ameghinoi			I	+	I	r
		Polemonium antarcticum			I	r	I	r
4	1	*Stipa humilis*			V	10	III	+
		Stipa speciosa			V	15	III	+
		Doniophyton patagonicum			V	+	III	+
	2	*Adesmia lotoides*	I	r	V	+	III	r
	3	*Senecio filaginoides*	II	r	V	5	V	+
		Poa ligularis	II	1	V	5	IV	+
		Epheara frustillata	I	1	III	r	II	+
	4	*Stipa neaei*			I	+	I	1
		Huanaca acaulis			I	r	I	r
		Schismus sp.			I	1	I	1
5	1	*Chuquiraga aurea*					V	1
	2	*Nassauvia glomerulosa*			I	1	V	5
		Hoffmansegia trifoliata			I	1	IV	+
		Azorella monantha			I	+	IV	+
		Chuquiraga kingii			I	+	IV	1
		Haplopappus diplopappus			I	r	IV	+
		Mulinum microphyllum			II	+	III	+
		Stipa ibari			I	+	III	1
		Brachyclados lycioides			I	r	III	1
		Tetraglochin caespitosus			I	+	III	+
	4	*Erysimum repandum*					II	1
		Brachyclados caespitosus					II	r
	5	*Verbena tridens*			I	5	II	5
		Perezia lanigera			I	+	II	+
		Stipa chrysophylla			I	1	I	1

Group	Sub-Group	Species	A Const.	A Cov. %	B Const.	B Cov. %	C Const.	C Cov. %
6	1	*Hordeum comosum*	V	1	V	+	V	+
		Cerastium arvense	V	1	IV	+	III	+
		Microsteris gracilis	IV	1	IV	1	III	+
		Arjona patagonica	III	+	III	+	III	+
		Polygala darwiniana	III	+	II	1	III	1

Note:
A: *Festuca pallescens, Rhytidosperma picta* and *Lathyrus magellanicus* community (17 relevés)
B: *Stipa speciosa, S. humilis, Adesmia campestris, Berberis heterophylla* and *Poa lanuginosa* community (23 relevés)
C: *Nassauvia glomerulosa, Chuquiraga aurea* and *Ch. kingii* community (12 relevés)
Const.: Braun-Blanquet constance classes
Cov. %: Braun-Blanquet abundance classes (r + 1) and cover classes at 5% intervals
Subgroup 1: Constance in the relevé group/s > 40%, 0% outside them
2: Constance in the relevé group/s > 40%, < 20% outside them
3: Constance in the relevé group/s > 40%, > 20% outside them
4: Constance in the relevé group/s < 40%, 0% outside them
5: Constance in the relevé group/s < 40%, < 20% outside them

surfaces or special soil conditions prevail (Golluscio et al., submitted). This unit is dominated by *Nassauvia glomerulosa* and *Nassauvia ulicina*. These two species are small shrubs with very small verrucose leaves. Other variants of the shrub steppe, of higher floristic richness than the "erial", exhibit alternatively dense populations of *Nassauvia axilaris, Anarthrophyllum rigidum, Corynabutilon bicolor, Verbena tridens* and *Nardophyllum obtusifolium* (Soriano 1956a).

3.4. Central district

The Central district, the most extensive in Patagonia, encompasses the most arid portion of the region, with less than 200 mm average annual rainfall. It extends from Maquinchao in Rio Negro Province to the Coyle River in Santa Cruz (Fig. 3). Shrub and low shrub steppe or semi-desert ("eriales") are the most frequent vegetation types in the Central district. The shrub steppe, restricted to hilly areas, often has a significant grass cover. Halophytic steppe and desert and coastal halophytic scrub are also characteristic of this district (Movia et al. 1987). Two subdistricts have been defined: the northern or Chubutian subdistrict, where most of the communities are dominated by *Chuguiraga avellanedae*, and the Southern or Santacruzian, where *Verbena tridens*, a rare species in the Chubutian subdistrict, becomes one of the dominants (Soriano 1956a).

3.5. Chubutian subdistrict

An open shrub steppe of *Chuquiraga avellanedae* and /or *Colliguaya integerrima* is one of the most conspicuous physiognomies of the Chubutian subdistrict. Generally it presents two open shrub layers and a total cover ranging from 50 to 70%. The upper plant layer, higher than 100 cm along with the above-mentioned species includes *Lycium ameghinoi, L. chilense, Verbena alatocarpa, Prosopis denudans* and sometimes *Schinus polygamus*. In the lower layer, (15–20 cm high) *Acantholippia seriphiodes, Nassauvia ulicina, Pleurophora patagonica* and the grasses *Stipa humilis* and *Poa lanuginosa* are found. On low alluvial plains, *Distichlis scoparia* and *Juncus balticus* are also present (Bertiller et al. 1981b).

Similar physiognomies have been described in large areas of the Rio Negro Province of Argentina, in the northern extreme of the subdistrict (Speck 1982). Some of these have a relatively important presence of *Mulinum spinosum* or *Atriplex lampa*.

Sandy areas (i.e. near Lake Colhue Huapi) are dominated by *Prosopis denudans* and *Lycium chilense*, together with *Atriplex sagitifolium, Senecio filaginoides* and *Sporobolus rigens* (Bertiller et al. 1981b). In badland areas, almost completely barren, very dispersed individuals of *Ameghinoa patagonica* and *Nicotiana ameghinoi* occur (Soriano 1956a).

The low shrub steppe is as frequent as the vegetation type described above. It has a low cover, usually less than 50%, consisting of grasses and small shrubs, often of cushion-like form. This extremely xeric physiognomy has received several names: semidesert, dwarf scrub or, locally, "peladal" or "erial". The latter denomination, employed by Castellanos and Perez Moreau (1944) and Frenguelli (1941), is also used in this chapter because it emphasizes the individuality of this vegetation type.

In south-central Chubut Province several communities have been defined in the "erial". Their common characteristic is the dominance of *Nassauvia glomerulosa, Nassauvia ulicina* and *Chuguiraga aurea*, and the presence of some of the following companion species: *Chuguiraga avellanedae, Ch. kingii, Hoffmansegia trifoliata, Acantholippia seriphiodes, Brachyclados caespitosus, Lycium chilense, Acaena caespitosa, Pleurophora patagonica, Perezia lanigera, Stipa humilis, S.ibari, S.ameghinoi, Schinus polygamus*, etc. (Bertiller et al. 1981b). In the Senguerr River plateau region, west of Sierra de San Bernardo, Golluscio et al. (1982), phytosociological defined an "erial" community named after "*Nassauvia glomerulosa, Chuguiraga aurea* and *Chuguiraga kingii*", with 17% of mean total cover and a floristic richness of 19 species (Table 1).

3.6 Santacruzian subdistrict

The "erial" physiognomy is widely represented in the Santacruzian subdistrict (Movia et al. 1987) with similar plant communities to those described above. They extend in areas with stony soils topped by clay-loam crusts. Grasses in these communities are only found in sites with sandy accumulation. A number of cushion-like species are common components of them: *Acantholippia seriphiodes, Chuguiraga aurea, Petunia patagonica, Brachyclados caespitosum* and *Azorella caespitosa* together with the small shrubs *Nassauvia glomerulosa, Chuguiraga kingii, Mulinum microphyllum* and *Frankenia* sp.. Soriano et al. (1983) have postulated that this vegetation type is a degradation stage of the *Nassauvia glomerulosa* semidesert, which covers a small area in this region. In this vegetation unit, *Stipa humilis, S.chrysophyla, S.ibari* and with less frequency *S.neaei, S.psylantha, S.subplumosa, Poa liqularis, Alstroemeria patagonica, Ephedra frustillata, Polygala darwinii, Nassauvia ulicina, Cerastium arvense* and *Carex argentina* are also present. In old drainages and basins on flat-bottom lagoons with temporary water, relatively dense populations of *Verbena tridens* can be found, and give a unique aspect to the landscape.

The most broadly represented shrub steppes in the Santacruzian subdistrict are those dominated by *Verbena tridens* commonly found over the plateaus, those of *Berberis heterophylla* and *Senecio filaginoides* in present alluvial river plains (i.e. the Chico River) and the halophytic steppe of *Lepydophylum cupressiforme* in salinized depressions, stream beds, and estuary areas.

The *Verbena tridens* shrub steppe is around 70 cm high and covers 60% of the soil surface. Few grass species are present, the most important ones being *Stipa ibari, S. neaei, S.speciosa* and *Festuca pyrogea*. In this lower layer where grasses live there are also found *Nassauvia darwinii, Acaena poeppigiana*, and *Azorella caespitosa*. The *Berberis heterophylla* and *Senecio filaginoides* shrub steppe, with plants 150–200 cm high, also includes *Lycium chilense, Verbena tridens, Schinus polygamus*, and the less constant *Mulinum spinosun*. The halophytic steppe, is lower in both total cover and species richness. *Chuguiraga aurea, Puccinellia* sp., *Atriplex sagittifolia* and *A.rosea* are among the dominant species.

Two vegetation types are common to both subdistricts. In basaltic landslides ("escoriales"), a tall scrub with *Verbena tridens, Schinus polygamus, Lycium chilense, Berberis heterophylla, Nardophyllum obtusifolium, Verbena ligustrina, Adesmia boroniodes* and *Anartrophyllum rigidum* can be found. In endorreic watersheds, valley bottoms or salinized meadows, with clay-loam crust and beaches during most of the year, true deserts can develop that only have vegetation cover in wet years. Such cover includes ephemerals like *Hallophytum ameghinoi, Suaeda patagonica* and several species of *Atriplex, Chenopodium* and *Polygonum* (Movia et al. 1987).

3.7. San Jorge Gulf district

This district extends from Cabo Raso (Chubut) to Punta Casamayor (Santa Cruz), associated with the high plateaus surrounding Gulf San Jorge and the valleys and hillside areas between the plateaus and the Atlantic Ocean. Grass steppe and scrub are here the common vegetation types. Grass steppe is present on flat areas, locally named "pampas" at 700 m above sea level (Mesetas de Montemayor and the Espinosa and Pampa del Castillo), while scrub covers hillsides and valleys (Soriano 1956a).

Grass steppes have a total cover of about 80% and are 25–40 cm high. Dominant species are two tussock grasses: *Festuca pallescens* and *F.argentina*; and several shrubs and small shrubs: *Senecio filaginoides, Nardophyllum obtusifolium, Mulinun spinosum, Adesmia campestris, Verbena thymifolia* and *Acaena platyacantha*. Important associates are *Nassauvia darwinii, Mulinum halei, Perezia patagonica, Adesmia lotoides* and several *Azorella* species (Bertiller et al. 1981b). The grass steppe shows great uniformity, and prevail over wide areas only interrupted by patches of *Verbena tridens* associated with flat depressions.

Two types of scrub are found in this district. In both *Stipa humilis* and *S. speciosa* are dominant components in the grass layer and *Colliguaya integerrima* dominates in the shrub one. In the first scrub type those species are associated with *Senecio filaginoides, Grindelia chilensis, Baccharis darwinii, Perezia recurvata* ssp. *beckii* and *Nassauvia ulicina*. Vegetation is less than 80 cm high and has only one open stratum with a total cover of 40–50%. *Poa lanuginosa, Phacelia magellanica* and *Mutisia retrorsa* are present as companion species. This scrub prevails on north exposed slopes descending from the eastern watershed plateaus (Bertiller et al. 1981b). This landscape unit is the most arid in the district. It also covers the lower level of the western hillslopes and forms a boundary with the Central district. A higher portion of these hillslopes support an open scrub of *Anarthrophyllum rigidum* with *Senecio filaginoides* and *Mulinum spinosun* and a grass layer similar to that in the grass steppe on the plateaus (Bertiller et al. 1981b).

In the second scrub type, plants of *Colliguaya integerrima* reach an average height of 200 cm (up to 300 cm, Soriano 1956a) and cohabit with *Trevoa patagonica*, a quasi-aphyllous species of the Rhamnacea armed with spring stems. This scrub has *Senecio bracteolatum* as a codominant shrub in the lower layer and widely scattered individuals of *Acantholippia seriphiodes* and *Acaena platyacantha*. The associated species in the lowest layer are *Festuca argentina, Stipa neaei, Phacelia magellanica* and *Erodium cicutarium*. This is considered to be the most productive community in the district an observation consistent with its being located on southern exposed, eastern hillsides (areas with the most favorable water balance).

Several species of the Monte Phytogeographical Province attain their southernmost distribution in the San Jorge Gulf district: *Stipa tenuis* among the grasses and *Prosopis denudans* among the shrubs. A typical Monte genus, *Larrea*, occurs in this district only as a species with a woody carpet form: *Larrea ameghinoi* (Soriano 1956a).

3.8. Subandean district

Grass steppes of this district are the contact between semiarid Patagonia and the Subantartic

Province (Fig. 3). The Subandean district extends over regions with mean annual rainfall higher than 250 mm. It is replaced by *Nothofagus* Deciduous Forest wherever the precipitation reaches 500 mm. It extends as a narrow continuous north–south strip between 71° and 71°30′W, from 43°30′S (Tecka) to 46°S (Gengel River). South of the 50°30′S parallel the Subandean district widens to the east, reaching the Atlantic Ocean up to the border of the Magellan Strait. Between 46°S and 50°30′S, and from north of Tecka to Neuquén the district occurs as isolated grassland stands. They are located in those topographic positions with a more favorable water balance because of its altitude or exposure (Soriano 1983). In Chubut Province a broad ecotone links this district with the Occidental district (Soriano 1956a; León and Facelli 1981).

This vegetation unit is often found in fluvioglacial–modelled landscapes as in the El Coyte, Río Pico and Alto Río Mayo regions in Argentina or in the upper valleys of the Cisnes and Nireguao Rivers in Chile. This affinity with a definite landscape is more evident in the southern portion of the district. South of Río Gallegos the landscape consists of rolling plains modelled on fluvioglacial sediments by the Turbio-Gallegos watershed tributaries (Scalabrini et al. 1985). Isolated grasslands are found on plateaus, as in Tepuel, Tecka or the interfluvium of the Santa Cruz and Chalía Rivers (Movia et al. 1987).

A remarkable feature of the Subandean district is its high physiognomic homogeneity characterized by a grass steppe, with high total cover (more than 60%) and few shrubs, except in overgrazed stands (León and Aguiar 1985; Borelli et al. 1985; Anchorena 1985). Recent phytosociological studies in two areas, are consistent with the original physiognomic and floristic description of the district (Soriano 1956a): (1) the southwest of Chubut (Golluscio et al. 1982); and (2) south of Santa Cruz (Roig et al. 1985).

The "Festuca Grassland" (Auer 1951) or "coirón blanco grassland" (Soriano 1956a) in southwest Chubut was identified by Golluscio et al. (1982) as the "*Festuca pallescens, Rhytidosperma picta* and *Lathyrus magellanicus* community". In addition to these three, several other species are restricted to this community. Among them *Nassauvia aculeata, Relbunium ricchardianum, Senecio sericeonitens, Erigeron andicola, Luzula chilensis, Festuca magellanica* and *Vicia biyuga* can be found with constancies of 60% or more (Table 1). The group headed by *Agoseris coronopifolia*, although with low constancies, is also restricted to this community (Table 1). Several species of *Acaena* and *Koeleria* are also important. Species found in more xeric Patagonian communities are also present, including *Bromus setifolius, Hordeum comosum, Cerastium arvense*, and *Mulinun spinosum*.

Festuca pallescens, the dominant species accounts for 44% of the mean total cover and 69% or relative cover. Other grasses increase the range value of this community: *Festuca ovina, F.pyrogea, Deschampsia elegantula, D.flexuosa, Phleum commutatum, Elymus patagonicus* and *Rytidosperma virescens*. *Festuca argentina* (huecú) and *Astragalus* species have been mentioned as poisonous for domestic grazers (Soriano 1956a).

In the central part of Santa Cruz patches of a grassland with this same floristic composition appear at special locations. In high plateaus above 700 m the grass steppe becomes rich in cushion-like habit shrubs such as *Nardophyllum obtusifolium* and *Verbena* sp. (Movia et al. 1987).

South of the Coyle River *Festuca gracillima* replaces *F. pallescens* in climax grasslands. However, in azonal environments, such as meadows or valley bottoms, *Festuca pallescens* continues to be the dominant grass. At this latitude Roig et al. (1985) distinguish two different types in the *Festuca gracillima* steppe. The more humid of the two has *Gamochaeta nivalis* and the more xeric one includes *Nardophyllum brioides* or *Burkartia lanigera*. The first type occurs in the western portion, within the limits of the *Nothofagus antarctica* forest and interspersed with patches of a dwarf scrub dominated by *Empetrum rubrum* locally known as "murtillar". The more xeric community reaches the Atlantic Ocean with variants which may be considered an ecotone with the Central district. For example, this is the case of the xeric grassland of *F.gracilllima, Stipa ameghinoi* and *Nassauvia ulicina* in the Coyle River region (Seibert, 1985).

4. AVHRR/NOAA DATA IN VEGETATION STUDIES IN PATAGONIA

The selected study area covers the middle part of the Patagonia region, between 44° and 47°S and between 66° and 72°W (Fig. 3). The Subandean, Occidental, Central and San Jorge Gulf districts of the Patagonia Phytogeographical Province (Soriano 1956a) and the Deciduous Forest district of the Subantartic Province (Cabrera 1971) are represented in this territory.

The spectral data used were obtained from the only six cloud-free NOAA 6 and NOAA 8 images available in the period October–December of 1985 (namely 2/10, 7/10, 25/10, 26/11, 5/12 and 29/12). A grid of 127 points was drawn over each image (Fig. 4). For every point in the grid six Normalized Difference Vegetation Index (Curran 1981) values were calculated using

$$NDVI = (C2 - C1) / (C2 + C1)$$

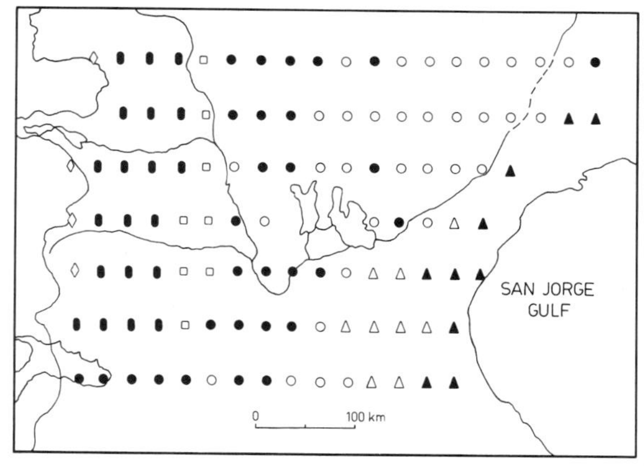

Fig. 4. Location of the sampling points in the study area. Symbols indicate different phytogeographical units. Subandean district (◊); Occidental–Subandean ecotone (▮); Occidental district (□); Central district, western area (●) and northeast area (○); San Jorge Gulf district, *Festuca-Poa* steppe (△) and *Colliguaya-Trevoa* tall scrub (▲).

where C1 and C2 are, respectively, the radiometric measures in the visible (0.58–0.68 μm) and in the near infrared (0.75–1.1 μm) spectral channels. This resulted in a data matrix containing the NDVI values of 127 one-km^2 sample points for six dates spread along the growing season. Those values were linearly transformed

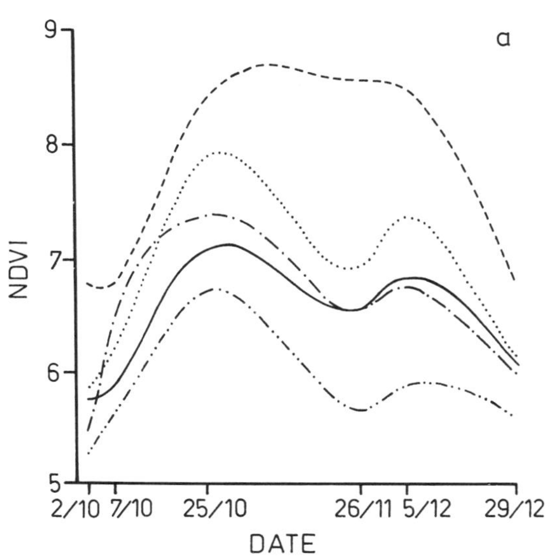

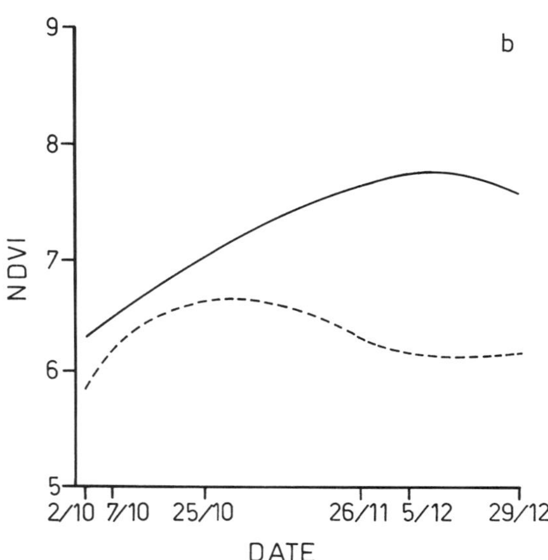

Fig. 5. Smoothed averaged NDVI change along the sampling period October–December. (a) Subandean district ----, Occidental–Subandean ecotone ····, Occidental district –·–·, Central district (W) —, Central district (NE) –··–··; (b) San Jorge Gulf: *Colliguaya-Trevoa* scrub — and *Festuca–Poa* grass steppe --- (adapted from Aguiar et al. 1988).

onto a scale ranging from 50 to 150 for the analysis. Each sample point was considered an observation unit with its own seasonal biomass dynamics described by the corresponding vector of six NDVI values. Previous phytogeographical maps (Soriano 1956a; Cabrera 1971; Bertiller et al. 1981b) were used to determine the phytogeographic unit in which each sample point lay (Fig. 4).

The successive steps of the spectral data analysis attempted to answer the following questions, previously formulated more generally:

1. What are the main directions of variation of NDVI seasonal curves among sample points at the regional scale?
2. Do the phytogeographic units represented in the study area differ in their average NDVI curve? If so, what are the main directions of variation among the units?
3. Are the differences between phytogeographic units the main source of variation in the seasonal NDVI curves?

The pattern of NDVI change along the growing season (Fig. 5) suggests a high degree of matching between the vegetation dynamics and gross environmental features (e.g. rainfall, temperature). In the spring, green biomass shows a rapid increase following temperature rise and soil moisture recharge (Barros et al. 1979). The subsequent sudden decrease of NDVI that occurs in western units (Fig. 5) may be associated with soil water depletion (Sala et al. 1989). For the western units, a longer duration of the peak of the Subandean district seems to be correlated with greater soil water availability.

Two approaches were used to answer the first question. Aguiar et al. (1988) analyzed the primary matrix via Polar Ordination (Bray and Curtis 1957; Cottam et al. 1973). Similarity between samples was measured by Czekanovski's Community Coefficient (Goodall 1973) and samples were arranged along two axes.

Most of the data variability was explained by the first ordination axis (Fig. 6). This axis would be associated with difference in NDVI, hence in

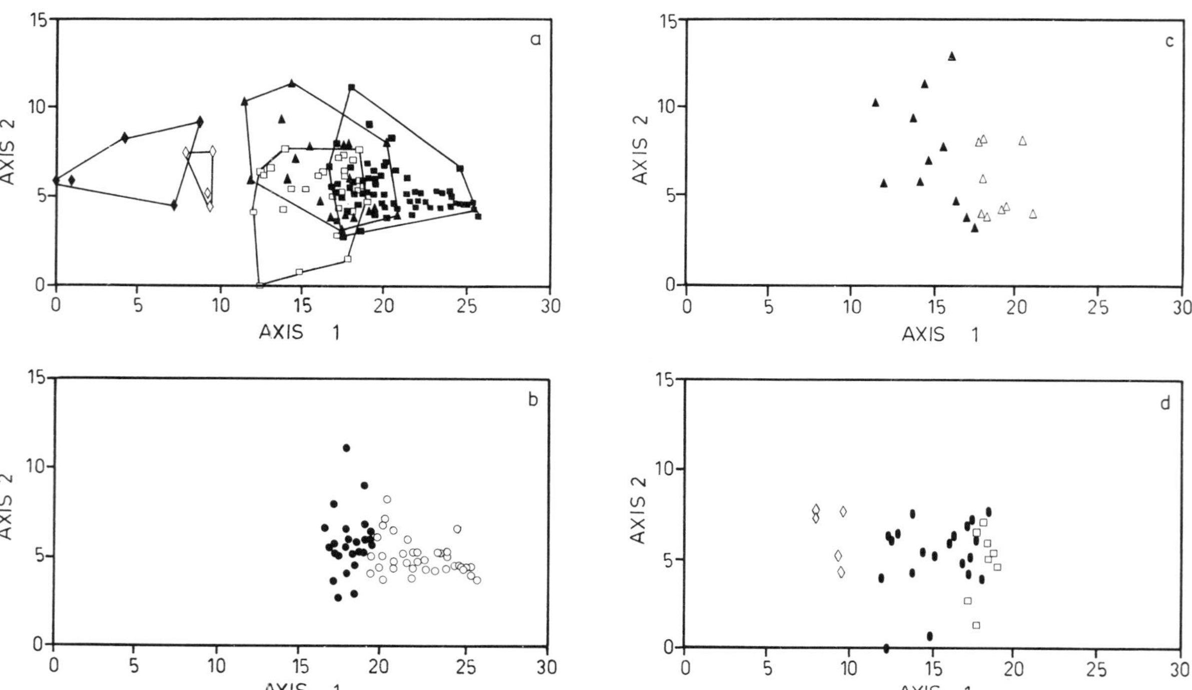

Fig. 6. Polar ordination plots of: (a) all points; (b) Central; (c) San Jorge Gulf, and (d) Occidental to Subandean district sample points. Symbols indicate different phytogeographical units. Symbols in figure (a): Deciduous forests (♦), Subandean district (◊), Occidental district (□), San Jorge Gulf district (▲) and Central district (■). For figures (b), (c) and (d) the symbols are the same as in Fig. 4.

green biomass, among floristic districts. Since the second axis did not separate the sample points and thus not show any interpretable data structure, the analysis continued based only on positions along the first axis.

The lowest values on the first axis correspond to Deciduous Forest district sample points (Fig. 6a) followed by those of the Subandean district. Sample points from the Occidental and San Jorge Gulf districts were scattered over the middle part of the ordination axis, while those from the Central district, the most arid part of Patagonia (Soriano 1956a) presented the highest scores.

Three vegetation types are spread along the Polar Ordination: forest (*Nothofagus* deciduous forest of the Subantartic Province), grass steppe (Subandean district) and shrub steppe or scrub (Occidental, Central and San Jorge Gulf districts). Samples from the *Nothofagus* Deciduous Forest, and from the *Festuca pallescens* grass steppe (Subandean district), are clearly set apart over the axis, while the points of the remainder phytogeographic units overlap partially. All of these patterns are correlated not only with great differences in average green biomass but also to differences in physiognomy.

León and Facelli (1981) judged that a broad ecotone marks the transition from the Occidental to the Subandean district. Actually, the sample points belonging to this ecotone showed lower scores than those of the Occidental district although they displayed a partial overlap (Fig. 6d). This may be the result of gradual changes in NDVI values within the shrub steppe vegetation type that follows the regional rainfall trend.

Scores of the Central district sample points along the ordination axis correlated closely with their geographic location. Lower values came out in the western area of the Chubutian subdistrict, while the highest were in the northeast and eastern parts of this unit (Fig. 6b). Although detailed vegetation descriptions of these areas are lacking, León (pers. commun.) suggested an increase of *Nassauvia ulicina* and *Nassauvia glomerulosa* towards the northeast of the area corresponding to semideserts ("eriales").

In the San Jorge Gulf district, sample points with lower scores over the first axis were located near the coast, while those with higher values corresponded to inland locations (Fig. 6c). Distribution of these two point groups conforms to that of the two distinct vegetation types: dense scrub of *Colliguaya integerrima* and *Trevoa patagonica*; and grass steppe of *Festuca pallescens, F.argentina* and *Poa ligularis* (Soriano 1956a; Bertiller et al. 1981b). Polar Ordination revealed a significant separation in NDVI dynamics between the western part of the surveyed territory (Deciduous Forest and Subandean districts) and the rest.

A second approach used to answer the first question was to perform a Principal Components Analysis (PCA; Ksirasagar 1972) to reveal the inner structure of the central and eastern parts. For this analysis the Deciduous Forest and Subandean districts sample points were excluded, as were sample points with missing values. The calculation of the PCA was done from the variance-covariance matrix estimated from the resulting data set.

The first two components accounted for 66.74% of the total NDVI variance. The first, explaining 48.57% of variance, resulted from the weighted sum of the six NDVI values with similar positive loadings for each one. The second component, accounting for the remaining 18.16% of the variance, can be roughly seen as a contrast between the sum of the second and third NDVI values and the sum of the fourth and sixth ones (Fig. 7).

Each PCA axis (Fig. 7) could be objectively interpreted because each component is a linear combination of the input variates. The first Principal Component was closely related with the integrated NDVI, corresponding to the area under the NDVI seasonal curve (Fig. 5). As the former, it results from a linear combination of NDVI values with positive loadings for each date. The integrated NDVI value could be associated with average green biomass during the period under study (Tucker et al. 1984a). The first axis separated the high biomass Subandean–Occidental ecotone and *Colliguaya integerrima* scrub from the most xeric phytogeographic units reasonably well (Fig. 7). Within the Central dis-

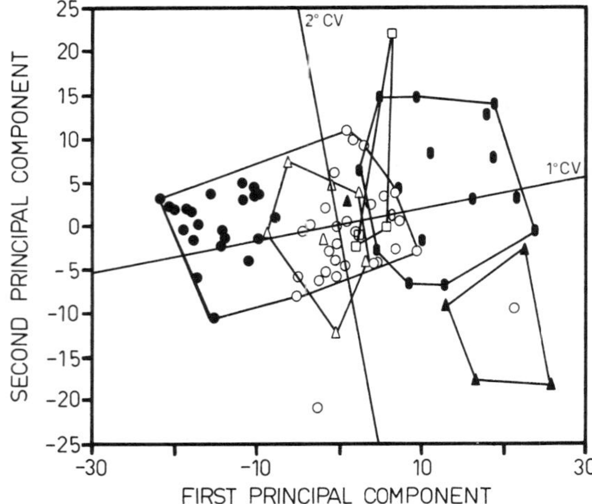

Fig. 7 Ordination plot of the sample points in the plane of the first two principal components. Symbols as Fig. 4. 1° and 2° CV show the directions of the first and second canonical variates derived by Canonical Correlation Analysis ($r1^2$ = 0.898, $r2^2$ = 0.316) between this and the Discriminant Coordinates ordination in Fig. 8. Eigenvectors (loadings) are 0.208, 0.288, 0.358, 0.577, 0.554, 0.323 and 0.162, 0.466, 0.586, −0.473, 0.057, −0.431, for the first and second component respectively.

trict it discriminated between the northeast and western subunits already recognized in the Polar Ordination (Fig. 6), supposedly reflecting differences in average standing green biomass. In spite of the structural and floristical differences distinguishing the Central and Occidental districts, they showed similarly shaped NDVI curves (Fig. 5) and a high overlap in both the Polar Ordination (Fig. 6a) and the PCA (Fig. 7). This result may be due to the similar amount of green biomass present in both units (Bertiller 1984; Fernandez, A. 1986).

Unlike the Polar Ordination, the second PCA axis allowed an objective interpretation. This component measured the decrease of NDVI values towards the end of the study period. The samples of the *Colliguaya integerrima* and *Trevoa patagonica* scrub from the San Jorge Gulf district were clearly separated from the rest along this axis. This reflected the continuous increase in their NDVI values along the studied period (Fig. 5). The longer growing season in the San Jorge Gulf district was also suggested by the corresponding average NDVI curve (Fig. 5). Paruelo et al. (submitted) suggested that this could be due to an ocean influence in the sense of reducing water vapor pressure deficit.

In relation to the first question, two main directions of variation came out in the NDVI curves. The first one highlighted both in Polar Ordination and PCA can be roughly associated with differences in the average NDVI along the sample period. The second direction shown by PCA seems to be related to the slope of NDVI changes between the beginning and the end of the observation period. These two principal directions of data variation would likely reflect differences in average green biomass and biomass dynamics along the growing season among sample points.

The second question concerned the existence and nature of significant differences in NDVI dynamics among districts. Our first attempt was to compare graphically the smoothed average NDVI curves corresponding to the different phytogeographic units (Aguiar et al. 1988). This approach suggested the existence of differences among units in the seasonal NDVI maximum and in the pattern of seasonal NDVI change. Two patterns could be recognized (Fig. 5). The first, corresponding to the western units (Subandean, Occidental and Central districts), showed a sudden NDVI rise at the beginning of spring and a rapid decrease towards the end of the growing season. The second pattern, typical of the *Colliguaya integerrima* and *Trevoa patagonica* scrub, presented a slower NDVI increase and a delayed peak, suggesting a growing season longer than the observation period.

Another approach to answer the second question (Batista 1988) was the use of Discriminant Coordinates Analysis (DCA; Gnanadesikan 1977; Seber 1984). This technique allowed a formal test of the previous observations by displaying the principal differences among classes (phytogeographic units) in relation to a given group of variates (NDVI × date). It has already been used to analyze floristic differences between otherwise defined vegetation units (Norris and Barkham 1970; Grigal and Goldstein 1971; Goldstein and Grigel 1972; Matthews 1979) and to display vegetation–environmental relationships (McCune and Allen 1984; Tisdale and Bramble-Brodahl 1985; Gerdol et al. 1985).

The application of DCA to the same data set subjected to PCA, revealed a four-dimensional discriminant structure where the first two DCA axes accounted for 88.2% of the total among-units variation in the average NDVI curve. Consequently it was assumed that the between-classes differences could be reasonably well approximated by the plane spanned by this two leading directions.

According to the loadings (eigenvectors) for the variates the first axis (67.65% of the variance), could be described as a sum of the NDVI values in the four central dates, with special weights for the third and fifth, and slightly corrected by the first and last values (Table 2). Hence, it could be interpreted as an index of the height and convexity of the NDVI curve, a magnitude that can be associated with the maximum seasonal within NDVI in each unit.

The second axis, explaining 20.56% of the among units variance, could be regarded contrast between the sum of the second and the sixth NDVI with the third (Table 2). Sample points showing little change between the second and third dates and high NDVI at the end of the sampling period would get high scores on this coordinate.

The plot of the first two discriminant coordinates (Fig. 8) showed the Subandean–Occidental ecotone at the top of the first axis, reflecting its higher average NDVI and/or a more convex seasonal curve (Fig. 5). The *Colliguaya integerrima*

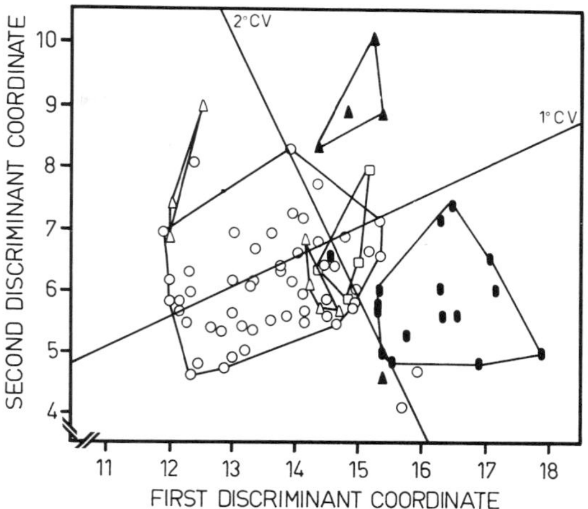

Fig. 8. Ordination plot of the sample points in the plane of the first two discriminant coordinates. Symbols as in Fig. 4. 1° and 2° CV show the directions of the canonical variates derived by Canonical Correlation Analysis ($r1^2$ = 0.898, $r2^2$ = 0.316) between this and the Principal Components ordination (Fig. 7).

scrub had the highest score for the second coordinate. The average seasonal NDVI curve for this unit (Fig. 5) corresponded to the second pattern defined above by showing a small change between the second and third dates and high NDVI in the sixth. The Central district had lower scores than the Subandean–Occidental ecotone on the first axis and a similar position along the second one. The average seasonal NDVI curves of this unit (Fig. 5) fit the first pattern, being lower than the Subandean–Occidental ecotone curve but similarly shaped. The Occidental sample points showed an intermediate behaviour between the Central district

Table 2. Results of Discriminant Coordinates Analysis (adapted from Batista 1988).

Discriminant coordinate	I	II	III	IV	V	VI
Eigenvalue	1.108	0.337	0.143	0.046	0.0	0.0
% Total sq. distance explained	67.65	20.56	8.9	2.80	0.0	0.0
Eigenvectors (loadings)	− 0.042	0.012				
	0.021	0.089				
	0.133	− 0.120				
	0.035	0.014				
	0.104	0.029				
	− 0.031	0.101				
Associated F Statistic (4 & 79 d.f.)	21.88	6.6	2.88	0.91	0.0	.0.

and the Subandean–Occidental ecotone. The *Festuca pallescens* and *Poa ligularis* steppe sample points appeared as two clusters in the plot, one of them merged with those of the Central district. This could be due to problems of geographic location of sample points of this unit which stretches as a narrow belt between the scrub (*Colliguaya integerrima* and *Trevoa patagonica*) and the Central district.

The clearcut discrimination found among the Central district, the Subandean-Occidental ecotone and the *Colliguaya integerrima* and *Trevoa patagonica* scrub indicates significant differences in seasonal NDVI curves, hence in seasonal green biomass changes, among phytogeographic units. The major among-unit differences seem related to total green biomass growth during the observed period and to differences between the two patterns of seasonal curves describing the length of growing season, rate of initial growth and date of the green biomass peak as assessed by NDVI.

The third question, concerning the degree to which the between-district differences explained the global regional variation in the NDVI curves, was addressed by comparing the PCA and DCA ordinations. The comparison was performed informally by contrasting the ordination plots and variates loadings and formally by applying Procrustes Analysis (Hurley and Cattel 1962) and Canonical Correlation Analysis (Sibson 1978). The agreement between the two configurations of sample points in the ordination spaces was to be considered a sign that the variation in seasonal NDVI curves could be mainly attributed to the between phytogeographic units differences.

The relative positions of the different phytogeographical units on the PCA and DCA plots (Figs 7 and 8) suggested certain degrees of correspondence between both types of analyses. The interpretation of the axes could also be related in terms of vegetation. The average green biomass (first PC axis) and the height and convexity of green biomass curve, (first DC) were expected to be highly associated vegetation attributes in the markedly seasonal Patagonian environments. Only for the longer growing *Colliguaya integerrima* and *Trevoa patagonica* scrub did the average NDVI during the observed period not resemble the height and convexity of the corresponding curve; hence the different relative positions along both first axes. Both the second PCA axis and the second DCA axis related to the shape of the NDVI curves. Although they formally measure different and not necessarily related aspects of this shape, for the present study both distinguished between the two patterns of seasonal growth postulated by Aguiar et al. (1988).

The above observations were supported by the formal comparison of both ordinations by Procrustes analysis and Canonical Correlation Analysis. After rigid axes rotation reflection and uniform scale transformation, the distance between configurations resulted in 39.1% of their total variation ($\alpha 2$ = 0.391, Sibson 1978). Hence 60.9% of the NDVI curves total interpreted variation was attributed to differences among phytogeographical units.

Canonical Correlation Analysis showed that the intermediate degree of coincidence between configurations, measured by Procrustes Analysis, resulted from averaging high correlation in the direction of the first canonical variates (r^2 = 0.898) with low correlation in the direction of the second ones (r^2 = 0.316). The first canonical variates were moderate rotations of the first axes of both ordination (Figs 7 and 8). Their high correlation gives support to the hypothesis that differences among phytogeographic units in the height and convexity of the growth curve would explain the regional (between points) variability in the average green biomass along the observed period. Paruelo et al. (submitted) have related the among phytogeographic units differences to the mean annual rainfall. The second canonical variates were moderate rotations of the second axes (Figs 7 and 8). Their low correlation would indicate that the slope of the NDVI change between the first and second parts of the growing season (second PC axis) would not be explained by the differences between phytogeographic units. Reciprocally, the principal difference among the shapes of the different phytogeographic units average NDVI curves (second DC axis) would not be a principal direction of the total variation. This result suggested the existence of within phytogeographic unit factors responsible for the variability (altitude, exposure, soils, etc.).

5. CONCLUSIONS

Use of multivariate methods in vegetation studies is common at the community level, but it also can be used profitably on a global scale (regional, continental). In this chapter, we presented a vegetation analysis of Extra-andean Patagonia at the regional level. This study is based on the Normalized Difference Vegetation Index (NDVI), a variate highly correlated with green biomass and calculated from spectral data provided by NOAA/AVHRR satellites.

Six satellite images from six dates spread over the growing season were sampled at the study area in central Patagonia, a territory in which five different phytogeographic units have been identified. Each sample point was characterized as a vector with six NDVI values along the observation period. This vector was taken as a description of the seasonal NDVI curve. Three questions concerning use of spectral data were posed and the following conclusions drawn:

(1) The results of Polar Ordination and PCA suggest that at a regional scale the NDVI curves were different in two main directions, the first roughly associated with the average NDVI during the growing season and the second with the slope of NDVI change between the beginning and the end of this period. This two-directions-of-data variation indicate differences in green biomass and its change along the season among sample points.

(2) The clear discrimination among Central, Subandean–Occidental ecotone and *Colliguaya integerrima* scrub found in the DCA suggests significant differences exist in NDVI curves between phytogeographic units. Major differences seem related to the height and convexity of the NDVI curve (DCA axis 1), and to the extent of initial growth and the occurrence of NDVI peaks in the growing season (DCA axis 2).

(3) Procrustes and Canonical Correlation analyses indicated that a high portion of the variability of the first component may be explained by phytogeographic differences. Associated with the average green biomass of sample points, this variation is correlated with changes in mean annual rainfall (Paruelo et al., submitted). The second component of the variation between NDVI curves, associated with its shape, would be related to intra-district rather than to among-district differences.

ACKNOWLEDGEMENTS

We would like thank A. Soriano, S. Burkart, O.E. Sala and T. Crovello for they useful comments. Data used in this chapter was provided by the Servicio Meteorológico Nacional. Data processing was performed with the aid of Gloria Pujol. This work was supported by grant of the Universidad de Buenos Aires (No. AG 18) and of CONICET (PID No 424/89).

REFERENCES

Aguiar, M.R., Paruelo, J.M., Golluscio, R.A., León, R.J.C., Pujol, G. and Burkart, S. 1988. The heterogeneity of the vegetation in arid and semiarid Patagonia: An analysis using AVHRR/NOAA satellite imagery. Annal. di Botanica 46, 103–114.

Allen, T.H.R. and Starr, T.B. 1982. Hierarchy: perspective for ecological complexity. Univ. of Chicago Press, Chicago.

Ancibor, E. 1980. Estudio anatómico de la vegetación de la Puna de Jujuy. II. Anatomía de las plantas en cojín. Bol. Soc. Arg. Bot., 19, 157–202.

Anchorena, J. 1985. Cartas de aptitud ganadera (dos ejemplos para la región magallánica). In Boelcke, O., Moore, D.M., Roig, F.A. (eds), Transecta Botánica de la Patagonia Austral CONICET (Argentina), Buenos Aires, pp. 520–540.

Asrar, G., Fuchs, M., Kanemasu, E.T. and Hatfield, D.L. 1984. Estimating absorbed photosynthetic radiation and leaf area index from spectral reflectance in wheat. Agro. J., 76, 300–306.

Auer, V. 1951. Consideraciones científicas sobre la conservación de los recursos naturales de Patagonia. Inf. Invest. Agric. No. 40–41.

Barros, V. R., Scian, B. V. and Mattio, H. F. 1979. Mapa de precipitaciones de la provincia de Chubut. Centro Nacional Patagónico.

Batista, W. B. 1988. Relating new information to a previous vegetation classification: a case of discriminant coordinates analysis. Vegetatio, 75; 153–158.

Bertiller, M. B. 1984. Specific primary productivity dynamics in arid ecosystems: a case study in Patagonia, Argentina. Acta Oecologica. Oecologia Generalis, 5, 365–381.

Bertiller, M. B., Beeskow A. M., and Irrisarri, P. 1981a. Caracteres fisonómicos y florísticos de la vegetación del Chubut. 1. Sierra San Bernardo. Contribución No. 40. CONICET. Centro Nacional Patagónico.

Bertiller, M. B., Beeskow, A M. and Irrisarri, P. 1981b. Caracteres fisonómicos y florísticos de la vegetación del Chubut. 2 La península de Valdés y el istmo Ameghino. Contribución No. 41. CONICET. Centro Nacional Patagónico. 20 pp.

Boelcke, O. 1957. Comunidades herbáceas del Norte de la Patagonia y sus relaciones con la ganadería. Rev. Inv. Agric., 11 (1), 5–98.

Bonvissuto, G., Moricz de Tecso, E., Astibia, O. and Anchorena, J. 1983. Resultados preliminares sobre los hábitos dietarios de ovinos en un pastizal semidesértico de Patagonia, IDIA 36 (Supl.), 243–253.

Borelli, P. R., Chappi, C. A., Jacomini, M. H. and Ramstrom, A. 1985. Condición de pastizales en el sitio Terraza de Río Gallegos. Rev. Arg. Prod. Animal, IV (9), 879–897.

Bray, J. R. and Curtis, J. T. 1957. An ordination of the upland forest communities of Southern Wisconsin. Ecol. Monogr., 27, 325–349.

Cabrera, A. L. 1971. Fitogeografía de la República Argentina. Bol. Soc. Arg. Bot., 14 (1–2), 1–42.

Cabrera, A. L. 1976. Regiones fitogeográficas argentinas. Enciclopedia Argentina de Agricultura y Jardinería (2da. ed). Tomo II, Fasc. 1 ACME, Bs. As. 85 pp.

Castellanos, A. and Perez Moreau, R. A. 1944. Los tipos de vegetación de la República Argentina. Monografías del Instituto de Estudios geográficos 4. Universidad Nacional de Tucumán. 154 pp.

Coldwell, J. E. and Hicks, D. R. 1985. NOAA Satellite Data: a useful tool for macro inventory. Environmental Management, 9, 463–470.

Correa, N. M. 1969. Flora Patagónica (Rep. Argentina). Tomo VIII Parte II Monocotyledonae (excepto Graminae). Colección Científica INTA. 219 pp.

Correa, N. M. 1971. Flora Patagónica (Rep. Argentina) Tomo VIII parte VII. Compositae. Colección Científica INTA. 451 pp.

Correa, N. M. 1978. Flora Patagónica (Rep. Argentina) Tomo VIII parte III. Graminae. Colección Científica INTA. 563 pp.

Correa, N. M. 1984a. Flora Patagónica (Rep. Argentina) Tomo VIII parte IVa. Dicotyledoneas dialipétalas (Salicáceas a Crucíferas) Colección Científica INTA. 559 pp.

Correa, N. M. 1984b. Flora Patagónica (Rep. Argentina) Tomo VII parte IVb. Dicotyledoneas dialipétalas (Droseaceae a Leguminosae) Colección Científica. 309 pp.

Cottam, G., Goff, F. G. and Whittaker, R. H. 1973. Wisconsin comparative ordination. In: Whittaker, R. H. (ed.), Handbook of Vegetation Science, Part 5, pp. 195–221. Junk, The Hague.

Curran, P. J. 1981. Biomass and productivity. In: Plants and the daylight, spectrum (ed. by H. Smith). Academic Press. 508 pp.

Facelli, J. M. and León, R. J. C. 1986. La diversidad específica de pastizales patagónicos subandinos sometidos al pastoreo. Turrialba, 36 (4), 461–468.

Fernández A., R. J. 1986. Estimación de la productividad primaria neta aérea de pastos y arbustos en la estepa árida patagónica. Trabajo de Intensificación, Facultad de Agronomía (UBA).

Frenguelli, J. 1941. Rasgos principales de la Fitogeofrafía Argentina. Rev. Museo de la Plata III Botánica No. 13, 65–181.

Gerdol, R., Ferrari, C. and Piccol, F. 1985. Correlation between soil and forest types: a study in multiple discriminant analysis. Vegetatio, 60, 49–56.

Gnanadesikan, R. 1977. Methods for statistical data analysis of multivariate observation. Wiley, New York.

Golluscio, R. A., León, R. J. C. and Perelman, S. B. 1982. Caracterización fitosociológica de la estepa del oeste del Chubut. Su relación con el gradiente ambiental. Boletín de la Sociedad Argentina de Botánica, 21 (1–4), 299–324.

Goodall, D. W. 1973. Sample similarity and species correlation. In: Whittaker, R. H. (ed.), Handbook of Vegetation Science, Part 5, pp. 105–156. Junk, The Hague.

Goward, S. N., Tucker, C. S. and Dye, D. G. 1985. North American vegetation patterns observed with the NOAA–7 advanced very high resolution radiometer. Vegetatio, 64, 3–14.

Grigal, D. F. and Goldstein, R. A. 1971. An integrated ordination classification analysis of intensively sampled oak-hickory forest. J. Ecol., 59, 481–492.

Goldstein, R. A. and Grigal, D. F. 1972. Definition of vegetation structure by canonical analysis. J. Ecol., 60, 277–284.

Hager, J. 1986. Zur Verbreitung der Polsterpflanzen in der patagonischen Zwergstrauch-Halbwüste. Ein Beitrag zum ökologischen Verständnis der Wuchsform. Bot. Jahrb. Syst., 106 (4), 511–540.

Haralick, R. M. and King Su Fu, 1983. Pattern recognition and classification. In: Manual of remote sensing (2nd edn), ed. by Coldwell, R. N. Fallschurch, Virginia, American Society of Photogrammetry. 2111 pp.

Hatfield, J. L. 1983. Remote sensing estimates of potential and actual crop yield. Remote Sensing of Environment, 13, 301–311.

Hauman, L. 1920. Un viaje botánico al Lago Argentino. An. Soc. Cient. Arg., 89, 179–281.

Hauman, L. 1926. Etude phytogéographique de la Patagonie. Bull. Soc. Roy. Bot. Belg., 58, 105–180.

Holben, B. N. and Justice, C. O. 1981. An examination of spectral band rationing to reduce the topographic effect on remotely sensed data. Remote Sensing, J. 2, 115.

Horler, D. N. and Barber, J. 1981. Principles of remote sensing of plants. In: Plants and the daylight spectrum (ed. by Smith, H.). Academic Press. 508 pp.

Hueck, K. and Seibert, P. 1981. Mapa de la vegetación de América del Sur, Stuttgart, New York, 2 Aufl.

Hurley, J. R. and Cattel, R. B. 1962. The Procrustes program: producing direct rotation to test a hypothesized factor structure?. Behav. Sci., 7, 258–262.

Jackson, R. D., Slater, P. N. and Pinter, P. J. 1983. Discrimination of growth and water stress in wheat by various vegetation index through clear and turbi atmospheres. Remote Sensing of Environment, 15–187.

Justice, C. O., Townshend, J. R. G., Holben, B. N. and Tucker, C. J. 1985. Analysis of the phenology of global vegetation using meteorological satellite data. Int. J. Remote Sensing, 61, 271–1318.

Kidwell, K. B. 1984. NOAA Polar orbital users guide (Tiros-N, NOAA 6, 7, 8). NOAA, NOAA National Climate Center, Washington, DC.

Kshirsagar, A. M. 1972. Multivariate Analysis. Marcel Dekker, New York. 534 pp.

Kumar, M. and Monteith, J. L. 1981. Remote Sensing of crop growth. In: Plant and the daylight spectrum (ed. H. Smith). Academic Press, 508 pp.

León, R. J. C. and Aguiar, M. R. 1985. El deterioro por uso pasturil en estepas herbáceas patagónicas. Phytocoenologia, 13; 181–196.

León, R. J. C. and Facelli, J. M. 1981. Descripción de una coenoclina en el SW del Chubut. Rev. Fac. Agron. 2: 163–171.

Lores, R. D., Ferrereira, C. A., Anchorena, J., Lipinski, V. and Marcolin, A. 1983. Las unidades ecológicas del campo Experimental Pilcaniyeu (Río Negro) su importancia regional. Gaceta Agronómica IV (16): 660–690.

Malingreau, J. P. 1986. Global vegetation dynamics: observations over Asia. Int J. Remote Sensing, 7, 1121–1146.

Matthews, J. A. 1979. A study of the variability of some successional and climax plant assemblage-types using multiple discriminant analysis. J. Ecol., 67, 255–271.

Mc Cune, B. and Allen, T. F. H. 1984. Will similar forests develop in similar sites? Can. J. Bot., 63, 367–376.

Morrison, J. J. 1917. La ganadería en la región de las mesetas australes del territorio de Santa Cruz. Tesis, Facultad de Agronomía y Veterinaria. Buenos Aires, No. 55, 172 pp.

Movia, C. P. 1972. Formas de erosión eólica de la Patagonia. Photointerprétation No. 6/3 (Editions Technip, Paris).

Movia, C. P., Soriano, A., and León, R. J. C. 1987. La vegetación de la cuenca del Río Santa Cruz. Darwiniana, 28 (1–4), 9–78.

Myers, V. I. 1983. Remote sensing applications in agriculture. Ch. 33. In: R. N. Colwell (ed.), Manual of remote sensing (2nd edn), Vol. II. Interpretation and applications. American Society of Photogrammetry. 2111 pp.

Norris, J. M. and Barkham, J. P. 1970. A comparison of some Cotswold beechwoods using multiple discriminant analysis. J. Ecol., 58, 603–619.

Norwine, J. and Greegor, D. M. 1983. Vegetation classification based on AVHRR satellite imagery. Remote Sensing of Environment, 13, 69–87.

Ragonese, A. E. and Piccinini, B. C. 1969. Límite entre el Monte y el Semi-desierto Patagónico en las provincias de Río Negro y Neuquén. Bol. Soc. Arg. Bot., 11 (4), 299–302.

Roig, F. A. 1972. Bosquejo fisonómico de la vegetación de la provincia de Mendoza. Bol. Soc. Arg. Bot., XIII (suplemento), 49–80.

Roig, F. A., de Marco, G. and Wuilloud, C. 1980. El límite entre las provincias fitogeográficas del Monte y de la Patagonia en las llanuras altas de San Carlos, Mendoza. Bol. Soc. Arg. Bot., XIX (1–2), 331–338.

Roig, F. A., Anchorena, J., Dollenz, O., Faggi, A. M. and Mendez, E. 1985. Las comunidades vegetales de la Transecta Botánica de la Patagonia Austral. La vegetación del área continental. In Boelcke, O. Moore, D. M. and Roig, F. A. (eds), Transecta Botánica de la Patagonia Austral CONICET (Argentina), Buenos Aires, pp. 350–456.

Ruiz Leal, A. 1972. Los confines boreal y austral de las provincias Patagónica y Central respectivamente. Bol. Soc. Arg. Bot., XII (suplemento), 89–118.

Ruthsatz, B. 1978. Las plantas en cojín de los semidesiertos andinos del NO argentino. Darwiniana, 21 (2–4), 491–539.

Sala, O. E., Golluscio, R. A., Lauenroth, W. K. and Soriano, A. 1989. Resource partitioning between shrubs and grasses in the Patagonian steppe. Oecologia, 81, 501–505.

Scalabrini, O., Spikerman, J. and Medina, F. 1985. Geologia y geomorfología de Santa Cruz entre los paralelos 51 y 52 de lat. Sur. In Boelcke, O. Moore, D. M. and Roig F. A. (eds), Transecta Botánica de la Patagonia Austral CONICET (Argentina), Buenos Aires, pp. 41–48.

Seber, G. A. F. 1984. Multivariate Observations. Wiley, New York. 210 pp.

Schlichter, T. M., León, R. J. C. and Soriano, A. 1978. Utilización de índices de diversidad en la evaluación de pastizales naturales en el centro-oeste de Chubut. Ecología, 3, 125–132.

Seibert, P. 1985. Ordenamiento fitogeográfico y evaluación territorial. In Boelcke, O. Moore, D. M. and Roig, F. A. (eds), Transecta Botánica de la Patagonia Austral CONICET (Argentina), Buenos Aires, pp. 520–540.

Sibson, R. 1978. Studies in the robustness of multidimensional scaling: procrustes statistics. J.R. Stat, Soc. B. 40, 234–238.

Smith, W. L., Bishop, W. P., Dvorak, V. F., Hayden, C. M., McElroy, J. H., Mosher, F. R., Oliver, V. J., Purdom, J. F. and Wark, D. Q. 1986. The meteorological satellite: overview of 25 years of operation. Science, 231, 455–462.

Soriano, A. 1949a. El límite entre las provincias botánicas Patagónica y Central en el Territorio del Chubut. Lilloa, 20, 193–202.

Soriano, A. 1949b. El pastoreo en la parte semidesértica del Chubut. IDIA, 2 (21) 8–14.

Soriano, A. 1956a. Los distrítos florísticos de la provincia Patagónica. Rev. Inv. Agr., 10, 323–347.

Soriano, A. 1956b. Aspectos ecológicos y pasturiles de la vegetación patagónica, relacionados con su estado y capacidad de recuperación. Rev. Inv. Agr., 10, 349–372.

Soriano, A. 1959. Síntesis de los resultados obtenidos en las clausuras instaladas en Patagonia en 1954 y 1955. Rev. Agron. Noroeste Arg., 3 (1–2), 163–176.

Soriano, A. 1981. Ecología del pastizal de coirón amargo en el sudoeste de Chubut. Producción Animal, 8, 38–43.

Soriano, A. 1983. Deserts and semi-deserts of Patagonia. Temperate Deserts and Semi-Deserts (ed. by N. E. West). Elsevier Scientific, Amsterdam, The Netherlands.

Soriano, A. 1986. Relaciones entre los métodos de uso de los recursos y la oferta de los sistemas ecologicos en la Patagonia. Anal. Acad. nac. Cs. Ex. Fis. Nat. Tomo, 38, 138–144.

Soriano, A. and Brun, J. 1973. Valoración de campos en el centro-oeste de la Patagonia: desarrollo de una escale de puntajes. Rev. Inv. Agrop. Serie 2, 10 (5), 173–185.

Soriano, A., Alippe, H. A., Sala, O. E., Schlichter, T. M., Movia, C. P., León, R. J. C., Trabucco, R. and Deregibus, V. A. 1976. Ecología del pastizal de coirón amargo (*Stipa* spp.) del Sudoeste de Chubut. Acad. Nac. Agric. Vet., 30, 1–13.

Soriano, A., Sala, O. E. and León, R. J. C. 1980. Vegetación actual y vegetación potencial en el pastizal de coirón amargo (*Stipa* spp.) del SW de Chubut. Bol. Soc. Arg. Bot., 19, 309–314.

Soriano, A. and Sala, O. E. 1983. Ecological strategies in the Patagonian Arid Steppe. Vegetatio, 56, 9–15.

Soriano, A., Movia, C. P., León, R. J. C. 1983. Deserts and semideserts of Patagonia. In: West, N. E. (ed.) Temperate Deserts and Semideserts, pp. 440–453. Elsevier Publ. Co.

Soriano, A. and Movia, C. P. 1986. Erosión y desertización en la Patagonia. Inteciencia, 11, 77–83.

Soriano, A. and Sala, O. E. 1986. Emergence and survival of *Bromus setifolius* seedlings in different microsites of Patagonian Arid Steppe. Israel J. Botany, 35, 91–100.

Speck, N. H. 1982. Vegetación y pasturas de la zona Ingeniero Jacobacci-Maquinchao. In INTA (ed.), Sistema Fisiográfico de la Zona Ingeniero Jacobacci-Maquinchao (Prov. Río Negro). Buenos Aires, pp. 157–208.

Swain, P. H. and Davis, J. M. 1978. Remote Sensing: The Quantitative Approach. McGraw-Hill, New York. 320 pp.

Tisdale, E. W. and Bramble-Brodahl, M. 1983. Relationships of site characteristics to vegetation in canyon grasslands of west central Idaho and adjacent areas. J. Range Manage., 36, 775–778.

Tucker, C. J. 1979. Red and photographic infrared linear combination for monitoring vegetation. Remote Sensing Environ., 8, 127–150.

Tucker, C. J., Holben, B. N., Elgin, J. H. and McMurtrey, J. E. III. 1981. Remote sensing of total dry matter accumulation winter wheat. Remote Sensing Environ., 11 (3), 171–190.

Tucker, C. J. and Sellers, P. J. 1986. Satellite remote sensing for primary production. Int. J. Remote Sensing, 7 (11), 1395–1416.

Tucker, C. J., Townshend, J. R. G. and Goff, T. E. 1985. African land cover classification using satellite data. Science, 227, 369–375.

Tucker, C. J., Gatlin, A. and Schneider, S. R. 1984a. Monitoring vegetation in the Nile Delta with NOAA6 and NOAA7 AVHRR photogram engng. J. Remote Sensing, 50, 53–61.

Tucker, C. J., Vanpraet, C. L., Boerwinkle, E. and Gaston, A. 1984b. Satellite remote sensing of total dry matter accumulation in the Senegalese Sahel. Remote Sensing Environ., 13, 461–474.

Townshend, J. R. G., Justice C. O. and Kalb, V. 1987. Characterization and classification of South American land cover types using satellite data. Int. J. Remote Sensing, 8 (8), 1189–1207.

Volkheimer, W. 1983. Geology of extra-Andean Patagonia. In Deserts and semideserts of Patagonia. In: West N. E. (ed.), Temperate Deserts and Semideserts, pp. 440–453. Elsevier Publ. Co.

7. DISTRIBUTION PATTERNS, ADAPTIVE STRATEGIES, AND MORPHOLOGICAL CHANGES OF MOSSES ALONG ELEVATIONAL AND LATITUDINAL GRADIENTS ON SOUTH PACIFIC ISLANDS

DALE H. VITT

Department of Botany, The University of Alberta, Edmonton, Alberta, Canada T6G 2E9

1. INTRODUCTION

One of the characteristics of polar landscapes is the relative abundance of bryophytes. These plants are especially well developed on organic soils and are largely associated with areas of local ground water flow (Vitt and Pakarinen 1977). In contrast to these terrestrial communities in both northern and southern polar regions, one of the defining features of tropical montane forests is the abundance of epiphytic mosses, hepatics, orchids and in the New World, bromeliads. This component is particularly well developed in areas where moisture-laden air masses are orographically raised and result in permanent, higher elevation mist zones. In both these situations, species richness of vascular plants is relatively low and overall community development is reduced. In general, animal diversity decreases along both elevational and latitudinal gradients as well (e.g. Terborgh 1971, 1977; Kikkowa and Williams 1971).

Despite the recognition by several authors that the bryophyte component becomes more conspicuous in polar regions, few have considered these plants in any context other than that of an "inert" carpet in which vascular plants grow (Savile 1972). Likewise in tropical rainforests, the abundant epiphytic bryophyte component has long been recognized as being evident; however, in almost all cases, the literature suggests that these epiphytic plant assemblages are no more than "large inanimate sponges". In polar regions, these plants were thought of only as a medium for rooting of other "higher" plants, while in tropical forests they were thought only to reduce incident radiation at leaf surfaces.

Although the view that ascribes to the limited importance of the bryophyte component in both of these contrasting systems is still prevalent in the current literature, there are a number of recent studies that suggest a very important role of these organisms. In high arctic meadows, high biomass, considerable productivity sometimes equaling that of above ground vascular plants, and early season growth (Vitt and Pakarinen 1977) suggest that bryophytes act to immediately uptake seasonal nutrients, especially those available from nutrient flushes after snow melt, when bryophytes show their highest seasonal activity and vascular plants are still dormant. Subsequent gradual and steady release of these nutrients from decomposing bryophytes make nutrients available to vascular plants later in the season. Damman (1978) and others have shown that much of the potassium and manganese released in peatland systems from decomposition processes is actively transported to the capitula of *Sphagnum* plants. As well, vascular plant roots are largely confined to the zone of active decomposition of the ground layer. More recently, Bayley et al. (1987) have presented data indicating that when nitrogen (in the form of NO_3) is added to a boreal peat forming system, within 24 hours over 90% is taken up by the *Sphagnum* lawn, with an increase in *Sphagnum* growth over the course of the year. No increase in the vascular plant component can be demonstrated after five years of the experiment (Vitt and Bayley, unpublished data). All of these data indicate that

P. L. Nimis and T. J. Crovello (eds): Quantitative Approaches to Phytogeography, pp. 205–231.

the bryophyte component can play a highly significant role in trapping and subsequently releasing nutrients for further use within the ecosystem.

Similarly, recent studies in tropical rainforests have indicated the importance of the epiphytic component in the active intervention of water (Pócs 1976, 1980), while studies in more northern ecosystems indicate the intervention of cryptogams in nutrient retention (Skre et al. 1983; Weber and Van Cleve 1984). As well, the potential for nutrient release owing to repeated wetting/drying cycles has been shown to be of significance (Dudley and Lechowicz 1987). These studies provide the base for suggestions that the epiphytic bryophyte community may play a critical role in first interception of nutrients, and subsequent release of these for use by vascular plant members of the forest.

In both polar and tropical ecosystems, bryophyte components comprise a high proportion of the total biomass; in both, nutrient limitations are commonly perceived as a major constraint to vascular plant growth, and it is beginning to appear that nutrient release by bryophytes is a critical component in the nutrient flow of these communities. Further studies in both ecosystems are needed in order to bring this apparent role of bryophytes more into focus.

Early studies on growth forms of mosses are available in the work of Meusel (1935). Subsequently, Gimingham and Birse (1957) outlined a classification of growth forms, and related these to particular habitats and micro-environments, while Vitt (1979) correlated elevation to growth form and higher taxonomic groups and Mägdefrau (1982) reviewed life forms of mosses. However, until very recently, few studies have examined either latitudinal or tropical elevational gradients in any degree of detail. In 1910, Geisenhagen described moss growth forms in Malaysian rain forests, while Seifriz (1924) reported on the elevational distribution of cryptogams in Java. Our knowledge of tropical bryophyte ecology was effectively summarized by Richards in 1932. Considerable detail to our view of these ecosystems was added by 1952 when Richards' treatment of tropical rainforests was published. Since then, the theory that nutrient cycling in cloud forests largely excludes the soil profile has become widely accepted. The importance of rainforest bryophytes as effective rainfall interceptors was established by Pócs (1976, 1980). Elevational increase in bryophyte biomass was found to be about six times higher in Tanzanian elfin forests as compared to lower submontane rainforests (Pócs 1982). Van Reenan and Gradstein (1983) reported that total bryophyte cover along an elevational gradient in the Andes increases 10-fold between lowland and montane rainforests, and Frahm (1987) found similar trends along a transect completed in Peru. As well as increases in biomass and cover, species richness increases with elevation in tropical rainforests as shown by Gradstein and Frahm (1987). More detailed studies by Van Reenen (1987) showed that not only epiphytic cover increases with elevation, but also terrestrial cover. Much of the biogeography of tropical bryophytes is reviewed by Gradstein and Pócs (1989).

These studies, perhaps beginning with the observations of Spruce (1908) and the detailed, quantitative data gathered by later authors indicate that bryophyte cover, biomass, and species richness increase with elevation. Reasons for this have ranged from high humidity (Giesenhagen 1910), humidity and light (Seifriz 1924), cooler climate and constant humidity (Richards 1952), precipitation and fog (Pócs 1982), to influences of temperature (Richards 1984). However, only Frahm (1987) has presented experimental data, from which he concluded that montane rainforest species are physiologically excluded from lower elevation forests and that the high biomass found at higher elevations is largely due to light intensity combined with low temperature.

Biogeography in the Pacific has had a rich history, from Wallace's (1869, 1880) work through MacArthur and Wilson's (1967) ideas on Island Biogeography. Many of our present-day concepts can be traced to the works of early biogeographers who spent time in the South Pacific. Such fundamental ideas as long-distance dispersal (Hooker 1860) date from J. D. Hooker's travel to isolated South Pacific islands; our thinking on the origins of island biota are largely predicated on works by Zimmerman (1984), Darlington (1965), Carlquist (1974) and of course Darwin, all of whom formulated their suggestions while working on or visiting Pacific islands. Against this wealth of biogeography it is somewhat surprising to find little on the geography of Pacific bryophytes.

Basic to any biogeographic study is a sound taxonomic base. For the South Pacific, this exists only for limited areas. Bartram's floras of the Hawaiian Islands (1933) and the Philippine Islands (1939) serve as early, but successful treatments for the northern Pacific regions. H. A. Miller's long-term collecting throughout the Pacific has resulted in a comprehensive base of specimens. Several florules have been published largely based on these, including those for Guam and northern Micronesia (Miller 1968) and for the Micronesian Atolls (Whittier et al. 1963). Whittier's (1976) full fledged flora of the Society Islands is the only one available for the tropical southern Pacific area. Whittier et al. (1983) unpublished computerized data base for 108 tropical Pacific islands will be available in the future. Treatments for some genera are available for Melanesia as part of Koponen and Norris (1983) current floristic treatment of that area. Farther south, the temperate elements of both New Zealand and Australia are well documented in floras by Sainsbury (1955) and Scott and Stone (1976).

Despite these deficiencies, some helpful literature is available. Miller et al. (1978, 1983) two Prodromus' effectively bring together all the literature on South Pacific bryophytes; Ramsay (1984) has summarized the bryoflora of Lord Howe Island; and Vitt (1974, 1979) has presented a key and annotated list of Campbell and the Auckland Islands.

Studies on the bryogeography of the South Pacific are limited. None deals with the bryofloras of the South Pacific tropical islands. However there is a widely scattered series of papers that relate in a variety of ways to this region. From a regional point of view, reviews of bryophyte disjunctions clarify some individual species distributions. These include papers by Schofield and Crum (1972), Schofield (1974), Gradstein et al. (1983), and Schuster (1969, 1983). Monographic work has contributed much to the understanding of South Pacific species ranges, in particular monographs by Touw (*Hypnodendron*, 1971), Salazar A. (*Leucophanes*, 1986), During (Garovaglia, 1977), Nowak (*Mitthyridium*, 1980), and Reese (*Syrrhopodon* 1987a and *Calymperes*, 1987b).

Perhaps more relevant to the present work, are the local bryogeographic analyses. Although few, these present a historical view of significance to biogeography. These include works by Hoe (1979) on the Hawaiian Islands, Smith (1969) on Micronesia, and Fife (1985) for New Zealand. Van Zanten's (1978) elegant work on spore viability and long-distance dispersal is highly relevant to our understanding of distribution patterns and dispersal. Whereas in some cases, data appear to support the view that long-distance dispersal is important (Van Zanten and Pócs 1981) in other cases, species distributions seem to be largely controlled by the movement of continents and plate tectonics (Schuster 1972, 1976). In some genera, both processes are seemingly important in the evolution of present-day distribution patterns (Vitt 1976, Vitt and Marsh 1988).

In 1979, working in the southern temperate area Vitt proposed that regional moss floras can be characterized on both taxonomic and morphological grounds. He suggested that on a gradient from the tropics to the poles, several trends were apparent from his data set, including an increase in acrocarpy, and a decrease in Isobryalean and Hypnobryalean species. Tropical, southern temperate, northern temperate, and polar moss assemblages each have distinctive components of these factors. As well, he showed that along a gradient from New Zealand southward (with additional data from the northern hemisphere) the ratios of species to genera decreased, while the ratios of moss species to vascular plant species increased. Total moss species richness decreased, but at a much lower rate than did richness of vascular plants.

Based on these results, I decided to expand these analyses to include intensively sampled subtropical and tropical islands of similar size and age and to expand the methodology employed to use quantatitive biogeographic methods. The questions asked in this study are:

(1) Are assemblages of mosses along elevational gradients on a tropical South Pacific island different from those elevational gradients on subtropical, temperate, and subantarctic islands?

(2) How are such differences manifested – in structural features?, in taxonomic components?, or in species richness?

(3) What adaptive strategies are present along these elevational and latitudinal gradients?

(4) Can broad evolutionary patterns in the Bryopsida be interpreted in light of differences that are found to be present?

These specific questions lead to the recognition of a broader biological question in community ecology – that is, are there specific attributes present in community pattern and, if so, to what extent are these attributes relatable to adaptive strategies and evolutionary change? This study addresses these questions for the plant group Bryopsida (mosses – Fig. 4a-c) on four South Pacific islands that are arranged along a latitudinal gradient.

2. STUDY AREA

Locations of the four South Pacific islands are

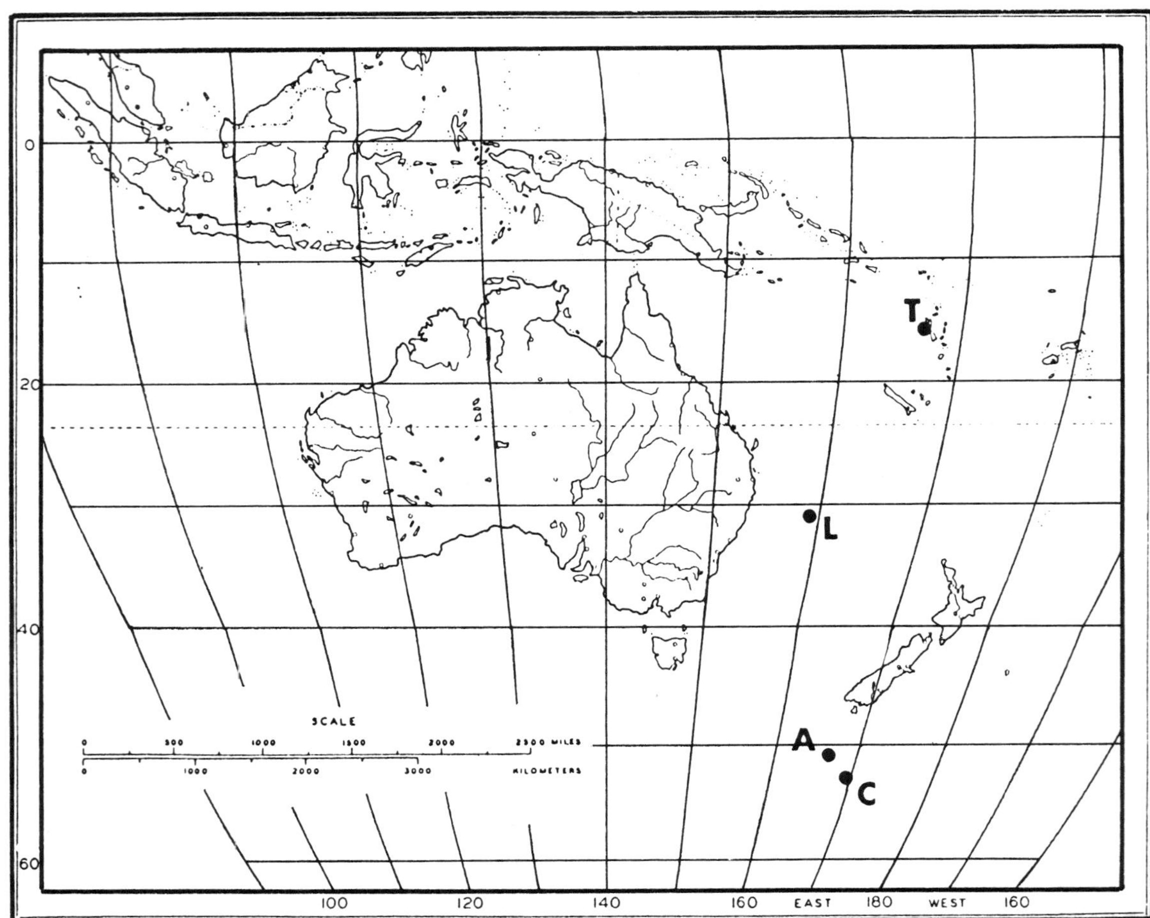

Fig. 1. Map of Australasia and the western South Pacific showing the location of four islands studied. C, Campbell I.; A, Auckland Is.; L, Lord Howe I.; T, Mt Tabwemasana on the western coast of Espirito Santo (Vanuatu).

shown in Figure 1. All four islands occur between 159° and 170° east longitude. The most southern, Campbell Island, lies about 480 km southwest of the Auckland Islands. Lord Howe Island is about 2400 km north-northwest of the Aucklands, while Mt Tabwemasana on the western coast of Espiritu Santo (Vanuatu, formerly the New Hebrides) is 1800 km north-northeast of Lord Island and almost directly north of the Auckland Islands. These four islands form a gradient from high subantarctic Campbell Island (Bliss 1979) through the low subantarctic (with trees) or south temperate Auckland Islands, subtropical Lord Howe Island, to tropical Espiritu Santo. All are located

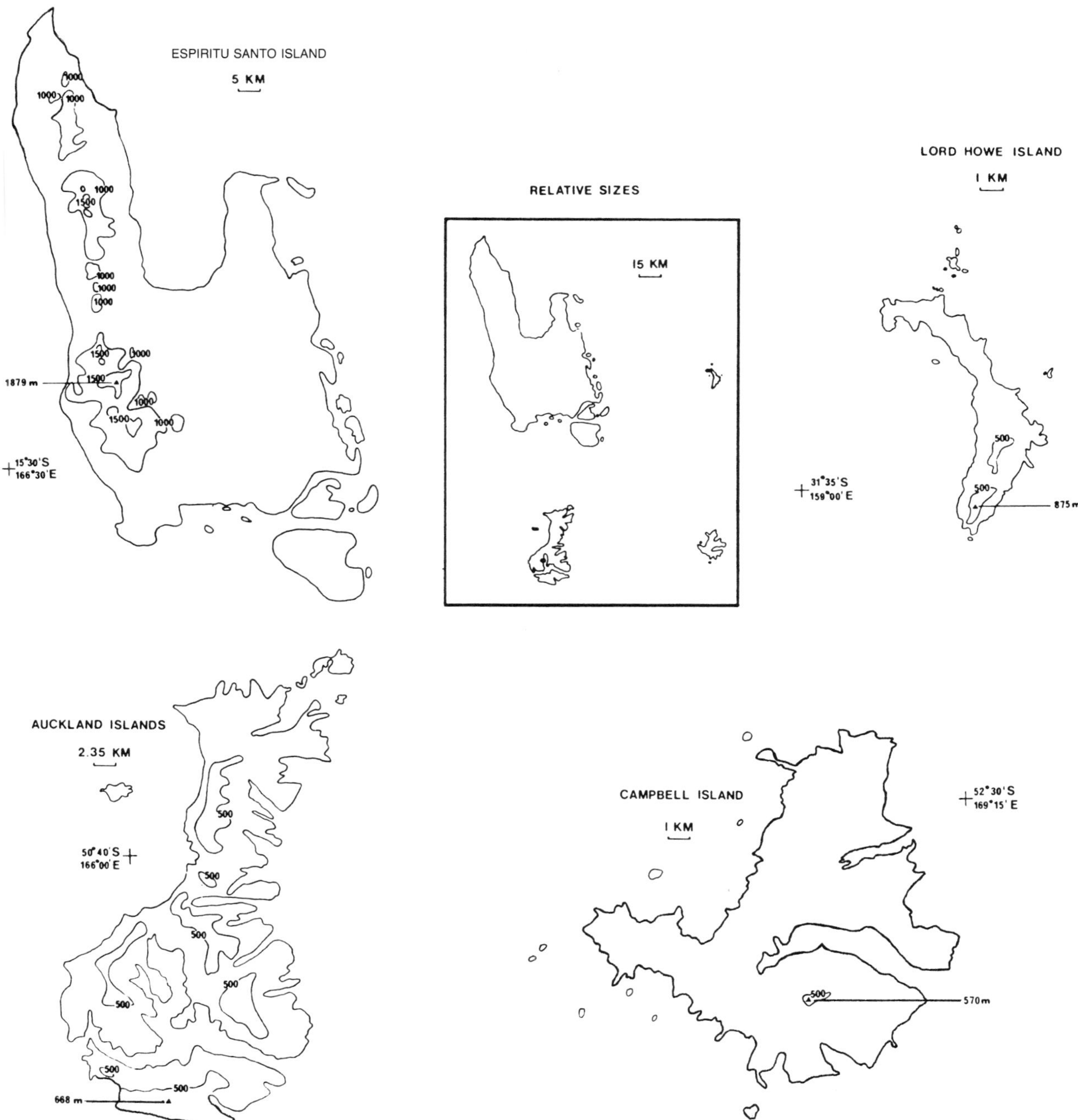

Fig. 2. Comparative area, elevation, and shape of the four islands.

in the South Pacific Ocean within a 950 km east–west band and a distance of 4200 km between the most southern and northern sites. Figures 2 and 3 give size and elevational comparisons of the four island areas.

Campbell Island is located at 52°33S latitude and 169°09E longitude, about 670 km south of Invercargill, New Zealand. It covers 115 square km and is approximately 16 km long north and south and about the same east and west. Mt Honey reaches 433 m elevation.

The climate of the island is characterized by high winds, slight diurnal and annual variation, humid cloudy conditions, and cool temperatures. Relative humidity averages around 87%. Mean annual precipitation is 139 cm and occurs evenly throughout the year. Rainfall occurs on most days of the year. Mean hours of sunlight recorded for a five-month period was 421. Annual mean temperature is about 6 to 7 °C. Snow falls on approximately 40 days per year.

Extensive blanket peat covers most of the island. The origin of Campbell Island, from volcanic activity, has been suggested as dating from late Miocene or earliest Pliocene (Marshall 1909, Fleming 1975). Wind erosion has removed much of the original land mass while calcareous rock is exposed in only a few coastal areas. Valley glaciers were present during the Pleistocene glaciations.

The vegetation of Campbell Island can be

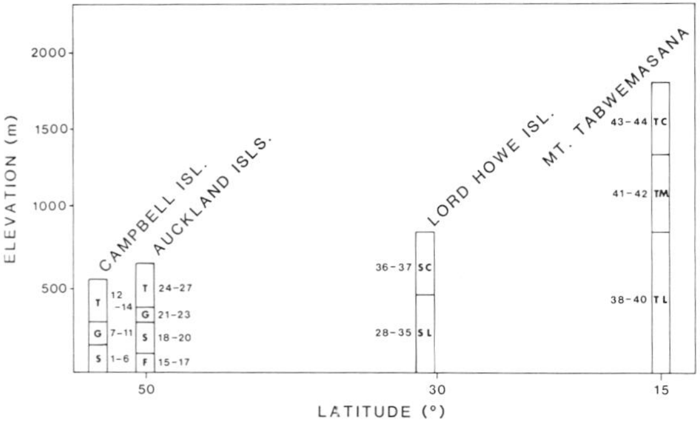

Fig. 3. Comparative elevation, latitude and physiography of the four islands. Numbers to side of bars are stand numbers (see Table 2). Abbreviations: S, scrub; G. grassland; T, tundra; F, forest; SL, seasonal subtropical forest; SC, mossy subtropical (non-seasonal) forest; TL, seasonal tropical forest; TM, montane tropical forest; TC, mossy tropical forest.

divided into three physiographic zones. No forest exists on the island. At sea level to about 150 m elevation, a high, dense scrub thicket occurs (Fig. 4h – scrub here used in the sense of closed-canopy shrubland). Dominant here are *Dracophyllum scoparium, D. longifolium, Myrsine divaricata*, and *Polystichum vestitum*. As elevation increases, the scrub vegetation diminishes and intergrades with tussock-grassland. On Campbell Island, the grassland occurs between 150 and 350 m elevation (Fig. 4d). Due to extensive grazing by sheep, *Poa litorosa* and *Chrysobactron rossii* are now abundant, as well as the original dominant – *Chionochloa* (*Danthonia*) *antarctica*. Above 300 to 350 m elevation, cushion plant-moss tundra occurs. Cliffs and rock debris are plentiful, however, the occurrence of extensive peat deposits extends into the alpine and gives the zone a particularly lush appearance. Herbfields have been severely reduced due to overgrazing. *Sphagnum australe* is a dominant component of the mires of lower and mid elevations.

The **Auckland Islands** form a small archipelago located about 420 km south-southwest of Stewart Island, New Zealand at 50°40′S latitude and 166°00′E longitude. The group of two major and four smaller islands is about 50 km long and 26 km at the widest point, with an area of about 560 square km. Highest elevations are Mt Dick at 668 m on Adams Island and Mt Raynal at 645 km on Auckland Island.

The Auckland Islands have a highly oceanic climate. Relative humidity averages about 85%. Mean annual precipitation varies between 152 cm in the north to 210 cm in the south. Rainfall occurs on between 311 days in the north (Port Ross) to 331 days to Carnley Harbour in the south. Rainfall occurs in a relatively constant pattern over the year. Mean hours of sunlight recorded for a five month period were 610. Annual mean temperature is about 8 °C. Snow falls on about 23 days per year.

Peat covers much of the basaltic rock of the islands. Although the historical geology of these islands is not well known, the volcanic activity centered around two basaltic volcanoes of around middle to late Tertiary age. Calcareous rock is rare on the islands. Valley glaciation is evident, dating from the Pleistocene.

Four physiographic vegetation zones can be recognized on the Auckland Islands (Cockayne 1958; Godley 1965; Vitt 1979). At elevations less that 50–60 m, forest occurs, dominated by *Metrosideros umbellata* (Fig. 4f). These trees have large, prostrate to semi-erect trunks with numerous, erect branches. The forest floor is

Fig. 4. Mosses and Vegetation: (a) Isobryalean moss with erect, stiff, epiphytic growth form; (b) Isobryalean moss with pendent, lax, epiphytic growth form; (c) Bryalean moss with dendroid, terrestrial growth form; (d) tussock-grassland zone on Campbell Island (with Royal Albatross); (e) tundra zone on Auckland Island; (f) coastal forest on the Auckland Islands; (g) Mt Gower (right) and Mt Lidgbird (left), covered by high-elevation cloud on Lord Howe Island; (h) low-elevation scrub of *Dracophyllum* on Campbell Island; (i) low-elevation seasonal subtropical forest on Lord Howe Island; (j) abundant cryptogamic epiphytes on canopy branches of higher elevation non-seasonal subtropical forest on Lord Howe Island.

densely shaded and the canopy at 6–9 m height is composed of *M. umbellata*, as well as *Pseudopanax simplex, Dracophyllum longifolium*, and *Coprosma foetidissima*.

Between 100 and 300 m, an extensive scrub zone occurs. It usually gradually merges with the lower elevation forest and thins with elevation, intergrading with tussock grassland. The divaricating shrubs dominating this nearly impenetrable zone are *Myrsine divaricata, Cassinia vauvilliersii, Dracophyllum longifolium*, and stunted *Metrosideros umbellata*. At lower elevations, the scrub thicket may reach 2–3 m in height, whereas at higher levels it generally attains only 0.5–2 m. Tussock-grassland, dominated by *Chionochloa antarctica* to heights above 1 m, occurs between elevations of 300 and 500 m.

Cushion plant-moss-herb tundra occupies elevations above 500–600 m (Fig. 4e). Tussocks of grass are absent or confined to bases of rock outcrops. Cushion plants, including *Colobanthus hookeri* and *Phyllachne clavigera*, are present and herbaceous species are often abundant including *Pleurophyllum hookeri* and *Celmsia vernicosa*.

In addition to these four zones, herbfields (meadows) and mires occur scattered at the lower and mid elevations. These are dominated either by megaherbs (*Pleurophyllum* spp., *Stilbocarpa polaris* and *Anisotome latifolia*) or by large hummocks of the cushion plants *Oreobolus pectinatus* and *Phyllachne clavigera*. *Sphagnum* is not an important part of the organic terrain of the island.

Lord Howe Island, at 31°30′S, 159°05′E, is a small isolated island located some 680 km east of Australia, between Sydney and Brisbane. This crescent-shaped island is about 10 km long and from 0.3 to 0.2 km, with an area of 15.2 square kilometres. Mt Gower at 875 m and Mt Lidgbird at 777 m are the highest elevations (Fig. 4g).

The island has a subtropical climate. Relative humidity averages about 70%. Mean annual precipitation is 168 cm. Rainfall occurs, on the average of 185 days per year, with the maximum number of rain days in May (20), June (21), and July (22), and the minimum number in December (11), January (11), and February (11). Annual mean temperature is 19.1° C. Frost has never been recorded. (Data from Pickard 1983.)

Lord Howe Island is an erosional remnant of a series of volcanic eruptions beginning in the Oligocene at about 30 million years ago, with subsequent activity during the Miocene at about 13 million years ago. Pleistocene or younger age sediments of wind blown coral sand are present on about 20% of the island at low elevations. The island has not been glaciated.

Pickard (1983) has described and classified the vegetation of the island. He recognized eight formations and 26 associations. Closed forest covers 76% of the island's surface. Below 400 m, the dominant vegetation is evergreen closed rainforest, dominated by *Drypetes australasica* and *Cryptocarya triplinervis* (Fig. 4i). Also abundant below 400 m is evergreen closed rainforest dominated by *Cleistocalyx fullagarii* and *Chionanthus quadristamineus*. Both these mixed lowland forests are here considered seasonal rainforests, with pronounced wet and dry seasons. Additionally, common at the lower elevations are palm-dominated forests; especially abundant are those with *Howea forsteriana*. Above 400 m, but especially at elevations greater than 750 m, mossy forest becomes predominant (Fig. 4j). I include three of Pickard's alliances in these higher elevational non-seasonal, mossy forests. This high elevation vegetation is dominated by *Dracophyllum fitzgeraldii, Bubbia howeana, Hedyscape canterburyana*, and *Metrosideros villosa*. Tree ferns and epiphytes, particularly filmy ferns mosses and hepatics are abundant.

Espiritu Santo (Vanuatu), the largest of the New Hebridean islands, is located at 15°30′S and 166°30′E longitude. Mount Tabwemasana, on the western side of the island, rises to 1879 m elevation and is the site of the present study. The island has an area of 3947 square km, with the Mt Tabwemasana study area limited to 50 sq km of the mountain and coast area to the west, near the village of Kerepoa.

The climate is tropical. Relative humidity is about 85% on the average. Mean annual precipitation is 283 cm (data from Espiritu Santo-Pekoa at 42 m elevation). Rainfall occurs on 214 days of the year. The maximum number of rain days occur in January (23), February (19) and March (23); the minimum in August (15), September (13) and October (16). Mean hours of sunlight per year is 2206; annual mean temperature is 25.2° C.

The islands of the New Hebrides are of volcanic origin and date from the Tertiary. The oldest activity is of Miocene age, while some islands are Pleistocene. Volcanic activity is still present on the more southern islands of Ambrym, Tanna, Lopevi, Aoba and Gaua. Espiritu Santo is one of the older islands of the group and consists of not only basaltic rocks, but also raised calcareous coral reefs.

The vegetation of the west coast of Espiritu Santo is tropical rainforest. At elevations from sea level to 800 m, the rainforest is seasonal, with a definitive wet and dry season. Mid-elevation forests (from 800 to 1400 m) appear much less affected by seasonality of precipitation. These montane rainforests have more epiphytes than the low elevation forests. As well, the inorganic soil is largely covered by vegetation whereas at lower elevations, the forest floor is largely covered by leaf litter. Vines are abundant at both lower and mid-elevations. At 1400 m, mossy forest, with abundant epiphytes, few vines and largely organic soil formation, appears and continues to the summit area of Mt Tabwemasana. At the highest elevations, the dominant trees are *Metrosideros* sp. and *Vaccinium whiteanum*.

3. METHODS

Expeditions were made to Campbell Island (21 December 1969–23 January 1970); Auckland Islands (11 December 1972–12 January 1973); Lord Howe Island (15 November 1981–21 November 1981); Mt Tabwemasana on Espiritu Santo (15 August 1985–18 August 1985). The numbers of collections from each area are given in Table 1. Collections of all species of mosses were made at each site visited; elevation and dominant vascular plant vegetation were recorded. Voucher specimens are deposited in ALTA for all localities, except for Campbell Island, where the first set of specimens are in MSU. Elevation, comparative size, and geographic location for the four islands are shown in Figures 1 and 2. The four study areas form a latitudinal gradient from subantarctic Campbell Island, south temperate Auckland Islands, subtropical Lord Howe Island to tropical Espiritu Santo. All islands are within 11 degrees longitude and span 37 degrees of latitude in the South Pacific Ocean. Elevations range from 433 to 1879 m, while areas of the islands studied vary between 15 and 570 square kilometres (Table 1).

3.1. Data processing

On each island, the elevational gradient was divided into categories, ranging from 14 on Campbell Island to 7 on Espiritu Santo. Table 2 gives the elevational ranges of each elevational category. Estimates of occurrence and abundance for all moss species were calculated by recording each collection in these elevational categories. These elevational categories from here on are termed stands. For the purposes of this paper, the term occurrence and abundance are defined as follows. Abundance of a species in a stand or

Table 1. Floristic and physiographic synopsis of Campbell, Auckland, and Lord Howe islands and the Mt Tabwemasana area.

Feature	Campbell I.	Auckland I.	Lord Howe I.	Mt Tabwemasana
Elevation (m)	433	668	875	1879
Latitude (°)	52	50	32	15
Area (km^2)	115	570	15	50
No. of genera	69	75	52	57
No. of species (varieties)	119 (4)	145 (7)	105	110
No. of moss collections made	1100	1585	449	362
No. of elevational categories	14	13	10	7

Table 2. Elevational ranges, physiography (Physiog.), and DCA stand grouping of 44 stands. See Fig. 3 for abbreviations of physiographic categories.

CAMPBELL ISLAND				AUCKLAND ISLANDS			
Stand no.	Elevational range	Physiog.	DCA stand group	Stand no.	Elevational range	Physiog.	DCA stand group
1	0–10	S	4	15	0–10	F	3
2	10–25	S	4	16	10–25	F	3
3	25–40	S	4	17	25–40	F	3
4	40–70	S	4	18	40–70	F–S	4
5	70–100	S	4	19	70–100	S	2
6	100–150	S	4	20	100–235	S	2
7	150–175	G	1	21	235–330	S–G	2
8	175–200	G	1	22	330–375	G	2
9	200–235	G	1	23	375–435	G	2
10	235–280	G	1	24	435–500	G–T	2
11	280–350	G	1	25	500–550	T	2
12	350–410	T	1	26	550–625	T	2
13	410–470	T	1	27	625–670	T	2
14	470–550	T	1				
LORD HOWE ISLAND				MT TABWEMASANA			
Stand no.	Elevational range	Physiog.	DCA stand group	Stand no.	Elevational range	Physiog.	DCA stand group
28	0–10	SL	6	38	0–400	TL	7
29	10–20	SL	6	39	400–700	TL	7
30	20–30	SL	6	40	700–900	TL	7
33	125–175	SL	6	43	1200–1600	TC	7
34	175–225	SL	5	44	1600–1800	TC	7
35	225–350	SL–SC	5				
36	350–550	SC	5				
37	550–825	SC	5				

stand group is calculated as the total number of sites from which a species was collected, adjusted to values ranging from 0 to 5 (see below). Occurrence of a species is the presence of a species in a stand or stand group. Values of species abundance were used in the ordinations.

The elevational categories (stands) were ordinated using DECORANA, a detrended correspondence analysis (Hill, 1979a; Hill and Gauch 1980, program CEP-40 in the Cornell Ecology Program Series), while species and stand clusters were arranged using TWINSPAN, a polythetic divisive method of classification (Hill 1979b, program CEP-41 in the Cornell Ecology Program Series).

Intensive collecting at easily accessible low elevation sites, especially on the Auckland Islands, yielded some relatively high abundance values. Individual species abundance values in individual stands range from 1 to 10, however, above 6 represents over-collecting in some stands; this was compensated for by delimiting cut levels in the ordinations as 0, 1, 2, 3, 4 and 5, thus all occurrence values over 5 were considered equal. Frequency distributions of species groups were adjusted to equal block size using the methods of Lausi and Nimis (1985), Orlóci and Kenkel (1985), and Vitt et al. (1986).

Species concepts and nomenclature largely follow those of Vitt (1979), except where more recent taxonomy has shown otherwise. Taxonomy of tropical species mostly follows treatments and determinations of current specialists (see Acknowledgments). Many Espiritu Santo determinations must remain tentative at the present time.

4. RESULTS AND DISCUSSION

Of significance to discriptive plant sociology has been the description of plant communities. From these descriptions, associations of species can be found that characterize particular communities. One of the important goals of plant ecology is to determine environmental factors that influence the distribution of individual species. Environmental factors exist as gradients and it is along individual gradients that each species will have specific tolerances. As a result, species are limited to their fundamental niches by tolerances to particular environmental gradients.

Evolution of plants to new situations requires increases in tolerances to environmental gradients that exist in a given area. These evolutionary changes take the form of specific adaptations, and are dependent on the genetic capability of an organism to evolve specific characteristics. Particular adaptations, both morphological and physiological, then may be important for existence along specific portions of environmental gradients. It seems reasonable to examine relationships between (1) the number of species that can exist at particular points on a gradient, and (2) the specific morphological features found in plant groups along a gradient. If these occur non-randomly along a gradient, then it may be cause to consider these as features associated with particular segments of a gradient. These then can be adaptations to the gradient and represent part of a complex adaptive strategy. Finally the pattern of occurrence of adaptations among members of the plant group can be examined. That is, are the adaptations limited to a few closely related members of a plant group or have they occurred numerous times in unrelated members of the group? If the latter is true, then the success of a plant group may be partially dependent on particular adaptations, and allow its members to be an important component of the ecosystem.

Elevation and latitude form complex gradients along which numerous environmental factors change dramatically. Included in these factors are physiographic changes of the vegetation itself. On the four islands studied this is especially apparent. Forest is not present on Campbell Island, while tundra, grassland, and scrub zones are not present on Lord Howe Island or on Mt Tabwemasana. Since moss habitats are in part controlled by the vascular plant physiography, which creates microhabitats suitable for moss species to exist, it is important to recognize the physiographic zonation present on the four islands as shown in Figure 3.

Whereas the vegetation physiography of the four islands is considerably different along both elevational and latitudinal gradients and suggests significant differences in environmental factors, age of the islands, origin, area, geology, and longitude are all nearly alike. Since these latter factors are relatively similar, they should not influence the relationships brought out below. Distance from harmonious, continental floras is also important. The Auckland Islands are 470 km from New Zealand, Campbell Island is 670 km from New Zealand, Espiritu Santo is 600 km from New Caledonia, while Lord Howe Island is 680 km from Australia. Bougainville in the northern Solomon Islands is 1700 km northeast of Espiritu Santo and has peaks reaching 2750 m. These distances are relatively similar and are not important influences in species patterns among the four island floras.

4.1. Species richness

The total number of species on each island (Table 1) varies between Lord Howe Island with 105 to the Auckland Islands with 145. Even though more collections were obtained from the Aucklands, few new species were collected after the initial 500 collections. The total number of species is probably fairly accurately represented in these numbers; however, if there is error, it is most probably in an underestimate of the species richness of Mt Tabwemasana.

The larger number of species on the Auckland Islands is due to the presence of four physiographic zones, as opposed to two or three on the

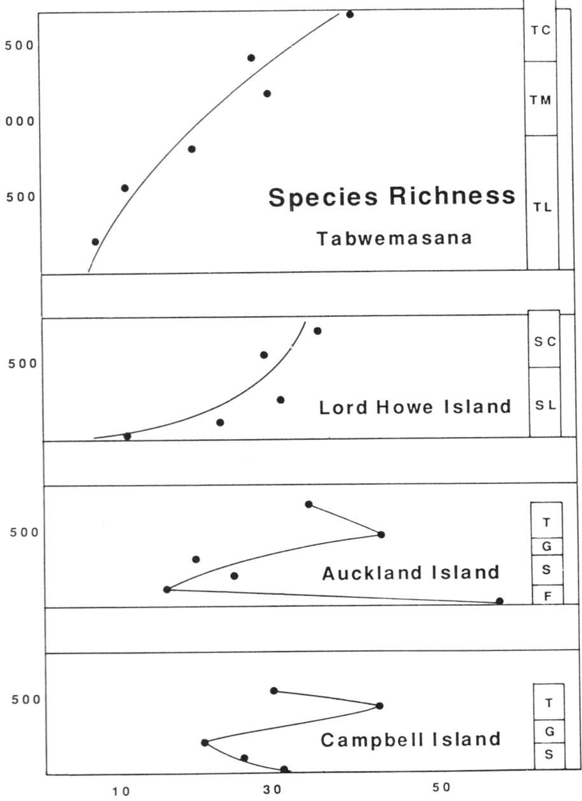

Fig. 5. Species richness along elevation gradients for each of the four islands. Points are derived from means of nearest 1–3 stands. Abbreviations are given in Figure 3. Vertical scale in metres; horizontal scale is number of species.

remaining islands. Also, the Aucklands have a somewhat greater area (Table 1). As will be seen later, this increase in Auckland Island species richness is due to the high percentage of southern temperate and austral/cosmopolitan species restricted mostly to the forests of the island group. Vitt (1979) listed 16 species restricted to Auckland forests and not found on Campbell. Otherwise, total species richness is similar between the four islands and does not vary with latitude. The number of genera ranges from 52 to 75 in a pattern similar to that of species richness. The mean number of species per genus varies from 1.7 on Campbell Island to 2.0 on Lord Howe Island. No trend is evident.

Elevational change in species richness is shown in Figure 5 for the four islands. In general, number of species increase with increasing elevation. This is especially so on Lord Howe and on Mt Tabwemasana in association with changes in seasonal to non-seasonal rainforests. Elevational increase in richness is more dramatic on Lord Howe than on Mt Tabwemasana, probably due to the greater influence of orographic precipitation at higher elevations on Lord Howe Island.

Auckland and Campbell Islands have nearly identical patterns. The Aucklands have maximum species richness in the forest zone; it is reduced in scrubland and gradually increases through grassland to a high number in tundra. A high elevation decrease is present owing to extreme conditions (particularly wind) at the island summits. On Campbell Island a similar pattern is evident, with coastal scrubland having intermediate richness values due to salt-tolerant species and a few more hardy forest species increasing the normally low scrubland diversity values. Gradual increases occur and result in nearly identical higher elevational species richness as on the Auckland Islands.

These patterns indicate that species richness is not markedly different along latitudinal gradients, but generally increases with elevation. At the higher latitudes, this general increase is modified by increase in diversity at low elevations due to the presence of salt-tolerant species along with temperate forest species overlapping ranges with more subantarctic ones. Salt-tolerant species appear to not be as frequent on the more tropical islands as compared to temperate ones (Vitt and Glime 1984). At higher elevations, habitat limitations are present due to extreme summit conditions, whereas more moderate conditions prevail at the lower latitudes.

The presence on Campbell and Auckland Islands of forest-scrub-grassland- and tundra physiographic zones is an opportunity to examine elevational species patterns in more detail and a subset of the data is utilized to explore these relationships.

4.2. Elevational patterns on Campbell and Auckland Islands

Abundance values for all moss species from the 27 stands (Table 2) on Campbell and the Auckland Islands were used to determine stand

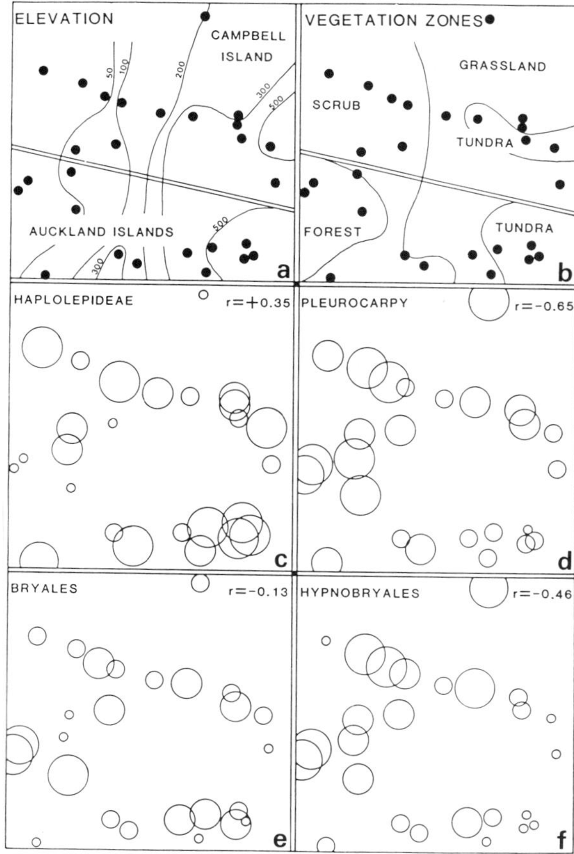

Fig. 6. DCA ordination of 27 stands from Campbell and Auckland islands. Eigenvalue for axis 1 (x) is 0.43; for axis 2 (Y) 0.22. (a) Elevational distribution of stands, values in metres; (b) vegetation zones based on physiography and their relationship to stands; (c) percentage of Haplolepideae in flora of each stand, circle size from largest to smallest >40%, 32–39%, 26–31%, <25%; (d) percentage of pleurocarpous species found at each stand, circle size from largest to smallest 40–50%, 30–39%, 20–29%, <19%; (e) percentage of Bryales (including Funariales) in flora of each stand, circle size from largest to smallest 75–90%, 40–75%, 20–40%, <20%; (f) percentage of Hypnobryales in flora of each stand, circle size from largest to smallest 25–30%, 20–24%, 15–19%, 14%.

similarities. Based on these species abundance similarities, stand distribution is shown by the DCA ordination in Figure 6a. The first axis (x) is associated with elevation (Fig. 6a) and also with vegetation physiography (Fig. 6b), while the second axis (y) is related to the specific island location of the stands (Fig. 6a, b). No correlation exists between elevation and the species richness of the 27 stands ($r = -0.13$) due to high species richness in Auckland Islands, forests and in tundra on both islands (Fig. 5).

However, there as distinctive patterns present between elevation and higher taxonomic groups of the Bryopsida. The abundance of the Haplolepideae (classification follows and is simplified from that in Vitt 1984) is significantly correlated ($p < 0.05$) with elevation; members of this group increase in abundance as elevation increases (Fig. 6c). The Bryales show no relationship with elevation, although they are more abundant in forests on the Auckland Islands (Fig. 6e). Abundance of Hypnobryales is significantly correlated with elevation, however these species decrease in abundance as elevation increases (Fig. 6f). These quantitative abundance data support floristic occurrence data of Vitt (1979), who suggested that Hypnobryales are prevalent at lower elevations. Crosby (1980) and Vitt (1984) both proposed that the ancestors of modern mosses should have characteristics that include acrocarpy and a diplolepideous peristome. These features are found in members of the Bryales and Funariales (here considered together in a broad sense as the Bryales). From this acrocarpous diplolepideous (Bryalean) ancestor, evolution proceeded in two directions: (1) toward xeromorphy with the resultant groups being acrocarpous and haplolepideous (Haplolepideae), and (2) toward mesomorphy, with the groups being pleurocarpous and diplolepideous (Hypnobryales and Isobryales – here including the Hookeriales).

Figure 6d graphically illustrates the abundance pattern of pleurocarpous species along the elevational gradient (the acrocarpous pattern would be the opposite of that shown). There is a highly significant ($p < 0.01$) correlation, indicating a marked decrease in pleurocarpous species with an increase in elevation. Whereas the acrocarpous and less derived Bryales show no elevational tendencies (but occur most abundantly in the south temperate forest zone), the acrocarpous Haplolepideae are strongly related to high elevation habitats (especially tundra). The acrocarpous growth form couples production of perichaetia with growth. When perichaetia are present and if fertilized and sporophytes produced, then growth can only continue by subsequent differentiation of cells below the apical area. Thus sexual reproduction and sporophyte development directly interferes with growth of the gametophyte. Many acrocarpous species rarely branch, and have limited life spans. Pleurocarpy makes available to the moss plant the ability to be a long-lived perennial, with perichaetia produced laterally and separate from the apical cell. Here, growth by the apical cell can continue indefinitely without hindering female sex organ production. Pleurocarpy then is theoretically an adaptive strategy for life in mesic, relatively stable habitats, for example in forest and scrub vegetation. These species would be placed well into the K adaptive strategy, whereas many acrocarps could be considered *r*-adapted species.

In summary, these abundance data indicate distinct taxonomic and structural patterns relative to elevation, even though overall species richness is not correlated to this gradient. Pleurocarpy, found in the Hypnobryales (and Isobryales that show a similar pattern not presented here), is an adaptive strategy that enabled mesic forest species to develop long-lived perennial life styles. Acrocarpy is associated with non-treed higher elevation habitats; these much more open to disturbance. These elevational changes reveal seemingly important patterns in moss distributions, and they lead to the larger question of how elevational changes relate to latitudinal ones.

4.2.1. Species Abundance

The relationship of species abundance (commonness) to number of species is shown in Figure 7. The number of rare species greatly outnumbers the number of common ones. The patterns are similar for all four islands (Fig. 7e). Similar patterns of species abundance have been noticed previously by several outhors, including Williams (1953) and Preston (1962). The statement of Simberloff (1986) "that lots of species are common nowhere" is particularly evident in these data. Rarity in mosses is especially noticeable (Slack 1980; Vitt et al. 1986). Although comparable data from Tracheophytes are not available, the pronounced rarity of many mosses is readily shown here.

Drury (1974) and Rabinowicz (1981) observed that rarity may result from geographic range, habitat breadth, or local population size. Few mosses are narrowly endemic; most occur disjunctively or are continuously widespread, yet many, if not most are locally rare as in this data set. Most species occur as small, local populations, widely spaced from neighbouring ones. This appears to result from these species having specialized habitat requirements, together with the habitats themselves being rare (spatially or temporally). Thus rarity in mosses is not so much determined by restriction of geographic range, but more dependent on habitat occurrence. Local population abundance may be determined largely by the extent of the specific habitat available for individual species. These observations support the suggestions of Rabinowicz (1981) where she found few rare species having either narrow geographic ranges or broad habitat ranges.

Local rarity of moss species appears to occur in a very large percent of the flora of a given area. It does not appear to be influenced by latitude. Rare species of mosses exist largely as a result of small, local populations dispersed throughout a broad geographic range. Persistence of these species may be a matter of relative effective colonization and establishment rates, coupled with high extinction rates, rather than long-term persistence in locally stable environments. (Here I am not advocating effective long-distance dispersal, rather effective local dispersal capabilities.)

The presence of high species richness (Fig. 5) in areas having more frequent unstable microhabitats (e.g. tundra, cliff faces, sea coast, and epiphytic substrates) support this view of a dynamic population structure of moss species.

4.3. Abundance patterns along a latitudinal gradient

The 44 stands and 293 species were arranged hierarchically into stand groups and species

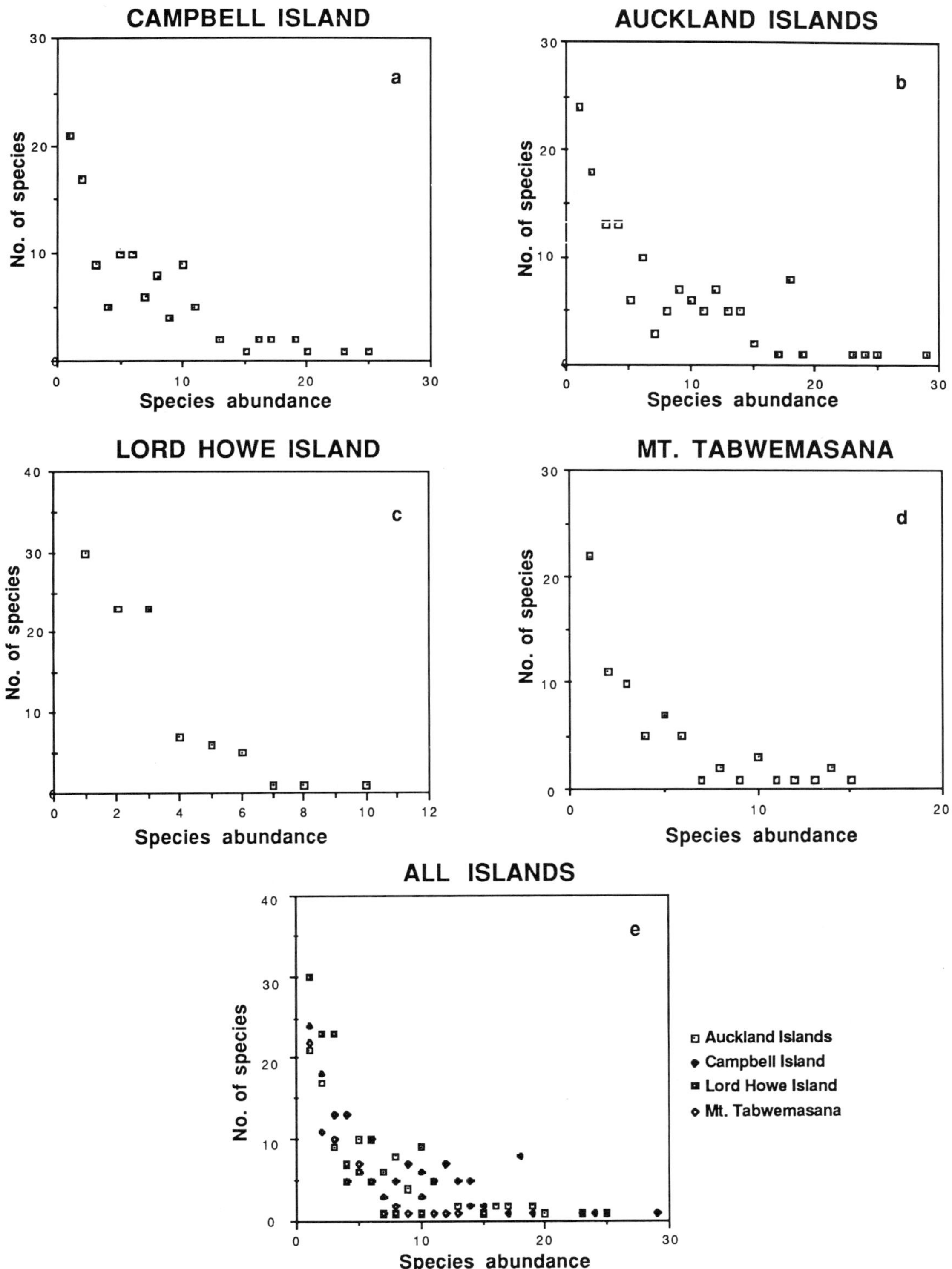

Fig. 7. Patterns of species abundance, presented as number of species plotted against species abundance in each stand (see methods for abundance calculation). (a) Campbell Island $r_s = -0.90$; (b) Auckland Islands $r_s = -0.84$; (c) Lord Howe Island $r_s = -0.98$; (d) Mt Tabwemasana $r_s = -0.74$; (e) All islands combined $r_s = -0.96$. Spearman coefficients of rank correlation (r_s) are all significant at $p = 0.001$.

Table 3. Distribution of the 61 most frequent species arranged in species groups, and their abundance in the seven stand groups given as mean stand group abundance (species abundance (0–5) in each stand/No. of stands in stand group).

	Number of occurrences	Stand groups						
		1	2	3	4	5	6	7
1. **Tundra** (42 species)								
Andreaea subulata	14	1.3	1.2	1.3	–	–	–	–
Bartramia robusta	13	1.1	1.1	0.7	0.1	–	–	–
Blindia contecta	10	0.5	1.3	0.3	–	–	–	–
Conostomum pentastichum	12	1.0	1.3	–	–	–	–	–
Ditrichum strictum	12	1.1	1.2	1.3	–	–	–	–
Rhacocarpus purpurescens	11	0.6	1.2	0.7	–	–	–	–
Racomitrium crispulum	21	2.2	2.2	3.0	0.3	–	–	–
Dicranoweisia antarctica	10	2.0	0.4	–	0.1	–	–	–
2. **Subantarctic** (22 species)								
Campylopus clavatus	16	0.5	1.4	1.6	0.3	0.3	0.2	–
Bryum laevigatum	11	0.6	0.8	1.3	–	–	–	–
Hypopterygium novae-seelandiae	10	0.6	0.8	2.7	0.1	–	–	–
Isopterygium limatum	11	0.3	1.2	2.3	0.3	–	–	–
Weymouthia cochlearifolia	18	1.0	1.1	3.0	0.3	–	–	–
Anisothecium persquarrosum	13	1.0	0.6	1.0	0.4	–	–	–
Zygodon intermedius	12	1.1	0.4	0.7	0.6	–	–	–
Camptochaete gracilis	19	1.8	0.8	3.3	1.3	–	–	–
Macromitrium longirostre	18	2.0	1.1	4.7	1.3	–	–	–
3. **Widespread South Temperate** (39 species)								
Thuidium furfurosum	14	1.0	0.2	2.7	1.6	–	–	–
Breutelia pendula	18	0.9	1.2	4.7	1.1	–	–	–
Bryum blandum	13	0.3	0.6	2.7	0.6	–	–	–
Dicranoloma billardieri	17	0.1	1.1	2.7	1.6	0.3	–	–
Dicranoloma robusta	16	0.9	0.7	4.0	1.0	–	–	–
Achrophyllum dentatum	16	0.4	0.6	4.3	0.4	0.8	–	–
Bryum billardieri	13	0.4	0.2	1.7	1.0	–	0.3	–
Hypnum chrysogaster	16	0.4	0.3	5.0	1.4	0.3	0.2	–
Ptychomnion aciculare	13	0.5	0.2	4.0	1.0	0.5	–	–
Holomitrium perichaetiale	15	1.3	0.3	0.3	1.3	0.5	–	–
4. **Forested South Temperate** (48 species)								
Campylopus introflexus	14	–	0.3	2.7	1.6	0.3	–	–
Campylopus pallidus	6	–	–	3.0	0.7	–	–	–
Calyptrochaeta apiculata	7	0.1	0.1	3.3	0.6	–	–	–
Pohlia wahlenbergii	7	0.3	–	1.3	0.9	–	–	–
Fissidens leptocladus	10	0.4	–	4.0	0.7	–	0.2	–
Distichophyllum pulchellum	6	–	–	3.7	0.4	–	–	–
Distichophyllum rotundifolium	7	0.1	0.1	2.7	0.3	–	–	–
Rhaphidorrhynchium tenuirostre	8	–	0.2	3.7	0.3	–	–	–
Rhizogonium novae-zeelandiae	6	–	0.1	4.3	0.3	–	–	–
Rhizogonium pennatum	6	–	0.2	2.3	0.1	–	–	–
5. **Austral/Cosmopolitan** (5 species)								
Bryum erythocarpum	8	–	0.3	3.3	–	0.3	0.2	–
Ceratodon purpureus	7	–	–	2.3	0.3	–	0.3	–
Cyathophorum bulbosum	7	–	0.1	2.3	0.1	0.5	–	–
Echinodium hispidum	12	0.8	1.0	1.0	0.1	0.3	0.3	–
6. **Subtropical** (62 species)								
Bryum dichotomum	10	0.1	–	1.0	0.9	0.5	0.5	–
Fissidens tenellus	7	0.1	–	1.3	0.3	0.3	0.3	–
Rhynchostegium tenuifolium	9	–	0.1	1.7	–	0.8	0.5	–

	Number of occurrences	Stand groups						
		1	2	3	4	5	6	7
Euptychium mucronatum	5	–	–	–	–	0.8	0.3	–
Fissidens oblongifolius	5	–	–	–	–	0.5	1.0	–
Marcomitrium periaristatum	6	–	–	–	–	0.8	0.5	–
Ptychomitrium australe	5	–	–	–	–	0.8	0.5	–
Tortella calycina	5	–	–	–	–	0.8	0.5	–
Tortula papillosa	5	–	–	0.7	–	0.3	0.5	–
Bryum argenteum	5	0.1	–	0.3	–	0.5	0.2	–
7. **Widespread Tropical** (9 species)								
Isopterygium albescens	5	–	–	–	–	0.3	0.3	0.3
Calomnium n.sp.	7	–	–	–	–	0.8	–	–
Hypopterygium muelleri	9	–	–	–	–	0.5	0.5	1.1
Leucobryum candidum	7	–	–	–	–	0.3	0.5	0.6
8. **Restricted Tropical** (66 species)								
Acroporium hermaphoditum	5	–	–	–	–	–	–	1.3
Aerobryopsis longissima	5	–	–	–	–	–	–	2.1
Hypnodendron dendroides	5	–	–	–	–	–	–	2.0
Macromitrium sp.	5	–	–	–	–	–	–	1.6
Mitthyridium samoanum	5	–	–	–	–	–	–	0.9
Rhizogonium spiniforme	5	–	–	–	–	–	–	1.7
Number of Stands		8	9	3	7	4	6	7

groups. From these, eight species groups and seven stand groups were selected as representing assemblages with biological meaning.

4.3.1. Species groups

Species groups represent clusters of species with similar distributions among the stand groups. In turn, each species group can be characterized geographically by comparisons of individual species distributional ranges and their ecological occurrences. Table 3 presents the most abundant species in each species group and their occurrences in the seven stand groups. The complete association table is available on request from the author. Species ranges in the Southern Hemisphere were used to examine the phytogeographical affinities of the eight species groups. The distributions of individual species are sometimes not well known, however these approximations can be interpreted to depict the major phytogeographical trends.

Most species of group one are restricted to alpine and subalpine *Tundra* of the subantarctic zone. They occasionally are found on the higher mountains of southern Australasia and southern South America. Only rarely do they occur at low elevations, where they occur on exposed cliff faces (e.g. *Andreaea subulata*).

Subantarctic species are those that geographically are mostly restricted to the subantarctic islands and southern South America and Australasia, however their habitat preferences include a broad range of scrubland, open forest, and mesic tundra situations. These species would occur along a broad range of elevations in New Zealand and Tasmania (e.g. *Macromitrium longirostre*).

Widespread South Temperate species have distribution ranges extending throughout the south temperate zone, occurring in low and mid-elevational habitats. This temperate element is not restricted to forests, occurring also, and often more frequently in meadows, on mesic exposed rock faces, and in peatlands. These species are mostly excluded from exposed tundra environments (e.g. *Thuidium furfurosum*).

The *Forested South Temperate* species are Australasian or sometimes circumpolar species that are restricted to shaded, forested habitats.

They are all mesophytic species and are much more limited in their distribution than the previous group (e.g. *Rhizogonium pennatum*).

The *Austral/Cosmopolitan* species have broad ranges, some extending to a worldwide distribution, however they are excluded from the tropical stands in this study (e.g. *Echinodium hispidum*).

Subtropical species are mostly excluded from temperate rain forests, but occupy a zone from mid-Australia, northern New Zealand and a few extend to middle South America. Most of the species of this group are Australasian in distribution (e.g. *Ptychomitrium australe*).

Widespread tropical species mostly are those that occur in both subtropical to tropical areas. They are often pan tropical or have a wide distribution in the old world tropics (e.g. *Leucobryum candidum*).

Species restricted to tropical latitudes are placed in group 8; these *Tropical* species were not found at more southern latitudes. Some of them are widespread tropical species (e.g. *Rhizogonium spiniforme*); others are restricted to the South Pacific (e.g. *Mitthyridium samoanum*).

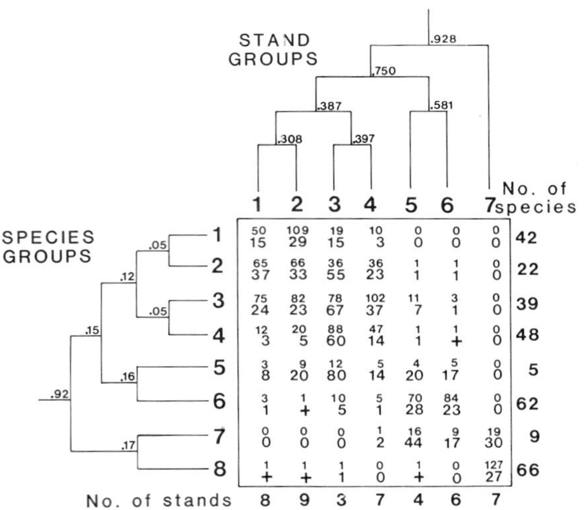

Fig. 8. Dendrograms of species groups and stand groups as delimited by TWINSPAN. Numbers of species in species groups given in right column, number of stands in stand groups given by bottom row. Small, upper number is number of species occurrences in stand group, large lower number is number of species occurrences adjusted to equal block size. Figure 9 is based on the percent of the latter figure. Eigenvalues for successive dichotomies given for stand and species group hierarchies.

Species groups with tropical species (7, 8) are separated at a much higher level than any others (Fig. 8). Subtropical (6) and Austral/Cosmopolitan (5) species groups are placed together, while Tundra (1), Subantarctic (2), and South Temperate (3, 4) groups form a related dichotomy. Finally Tundra–Subantarctic groups (1, 2) are separated from South Temperate ones (3, 4).

4.3.2. Stand groups

Similarities between stand groups (Fig. 8) are greatest between stands found on the Auckland (2, 3) and Campbell Islands (1, 4). Within these four stand groups, similarities are between high elevation (stands 1, 2) and low elevation (3, 4). Stand group 7, representing all stands from Mt Tabwemasana, is least similar to the remainder, while groups 5, 6 that include Lord Howe Island stands are classified separate from the more southern ones (Fig. 8). This stand classification indicates that when similarities of species abundances are used to classify the moss associations, latitude is the primary factor, but that within broad latitudinal categories, elevation becomes of significance. Thus, on Campbell and Auckland Islands, there is greater similarity within high-elevation and also within low-elevation stands from both islands, then within the vegetation along the elevational sequences.

The relationships between species groups and stand groups are shown in Figure 8. Based on these data, relative importance of species groups can be determined for individual stand groups. A synthesis of these relationships is given in Figure 9, where the phytogeographic category is provided for each species group (based on a study of the phytogeographic elements in each group). In this figure, the centroid distances between stand groups shown in Figure 11 (and discussed below) form the horizontal axis.

Tundra-grassland stands from Campbell Island have a dominance of Subantarctic species, while Tundra species dominate the higher elevations of the Auckland Islands. This difference may relate to the somewhat more harsh conditions on

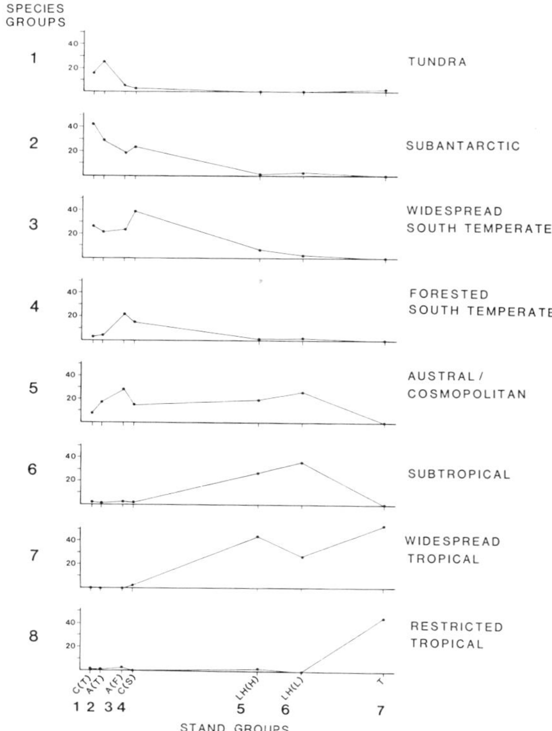

Fig. 9. Frequency distribution of species belonging to eight principle geographical distribution patterns plotted against the relative DCA centroid distances of stand groups (Fig. 11). Species group distributions adjusted to equal block size. Species groups named by comparisons of overall geographic ranges. Abbreviations: C, Campbell Is.; T, tundra-grassland; A, Auckland Isls.; F, forest; S, scrub; LH, Lord Howe Isl.; (H), high elevation; (L) low elevation; T, tropical forest of Mt Tabwemasana.

Campbell Island that limits the occurrence of the more restricted tundra species, while the more tolerant and more broadly distributed Subantarctic species are more abundant.

Low elevations on Campbell Island (scrubland) are dominated by Widespread South Temperate species, whereas the forested lower elevations of the Auckland Islands are characterized by Forested South Temperate species as well as by Austral/Cosmopolitan species. The Forested South Temperate species are largely excluded from or limited in occurrence on Campbell Island. This unique group of species, restricted to Auckland Islands forests, was noted earlier by Vitt (1979) as a critical difference in the moss floras of these two islands.

Whereas Tundra species (group 1) are restricted to high elevations, Subantarctic species are broadly distributed throughout the elevational ranges of these two islands. The main phytogeographic features of the lower elevations of these two islands is the high frequency of Widespread South Temperate elements, while the higher elevations are characterized by Subantarctic species (Fig. 9).

Subtropical lowland forests (seasonal) are characterized by species in group 7 (Widespread Tropical) while high elevation mossy forests on Lord Howe Island have a dominance of species from species group 6 (Subtropical). The high-elevation subtropical mossy forests have many more species with a restricted distribution, whereas the low-elevation stands have a higher frequency of broadly distributed species. Lord Howe Island high-elevation stands have few species common with either Campbell or the Auckland Islands, except for the Austral/Cosmopolitan element (species group 5 in Fig. 9). However there are significant similarities in species group 7 between Lord Howe Island stands and those from Mt Tabwemasana. The limited relationships between the tropical stands (stand group 7) and all others is shown by the restriction of nearly all species in species group 8 (66 species) to those stands on Mt Tabwemasana. Based on these data, tropical forests have little in common with subtropical forests or with areas farther south. Subtropical stands have more in common with those of the tropics, than with those of the south temperate area. Of the 293 species in this study, only one is found in stand groups 1 and 7 (*Leptotheca gaudichaudii*). Species such as *Bryum argenteum* (sp. grp 6) *Ceratodon purpureus* (sp. grp 5), and *Funaria hygrometrica* (sp. grp 5), commonly thought of as cosmopolitan were not found on Mt Tabwemasana. These species, common in temperate and subtropical areas, may not be as common in the tropics (in undisturbed areas) as they are in the cooler regions of the world and in fact they may not be as widespread in distribution in natural vegetation as is commonly thought.

The results of the DCA, utilizing the same data set as that in the species–stand association

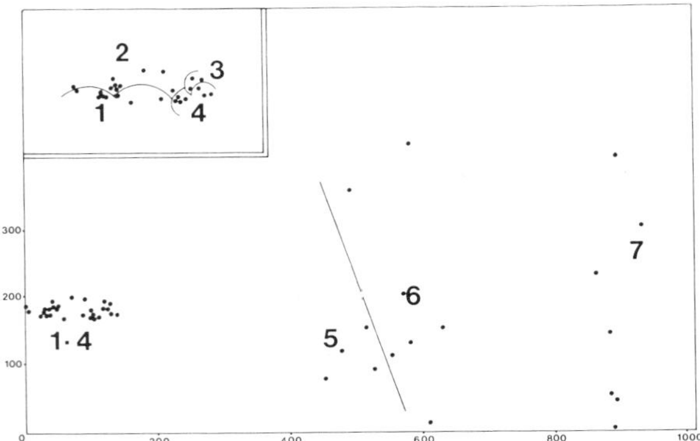

Fig. 10. Dispersal of 44 stands against the first (x) and second (y) axes of the DCA ordination. Eigenvalues are 0.83 for x axis, 0.37 for y axis. TWINSPAN classification of seven stand groups shown, details of stand pattern in groups 1–4 enlarged at upper left.

table (Table 3), are shown in Figure 10. The 44 stands are distributed along the first axis (x) in an increasingly broad pattern from left to right. The sequence of stands is from those in stand groups 1–4, 5–6, and finally those in group 7 to the far right. Figure 11 displays the calculated centroid positions for each of the seven stand groups. These partly correspond to elevational categories on individual islands. A clear latitudinal trend is evident along the x axis, represented by a decrease in latitude along the x axis. A less obvious trend of elevation is present along the second axis (y). The relatively high eigenvalue (0.83) of the x axis indicates that latitude is the principle factor accounting for the stand distribution. Elevational differences in floristic composition on individual islands, even between south temperate forest and tundra, are less pronounced than differences between south temperate, subtropical, and tropical forests. Stand distribution shown in Figure 10 emphasizes this point. Greater stand separation is present between stands on the seasonally dry to moist tropical elevational gradient on Mt Tabwemasana than on the forest-scrub tundra elevational gradients on Campbell and Auckland Islands. These trends are probably explained best by the lower overall elevations of the southern islands, causing a concentration effect of species, these overlapping one another along a compressed elevational gradient. Whereas species from species groups 2 and 3 have broad tolerances to all elevational zones on the Auckland and Campbell Islands causing these stand groups to show close similarity, subtropical and tropical species from species groups 5, 6 and 7 show relatively high fidelity to both their stand groups and individual stands, with less overlap of species along the elevational gradient.

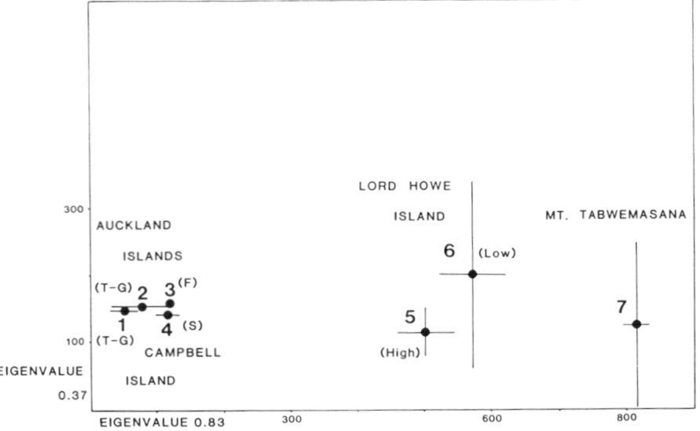

Fig. 11. Centroid locations of the seven stand groups (1–7) as delimited by TWINSPIN. Vertical and horizontal lines are centroid standard deviations. See Figure 9 for abbreviations.

Stating this somewhat surprising conclusion in other words, based on Figures 10 and 11, sub-antarctic tundra stands on Campbell Island have more in common with south temperate forested stands on the Auckland Islands, than the moss assemblages from low-elevation seasonal forests on Lord Howe Island have with high-elevation non-seasonal forests on the same island. Whereas latitude is largely a temperature gradient, these latter differences suggest the sensitivity of mosses to precipitation patterns. Within a common temperature regime, precipitation has an overriding influence.

4.4. Taxonomic patterns

When compared on a latitudinal basis, the higher taxonomic groups of the Bryopsida occur in distinctive patterns. However these latitudinal patterns are somewhat different to the elevational ones presented earlier. Figure 12 presents the changes in eight taxonomic groups along both elevational and latitudinal gradients as represented by the four islands used in this study.

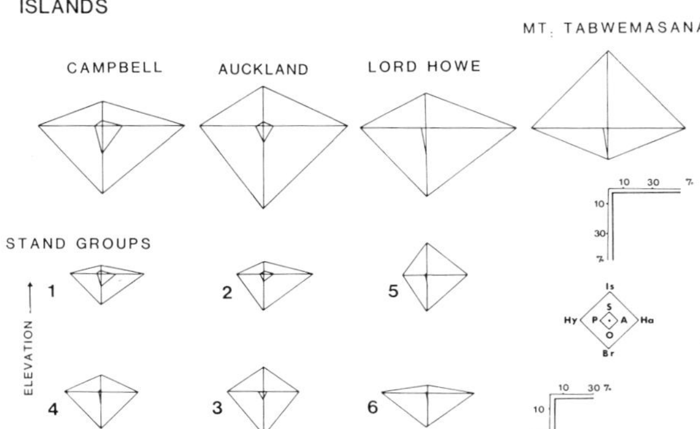

Fig. 12. Behavior of principle higher taxonomic categories of Bryopsida on latitudinally and elevationally correlated stand groups (1–6). Data in percent, and shown as line length from center cross. Scales given at right for top four islands, at lower right for elevational groups. Abbreviations: Is, Isobryales (+ Hookeriales); Ha, Haplolepideae including suborders 9–15 Vitt (1984); Br, Bryales (+ Funariales); Hy, Hypnobryales; S, Sphagnidae; A, Andreaeidae; O, Orthotrichales; P, Polytrichales (classification based on that in Vitt (1984), with some groups recognized at the order rank).

Four of these groups occur consistently in less abundance, when compared to the remaining four that make up about 80–90% of the total moss flora.

Among the four less abundant groups, distinctive patterns exist. Sphagnidae (with only the genus *Sphagnum*) decreases with decreasing latitude, and this genus is a characteristic component of Subantarctic–south temperate floras, although never as abundant as it is in the Northern Hemisphere boreal forest where *Sphagnum* may be a dominant component (Vitt 1979). Andreaeidae, also a relatively small group, characterizes the higher elevations of the Auckland and Campbell Islands and contains species that are common in tundra-grassland stands (Fig. 12, stand grps 1–2). As opposed to the changes in these two groups, the Polytrichales and Orthotrichales have constant patterns along both elevation and latitude. As well, the ratios between these latter two groups is constant, with the Orthotrichales having 3–4 times the number of species as the Polytrichales.

Four major higher groups of the Bryidae are treated. The Hypnobryales, Isobryales (including the Hookeriales), Bryales (including the Funariales), and the Haplolepideae (including Dicranales, Pottiales, Fissidentales, Seligeriales, and Grimmiales).

Hypnobryales: This pleurocarpous group remains constant with latitude, however decreases are evident with increasing elevation to all three islands analyzed; of these the elevational decrease in species in most evident on Lord Howe Island (between seasonal and non-seasonal rainforest).

Isobryales: Latitudinally, there is a dramatic increase in frequency of Isobryalean species on Mt Tabwemasana. The slight increase on the Auckland Islands, when analyzed elevationally, is shown to be due to increases in stand group 3 (or in the forest component). Likewise, on Campbell Island the Isobryales are less common at higher elevations (stand group 1 *versus* 4). Even more pronounced is the elevational difference in the Isobryalean component on Lord Howe Island, where there is a marked increase in the group at higher elevations.

Bryales: The Bryales show no changes in comparative species numbers between Campbell, Auckland, and Lord Howe Islands, including limited changes with elevation – only a slight increase can be detected for stand group 3 (forested Auckland Island stands). However, a decided decrease in species of Bryales is evident on tropical Mt Tabwemasana.

Haplolepideae: As presented earlier (Fig. 7c), this group of species increases with elevation on Campbell and the Auckland Islands (also compare Fig. 12 stand grps 1 with 4 and 2 with 3). However, this trend is not evident on Lord Howe Island (stand grp 5 with 6); nor is a trend present for latitude. Although Mt Tabwemasana has a slightly less percentage of Haplolepideae, the trend is not of relative importance.

In summary, whereas *Sphagnum* and *Andreaea* decrease with decreasing latitude and characterize south temperate and subantarctic floras, the Orthotrichales and Polytrichales remain constant. (Note: however within these orders, generic differences along latitudinal gradients can be significant.) Along the latitudinal gradient (discussed in terms of decreasing latitude), Isobryales increase, while Bryales decrease; both Haplolepideae and Hypnobryales remain relatively constant. Along the elevational gradients the Bryales remain constant, while Isobryales increase on subtropical Lord Howe Island, and decrease at higher (non-treed) habitats of more southern islands. The Haplolepideae increase with elevation on southern islands (Campbell and the Aucklands), while Hypnobryales decrease with elevation on all three islands.

Historically, the Sphagnidae and Andreaeidae have long been considered highly specialized, isolated groups with many primitive features; the Bryales have been considered the group nearest the ancestor of the Bryidae. Crosby (1980) and Vitt (1982, 1984) considered the Funariales to include the most primitive Bryidae (this relatively small order is here considered with the Bryales). These results suggest from an evolutionary point of view, that primitive groups are most abundant at higher latitudes; while tropical floras in particular have larger numbers of advanced Bryidae, especially Isobryales.

Along the elevational gradients, primitive Bryidae occur in moist non-seasonal rainforests and become less frequent in non-treed high-elevation as well as seasonal, low-elevation forests. High-elevation non-treed areas have a larger number of Haplolepideae, a relatively advanced lineage of Bryidae that seemingly adapted to xeric habitats early in its evolution (Vitt 1984).

4.5. Structural patterns

Crosby (1980), Vitt (1984), and Vitt and Buck (1986) have each suggested patterns of character evolution in the Bryopsida. Although some differences are present, overall there are similarities of thought. Here, four characters are chosen as representing general evolutionary trends in the Bryidae. These are as follows:

(1) *Papillosity*: the presence of ornamented upper leaf cells is considered to be apomorphous; the lack of these is plesiomorphous.

(2) *Peristome reduction*: The presence of a peristome is a defining synapomorphy of the Bryidae. The reduction, limited function, and eventual complete loss of the structure is considered to be a derived condition. Well-developed peristomes of any of the major types are considered plesiomorphous.

(3) *Pleurocarpy*: The use of the apical cell in archegonial production (acrocarpy) is plesiomorphous; when archegonia are produced from lateral buds, with the apical cell used only for apical growth, the growth form is pleurocarpous and considered apomorphous. Vitt and Buck (1986) showed that pleurocarpy can occur in a small group of the Bryales, and in all members of the Isobryales and Hypnobryales; in the latter two orders, the buds that give rise to perichaetia and to lateral branches are protected by pseudoparaphyllia.

(4) *Costa*: The presence of vascular and support tissue in a central, multi-stratose strand of cells running the length of the leaf is characteristic of primitive groups of mosses (Bryales, Funariales). Vitt (1984) showed the distribution of taxa without a costa to be associated with the most derived groups of mosses, and also to be generally distributed throughout the Bryidae. I consider lack of costa to indicate the loss of this structure and to be apomorphous; well-developed, single or double costae are considered plesiomorphous.

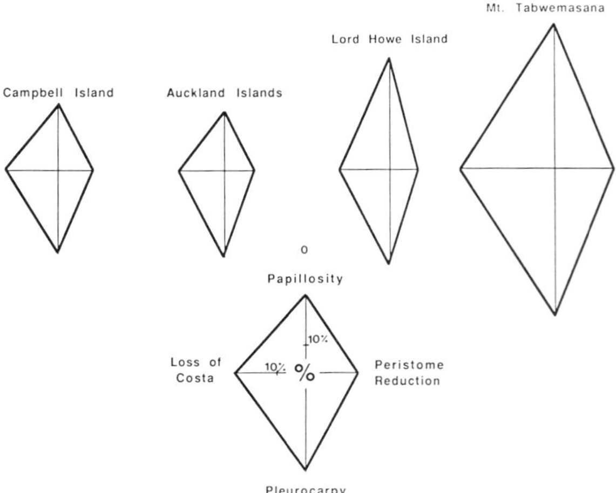

Fig. 13. Distribution of four major apomorphous character states on the four islands. Data in percent, and shown as line length from center cross.

Figure 13 shows the distribution of species having the four apomorphous states on the four islands. There is a definite trend of increasing frequency of species having apomorphous features from the subantarctic to the tropics. In all four characters, the tropical flora contains a considerably higher percentage of species with apomorphous states. Figure 14 expands this analysis to include the elevational component of Campbell and Auckland Islands. Several points are evident, with the four characters showing three separate patterns of behavior. Papillosity increases in even steps between the southern islands, subtropical and tropical ones. This linear increase relates directly to latitude; elevational differences on the southern islands are not evident. Loss of a costa and peristome reduction show similar patterns. Both are unaffected by elevation and both show increased frequency only in the tropical flora; the subtropical frequency being similar to that of the south temperate and subantarctic islands. Pleurocarpy decreases with elevation on the southern islands. It is similar between low-elevation Campbell and Auckland areas and on subtropical Lord Howe Island, and increases markedly in the tropical flora. These distribution patterns have been given in terms of percent number of species or relative species

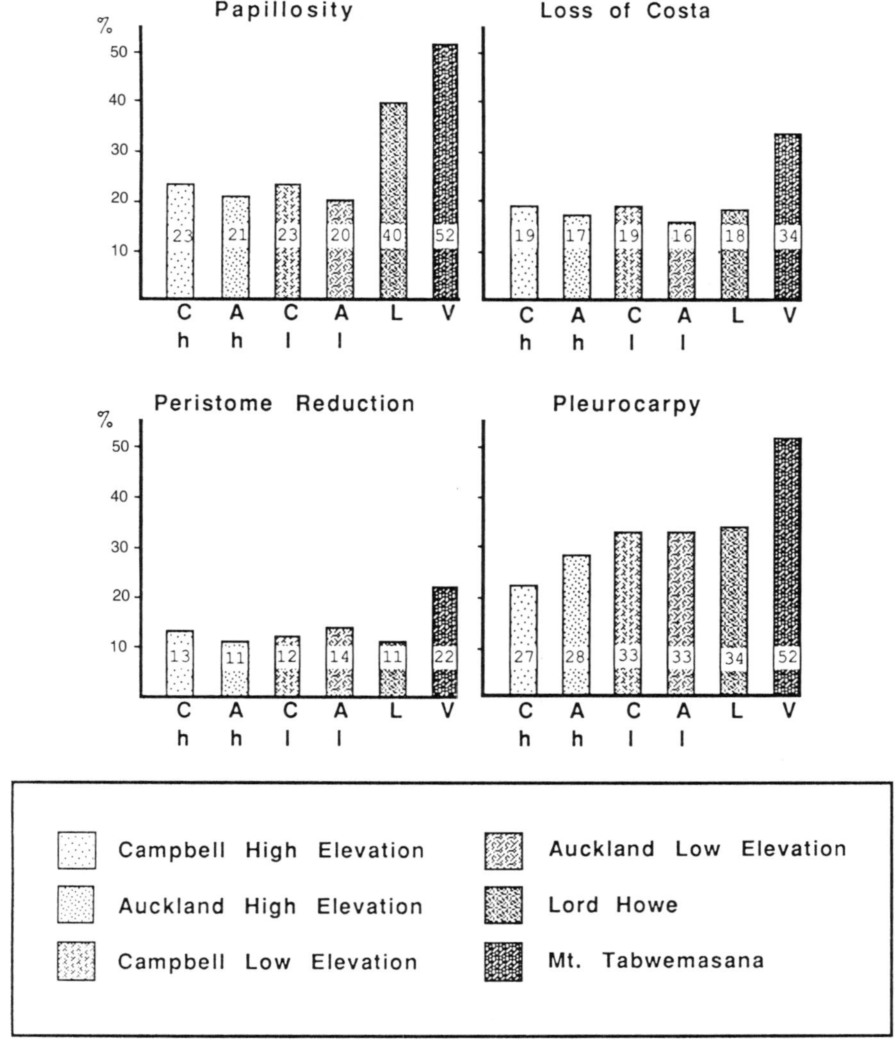

Fig. 14. Changes in percent frequency of four character states for six elevational and latitudinal stand groups. Numbers in bars are percent of species in group having the apomorphy.

Table 4. Pleurocarpous species patterning comparing percent of pleurocarpous species to the abundance of pleurocarpous species in the eight species groups. Abundance calculated by the sum of species abundance in each stand divided by sum of total species abundance (see methods for abundance measures). Correlation coefficient, $r = 0.91$.

Species group	Name	% number of pleurocarpous species	Abundance of pleurocarpous species in % of possible occurrences
1	Tundra	19	9
2	Subantarctic	36	41
3	Widespread South Temperate	36	38
4	Forested South Temperate	54	54
5	Austral/Cosmopolitan	40	50
6	Subtropical	37	36
7	Widespread Tropical	56	51
8	Restricted Tropical	50	56

richness. Species importance can also be examined in terms of abundance (here measured as frequency of possible occurrences, with values ranging from 0 to 5 in each stand – see methods). Relations are shown in Table 4 and indicate for pleurocarpy that there is a strong relationship between richness and abundance. Not only are pleurocarpous species more numerous in tropical moss assemblages, they are also an important component of the vegetation in terms of abundance. In summary, apomorphous features are more frequent in tropical moss floras whereas more temperate floras have a high percentage of plesiomorphous features. These trends are much less evident along elevational gradients.

Since species of Hepaticae are often a conspicuous component of epiphytic, tropical vegetation, but rarely are they as dominant as are mosses in the tundra, it would be of interest to determine if similar patterns are present in this plant group.

5. CONCLUSIONS

Clarification of pattern in plant communities is the first step towards understanding the complexities of community and ecosystem dynamics. Quantification of phytogeographical data allows detailed ecological interpretations of distribution patterns, adaptive strategies, and morphological changes as they occur along broad environmental gradients. These, in turn, allow us to generate hypotheses of both evolutionary and ecological relevance.

Moss species richness is constant over the latitude of this study, however there is a general increase in the number of species with elevation on individual islands. Exceptional richness is present in southern temperate forests on the Auckland Islands.

Most species are rare on a given island, however few have narrow geographic ranges. Rarity may be a function of temporal and spatial habitats, with local populations largely existing by highly dynamic rates of colonization and extinction of local populations.

Whereas the diversity of the Bryales shows no relationship with elevation, the Haplolepideae are more diversified in non-treed, higher elevation areas. Pleurocarpous Hypnobryales have their greatest diversity in forested, low-elevation habitats. While pleurocarpous Isobryales increase dramatically in the tropics and high-elevation subtropical forests, Bryales decrease in these forests. Moss species, generally considered among the least derived, occur in greatest abundance at the higher latitudes, while moss floras of tropical Pacific islands have a much greater number of derived species; apomorphous features occur with greater frequency in the tropical floras as well. Whereas tropical moss floras evolved along with diversification of the pleurocarpous Isobryales, high-latitude and tundra vegetation evolved in association with the diversification of the acrocarpous Haplolepideae. The cool temperate forests and low-elevation subtropical forests contain the largest number of Bryales, a group perhaps closest to the primitive Bryidae. Based on these patterns, it appears most reasonable to consider tropical floras as highly derived. It is unlikely that the tropical rainforest formed the primary habitat for the early evolution of the mosses. Furthermore, diversification and adaptation to this habitat occurred mostly in derived groups of mosses; this radiation was probably secondary, occurring after the major lineages of mosses were present.

If indeed a fundamental role of mosses (and bryophytes in general) in many polar and tropical communities is to intercept and release nutrients, then the stability and origins of these systems themselves are affected by these plants. The evolution of tropical moss vegetation appears to have been largely independent from that of polar and tundra vegetation. Adaptive strategies and structural modifications involved in the evolution of these contrasting communities were considerably different. However, in both tundra and tropical systems the strategies employed may allow different, largely unrelated, moss groups to play a similar ecosystem role. Further experimental and manipulative studies are needed to test these suggestions, but the quantification of phytogeographical data has been indispensable in their initial presentation.

ACKNOWLEDGMENTS

This paper was presented as a portion of a symposium entitled "Life Strategies in Bryophytes: Tropical Rain Forests" at the 1987 International Botanical Congress in Berlin, West Germany. This research was supported by grant A6390 from the Natural Sciences and Engineering Research Council of Canada (NSERC) for which I am grateful. Field work on Campbell and Auckland Islands was made possible by a National Science Foundation Grant (NSF–US) to Henry Imshaug (East Lansing) and to the New Zealand Department of Lands and Survey. On Lord Howe Island, field work was supported by a travel grant to D. H. Vitt from NSERC and from a grant to Helen Ramsay (Sydney) from the Australian Biological Resources Survey, while on Vanuatu field work was supported by NSERC and by an NSF grant (US) to H. A. Miller (Orlando). ORSTROM (France) also provided field support while on Espiritu Santo.

Lord Howe Island mosses were identified with the assistance of H. During (*Euptychium*), J.-P. Frahm (*Campylopus*), Z. Iwatsuki (*Fissidens*), N. Nishimura (*Ctenidium* and *Ectropothecium*), H. Ochi (*Bryum*), H. Streimann (*Papillaria*), and R. Zander (*Barbula*). W. Reese (*Syrrhopodon*) and J.-P. Frahm (*Campylopus*) identified critical material from Espiritu Santo. I wish also to thank especially H. A. Miller, H. Ramsay, and P.-L. Nimis (Trieste) for help in a variety of ways with this research. Additional taxonomic help with difficult Campbell and Auckland Islands specimens is much appreciated and is gratefully acknowledged in Vitt (1974) and (1979).

REFERENCES

Bartram, E.B. 1933. Manual of Hawaiian Mosses. B.P. Bishop Mus. Bull., 101, 1–275.

Bartram, E.B. 1939. Mosses of the Philippines. Philipp. J. Sci., 68, 1–437.

Bayley, S.E., Vitt, D.H., Newbury, R.W., Beaty, K.G., Behr, R. and Miller, C. 1987. Experimental acidification of a *Sphagnum*-dominated peatland: First year results. Can. J. Fisheries Aquat. Sci., 44, Suppl 1, 194–205.

Bliss, L.C. 1979. Vascular plant vegetation of the southern circumpolar region in relation to antarctic, alpine, and arctic vegetation. Can. J. Bot., 57, 2167–2178.

Buck, W.R. and Vitt, D.H. 1986. Suggestions for a new familial classification of pleurocarpous mosses. Taxon., 35, 21–60.

Cockayne, L. 1958. The Vegetation of New Zealand (3rd edn.). Engelmann-Cramer, Weiheim, Germany.

Carlquist, S. 1974. Island Biology. Columbia Univ. Press, New York.

Crosby, M.R. 1980. The diversity and relationships of mosses. In: Taylor, R.J. and Leviton, A.E. (eds), The Mosses of North America, pp. 115–129, Pac. Div. Amer. Assoc. Adv. Sci., Calif. Acad. Sci., San Francisco, California.

Damman, A.W.H. 1978. Distribution and movement of elements in ombrotrophic peat bogs. Oikos., 30, 480–495.

Darlington, P.J., Jr. 1965. Biogeography of the Southern End of the World: Distribution and History of Far Southern

Life and Land, With an Assessment of Continental Drift. Harvard Univ. Press, Cambridge, Massachusetts.

Drury, W. H. 1974. Rare species. Biol. Conserv., 6, 162–169.

Dudley, S. A. and Lechowicz, M. J. 1987. Losses of polyol through leaching in subarctic lichens. Plant Physiol., 83, 813–815.

During, H. J. 1977. A Taxonomical Revision of the Garovaglioideae (Pterobryaceae, Musci). Bryophytorum Bibliotheca, 12, J. Cramer Verlag, Vaduz.

Fife, A. J. 1985. Biosystematics of the cryptogamic flora of New Zealand: Bryophytes. New Zealand J. Bot., 23, 645–662.

Fleming, C. A. 1975. I. The geological history of New Zealand and its biota. In: Kuschel, G. (ed.), Biogeography and Ecology in New Zealand, pp. 1–86, Monog. Biol. Junk, The Hague.

Frahm, J.-P. 1987. Which factors control the growth of epiphytic bryophytes in tropical rainforests? In: Pócs, T., Simon, T., Tuba, Z. and Podani, J. (eds), Proc IAB Conf. Bryoecol, Part B, pp. 639–648. Symp. Biol. Hung., Budapest.

Giesenhagen, K. 1910. Moostypen der Regenwälder. Annls. Jard. Bot. Buitenzorg Suppl., 3, 711–790.

Gimingham, C. Y. and Birse, E. M. 1957. Ecology studies on growth form in bryophytes. I. Correlations between growth form and habitat. J. Ecol., 45, 533–545.

Godley, E. J. 1965. Notes on the vegetation of the Auckland Islands. Proc. New Zealand Ecol. Soc., 12, 57–63.

Gradstein, S. R., Pócs, T. and Váňa, J. 1983. Disjunct Hepaticae in tropical America and Africa. Acta. Bot. Hung., 29, 127–171.

Gradstein, S. R. and Frahm, J.-P. 1987. Die floristische Höhenghederung der Moose entlang des BRYOTROP-transektes in NO-Peru. Beih. Nova Hedwigia, 88, 105–113.

Gradstein, S. R. and Pócs, T. 1989. Biogeography of tropical rain forest bryophytes. In: Lieth, H. and Werger, M. J. A. (eds). Tropical Rain Forest Ecosystems (Series Ecosystems of the World Vol. 14A), pp. 311–325. Elsevier Sci. Publ., Amsterdam.

Hill, M. O. 1979a. DECORANA – A FORTRAN program for detrended correspondence analysis and reciprocal averaging. Ecology and Systematics. Cornell Univ., Ithaca.

Hill, M. O. 1979b. TWINSPAN – A FORTRAN program for arranging multivariate data in an ordered two-way table by classification of individuals and attributes. Ecology and Systematics. Cornell Univ., Ithaca.

Hill, M. O. and Gauch, H. G., Jr. 1980. Detrended correspondence analysis: an improved ordination technique. Vegetatio, 42, 47–58.

Hoe, W. J. 1979. The Phytogeographical Relationships of Hawaiian Mosses. Ph.D. Dissertation, Univ. of Hawaii, Manoa.

Hooker, J. D. 1860. On the origin and distribution of species: introductory essay to the flora of Tasmania. Amer. J. Sci. Arts Ser. 2, 29, 1–25, 305–326.

Kikkawa, J. and Williams, E. E. 1971. Altitudinal distribution of birds in New Guinea. Search, 2, 64–69.

Koponen, T. and Norris, D. H. 1983. Bryophyte flora of the Huon Peninsula, Papua New Guinea. I. Study area and its bryological exploration. Ann. Bot. Fennici, 20, 15–29.

Lausi, D. and Nimis, P.-L. 1985. Roadside vegetation in boreal south Yukon and adjacent Alaska. Phytocoenologia 13, 103–138.

MacArthur, R. H. and Wilson, E. O. 1967. The Theory of Island Biogeography. Monogr. Pop. Biol. No. 1. Princeton Univ. Press, Princeton, New Jersey.

Mägdefrau, K. 1982. Life-forms of bryophytes. In: Smith, A. J. E. (ed.), Bryophyte Ecology, pp. 45–58. Chapman and Hall, London.

Marshall, P. 1909. The geology of Campbell Island and the Snares. In: Chilton, C. (ed.), The Subantarctic Islands of New Zealand, Vol. II, pp. 680–704. John Mackay, Wellington, New Zealand.

Meusel, H. 1935. Wuchsformen und Wuchstypen der europäischen Laubmoose. Nova Acta Leopoldina-Abhandl Kaiserl. Leopold.-Carol. Deutsch. Akad. Naturfors. N.F., 3 (12), 123–277.

Miller, H. A. 1968. Bryophyta of Guam and northern Micronesia. Micronesica, 4, 49–83.

Miller, H. A., Whittier, H. O. and Whittier, B. A. 1978. Prodromus Florae Muscorum Polynesiae. Bryophytorum Bibliotheca, 16. J. Cramer Verlag, Vaduz.

Miller, H. A., Whittier, H. O. and Whittier, B. A. 1983. Prodromus Florae Hepaticarum Polynesiae. Bryophytorum Bibliotheca, 25. J. Cramer Verlag, Braunschweig.

Nowak, H. 1980. Revision der Laubmoosgattung *Mitthyridium* (Mitten) Robinson für Ozeanien (Calymperaceae). Bryophytorum Bibliotheca, 20. J. Cramer Verlag, Vaduz.

Orlóci, L. and N. C. Kenkel. 1985. Introduction to Data Analysis with Examples from Population and Community Ecology. Stat. Ecol. Monog. Vol. 1. Intern Co-op Publ. House; Fairland, Maryland.

Pickard, J. 1983. Vegetation of Lord Howe Island. Cunninghamia, 1 (2), 133–266.

Pócs, 1976. The role of the epiphytic vegetation in the water balance and humus production of the rain forests of the Uluguru Mountains, East Africa. Boissiera, 246, 499–503.

Pócs, T. 1980. The epiphytic biomass and its effect on the water balance of two rain forest types in the Uluguru Mountains (Tanzania, East Africa). Acta Bot. Hung., 26, 143–167.

Pócs, T. 1982. Tropical forest bryophytes. In: Smith, A. J. E. (ed.), Bryophyte Ecology, pp. 59–104. Chapman and Hall, London.

Preston, F. W. 1962. The canonical distribution of commonness and rarity: Part I: Ecology, 43, 185–215; Part II, 43, 410–432.

Rabinowicz, D. 1981. Seven forms of rarity. In: Synge, H. (ed.), The Biological Aspects of Rare Plant Conservation, pp. 205–217. Wiley, London.

Ramsay, H. P. 1984. The mosses of Lord Howe Island. Telopea, 2, 549–558.

Reenen, G. B. A. van 1987. Altitudinal bryophyte zonation in the Andes of Colombia: A preliminary report. In: Pócs, T., Simon, T., Tuba, Z. and Podani J. (eds), Proc. IAB Conf. Bryoecol., Part B, pp. 631–637. Symp. Biol. Hung., Budapest.

Reenen, G. B. A. van. and Gradstein, S. R. 1983. Studies on Colombian Cryptogams XX a transect analysis of the bryophyte vegetation along an altitudinal gradient on the Sierra Nevada de Santa Marta, Colombia. Acta. Bot. Neerl., 32, 163–175.

Reese, W. D. 1987a. World ranges, implications for patterns of historical dispersal and speciation, and comments on phylogeny of *Syrrhopodon* (Calymperaceae). Mem. NY

Bot. Gard., 45, 426–445.

Reese, W. D. 1987b. *Calymperes* (Musci: Calymperaeae): World ranges, implications for patterns of historical dispersion and speciation, and comments on phylogeny. Brittonia, 39, 225–237.

Richards, P. W. 1932. Ecology. In: Verdoorn, F. (ed.), Manual of Bryology, pp. 367–395. M. Nijhoff, The Hague.

Richards, P. W. 1952. The Tropical Rain Forest. Cambridge Univ. Press, Cambridge.

Richards, P. W. 1984. The ecology of tropical forest bryophytes. In: Shuster, R. M. (ed.), New Manual of Bryology, Vol. 2, pp. 1233–1270. Hattori Bot. Lab., Nichinan, Japan.

Sainsbury, G. O. K. 1955. A handbook of the New Zealand mosses. Roy. Soc. New Zealand Bull., 5, 1–490.

Salazar, A., N. 1986. A Revision of the Pantropical Moss Genus *Leucophanes*. Ph.D. Thesis, Univ. of Alberta, Edmonton.

Savile, D. B. O. 1972. Arctic adaptations in plants. Plants Res. Inst. Monog. No. 6, 81 pp. Ottawa.

Schofield, W. B. 1974. Bipolar disjunctive mosses in the southern hemisphere, with particular reference to New Zealand. J. Hattori Bot. Lab., 38, 13–32.

Schofield, W. B. and Crum, H. A. 1972. Disjunctions in bryophytes. Ann. Missouri Bot. Gard., 59, 174–202.

Schuster, R. M. 1969. Problems of antipodal distribution in lower land plants. Taxon., 18, 46–91.

Schuster, R. M. 1972. Continental movements, "Wallace's Line" and Indomalayan-Australasian dispersal of land plants: Some eclectic concepts. Bot. Rev., 38, 3–86.

Schuster, R. M. 1976. Plate tectonics and its bearing on the geographical origin and dispersal of angiosperms. In: Beck, C. B. (ed.), Origin and Early Evolution of Angiperms, pp. 48–138. Columbia Univ. Press, New York.

Schuster, R. M. 1983. Phytogeography of the Bryophyta. In: Schuster R. M. (ed.), New Manual of Bryology, Vol. 1, pp. 463–636. Hattori Bot. Lab., Nichinan, Japan.

Scott, G. A. M. and Stone, I. G. 1976. The Mosses of Southern Australia. Academic Press, London.

Simberloff, D. 1986. The proximate causes of extinction. In: Raup, D. M and Jablonski, D. (eds), Patterns and Processes in the History of Life, pp. 259–276. Springer Verlag, Berlin.

Skre, O., Oechel, W. C. and Miller, P. M. 1983. Patterns of translocation of carbon in four common moss species in a black spruce (*Picea mariana*) dominated forest in interior Alaska. Can. J. For. Res., 13, 869–878.

Slack, N. G., Vitt, D. H. and Horton, D. G. 1980. Vegetation gradients of minerotrophically rich fens in western Alberta. Can. J. Bot., 58, 330–350.

Smith, D. R. 1969. Mosses of Micronesia. Ph.D. Dissertation, Washington State Univ., Pullman, Washington.

Seifriz, W. 1924. The altitudinal distribution of lichens and mosses on Mt. Gedeh, Java J. Ecol., 13, 307–313.

Spruce, R. 1908. Notes of a Botanist on the Amazon and Andes. Wallace, A. R. (ed.), Vol. 1, 518 pp. Macmillan, London.

Terborgh, J. 1971. Distribution on environmental gradients: theory and a preliminary interpretation of distributional patterns in the avifauna of the Cordillera Vilcabamba, Peru. Ecology, 52, 23–40.

Terborgh, J. 1977. Bird species diversity on an Andean elevational gradient. Ecology, 58, 1007–1019.

Touw, A. 1971. A taxonomic revision of the Hypnodendraceae (Musci). Blumea, 19, 211–354.

Vitt, D. H. 1974. A key and synopsis of the mosses of Campbell Island, New Zealand. New Zealand J. Bot., 12, 185–210.

Vitt, D. H. 1976. A monograph of the genus *Muelleriella* Dusen. J. Hattori Bot. Lab., 40, 91–113.

Vitt, D. H. 1979. The moss flora of the Auckland Islands, New Zealand, with a consideration of habitats, origins, and adaptations. Can. J. Bot., 57, 2226–2263.

Vitt, D. H. 1982. Sphagnopsida and Bryopsida. In: Parker, S. P. (ed.), Synopsis and Classification of Living Organisms, pp. 305, 307–336. McGraw-Hill, New York.

Vitt, D. H. 1984. Classification of the Bryopsida. In: Schuster, R. M. (ed.), New Manual of Bryology, Vol. 2, pp. 696–759. Hattori Bot. Lab., Nichinan, Japan.

Vitt, D. H. and Pakarinen, P. 1977. The bryophyte vegetation, production, and organic components of Truelove Lowland. In: Bliss, L. C. (ed.), Truelove Lowland, Devon Island, Canada: A High Arctic Ecosystem, pp. 225–244. Univ. of Alberta Press, Edmonton.

Vitt, D. H. and Glime, J. M. 1984. The structural adaptations of aquatic Musci. Lindbergia, 10, 95–110.

Vitt, D. H., Glime, J. M. and LaFarge-England, C. 1986. Bryophyte vegetation and habitat gradients of montane streams in western Canada. Hikobia, 9, 367–385.

Vitt, D. H. and Marsh, C. 1988. Population variation and phytogeography of *Racomitrium lanuginosum* and *R. pruinosum*. Beih. Nova Hedwigia 90, 235–260.

Wallace, A. R. 1869. The Malay Archipelago: The Land of the Orangutan, and the Bird of Paradise. Harper, New York.

Wallace, A. R. 1880. Island Life, or the Phenomena and Causes of Insular Faunas and Floras. Macmillan, London.

Weber, M. G. and Van Cleve, K. 1984. Nitrogen transformations in feather moss and forest floor layers of interior Alaska black spruce ecosystems. Can. J. For. Res., 14, 278–290.

Whittier, H. O. 1976. Mosses of the Society Islands. Univ. of Florida Press, Gainesville.

Whittier, H., Miller, H. A. and Bonner, C. E. B. 1963, Musci, in Bryoflora of the atolls of Micronesia. Beih, Nova Hedwigia, 11, 7–41.

Whittier, H. O., Pringle, H., Miller, H. A. and Whittier, B. A. 1983. Matrix analysis for Pacific insular biogeography. Amer. J. Bot., 70 (5, 2), 9 (Abstract).

Williams, C. B. 1953. The relative abundance of different species in a wild animal population. J. Anim. Ecol., 22, 14–31.

Zanten, B. O. van. 1978. Experimental studies on trans-oceanic long range dispersal of moss-spores in the southern hemisphere. J. Hattori Bot. Lab., 44, 455–482.

Zanten, B. O. van. and Pócs, T. 1981. Distribution and dispersal of bryophytes. Adv. Bryol, 1, 479–562.

Zimmerman, E. C. 1948. The Insects of Hawaii, Vol. 1. Univ. of Hawaii Press, Honolulu.

8. PHYTOGEOGRAPHY OF SOUTHERN HEMISPHERE LICHENS

DAVID J. GALLOWAY

Department of Botany, British Museum (Natural History), Cromwell Road, London SW7 5BD, UK

1. INTRODUCTION

The Southern Hemisphere as widely understood by biologists and biogeographers refers in a broad sense to areas of land and sea bordering the vast ice-covered continent of Antarctica. For purposes of the present discussion I make an arbitrary definition of Southern Hemisphere as that area of land and sea radiating outwards from the South Pole as centre, to the Tropic of Capricorn encircling the globe at Lat. 23°26′30″S. The region so defined consists of large areas of open sea (Southern Ocean, South Pacific Ocean, South Atlantic Ocean, Indian Ocean and the Tasman Sea), with the landmasses of Australia, New Zealand, South America, South Africa and the various Subantarctic islands (Macquarie, Auckland, Campbell, Antipodes, Falkland, South Georgia, Gough, Tristan da Cunha, Bouvetøya, Marion, Prince Edward, Crozet, Kerguelen and Heard) disposed around the central mass of Antarctica (Fig. 1). This geographical area approximates to the Holantarctic Kingdom of Takhtajan (1986) which includes not only cold and temperate zones of the Southern Hemisphere but also part of the subtropical zone. Udvardy (1987) designates the southern part of this area as his Antarctic Realm, comprising four biogeographical provinces, *Neozealandia, Maudlandia* and *Marielandia* (eastern and western Antarctica respectively), and *Insulantarctica* (comprising the scattered islands of the cold southern oceans). Fleming's (1987) modification of Udvardy's biogeographical arrangement recognizes *Neozealandia* as a subrealm of an Australian Realm, and supports the concept of *Insulantarctica* as a province of the Antarctic Realm, characterized by dominant temperate grasslands and herbfields (Fig. 2). Climates in the cool temperate zone of the Southern Hemisphere vary from oceanic to hyperoceanic, being characterized by low seasonal temperature fluctuations, high rainfall and humidity, frequent cloud cover and strong westerly winds. In Tasmania, the South Island of New Zealand, and in southern South America, there are marked east–west gradients of rainfall across the mountain chains, with strongly oceanic climates and high rainfall to the west, and continental climates east, of the mountains (Galloway 1988a). There are also affinities in climate, vegetation and soils between upper timberlines of the southern cool temperate zone and the cool tropical highlands, with the prevailing life forms in both the Subantarctic and tropical high altitudes being cushion plants, dwarfed shrubs, and tussock grasses (Troll 1973).

1.1. An historical perspective

The vegetation of the Southern Hemisphere and in particular of the South Pacific–Antarctic sector of it has held a fascination for botanists since the eighteenth century when European navigators first opened the Pacific to geographic and scientific discovery. Although lichens were first collected from Tierra del Fuego in c. 1690 by George Handisyd, surgeon of the East Indiaman, *Modena*, it was not until the end of the

P. L. Nimis and T. J. Crovello (eds): Quantitative Approaches to Phytogeography, pp. 233–262.

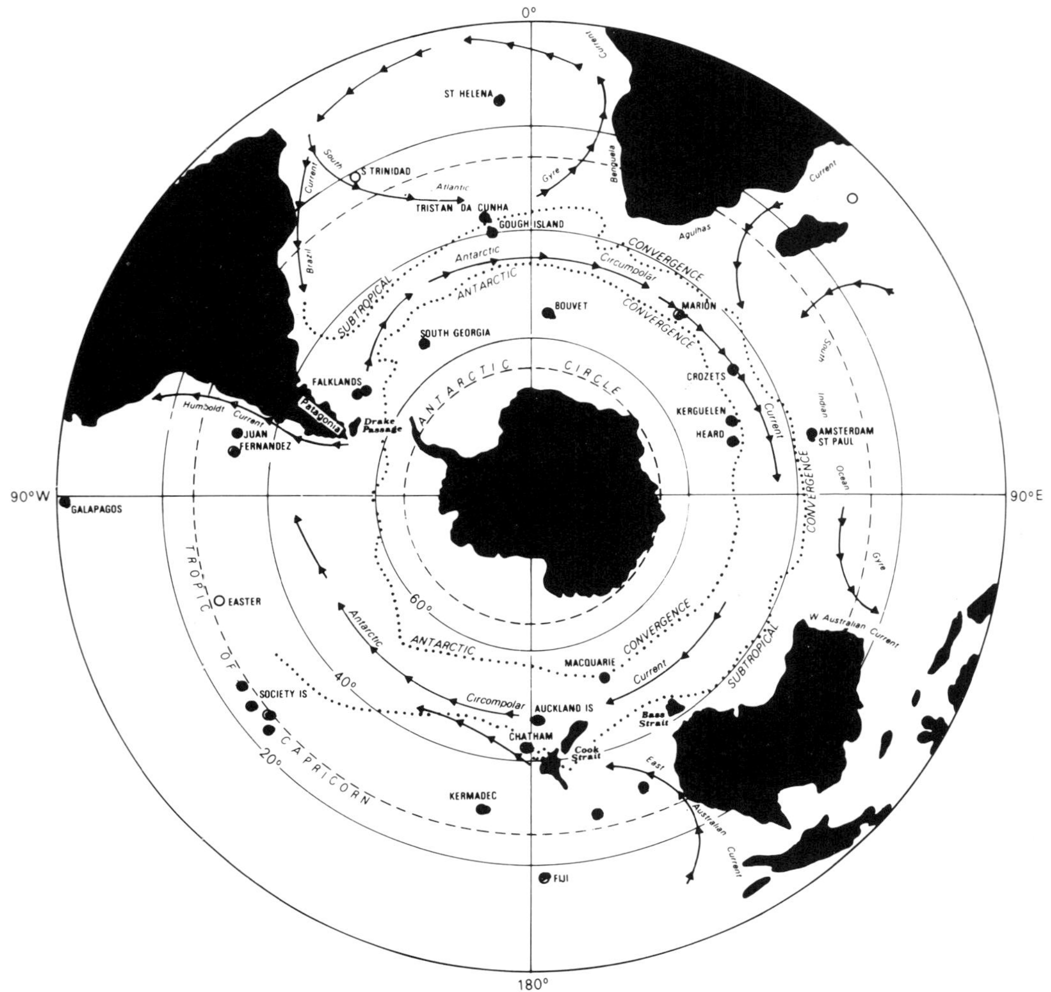

Fig.1 Southern Hemisphere showing major landmasses, principal islands, ocean currents, and convergence zones between water masses.

eighteenth century when the great European voyages of geographical and scientific discovery sailed into the largely uncharted Pacific that lichens from southern latitudes were seriously collected, described, illustrated and circulated in herbaria (Galloway 1985b). The first Southern Hemisphere lichens to be published were dramatic plants quite unlike anything then known from Europe. They were collected by George Forster from Tierra del Fuego and New Zealand on Cook's second voyage (1772–75), and by J. J. H. La Billardière in 1791 from Tasmania (Galloway 1988c). Olof Swartz (1781) described *Lichen filix* [*Sticta filix* (Sw.) Nyl.] from a New Zealand collection of Forster, a curious fern-like plant (Fig. 3) which was figured in colour by Hoffmann (1801) and Delise (1825), and La Billardière described and figured *Baeomyces reteporus* [*Cladia retipora* (La Bill.) Nyl.] from collections he made in Tasmania (Fig. 4). Of this latter lichen William Jackson Hooker wrote "... Nothing in nature can exceed the elegant lace-like appearance of this plant, a structure one would little expect to meet with in the humblest and least perfect part, as it is usually considered, of the vegetable creation – the lichens..." (Hooker 1842). The most extensive and important eighteenth century lichen collections from

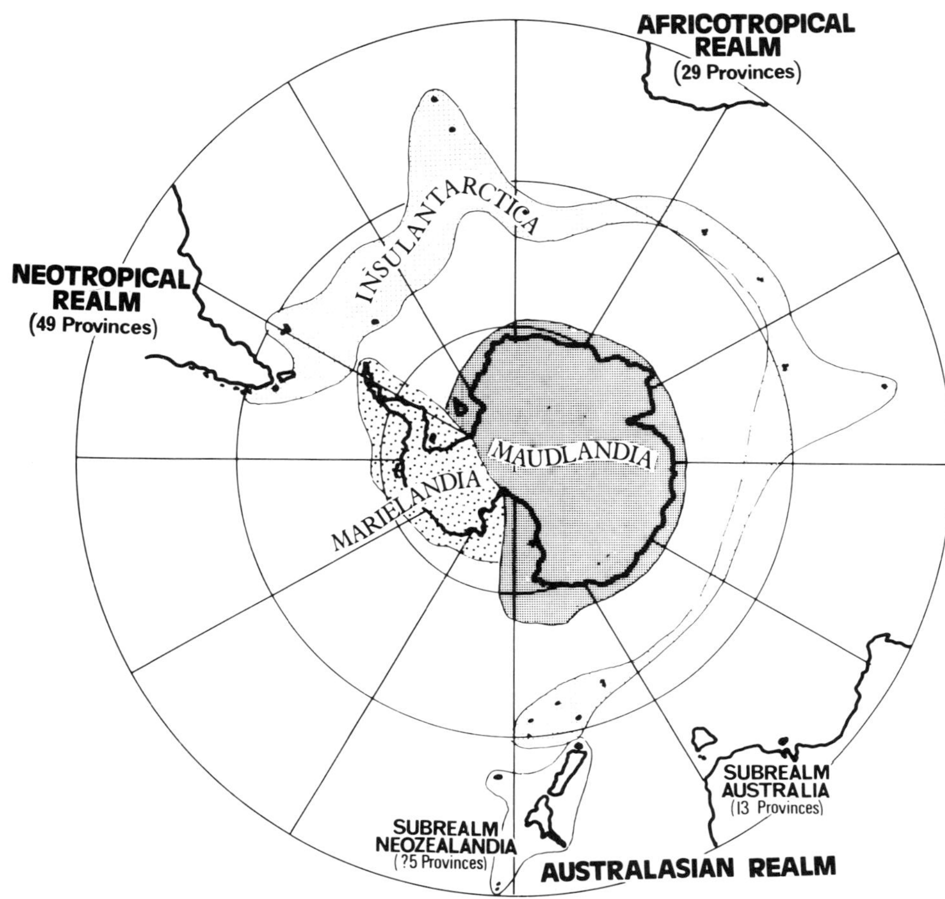

Fig.2. Southern Hemisphere biogeographical realms and provinces, from Fleming (1987).

Fig. 3. Lichen filix [*Sticta filix* (Sw.) Nyl.] from Swartz (1781).

the Southern Hemisphere were those made by the Scottish surgeon-botanist Archibald Menzies (1754–1842) who collected lichens during two major circumnavigations of the world, as surgeon on Captain Colnett's *Prince of Wales* expedition of 1786–89, and as naturalist on Vancouver's 1791–95 *Discovery* expedition (Galloway and Groves 1987). Menzies's Southern Hemisphere lichens came from Staten Island near Cape Horn, Dusky Sound, New Zealand, from Java and from the Cape of Good Hope. From his Staten Island gatherings sent to both Swartz and Acharius by J. E. Smith and by Menzies, Acharius described a number of new lichens (Galloway 1981c, 1985b, 1986, 1988b).

In the first half of the nineteenth century, expeditions from France, Britain, Germany and the United States visited the Southern Hemi-

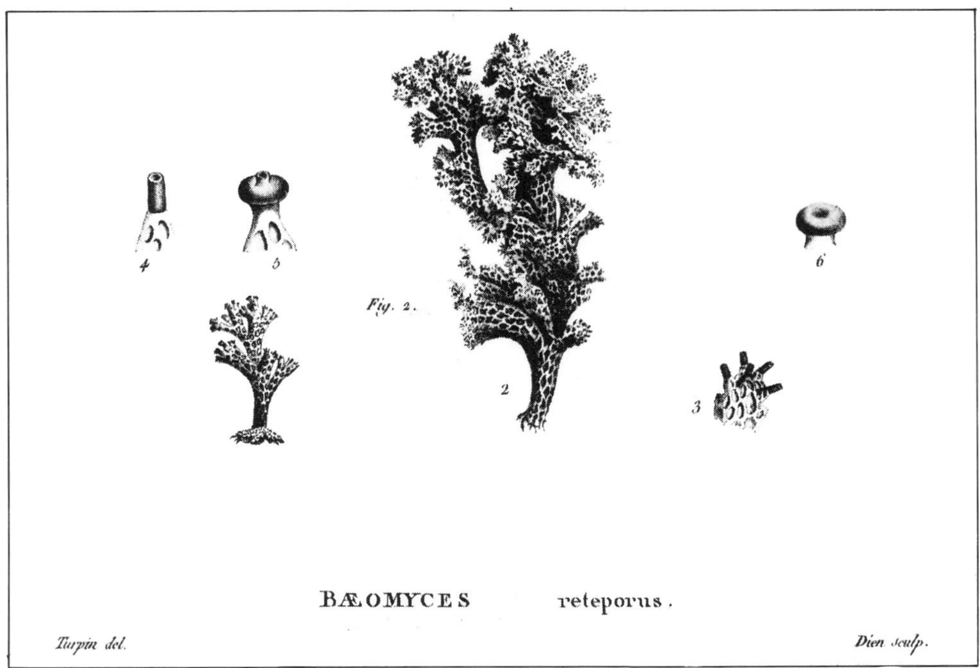

Fig. 4. Turpin's plate of Labillardière's *Baeomyces reteporus* [*Cladia retipora* (LaBill.) Nyl.] from Tasmania.

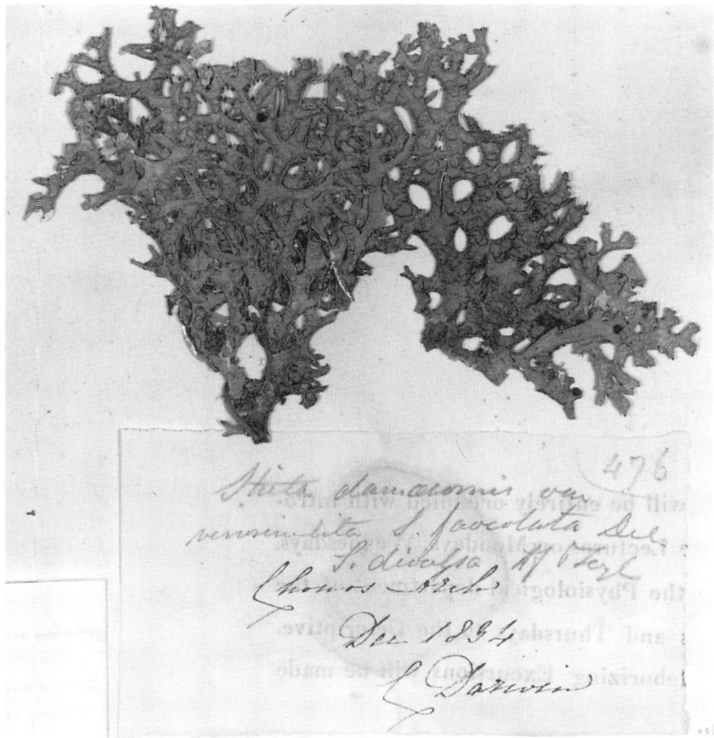

Fig. 5. Darwin's collection of *Pseudocyphellaria divulsa* (BM) from the Chonos Archipelago, southern Chile.

sphere and large collections of lichens were brought back to Northern Hemisphere herbaria to be named by lichenologists (see Galloway 1985b). Even Charles Darwin on the *Beagle* voyage (1832–36) was sufficiently impressed by the extensive growth of epiphytic lichens on the island of Chiloé and in other localities in Tierra del Fuego and southern Chile that he brought back to England fine collections of *Pseudocyphellaria* from which Thomas Taylor later described *Sticta nitida* [*Pseudocyphellaria nitida* (Taylor) Malme] and *Sticta divulsa* [(*Pseudocyphellaria divulsa* (Taylor) Imshaug], the latter gathered from *Nothofagus* forest in the Chonos Archipelago (Fig. 5). Of this area Darwin remarked

> "... Cryptogamic plants here find a most congenial climate. In the Strait of Magellan ... the country appears too cold and wet to allow of their arriving at perfection; but in these islands, within the forest, the number of species and great abundance of mosses, lichens, and small ferns, is quite extraordinary" (Darwin 1845).

J. D. Hooker gave considerable impetus to the study of Southern Hemisphere lichens with his important collections made at all landfalls during the Antarctic Voyage of 1839–43. He visited Kerguelen, Tasmania, the Auckland Islands, Campbell Island, the Bay of Islands, northern New Zealand, the Falkland Islands and Hermite Island near Cape Horn (Galloway 1985b) and accounts of these collections (Hooker and Taylor 1844; Taylor and Hooker 1845; Hooker and Taylor 1847; Babington 1855; Babington and Mitten 1859) are landmarks in Southern Hemisphere lichenology. The history of lichenological exploration in New Zealand, Australia, and Tasmania is discussed by Galloway (1985a), Filson (1976) and Kantvilas (1983) respectively.

In addition to opening up the riches of Southern Hemisphere lichenology to European botanists, Hooker also laid the foundations of modern biogeography in his introductory essays to *Flora Antarctica, Flora Novae Zealandiae, Flora Tasmaniae,* and in his later celebrated 1866 lecture on insular floras (Williamson 1984). In Brundin's words "... To do him full justice, Joseph D. Hooker has to be designated as the real founder of causal historical biogeography" (Brundin 1988). From his own extensive field observations and later analysis of his comprehensive collections of all plant groups, Hooker synthesized his ideas on relationships of the southern circumpolar flora in the following manner:

> "... Beyond the bare fact of the difficulty of accounting by any other means for the presence of the same species in two of the islands, there appeared nothing in the botany of the Antarctic regions to support or favour the assumption of a double creation, and hence I dismissed it as a mere speculation which, till it gained some support on philosophical principles, could only be regarded as shelving a difficulty; whilst the unstable doctrine that would account for the creation of each species on each island by progressive development on the spot, was contradicted by every fact. It was with these conclusions before me that I was led to speculate on the possibility of the plants of the Southern Ocean being the remains of a flora that had once spread over a larger and more continuous tract of land than now exists in that ocean; and that the peculiar Antarctic genera and species may be the vestiges of a flora characterized by the predominance of plants which are now scattered throughout the southern islands" (Hooker 1853: xxi),

and further,

> "... Enough is here given to show that many of the peculiarities of each of the three great areas of land in the southern latitudes are representative ones, effecting a botanical relationship as strong as that which prevails throughout the lands within the Arctic and Northern Temperate zones, and which is not to be accounted for by any theory of transport or variation, but which is agreeable to the hypothesis of all being members of a once more extensive flora, which has been broken up by geological and climatic change" (Hooker 1853 xxxvi).

1.2. Biogeographical frameworks and areas

Biogeography, simply defined as the study of the geographical distribution of organisms, both present and past, involves considerations both of patterns of distribution shown by organisms and also of the processes whereby such distribution have come about. For a concise overview of the aims, approaches and methods of biogeography see Rosen (1988). Problems of Southern Hemisphere biogeography have concerned botanists, zoologists and palaeontologists for many years with interpretations of past and present distribution patterns attracting considerable discussion and sometimes controversy when frameworks or methodologies of biogeographical analysis such as cladistic or phylogenetic biogeography (e.g. Humphries and Parenti 1986; Humphries et al. 1988; Brundin 1988), and panbiogeography (e.g. Craw 1985, 1988) are used to provide explanations for disjunct distributions of Pan-austral taxa such as the southern beeches (*Nothofagus*). Cladistic biogeography, the approach combining cladistics and vicariance biogeography "... has as its basic premise the search for patterns of relationships among areas of endemism" (Humphries and Parenti 1986), while Panbiogeography "... is a biogeographic method that focuses on spatiotemporal analysis of the distribution patterns of organisms" (Craw 1988). Although panbiogeographers criticize vicariance cladistic methodologies (e.g. Craw 1988), and are themselves subject to criticism (Seberg 1986), both frameworks provide useful insights into the problems of Circum pacific and Pan-austral biogeography.

In his attempt to provide a classification of the major biogeographical areas of the world Udvardy (1975) proposed eight areas or realms (equivalent to kingdoms of phytogeographers and regions of zoogeographers) of continental or subcontinental size unified by features of geography flora and fauna which he designated as Palaearctic, Nearctic, Afrotropical, Indomalayan, Australian, Oceanian, Antarctic and Neotropical. His later view of the Antarctic Realm (Udvardy 1987) encompassing four biogeographical provinces [Marielandia (West Antarctica), Maudlandia (East Antarctica), Insulantarctica (the Subantarctic islands) and Neozealandia (New Zealand)] was challenged by Fleming (1987) who suggested an amendment to the biogeographical realms of the Southern Hemisphere (Fig. 2) where New Zealand is included as a subrealm of the Australian Realm, and where the Antarctic Realm comprises Marielandia, Maudlandia and Insulantarctica. In the following account Southern Hemisphere lichen floras of the various component areas will be outlined in turn, supplemented by discussions of biogeographical affinities shown in these floras.

2. SOUTHERN HEMISPHERE LICHEN FLORAS

2.1 Australia

The lichen flora of Australia comprises 294 genera (R. W. Rogers *pers. comm.*) and possibly up to 2000 species (Rogers and Stevens 1981) though this is very much an approximate estimate. A checklist of Australian lichens is available (Filson 1988b) and an index to type specimens of Australian lichens (Filson 1986a). A concise account of the Australian environment as it affects the distribution of lichens in the continent is given in Stevens (1987), and a preliminary discussion of biogeographical affinities of the Australian lichen flora in Rogers and Stevens (1981). The distribution of lichen communities in Australia varies greatly according to climatic regions from tropical and subtropical areas which have a more or less year-round rainfall, to the "wet–dry" tropics which have an extreme winter drought and a relatively short wet summer season, to cool temperate areas in the south east (coastal southern New South Wales, Victoria and Tasmania). In tropical areas lichens show affinities with the Asian and Pacific tropics, and with Africa, in genera such as *Bulbothrix* (Hale 1976), *Canomaculina* (Elix and Hale 1987), *Heppia* (Filson 1988a), *Myelorrhiza* (Verdon and Elix 1986), *Pyxine* (Rogers 1986) and *Relicina*

(Hale 1975: Elix and Johnston 1986). Cool temperate, Southern Hemisphere affinities are found in southeastern Australia and especially in Tasmania, and so for present purposes it is the Tasmanian lichen flora which will be discussed further.

Tasmania is structurally and biogeographically part of Australia separated from the continental mainland by a recent narrow strait (Bass Strait). It has a well-developed Alpine flora (see Kirkpatrick 1986) and the best development of cool temperate rainforest in Australia (Thorne 1986, Kantvilas 1988) and in both habitats particularly, lichens are important components of the vegetation (Bratt 1976a, b, c; Bratt and Cashin 1975, 1976; Kantvilas 1985, 1987, 1988; Kantvilas and Elix 1987; Kantvilas and James 1987; Kantvilas et al. 1985). The greatest diversity and biomass of Tasmanian lichens occurs in cool temperate rainforest, and is predominantly epiphytic being dominated by austral cool temperate taxa from such genera as *Degelia, Leioderma, Menegazzia, Pseudocyphellaria, Psoroma, Psoromidium, Roccellinastrum, Sagenidium* and *Sphaerophorus* with major ecological factors influencing distribution of the rainforest lichen vegetation appearing to be availability of moisture and light, and characteristics of the host tree (Kantvilas 1988). The Tasmanian rainforest lichen flora is closely related in terms of genera represented to those found in similar forests in New Zealand and southern South America, although it is less diverse at the species level than either of these two latter areas (Kantvilas et al. 1985, Galloway 1988a). For instance in the genus *Pseudocyphellaria*, 15 taxa are recorded from Tasmania (Kantvilas 1985, 1988; Kantvilas and James 1987) compared with 48 species from New Zealand (Galloway 1988b) and 52 from South America (unpublished observations).

Kantvilas (1988) has described 11 epiphytic lichen communities from cool temperate rainforest at Little Fisher River; the communities are widespread and typical of a broad range of Tasmanian rainforest vegetation. He proposed interrelationships between the communities based on succession and changes in light and moisture regimes. The communities are given provisional names based on the most common, prolific and faithful species. These are: *Lecanactis abietina–Sagenidium molle* community; *Conotremopsis weberiana* community; *Sphaerophorus insignis* community; *Sphaerophorus ramulifer–Sphaerophorus tener* community; *Pseudocyphellaria dissimilis–Peltigera dolichorhiza* community; *Pseudocyphellaria multifida–Psoroma microphyllizans* community; *Pseudocyphellaria billardierei–Psoroma microphyllizans* community; *Pseudocyphellaria rubella–Pseudocyphellaria faveolata* community; *Parmelia testacea–Pertusaria truncata* community; *Opegrapha stellata–Coccotrema cucurbitula* community and *Pannoparmelia angustata* community. This phytosociological work which has ramifications also in the lichen vegetation of New Zealand and southern South America is the first detailed investigation of its kind in the Southern Hemisphere.

2.2. New Zealand

New Zealand as it exists today is only the emergent part of a much larger land mass, the

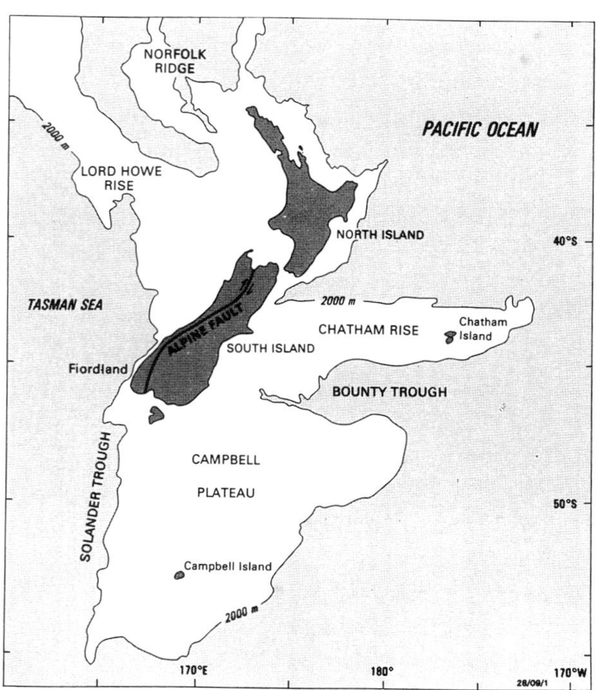

Fig. 6. New Zealand, a small emergent portion of a larger New Zealand microcontinent, from Ollier (1986).

New Zealand microcontinent much of which is now below the level of the sea (Fig. 6). Geologically New Zealand consists of an old part (Western Province) and a young part (Eastern Province) the two separated by the Median Tectonic Line (Korsch and Wellman 1988) with the oldest rocks in the Western Province being Precambrian, while the oldest rocks in the Eastern Province are Carboniferous. Perhaps the most important feature of the New Zealand region is that it straddles the boundary of the Pacific and Australia–India plates (Fig. 7). The plate boundary stretches in an almost straight line from south of Samoa at c. 15°S, to the triple junction (where Pacific, Antarctic and Australia–India plates meet) at c. 62°S, but varies a great deal in character along its length. To the north New Zealand is connected with New Caledonia and Norfolk Island by the long, narrow Norfolk Ridge System which is situated at the centre of a geologically complex region of ridges and basins between Pacific Basin ocean crust, and crust of continental Australia. There are close geological resemblances between New Caledonia and New Zealand (Waterhouse and Sivell 1987; Brothers and Lillie 1988) particularly between Permian and Mesozoic strata and faunas of the two regions. In South Island of New Zealand the Alpine Fault is a transform fault where the Pacific and Australia–India plates move laterally relative to each other and an important feature of the New Zealand region is the tectonism resulting from this interplate collision. Another important geological feature is the Campbell Plateau and evidence suggests that it is a remnant of Protopacific Gondwanaland split away from West Antarctica in the vicinity of Marie Byrd Land some 81 Ma (Adams 1986; Korsch and Wellman 1988). Fife (1986) suggests that the Campbell Plateau region may have provided a refuge for a cold-adapted flora forced out of West Antarctica by advancing continental ice-sheets developed in Antarctica in the Oligocene. Pleistocene glaciations led to great changes in patterns of land and sea in the New Zealand region and to the exctinctions there of many warm temperate and subtropical organisms (Stevens 1985). Further, the composite geological nature of New Zealand (Bishop et al. 1985; Howell 1985) has important consequences for New Zealand biogeography as emphasized by Craw (1985, 1988).

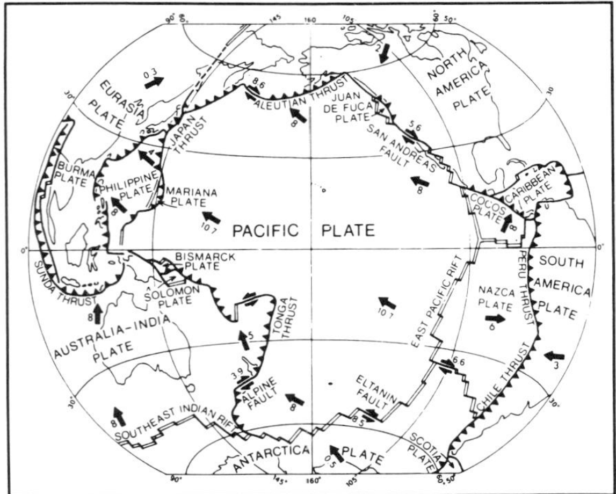

Fig. 7. New Zealand astride the boundary of the Pacific, and Australia–India plates, from Glasby (1985).

The New Zealand lichen vegetation and flora (Galloway 1979, 1985a, 1988b) is richly developed in the many diverse habitats available and comprises 1121 taxa distributed among 244 genera (Galloway unpublished). Major elements in the flora have the following affinities: Cosmopolitan; Australasian; Austral; Circumsubantarctic (Insulantarctic); Antitropical (bipolar); Indo-Pacific; Pantropical; Endemic. Similar elements are present in New Zealand and Australasian alpine bryophyte floras (Ramsay et al. 1986; Fife 1986). New Zealand's isolation and the persistence of moist, cool temperate conditions in both forest and shrubland habitats since the Cretaceous has given rise to many relict plants from angiosperms to mosses and hepatics (Mildenhall 1980; Fife 1986) as well as to many lichens. Although the level of endemism at the generic level is very low [only *Thysanophoron* in the Caliciales is endemic, although in recents accounts (Tibell 1984, 1987) it is placed in *Sphaerophorus*], at the species level the incidence of endemic taxa is much higher, with species from the following genera found only in the New Zealand region: *Anzia, Argopsis, Austroblastenia, Brigantiaea, Caloplaca, Chaenotheca, Collema, Cryptolechia,*

Dendriscocaulon, Haematomma, Hyperphyscia, Leproplaca, Lobaria, Megaloblastenia, Megalospora, Menegazzia, Mycoblastus, Neofuscelia, Pannaria, Parmeliella, Pertusaria, Phlyctis, Physcia, Placopsis, Polychidium, Psoroma, Rinodina, Sagenidium, Sclerophora, Siphula, Sphaerophorus, Steinera, Stereocaulon, Sticta, Thelotrema, Trapeliopsis, Usnea and *Xanthoria*. Other non-endemic elements in the New Zealand lichen flora are discussed below and in Galloway (1979, 1988b).

2.3. South America

South America in the Southern Hemisphere is generally considered to be that part of the continent south of Lat. 30°S to Cape Horn at 56°S and including Staten Island (55°S), the Falklands (52°S), South Georgia (54°S) and Juan Fernandez (33°S). A directly comparable latitudinal range in the South Pacific is that part of the New Zealand region from the Three Kings Islands (34°S) to Macquarie Island (54°30′S). This south-

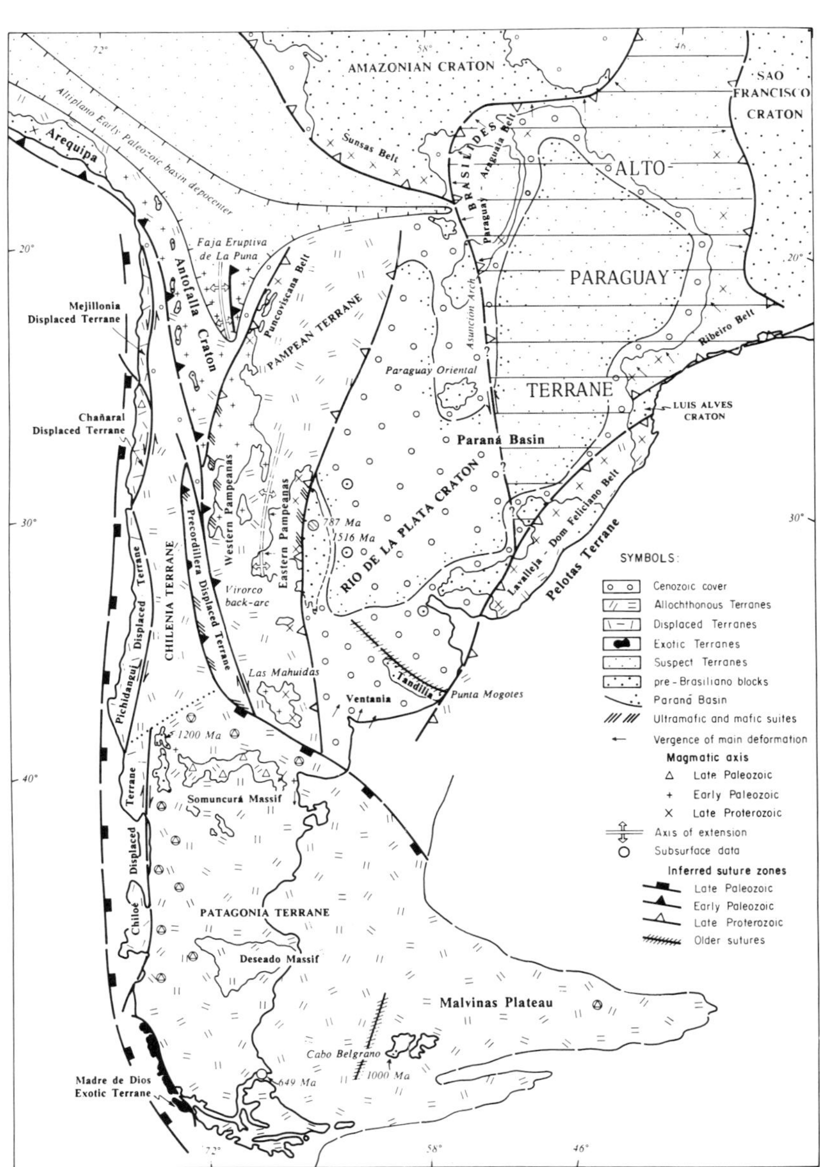

Fig. 8. Major terranes and cratonic blocks of southern South America, from Ramos (1988).

ern portion of South America is a mixture of cratonic blocks that were brought together along the southwestern margin of Gondwanaland in late Precambrian to early Palaeozoic times. The accretionary history of this region which is a complex collage of collisions and amalgamations between lithospheric plates, of cratonic blocks and composite terranes (Fig. 8) is discussed by Ramos (1988) and Hervé (1988). As in other fragments of Gondwanaland, the accretionary history of terranes comprising minor or major part of existing land areas in the Southern Hemisphere (Howell et al. 1985) has an important bearing on the distribution of the biota on those land areas.

Redon (1973, 1976) recognizes nine distinct elements in the Chilean lichen flora and gives characteristic lichens or lichen formations for each. The elements are:

1. Northern coastal element. Here saxicolous communities are dominated by species of *Roccellina* including *R. accedens, R. cerebriformis, R. chalybea, R. chilena, R. flavida, R. inaequalibus, R. limitata, R. luteola, R. lutosa, R. mahuiana, R. nigricans, R. obscura, R. suffruticosa, R. terrestris* (Tehler 1983). Epiphytic communities (mainly on cactus and *Euphorbia*) are dominated by delicate, filamentous taxa from the Roccellaceae adapted to desert fog conditions and include: *Darbishirella gracillima, Dolichocarpus chilensis, Ingaderia pulcherrima, Pentagenella fragillima, Roccellaria mollis.*

2. Fray Jorge–Peruvian–Brasilian element (here a number of warm temperate–tropical taxa are important features of the lichen flora and include: *Everniopsis trulla, Oropogon lorolobic, Pseudocyphellaria aurata, P. intricata, Roccellinastrum spongioideum*). See also Redon and Lange (1983).

3. Andean alteplano element.

4. Andean element (dominated by high Alpine taxa such as *Usnea patagonica, U. perpusilla* (Walker 1985), and species of *Pseudephebe* and *Umbilicaria* and various crustose species including *Omphalodina melanophthalma*).

5. Central Andean submontane element (the characteristic species of this element is the endemic monotypic *Xanthopeltis rupicola*).

6. Valdivian element. Here a rich diversity of southern cool temperate genera and species are characteristic and they also show close relationships with lichen floras in New Zealand and southeastern Australia (Redon 1972, 1974). Over 40 species of *Pseudocyphellaria* occur in this element with many endemic species developed viz. *P. coerulescens, P. compar, P. divulsa, P. encoensis, P. exanthematica, P. hirsuta, P. meyenii, P. nitida, P. nudata, P. obvoluta, P. pilosella, P. piloselloides, P. pluvialis, P. santessonii, P. scabrosa, P. subrubella, P. vaccina, P. valdiviana.* Other important taxa represented in this element include *Cladina laevigata, C. pycnoclada, Cladonia aspera, C. corniculata, C. lepidophora, C. tessellata* (Ahti and Kashiwadani 1984), *Coelocaulon epiphorellum, Erioderma leylandii, Lepolichen granulatus, Menegazzia albida, M. cincinnnata, M dispora, M. globulifera, M. magellanica, M. opuntioides, M. sanguinascens, M. valdiviensis, Metus pileatus* (Galloway and James 1987), *Nephroma analogicum, N. antarcticum, N. cellulosum, N. chubutense, N. microphyllum, N. papillosum, N. parile, N. skottsbergii* (White and James 1988), *Platismatia glauca, Protousnea alectorioides, P. poeppigii, P. teretiuscula, Psoroma caliginosum, P. dimorphum, P. hispidulum, P. pulchrum, Sphaerophorus dodgei, S. insignis, S. patagonicus, S. ramulifer, S. tener, Sticta caulescens, S. hypochra.*

7. Magellanic element. This occurs in the far south in high-rainfall areas between latitudes 48° and 56°S and from sea level to 1500 m and is characterized by Subantarctic tundra or Magellanic moorland (Godley 1960). Species of *Pseudocyphellaria* although obvious and at times prominent in certain habitats are much less diverse than in the Valdivian element and compromise *P. berberina, P. endochrysa, P. faveolata, P. freycinetti, P. granulata, P. lechleri, P. mallota* and *P. vaccina*. Other taxa represented are: *Brigantiaea fuscolutea, Cetraria islandica* ssp. *antarctica* (Kärnefelt 1979), *Cladina pycnoclada* (Ahti and Kashiwadani 1984), *Coelocaulon aculeatum, Hypogymnia lugubris, Ochrolechia frigida, Psoroma hypnorum*.

8. Juan Fernandez element. The Juan Fernan-

dez islands are intra-plate volcanic islands at Lat. 33°S in the SE Pacific. They rise from an east–west trending submarine ridge on the north side of the Challenger Fracture zone in the SE portion of the Nazca Plate. The two main islands are Robinson Crusoe (Mas a Tierra) which lies 660 km WSW of Valparaiso, and Alexander Selkirk (Mas Afuera) which is 180 km further west. Robinson Crusoe rises from sea floor dated at 37 Ma, whereas Alexander Selkirk is built on 32 Ma oceanic crust. The volcanic sequences of the two islands are interpreted as the products of a hot spot currently situated beneath Alexander Selkirk. Robinson Crusoe was positioned over the active hot spot some 3 Ma ago and has since been translated 180 km eastwards, corresponding with a spreading rate of 6 cm per annum (Baker et al. 1987). Robinson Crusoe (48 km) is a deeply dissected island with a highly irregular coastline of jagged promontories, high cliffs and broad embayments. The landscape is dominated by the massive anvil-shaped mountain Cerro Yunque (922 m) which rises abruptly from the densely forested interior. The main part of the island appears to be built around at least four volcanic centres with lavas dated between 3 and 4 Ma (Baker et al. 1987). Alexander Selkirk has an area of 52 km and has the form of an asymmetric dome or shield. Long straight valleys fan out from the summit area down towards the east coast. The island is severely reduced by marine erosion on the west coast, but in spite of this the general configuration of the original volcano is preserved. The evident youthfulness of landforms on Alexander Selkirk contrasts with those on Robinson Crusoe, and rock ages are about 1 Ma (Baker et al. 1987).

The lichen flora of Juan Fernandez (Zahlbruckner 1924; Redon and Quilhot 1977) is of great interest biogeographically as it has affinities with both South America and also with Australasia. Among endemic taxa may be mentioned: *Caloplaca selkirkii, Pertusaria hadrocarpa, Pseudocyphellaria berteroana, P. imbricatula, P. richardi, P. verrucosa, Psoroma angustisectum, P. cephalodinum, P. dasycladum, P. vulcanicum.* Affinities with the Chilean lichen flora are seen in such taxa as *Pseudocyphellaria berberina, P. crocata, P. flavicans, P. gilva, P. guillemini, P. intricata* and *P.mallota.* However, *Pseudocyphellaria dissimilis, P. mooreana,* and *P. physciospora* which all occur on Juan Fernandez are not found in Chile and show links with New Zealand (*P. dissimilis, P. physciospora*) and with eastern Australia (*P. mooreana*). The lichen flora has a pronounced austral character as evidenced by the presence of such taxa as *Brigantiaea chrysosticta, Leioderma pycnophorum, Nephorma australe, N. cellulosum, N. plumbeum, Parmeliella nigrocincta, Pseudocyphellaria glabra, Psoroma sphinctrinum.*

9. Antarctic element. See Redon (1985).

2.4. South Africa

Almborn (1987, 1988) gives a succinct introduction to the lichen flora and vegetation of South Africa and to the history of lichenological investigation of that country, stressing that many groups are still poorly known from a taxonomic standpoint, and predicting that a lichen flora of some 1800–2000 taxa is a distinct possibility by the end of the present century. He characterizes the region as

> "...an arid area, however, some areas, mainly the eastern slopes of the mountains have a high precipitation (800–2000 mm a year) and a considerable number of oceanic lichens many of them with a worldwide distribution such as... *Normandina*. Other parts of the area are dry, having steppe and desert vegatation with a most interesting lichen flora containing many endemics" (Almborn 1987).

Among the endemic taxa mention may be made of the following: *Almbornia cafferensis* (Esslinger 1981), *Lasallia capensis, L. dilacerata, L. glauca, L. membranacea, Neofuscelia* [of the 37 species of *Neofuscelia* known to occur in southern Africa, 25 are endemic (Esslinger 1986)], *Peltula marginata* (Büdel 1987), *Siphula verrucigera, Stereocaulon esterhuyseniae* and many species of *Xanthoparmelia*. Recently Brusse has described several endemic genera from South Africa including: *Corynecystis* [Heppiaceae]

(Brusse 1985); *Lithoglypha* [Acarosporaceae] (Brusse 1988a); and *Schizodiscus* [Porpidiaceae] (Brusse 1988c), as well as new taxa in *Parmelia* (Brusse 1987a); *Gonohymenia* (Brusse 1987b); *Thyrea* (Brusse 1987c); *Trapelia* (Brusse 1987a) and *Porpidia* (Brusse 1988b). Biogeographical affinities of the South African lichen flora are discussed by Almborn (1987, 1988).

2.5. Subantarctic islands (Insulantarctica)

The scattered islands of the cold southern oceans including the outer islands of the southern Chile archipelago (south of Lat. 48°S), the Falkland Islands, and the islands of the Scotia Arc, Tristan da Cunha, Marion, Prince Edward and the Crozet Islands, Kerguelen, and the islands of the Campbell Plateau (Antipodes, Auckland, Campbell and Macquarie Islands) have clear affinities with one another and, with respect to some groups of their biota (e.g. birds), with Antarctica as well. The virtual absence of forest and the presence of temperate grasslands and herbfields, moorland or tundra (Bliss 1979; French and Smith 1985) as the dominant eco-sytems, unite these scattered islands as a subrealm (Insulantarctica) of the Antarctic Realm (Udvardy 1987). Their climates are cool temperate and strongly oceanic with frequent rain and strong westerly winds, soil nutrients are affected by salt spray and often by seals and birds, and their isolation and in several cases relatively recent origin, has produced species-poor biotas (French and Smith 1985; Walker 1985). Pleistocene glaciations have caused major environmental changes in comparatively recent times, and in high-latitude oceanic islands, glaciations have caused massive extinctions (Moore 1979). Tussock grasslands and tundra formations of the shelf islands of the Campbell Plateau (Antipodes, Auckland, Campbell and Macquarie Islands) are closest ecologically to the snow tussock and high Alpine tundras of the central and eastern mountains of New Zealand's South Island (Bliss 1979), with Auckland and Campbell Islands having a more extensive lichen vegetation than any of the other islands of Insulantarctica.

The low scrub and coastal forest associations of the Auckland Islands comprising *Coprosma ciliata, Dracophyllum longifolium, Fuchsia excorticata, Metrosideros umbellata, Myrsine divaricata,* and *Pseudopanax simplex* have a diverse assemblage of lichens, many closely similar to those found in cool temperate, humid habitats in southern New Zealand, and including: *Coccotrema cucurbitula, C. porinopsis, Collema laeve, C. leucocarpum, Degelia duplomarginata, Leioderma amphibolum, L. pycnophorum, Menegazzia circumsorediata, M. subpertusa, Pseudocyphellaria billardierei, P. coronata, P. faveolata, P. glabra, P. multifida, P. physciospora, P. rubella, Psoroma athroophyllum, P. durietzii, P. leprolomum, P. microphyllizans, P. xanthomelanum, Psoromidium aleuroides, P. versicolor* and *Usnea xanthopoga.* In exposed grassland and tundra habitats *Pseudocyphellaria degelii, P. glabra* and *P. physciospora* are found, while on exposed rocks and in fellfield *Peltularia crassa* (Jørgensen and Galloway 1984), *Placopsis subgelida, Pseudocyphellaria glabra, Steinera radiata* ssp. *aucklandica, S. sorediata* (Hensen and James 1982), *Argopsis megalospora* and *Stereocaulon argus* (Galloway 1980), *Cladia aggregata, Hypogymnia lugubris* and *Knightiella splachnirima* (Galloway and Elix 1981) occur.

Lichen communities on other subantarctic islands are neither so prominent nor so diverse, being mainly confined to exposed rocky situations and tundras (see Walker 1985). Floristic or systematic accounts are recorded for various island groups viz: Tristan da Cunha (Jørgensen 1977, 1979), Marion and Prince Edward Islands (Lindsay 1977, Hertel 1984, Henssen and Büdel 1984), South Shetlands (Lindsay 1971, Hertel 1987b), South Georgia (Lindsay 1975); Huneck et al. 1984; Hertel 1984, 1987b), Bouvetøya (Engelskjøn and Jørgensen 1986; Jørgensen 1986; Øvstedahl and Hawksworth 1986), Kerguelen (Hertel 1984, 1987b), Macquarie Island (Filson 1981a, b, 1986b; Filson and Archer 1986), Auckland Islands (Fineran 1971;

Galloway 1985a, 1988b). Saxicolous lecideoid lichens of the Subantarctic islands are discussed by Hertel (1984, 1987a, b) who shows that both circumpolar and cosmopolitan taxa are represented.

2.6. Antarctica

The continent of Antarctica is part of a larger Antarctic plate bounded by mid-ocean ridges in the Pacific, Indian and South Atlantic oceans. The Antarctic plate comprises the continent and ocean floor to the mid-ocean ridge boundaries (Fig. 9). The continent itself is formed of two distinct geological provinces; East Antarctica, a Precambrian craton composed of mainly igneous and metamorphic rocks and forming a large stable continental area, and West Antarctica which, by contrast, is made up of a series of orogenic belts of early Palaeozoic to late Mesozoic age (Elliot 1985).

Antarctica was an integral part of Gondwanaland and the continental fits between Africa and South America (West Gondwanaland), and between Antarctica, Australia and India (East Gondawanaland) have been satisfactorily established. However, problems still remain with the microplate region of West Antarctica, the Pacific margin of Gondwanaland (Fig. 10) from southern South America, the Antarctic Peninsula, to the

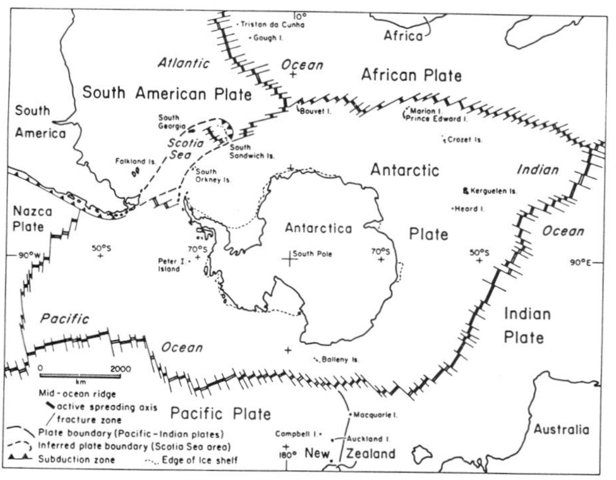

Fig. 9. The Antarctic plate, from Elliott (1985).

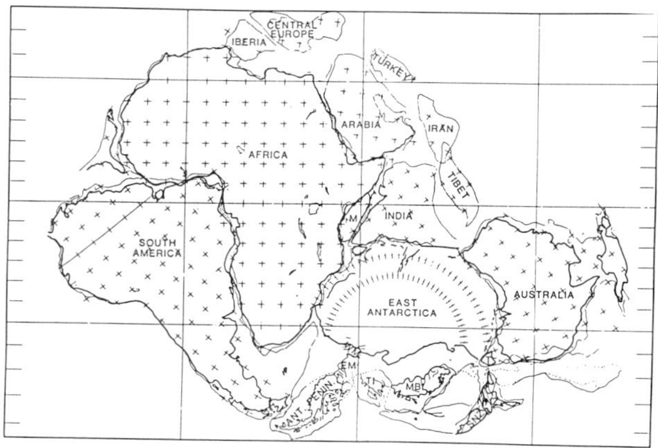

Fig. 10. Reconstruction of Gondwanaland, from Lawver and Scotese (1987).

New Zealand microcontinent (Dalziel and Elliot 1982; Dalziel and Grunow 1985; Lawver and Scotese 1987). It is this Protopacific or Panthalassic margin of Gondwanaland which is of profound significance in considerations of biogeographical relationships of the biota of the austral zone.

Before the breakup of Gondwanaland, Antarctic Gondwana was dominated geologically by the evolution of an active plate margin along the Protopacific border, and also climatically by a change from glacial to non-glacial temperate conditions. Glacial conditions are attributed to a polar position, but the arrangements of dry land and oceans differed very much from those obtaining in the region today. The glacial environment appears to have ended abruptly, being replaced by a temperate climatic regime allowing abundant plant growth, with eventually the formation of coal measures. Evidence from growth rings in fossil woods from Antarctica suggests a strongly seasonal temperate climate. The homogeneity of Permian floras points to a climatically uniform flora province, and conditions changed little during the Triassic with the vegetation being locally abundant and forming coal swamps (Elliot 1985). Antarctic climate during the late Mesozoic and early Cenozoic was equable and probably warm temperate as indicated by the common occurrence of angiosperm leaves and logs from large trees, and certainly seasonal as shown by well-developed growth rings in these logs (Elliot 1985). On the Antarctic Peninsula the most common fossil angiosperm wood is *Nothofagus*, and fossil remains of *Nothofagus* and the families Podocarpaceae and Araucariaceae indicate forest compositions very similar to those found today in South America and Australasia (Francis 1986; Mercer 1987). In the mid-Tertiary 25 to 30 Ma, Antarctica once again returned to glacial conditions, the history of the Antarctic ice-sheet and its response to changes in climate and sea level being extensively documented (e.g. Hall 1987; Robin 1988). This glaciation which persists to the present time, led to the extinction and elimination of the formerly richly diverse late Cretaceous–Tertiary plant cover, Pleistocene glaciations producing almost total ice cover over the whole continent and its offshore islands and also some Subantarctic islands.

Today the Antarctic terrestrial flora is almost completely cryptogamic and includes at least 200 lichens, 85 mosses, 28 macrofungi, 25 hepatics and two angiosperms, the grass *Deschampsia antarctica* and the cushion-forming *Colobanthus quitensis* (Longton 1985, 1988). General accounts of Antarctic lichens are given in Dodge (1973), Lindsay (1977b) and Redon (1985) while classification of Antarctic lichen communities is discussed by Longton (1985, 1988) and Smith (1988). Since over 97% of continental Antarctica is ice covered, niches available for lichen colonisation occur mainly in maritime regions, on islands and inland on nunataks and locally in dry valleys. Lichens (and vegetation generally) of the maritime Antarctic differ markedly from those found in continental Antarctica, the contrast being primarily in response to differences in climate and water availability. In favourable maritime Antartic localities the Antarctic herb tundra formation comprises eight subformations of which two are dominated by lichens, viz. a crustose lichen subformation, and a fruticose/foliose lichen subformation (Longton 1985). In the latter, the fruticose genera *Himantormia* (Lamb 1964) and *Neuropogon* (Walker 1985) are especially typical and often abundantly developed.

The lichen vegetation of the Antarctic including the Subantarctic islands (Insulantarctica), southern New Zealand, southern South America and Tasmania has a pronounced circumpolar and also a cosmopolitan aspect (Lindsay 1977a; Hertel 1987a). Although the Antarctic continent has a number of apparently endemic crustose species, apart from the genus *Austrolecia* (Hertel 1984), there are no other endemic genera present. Hertel (1987a, b) has shown that in Antarctic, saxicolous lecideoid taxa, widespread genera are represented in all southern regions (Fig. 11) with the following known from Antarctica itself: *Carbonea, Farnoldia* (formerly thought to be exclusively boreal), *Lecanora, Lecidea, Lecidella, Porpidia* and *Tremolecia*. In contrast to these there are several southern circumpolar saxicolous lecideoid genera which are not known from the Northern Hemisphere and these include: *Diomedella* (South America, South Georgia, islands of the southern Indian Ocean, New Zealand); *Notolecidea* (South Georgia, Prince Edward Island, Kerguelen); *Poeltiaria* (South America, Tasmania, New Zealand); *Poeltidea* (South America, South Georgia, islands of the southern

Fig. 11. Widespread genera of saxicolous lecideoid lichens in the Southern Hemisphere, from Hertel (1987a).

Indian Ocean, New Zealand); *Rhizolecia* (New Zealand); *Stephanocyclos* (South America, islands of southern Indian Ocean, Tasmania); and *Zosterodiscus* (islands of the southern Indian Ocean, New Zealand).

Hertel (1987a) considers that high humbers of genera and species in the saxicolous lecideoid lichen flora of continental Antarctica, together with the virtual absence of endemic genera, suggest that saxicolous lecideoid lichens in Antarctica are of recent age in contrast to those found in surrounding Insulantarctica, and that long-distance dispersal of propagules in the West Wind Drift has accounted for the patterns of distribution of many lichens in areas such as Marion, and Prince Edward Islands. West Wind Drift or neoaustral lichen distributions are discussed in Galloway (1987). The lichen flora of Antarctica appears therefore to be of fairly recent origin (post-Pleistocene) and it stands in contrast to the lichen vegetation developed on Insulantarctica where many older, relict elements currently occur. Cryptoendolithic lichens are also a feature of the lichen vegetation of Antarctica's dry valleys and are discussed by Kappen and Friedmann (1983) and Kappen (1988).

3. PHYTOGEOGRAPHICAL AFFINITIES OF SOUTHERN HEMISPHERE LICHEN FLORAS

3.1. New Zealand–Australia

Fleming (1987) has shown that there is much evidence that New Zealand's biogeographical relationships with Australia are closer than they are with the rest of Gondwanaland, and especially with the rest of Udvardy's (1987) Antarctic Realm, citing as evidence for his view, *inter alia*:

1. Australia and New Zealand were adjacent parts of Gondwanaland with a long history of contact before formation of the Tasman Sea beginning in the Upper Cretaceous.

2. In the Tertiary, New Zealand had a diverse assemblage of *Nothofagus* species of three pollen species groups (some with pollen identical to that found in the Australian Tertiary), and a diverse assemblage of Proteaceae of Australian–New Caledonian affinity.

3. The New Zealand ratite birds are more closely related to those of Australia than to those of other lands.

4. Many New Zealand invertebrates have Australian affinities.

5. The majority of New Zealand land and fresh water birds are of Australian origin.

6. In historic time (the last 150 years) New Zealand has continued to receive colonists from Australia, including plants, rusts, birds, insects, etc. (see also Close et al. 1978).

Many species in the New Zealand lichen flora also occur in Australia with most shared species being found in southeastern Australia and Tasmania indicating a close floristic and biogeographical relationship (Galloway 1979, 1985a, 1988a, b; Galloway and James 1985; Galloway and Jørgensen 1987). The cool temperate Australasian element is best seen in Alpine–Subalpine lichen vegetation in New Zealand, Tasmania and southeastern Australia. Climate in these areas is essentially similar being markedly oceanic, with humid, cloudy conditions prevailing and with frequent rainfall. Lichens of this element are also commonly associated with species of *Nothofagus*. Macrolichens of Tasmania and southern New Zealand are closely similar, with 80% of the Tasmanian species being common to New Zealand (Kantvilas and James 1987) but with the diversity in the New Zealand flora being much greater (Kantvilas et al. 1985).

Many of these disjunct Australasian taxa are presumed ancient relicts which evolved in Cretaceous times in cool temperate habitats around the Protopacific margin of Gondwanaland and which later became isolated in cool temperate habitats on the New Zealand microcontinent and in Tasmania and the uplands of southeastern Australia, after rifting and drift, consequent upon the opening of the Tasman Sea and Bass Strait. New Zealand being isolated from eastern Gondwanaland for a greater period of time than Tasmania (Veevers 1987) would have had greater opportunities for increased speciation events resulting in the evolution of more taxa in *Pseudocyphellaria, Placopsis* and *Sphaerophorus* for example in New Zealand than in Tasmania.

In addition to the relict or palaeoaustral element in the lichen floras of Tasmania and New Zealand (Galloway 1987, 1988b; Galloway and Jørgensen 1987) a distinctive and more recent relationship exists between Australian and New Zealand lichen floras (Galloway 1979, 1988a, b) with taxa derived from sources in Australia [such as *Chondropsis* (Elix and Childs 1987) and *Xanthoparmelia* (Galloway 1981b; Elix et al. 1986)] being transported to appropriate habitats in New Zealand by the prevailing westerlies (Close et al. 1978; Wardle 1978). Lichens disjunctly distributed between southeastern Australia (including Tasmania) and New Zealand include: *Austroblastenia pauciseptata* (Sipman 1983), *Bacidia buchananii, Baeomyces arcuatus, B. heteromorphus, Caloplaca cribrosa, Cladia fuliginosa, C. inflata, C. retipora, C. sullivanii* (Galloway 1977; Filson 1981c; Kantvilas and Elix 1987), *Cladonia bimberiensis, C. murrayi, C. neozelandica, C. subsubulata, C. weymouthii* (Archer 1985, 1988; Archer and Bartlett 1986), *Conotremopsis weberiana, Degelia durietzii* (Arvidsson and Galloway 1981), *Dendriscocaulon dendriothamnodes, Diploschistes muscorum* ssp. *bartlettii* (Lumbsch 1987), *Ephebe fruticosa, Hypogymnia billardierei, H. kosciuskoensis, H. mundata, H. turgidula* (Elix 1979), *Knightiella splachnirima* (Galloway and Elix 1981), *Lecidea coromandelica* (Hertel 1987b), *L. laeta, Leioderma amphibolum* (Galloway and Jørgensen 1987), *Megalospora atrorubicans, M. campylospora, M. subtuberculosa* (Sipman 1983, 1986), *Menegazzia aeneofusca, M. caliginosa, M. castanea, M. nothofagi, M. testacea, M. ultralucens, Metus conglomeratus* (Galloway and James 1987), *Miltidea ceroplasta, Neofuscelia loxodella, Nephroma australe, N. rufum* (White and James 1988), *Neuropogon acromelanus, N. ciliatus, N. subcapillacea* (Walker 1985), *Pannoparmelia angustata, P. wilsonii* (Galloway 1978; Yoshimura 1987), *Parmelia norcrambidiocarpa, P. salcrambidiocarpa, P. signifera, P. tenuirima, P. testacea* (Galloway and Elix 1983; 1984, Hale 1987), *Parmelina conlabrosa, P. labrosa, P. stevensiana* (Elix and Johnston 1987), *Pertusaria truncata, Placopsis trachyderma, Protoparmelia petraeoides* (Hertel 1985), *Pseudocyphellaria ardesiaca, P. billardierei, P. chloroleuca, P. cinnamomea, P. colensoi, P. coronata, P. haywardiorum, P. jamesii, P. multifida, P. neglecta, P. rubella* (Galloway 1988b), *Psoroma asperellum, P. caliginosum, P. durietzii, P. soccatum, Psoromidium aleuroides, P. versicolor* (Galloway and James 1985), *Ramalina inflata* (Stevens 1987). *Roccellinastrum neglectum* (Henssen et al. 1982), *Sagenidium molle, Siphula foliacea, S. fragilis, S. jamesii* (Kantvilas 1987), *Sphaerophorus imshaugii, S. insignis, S. macrocarpus, S. murrayi, S. notatus, S. scrobiculatus* (Tibell 1987), *Stereocaulon caespitosum, S. trachyphloeum* (Galloway 1980), *Teloschistes fasciculatus, T. sieberianus, T. xanthorioides, Trapeliopsis congregans, Usnea capillacea, U. contexta* and *Wawea fruticulosa* (Henssen and Kantvilas 1985).

3.2. New Zealand – Australia – South America

Biotas of the major austral land masses have many similarities with disjunct distributions observed between taxa in New Zealand and South America being analysed by both vicariance or cladistic biogeographical (e.g. Humphries and Parenti 1986; Brundin 1988; Seberg 1988), and panbiogeographical (Craw 1985, 1988) methodologies, although Seberg (1988) opines that "... the greatest obstacle to an effective analysis of the biogeography of the South Pacific is the lack of reliable taxon cladograms to base the area cladograms upon". Austral lichen floras also show these similarities and patterns of disjunct distributions with affinities being most noticeable at the generic rather than the species level (see Galloway 1979, 1985b, 1987b, 1988a, b; Jørgensen 1983).

Two major groupings are found in austral lichen floras (Galloway 1987b, 1988b).

1. *Palaeoaustral* lichens are considered to represent primitive Gondwanan groups poorly

adapted for transoceanic dispersal, and which derived from the Cretaceous (or earlier) when the Protopacific margin of Gondwanaland was available for colonization, where cool temperate conditions prevailed, and where a vegetated West Antarctica could link South American and South Pacific land masses. Palaeoaustral lichens show a number of shared features; they grow in cool temperate environments, often in forest dominated by species of *Nothofagus*, or else in subalpine shrubland or grassland habitats; they are often fertile and generally do not develop vegetative diaspores; they have been flexible enough to colonize or recolonize cool temperate habitats during and after periods of climatic deterioration; and they show disjunct distributions. Examples are *Degelia gayana, Leioderma pycnophorum* (Fig. 12), *Nephroma cellulosum, Pseudocyphellaria faveolata, Sphaerophorus patagonicus* (for additional taxa see Galloway 1987a).

2. *Neoaustral* lichens, are taxa dispersed after the fragmentation of Gondwanaland, mainly post-Pleistocene to the present. They are richly provided with vegetative propagules which allow long distance transport via birds, ocean currents or in the West Wind Drift (Fig. 13). Examples of such recently distributed taxa include *Pseudocyphellaria glabra, P. granulata* and *Stereocaulon corticatulum* (Galloway 1987b).

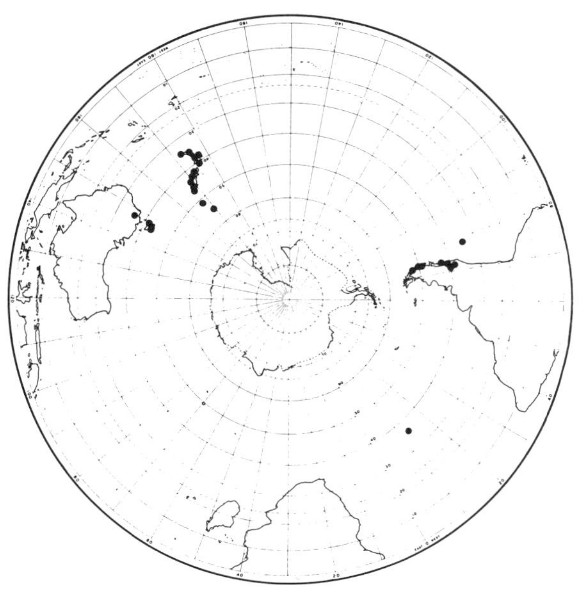

Fig. 12. Distribution of *Leioderma pycnophorum*, from Galloway and Jørgensen (1987).

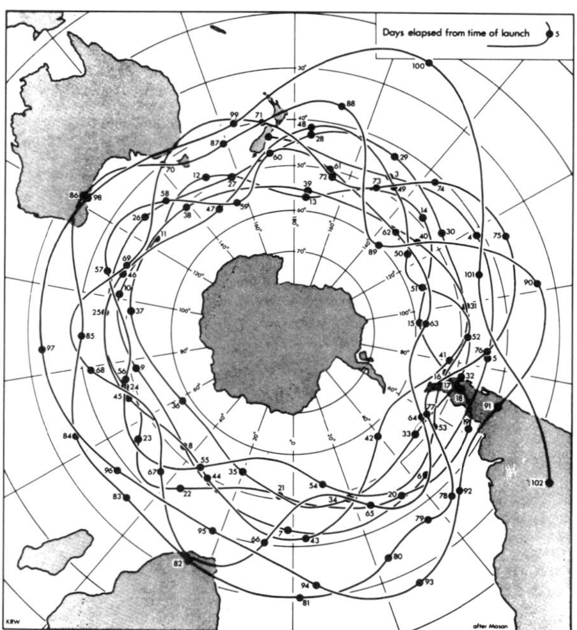

Fig. 13. West Wind Drift path of balloon released from Christchurch. New Zealand, making 8 circuits at c. 12 000 m during 102 days.

3.3. Australia – South Africa

A small number of species are shared by Australasia and South Africa. Degelius (1974) recorded that the cool temperate *Collema leucocarpum* which is distributed widely in New Zealand, Tasmania and southeastern Australia, also grows in cool temperate, humid sites at the Cape. *Parmelia kerguelensis* (Hale 1987) and *Cetraria chlorophylla* (Kärnefelt 1987) also occur at the Cape but are disjunct between Australasia, South America and the west coast of North America. Similar disjunctions in Afro-American hepatics are discussed in Gradstein et al. (1983).

However, in warmer drier habitats where continental climates prevail there are disjunctions observed between South African and Australian populations of *Dirinaria melanoclina, Eremestrella crystallifera* (Rogers and Stevens 1981; Jørgensen 1983; Almborn 1987), in *Neofuscelia*

(Esslinger 1977, 1986a, b), and especially in *Xanthoparmelia* (Knox and Brusse 1983; Elix et al. 1986; Elix and Johnston 1987b; Nash and Elix 1987). Of the 133 taxa presently known from Australia, at least 30 also occur in South Africa including: *Xanthoparmelia adhaerens, X. australasica*[*], *X. amplexula, X. constrictans, X. crateriformis, X. digitiformis, X. eradicata, X. exillima, X. filsonii, X. hypoleia, X. hypoprotocetrarica, X. incerta, X. iniquita, X. isdiigera*[*], *X. isidiosa, X. molliuscula*[*], *X. mougeotina*[*], *X. neorimalis, X. pertinax, X. pseudohypoleia, X. scabrosa*[*], *X. subdomokosii, X. suberadicata, X. subnigra, X. tasmanica*[*], *X. tegeta, X. terrestris, X. verruciformis* and *X. weberi*. Those taxa marked with an asterisk are present also in New Zealand, mainly in restricted semi arid areas of Central Otago, Canterbury and Marlborough. *Xanthoparmelia mougeotina, X. scabrosa* and *X. tasmanica* also occur more widely in the Pacific and North America (Elix et al. 1986).

3.4. Antitropical taxa

Antitropical, amphitropical or bipolar taxa have intrigued botanists and biogeographers for many years (see Humphries and Parenti 1986; Briggs 1987). Lichens also show distinctive antitropical disjunctions (see Galloway and Bartlett 1986 for discussion of earlier lichen literature), and from the Southern Hemisphere the following antitropical taxa are known: *Alectoria nigricans, Arthrorhaphis alpina, A. citrinella* (Fig. 14), *Aspicilia alpina, A. subsorediza, Brigantiaea fuscolutea, Caloplaca cinnamomea, Carbonea vorticosa* (Hertel 1985), *Cetraria delisei* (Kärnefelt 1979), *Coelocaulon aculeatum* (Kärnefelt 1986), *Lecanora polytropa, Lecidea lapicida, Pannaria hookeri, Parmelia sulcata* (Hale 1987), *Pertusaria dactylina, Placynthium nigrum, P. subradiatum* (Henssen 1984), *Pseudephebe minuscula, P. pubescens, Rimularia insularis* (Hertel 1985), *Solorina crocea, S. spongiosa, Xanthoria elegans* and *Zahlbrucknerella calcarea*. All these taxa are recorded from subalpine to high–alpine habitats.

3.5. Tropical taxa

All the major Southern Hemisphere lichen floras have a component showing affinities with those of tropical areas. Such tropical taxa in

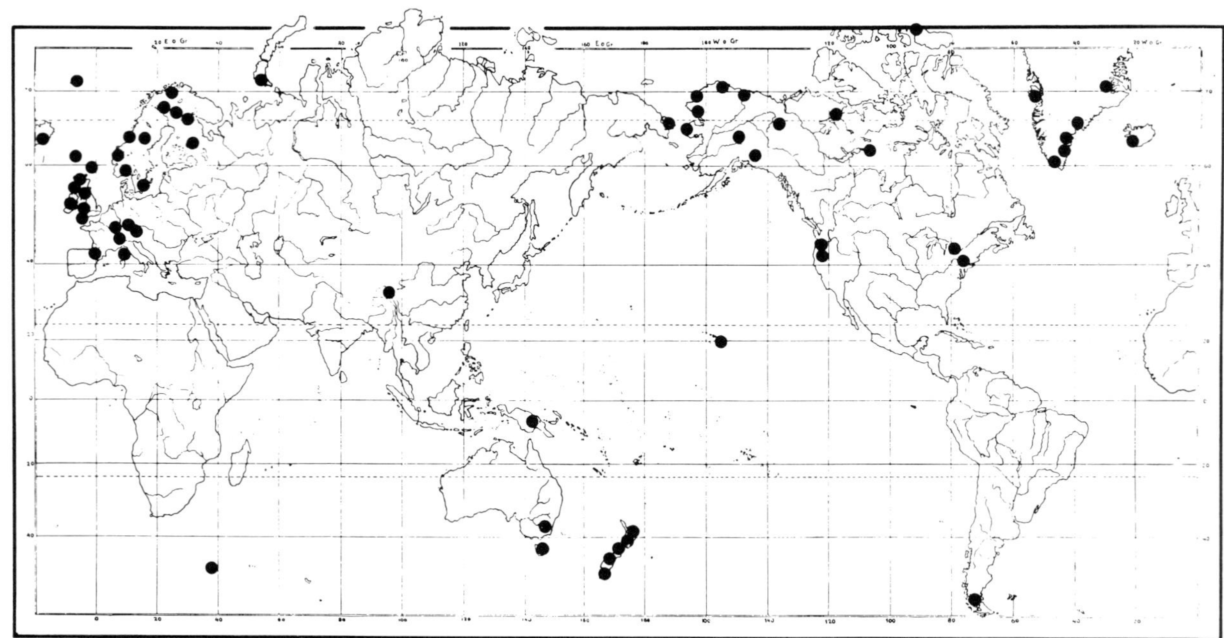

Fig. 14. Antitropical distribution of *Arthrorhapsis alpina* and *A. citrinella*, from Galloway and Bartlett (1986).

contrast to southern, cool temperate taxa, have warm temperate or Tethyan affinities and comprise those organisms shared with ancient Tethys (i.e. the vicinity of the modern Mediterranean) and with its seaway extension via the Arabian Peninsula, Himalayas, southwestern Asia, Indonesia and New Guinea to the islands of the Pacific (Galloway and Jørgensen 1987). Most of the lichens with tropical affinities found in Southern Hemisphere floras are lowland taxa, but it is also important to note cool temperate conditions exist at high elevations on tropical mountains and southern, cool temperate lichen taxa can survive in such habitats. For discussions on Indo-Australian and Australasian tropicalpine floras and their origins which have bearings on the distributions of lichens in the tropics see Holloway (1986) and Smith (1986a, b).

Tropical and/or Amphipacific taxa present in Southern Hemisphere lichen floras include: *Anzia madagascarensis, Bryoria indonesica* (Jørgensen and Galloway 1983), *Cetrelia braunsiana, Coccocarpia erythroxyli, C. palmicola, C. pellita* (Arvidsson 1983), *Coccotrema cucurbitula, C. porinopsis, Collema japonicum, C. subconveniens, C. subfragrans, Everniastrum sorocheilum, Heterodermia japonica, H. microphylla, Hypotrachyna ensifolia, H. thysanota, Leioderma duplicatum, L. erythrocarpum, L. sorediatum* (Galloway and Jørgensen 1987, see also Fig. 15), *Leprocaulon arbuscula, Lobaria retigera* (Galloway 1981a), *Lopadium subcoerulescens, Menegazzia eperforata, Pannaria fulvescens, Peltigera dolichorhiza, P. nana, Physma byrsaeum, Placopsis cribellans, Pseudocyphellaria argyracea, P. dissimilis, P. poculifera, P. sulphurea, Ramalina celastri* (Stevens 1987), *Sphaerophorus murrayi, Sticta sublimbata, Thelotrema weberi, Thysanothecium scutellatum* (Galloway and Bartlett 1983), *Usnea maculata. U. societatis* and *Xanthoparmelia scabrosa* (Elix et al. 1986).

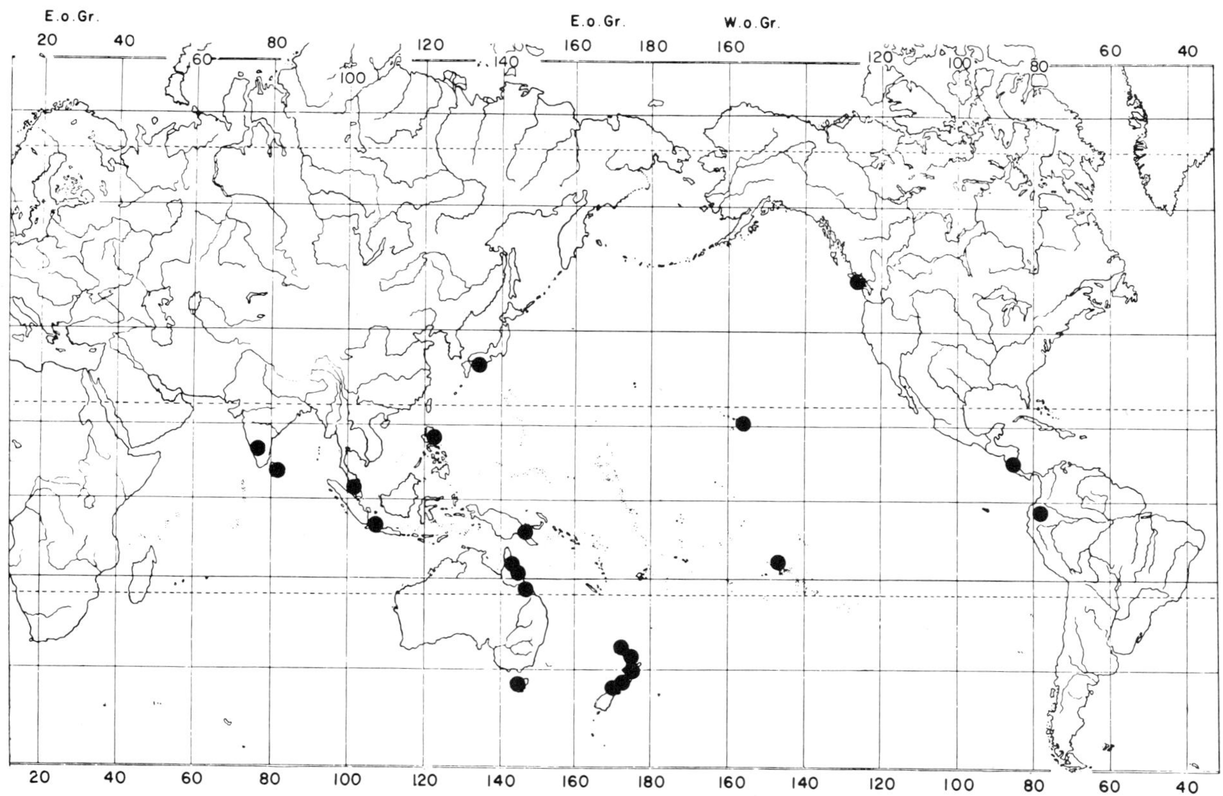

Fig. 15. Known distribution of *Leioderma sorediatum*, from Galloway and Jørgensen (1987).

3.6. Cosmopolitan taxa

The cosmopolitan element contains taxa which are generally widespread being present on all landmasses and on oceanic islands. The extremely wide distribution of members of the *Xanthorion*, especially *Xanthoria parietina, Physcia adscendens* and *P. stellaris* has already been noted (Galloway 1979) and is usually associated in New Zealand and South America at least, with introduced trees in urban areas, or decorticated wood such as fenceposts. The cosmopolitan, sterile alpine species *Thamnolia vermicularis* has a conservative morphology throughout its worldwide range yet exhibits two well-defined chemodemes and is widely distributed in Alpine and periglacial environments in both hemispheres and in the tropics. In the Southern Hemisphere it grows in environments of active mountain building which are geologically young and can scarcely be regarded as a remnant of an older lichen vegetation. Local spread of populations (as with certain species of *Siphula*, also sterile) in the absence of spores or asexual propagules, is dependent on thallus fragmentation and the action of wind, water, and mechanical trampling by animals and/or humans in certain areas. The mechanism of its wider spread is still not at all clear and is presumed to be one of long-distance dispersal. The order Caliciales has a number of taxa of cosmopolitan distribution, e.g. *Calicium abietinum, Chaenotheca brunneola, Cyphelium inquinans, Mycocalicium subtile* and *Sphaerophorus melanocarpus* (Tibell 1984, 1987), and the mainly cool temperate *Pseudocyphellaria* has three wide ranging species, *P. aurata, P. crocata, P. intricata*, all with soredia (Galloway 1988b).

3.7. Endemic taxa

An endemic element is present in the lichen floras of all southern landmasses and may consist of genera of very restricted distribution such as *Austrolecia* in Antarctica (Hertel 1987a), *Lethariella* and *Protousnea* in South America (Krog 1976), *Sarrameana* in Tasmania (Vezda and Kantvilas 1988), *Sagenidiopsis* in Australia (Rogers and Hafellner 1987), or of local endemic species from genera of much wider occurrence of which many examples may be cited for all areas. Most of the Southern Hemisphere endemics are of this latter category and include taxa from genera such as: *Argopsis* (Lamb 1974; Galloway 1980); *Brigantiaea* (Galloway 1985a); *Cladonia* (Archer and Bartlett 1986), *Bryoria* (Jørgensen and Galloway 1983); *Cladia* (Kantvilas and Elix 1987); *Collema* (Degelius 1974); *Degelia* (Arvidsson and Galloway 1981); *Haematomma* (Rogers and Bartlett 1986); *Leioderma* (Galloway and Jørgensen 1987); *Menegazzia* (Galloway 1985a); *Megalospora* (Sipman 1983, 1986); *Melanelia* (Esslinger 1977, 1978); *Neofuscelia* (Esslinger 1977, 1978); *Nephroma* (White and James 1988); *Parmelia* (Hale 1987); *Parmeliella* (Galloway and Jørgensen 1987); *Peltularia* (Jørgensen and Galloway 1984); *Phlyctis* (Galloway and Guzman 1988); *Phylisciella* (Henssen and Büdel 1984); *Pseudocyphellaria* (Galloway 1986, 1988b); *Sphaerophorus* (Tibell 1987); *Steinera* (Henssen and James 1982); *Stereocaulon* (Lamb 1977, Galloway 1980); *Xanthomaculina* (Hale 1985) and *Xanthoparmelia* (Elix et al. 1986).

Endemic taxa are of considerable biogeographical interest and may represent either the evolution of “new” genera or species from ancestors over a period of isolation, or else they reflect an “old” or relict distribution surviving after widespread extinctions as a consequence of geographic or climatic changes. For example the emergence of a profusion of species in *Xanthoparmelia* apparently adapted to local arid conditions in Australia, has resulted in a high proportion of endemic species of rather limited occurrence and with variations in morphology and especially in chemistry (Elix et al. 1986) produced, very likely as ecological adaptations to a northward rafting of Australia into lower and drier latitudes. A similar situation in the same genus is found in southern Africa. The long geographical isolation of New Zealand from Australia, Antarctica and South America has led to a great diversity of endemic species in such genera as *Menegazzia, Psoroma, Placopsis* and *Pseudocyphellaria* with many taxa being vicariants of those occurring in southern South America. For cool temperate rainforest elements in Tasmania Hill and Read (1987) propose three hypotheses to account for endemism which may also have a wider and more general application in considerations of Southern Hemisphere cool temperate endemics including lichens.

1. Some taxa evolved in southeastern Australia during the Tertiary in response to the changing climate. Some ancestral species still occur in temperate rainforest at lower latitudes in Australia.

2. Some species have remained essentially unchanged in Tasmania during Tertiary and Quaternary climatic changes.

3. Post glacial climatic changes (especially a decrease in rainfall) and human influence (land clearance, and fire) possibly combined to eliminate some cool temperate rainforest taxa from mainland Australia.

4. ORGANIC CHEMISTRY AND LICHEN DISTRIBUTION

Lichens produce a wide range of primary and secondary metabolites (Culberson 1969; Elix et al. 1984) the latter being widely used by lichenologists in systematic studies with chemical characters commonly invoked to support (or deny) variation in morphological characters and in geographical distributions of taxa (see Elix 1982; Brodo 1986; Egan 1986). W. L. Culberson (1986) has drawn attention to the importance of chemical characters in sibling speciation in lichen-forming fungi where chemical and ecological differentiation is observed in taxa having highly conservative morphologies, underlining the view that chemistry is a major marker of evolutionary change in the lichens. His view is further vindicated by the demonstration for the first time of gene flow in lichen-forming fungi by an analysis of secondary metabolites in the progeny of individuals from natural populations of mixed chemodemes of the *Cladonia chlorophaea* complex (Culberson et al. 1988). The increasing sophistication of techniques for detection of secondary metabolites in lichens (see Culberson et al. 1987) allows the elaboration of biosynthetic hypotheses which can be used in cladistic analyses of evolutionary relationships among taxa (Culberson 1986; Culberson et al. 1987) a field ripe for further investigation. Not only secondary metabolites from lichens but also an investigation of certain primary metabolites and metabolic pathways, may provide clues to biological and evolutionary relationships. Three broad groups of lichen metabolites are discussed in a very preliminary way in this regard below.

4.1. Triterpenoids

Terpenoids are compounds with varying numbers of carbon atoms, derived from C–5 units. They are widespread in the plant world and it is probable that more individual terpenes and terpenoids exist than any other group of plant products. Photosynthesis, the essential process on which all plant life depends, has a mandatory requirement for certain terpenoids and their derivatives (e.g. carotenoids and chlorophylls) and many plant hormones are terpenoids. Chemically all terpenes and terpenoids are derived from a basic 5-carbon isoprene skeleton and are classified according to the number of such units in the molecule. Triterpenoids have a skeleton of 30 carbon atoms derived from the acyclic hydrocarbon squalene, itself formed through the mevalonic acid pathway (Goodwin and Mercer 1983). As a group, triterpenoids occur widely in plants, many being well-known toxins such as saponins and cardiac glycosides. The comparative biochemistry of triterpenoids has been extensively studied in higher plants, bryophytes, algae and fungi, with taxonomic implications being drawn from their distribution patterns (Harborne and Turner 1984). Triterpenoids have rather complex structures (see below), with the main groups consisting of tetracyclic derivatives based on parent hydrocarbons such as lanostane, cycloartane and dammarane, and pentacyclic compounds derived from ursane, oleanane, lupane, hopane and related skeletons (Connolly and Overton 1972). Subsequent to polymerization and cyclization and the formation of the triterpenoid skeleton from its component isoprene units, inumerable rearrangements, oxidations, etc., are possible, with the formation of thousands of structurally unique terpenoids possible.

In the order Peltigerales triterpenoids are especially richly developed in the genera *Nephroma* (James and White 1987; White and James 1988), *Peltigera* (Tønsberg and Holtan–Hartwig 1983; Vitikainen 1985; Galloway 1985a; Holtan–Hartwig 1988) and *Pseudocyphellaria* (Galloway 1988b) and besides having an obvious use in separating taxa at the species level they may also prove to be of importance in phylogenetic studies in these genera.

Triterpenoids of the hopane series (see Galloway 1988b) are the most widely distributed of triterpenoids in lichens being known from *Heterodermia, Lobaria, Nephroma, Peltigera* and *Pseudocyphellaria* (Galloway 1988b), and *Physcia* (Elix et al. 1982). They are thought to be primitive phylogenetic precursors of sterols (Ourisson et al. 1979) and are also the most highly reduced of the triterpenoid series known from lichens. Four main hopanoid compounds are widely distributed in the Peltigerales, though at least 15 different compounds from the series are known. These major hopanes are: 7β-acetoxyhopane-22-ol, and hopane-15α, 22-diol, hopane-6α, 7β, 22-triol and hopane-6α, 22-diol (zeorin), and they occur in species specific patterns (see Wilkins and James 1979; Galloway et al. 1983; James and White 1987; Galloway 1988b).

Pseudocyphellaria has the highest chemical diversity of any genus in the Peltigerales and besides hopane triterpenoids, fernene, stictane, secostictane and lupane triterpenoids are known from species present in the Southern Hemisphere where the major areas of species diversity are New Zealand, South America, southeast Australia (including Tasmania) and the palaeotropics. Compounds from each of these latter triterpenoid classes are more highly oxidized than hopane triterpenoids and from a biosynthetic and also an evolutionary point of view, may be regarded as less primitive (apomorphic) compounds than hopane derivatives. It is noteworthy that these more highly oxidised triterpenoids all co-occur in species of *Pseudocyphellaria* which have the yellow pigments calycin, pulvinic acid and pulvinic dilactone present either in the medulla or in pseudocyphellae (Galloway 1988b). That hopanes may be considered pleisiomorphic to the other triterpenoid groups is supported by the fact that hopane-containing taxa in *Pseudocyphellaria* are both more numerous and also much more widely distributed geographically. Hopanes have also attracted considerable interest in the organic geochemistry of shales, coals, lignites, sediments and petroleum, and two triterpanes, 17α(H),-21β(H)-hopane and 17β(H), 21α(H)-hopane (moretane) have received wide attention as indicators of maturity, and to a lesser extent, sources of sedimentary organic matter (Czochanska et al. 1987). Their presence in sediments and oils is attributed to contributions from micro-organisms, algae and higher plant sources, and it is possible that they may even reflect a contribution from lichen fungus sources. In any event, the similarity of lichen hopanes and the triterpane biomarkers suggests the possibility of relating chemical structures to geological formations of known age which may have important implications in phylogenetic reconstructions of taxa in the Peltigerales. It is interesting to note in passing that taxa in both *Physcia* and *Pseudocyphellaria* produce hopanes and have thick-walled, brownish 1-septate spores of similar architecture.

The possible biological role of triterpenoids as distinct from their demonstrated taxonomic utility in lichens and especially in genera of the Peltigerales where they are particularly diverse and produced in some quantity (see James and White 1987; Galloway 1988b) invites considerable speculation. In plants generally, triterpenoids are well-known toxins (Goodwin and Mercer 1983; Harborne and Turner 1984) and are implicated, as are many other secondary compounds, in a variety of ecological roles as defensive agents in plant–plant (allelopathic), plant–herbivore, and plant–pathogen interactions. The considerable adaptive advantage to lichens in the production of secondary metabolites with these presumed functions points to a functional role and a co-evolutionary importance for such substances (see Harborne and Turner 1984; Lawrey 1984; Hawksworth 1988). For a recent review of plant–fungal interactions from a physiological perspective see Mayer (1989).

In cool temperate rain forests of the Southern Hemisphere species of *Nephroma, Pseudocyphellaria* and *Sticta* are richly developed, many species reaching a great size and representing a considerable epiphytic biomass. Since many have cyanobacterial photobionts and are presumed major contributors to forest nitrogen budgets (Galloway 1988a) they constitute a rich protein source for potential vertebrate and/or invertebrate herbivores. The fact that they are not significantly grazed in these areas (Rundel 1978; Galloway 1988b) suggests that an anti-herbivore defence system is in place in these genera. While it is tempting to suggest that triterpenoids may function as anti-herbivore compounds in *Nephroma* and *Pseudocyphellaria* for example, the absence of any secondary metabolites in species of *Sticta* and in two species of *Pseudocyphellaria* (*P. gretae* in New Zealand, and *P. nitida* in South America) makes it appear that other factors are involved in anti-herbivory, since species of *Sticta* reaching great size and being unattacked by herbivores, are frequently sympatric with taxa in *Nephroma* and *Pseudocyphellaria* which produce triterpenoids in quantity.

However, the investment in highly metabolized carbon that pools of triterpenoids represent in taxa in the Peltigerales especially is remarkable, and is undoubtedly connected with richly speciated groups in Southern Hemisphere cool temperate rainforests where they are such conspicuous components of the epiphytic and ground vegetation. The formation of complex molecules such as triterpenoids and their co-occurrence in closely related taxa in genera such as *Nephroma, Peltigera* and *Pseudocyphellaria* indicates conservation of biosynthetic pathways in phylogenetic lineages, with the products of biosyntheses (in this case triterpenoids (serving adaptive functions (see also Rodway 1987).

4.2. Proteins and enzymes

The extraction of proteins from lichens, and the visualization of protein and enzyme banding patterns by means of isoelectric focusing on polyacrylamide gels is a procedure having great potential as a taxonomic tool, and which to date has only been used infrequently in the Northern Hemisphere (see Fahselt and Jancey 1977; Fahselt 1985; Mattson and Kärnefelt 1986; Kilias 1987) and rarely for Southern Hemisphere material Although it is assumed that the major part of the protein extracted from lichen samples is of fungal origin, and as Mattson and Kärnefelt (1986) claim "... it is not necessary to isolate the different components with respect to taxonomic research as for that the lichen thalli are studied as a single unit", Kilias (1988) has recently shown in isolates of *Trebouxia* that focusing patterns of several enzymes are species specific and provide a basis for re-evaluation of species classification in this genus.

Lichen genera having no readily demonstrable chemistry such as *Sticta* for example may have considerable biogeographical and ecological importance, and the use of protein and isoenzyme banding patterns may well be useful in suggesting or confirming taxonomic relationships. *Sticta* speciates richly in tropical areas and in Southern Hemisphere cool temperate regions it forms interesting vicariant stalked taxa. However, it has proved somewhat difficult to resolve taxonomically because of its lack of confirmatory secondary chemical characters, but protein and isoenzyme patterns may eventually help provide additional clues to taxonomic and phylogenetic relationships.

4.3. Lipids

Apart from the occurrence of higher fatty acids and their lactones in lichens, a number of which (e.g. protolichesterinic and caperatic acids) have taxonomic significance in various genera (see Culberson 1969; Mosbach 1974) little work has been done on the characterization or metabolism of lichen lipids. Lipids (including triglycerides, free fatty acids and phospholipids) are universal primary metabolites, found in both metabolic and structural systems in all living cells. Being present in some diversity in both algae (Borowitzka 1988) and fungi they should also occur widely in lichens, indeed the highly reduced nature of lipid molecules would appear to offer the lichen symbiosis considerable potential for energy storage (as a pool of highly reduced carbon) under conditions of physiological and ecological stress, a proposition which has still not been properly tested.

Oil cells (a source of highly reduced carbon and hence a potential energy supply when the reduced carbon is oxidized in respiration) were noted in cells of various crustose lichens by several workers in the late nineteenth century (Smith 1921). Zukal in 1895, for example, regarded such oil globules as reserve products, of use at times of fruit development or in periods of drought and he observed they were formed when periods of luxuriant growth alternated with periods of starvation. Fünstuck in 1896 by contrast, considered oil body formation as an excretion of waste products of metabolism, and especially of excess carbon formed by the action of lichen acids on the limestone substrate (Smith 1921).

Lipid extraction procedures and silicic acid column chromatography fractionating total lipid extracts into various classes such as neutral lipids, complex lipids and free fatty acids are nowadays simple biochemical manipulations, and sophisticated detection by gas liquid chromatography (GLC), and high-performance liquid chromatography (HPLC) allows rapid resolution of fatty acid mixtures after hydrolysis and methylation. In a preliminary series of experiments (Galloway unpublished observations) on lipid extraction and fractionation, a range of New Zealand lichens from both alpine and lowland habitats (*Alectoria nigricans, Cetraria islandica* ssp. *antarctica, Cladina confusa, Hypogymnia lugubris, Pseudocyphellaria rufovirescens, Neuropogon ciliatus, Ramalina linearis, Stereocaulon ramulosum, Thamnolia vermicularis* and *Umbilicaria polyphylla*) were examined for triglycerides, phospolipids and free fatty acids. These lipid classes were found to be present in all lichens, but in widely varying amounts, with alpine lichens having relatively high amounts of phospholipids, but low amounts of neutral lipids and free fatty acids, while lowland, rapidly growing lichens such as *Pseudocyphellaria rufovirescens* had high values for all classes of lipids. When methylated fatty acids from hydrolysed lipid fractions were examined by GLC a diverse range of fatty acids from C 8 : 0 to C 21 : 0 were detected.

Conditions favouring lipid formation in algae appear to be: desiccation, increase in inorganic ion concentration, low oxygen tension, high light intensity, nitrogen depletion. Such environmental changes influencing lipid accumulation in algae are paralleled to some extent by many lichens in their natural habitats, in periglacial environments for example. One of the most striking successes of the lichenized state is the ability for bionts to remain viable for long periods in adverse habitat conditions. Lipids may in such cases confer a fitness for survival by sustaining a measure of resistance to desiccation, and further by being among the most highly reduced of primary metabolites they offer a greater energy potential than most other storage compounds.

The role of lipids in the lichen symbiosis is still a matter for conjecture and experiment and one deserving attention, for it may have bearings both on the ecological adaptation of lichens of particular environments, as well as on problems of taxonomy and phylogeny if patterns of fatty acids from simple and complex lipids are found to be useful taxonomic characters.

REFERENCES

Adams, C.J. 1986. Geochronological studies of the Swanson Formation of Marie Byrd Land, West Antarctica, and correlation with northern Victoria Land, East Antarctica, and South Island, New Zealand. NZJ Geol. Geophys., 29, 345–358.

Ahti, T. and Kashiwadani, H. 1984. The lichen genera *Cladia, Cladina* and *Cladonia* in southern Chile. In: Inoue, H. (ed), Studies on Cryptogams in Southern Chile, pp. 125–151. Kenseisha, Tokyo.

Almborn, O. 1987. Lichens at high altitudes in southern Africa. Bibliotheca Lichenol., 25, 401–417.

Almborn, O. 1988. Some distribution patterns in the lichen flora of South Africa. Monogr. Syst. Bot. Miss. Bot. Gard., 25, 429–432.

Archer, A. W. 1985. Two new lichens: *Cladonia bimberiensis* and *C. weymouthii*. Muelleria, 6, 93–95.

Archer, A. W. 1986. The chemistry and distribution of *Cladonia capitellata* (J. D. Hook and Taylor) Church. Bab. (Lichenes) in Australia. Proc. Linn. Soc. NSW, 108, 191–194 ["1985"].

Archer, A. W. 1988. The lichen genus *Cladonia* section *Cocciferae* in Australia. Proc. Linn. Soc. NSW, 110, 205–213.

Archer, A. W. and Bartlett, J. K. 1986. New species and distributions of the lichen genus *Cladonia* in New Zealand together with a revised key. NZJ Bot., 24, 581–587.

Arvidsson, L. 1983. A monograph of the lichen genus *Coccocarpia*. Opera Bot., 67, 1–96.

Arvidsson, L. and Galloway, D. J. 1981. *Degelia*, a new lichen genus in the Pannariaceae. Lichenologist, 13, 27–50.

Babington, C. 1855. Lichenes. In: Hooker, J. D. (ed), The Botany of the Antarctic Voyage. Flora Novae Zealandiae, Part II, pp. 266–311. Reeve, London.

Babington, C. and Mitten, W. 1859. Lichenes. In: Hooker, J. D. (ed), The Botany of the Antarctic Voyage. Flora Tasmaniae, Part II, pp. 343–354. Reeve, London.

Baker, P. E., Gledhill, A., Harvey, P. K. and Hawkesworth, C. J. 1987. Geochemical evolution of the Juan Fernandez islands, SE Pacific. J. Geol. Soc. Lond., 144, 933–944.

Bishop, D. G., Bradshaw, J. D. and Landis, C. A. 1985. Provisional terrane map of South Island, New Zealand. In: Howell, D. G. (ed), Tectonostratigraphic Terranes of the Circum-Pacific Region, pp. 515–521. Circum-Pacific Council for Energy and Mineral Resources, Houston.

Bliss, L. C. 1979. Vascular plant vegetation of the southern circumpolar region in relation to Antarctic, alpine and Arctic vegetation. Can. J. Bot., 57, 2167–2178.

Borowitzka, M. A. 1988. Fats, oils and hydrocarbons. In: Borowitzka M. A. and Borowitzka, L. J. (eds), Micro-algal Biotechnology, pp. 257–287. Cambridge Univ. Press, Cambridge.

Bratt, G. C. 1976a. Lichens of south west Tasmania. Part I – Lichens of the button grass areas. Tasm. Nat., 45, 1–4.

Bratt, G. C. 1967b. Lichens of south west Tasmania. Part II, Mountain peaks and plateaux. Tasm. Nat., 46, 1–4.

Bratt, G. C. 1967c. Lichens of south west Tasmania. Part III, Forests. Tasm. Nat., 47, 1–4.

Bratt, G. C. and Cashin, J. A. 1975. Additions to the lichen flora of Tasmania I. Pap. Proc. R. Soc. Tasm., 109, 17–20.

Bratt, G. C. and Cashin, J. A. 1976. Additions to the lichen flora of Tasmania II. Pap. Proc. R. Soc. Tasm., 110, 139–147.

Briggs, J. C. 1987. Antitropical distribution and evolution in the Indo-West Pacific Ocean. Syst. Zool., 36, 237–247.

Brodo, I. M. 1986. Interpreting chemical variation in lichens for systematic purposes. Bryologist, 89, 132–138.

Brothers, R. N. and Lillie, A. R. 1988. Regional geology of New Caledonia. In: Nairn, A. E. M., Stehli, F. G. and Uyeda, S. (eds), The Ocean Basins and Margins. 7B. The Pacific Ocean, pp. 325–374. Plenum Press, New York.

Brundin, L. 1988. Phylogenetic biogeography. In: Myers, A. A. and Giller, P. S. (eds), Analytical Biogeography, pp. 343–369. Chapman and Hall, London.

Brusse, F. 1985. Heppiaceae (Lichenes) *Corynecystis*, a new lichen genus from the Karoo, South Africa. Bothallia, 15, 552–553.

Brusse, F. 1987a. Two new brown subcrustose *Parmelia* species from southern Africa, Bothallia, 17, 25–28.

Brusse, F. 1987b. A new species of *Gonohymenia* from Etosha Pan limestone. Bothallia, 17, 35–37.

Brusse, F. 1987c. A new species of *Thyrea* from Otavi dolomite (Damara System). Bothallia, 17, 37–40.

Brusse, F. 1987d. A new species of *Trapelia* (Lichenes) from southern Africa. Bothallia, 17, 187–188.

Brusse, F. 1988a. (Acarosporaceae) *Lithoglypha*, a new lichen genus from Clarens sandstone. Bothallia, 18, 89–90.

Brusse, F. 1988b. (Porpidiaceae) A new species of *Porpidia* from the Drakensberg. Bothallia, 18, 93–94.

Brusse, F. 1988c. *Schizodiscus*, a new porpidioid lichen genus from the Drakensberg. Bothallia, 18, 94–96.

Büdel, B. 1987. Zur Biologie und Systematik der Flechtengattungen *Heppia* und *Peltula* im südlichen Afrika. Bibl. Lichenol., 23, 1–105.

Close, R. C., Moar, N. T., Tomlinson, A. I. and Lowe, A. D. 1978. Aerial dispersal of biological material from Australia to New Zealand. Int. J. Biometeorol., 22, 1–19.

Connolly, J. D. and Overton, K. H. 1972. The triterpenoids. In: Newman, A. A. (ed.), Chemistry of Terpenes and Terpenoids, pp. 207–287. Academic Press, London.

Craw, R. 1985. Classic problems of Southern Hemisphere biogeography re-examined, panbiogeographic analysis of the New Zealand frog *Leiopelma*, the ratite birds and *Nothofagus*. Zeitschr. Zool. Syst. Evol., 23, 1–10.

Craw, R. 1988. Panbiogeography: method and synthesis in biogeography. In: Myers, A. A. and Giller, P. S. (eds), Analytical Biogeography, pp. 405–435. Chapman and Hall, London.

Culberson, C. F. 1969. Chemical and botanical guide to lichen products. Univ. of North Carolina Press, Chapel Hill.

Culberson, C. F. 1986. Biogenetic relationships of the lichen substances in the framework of systematics. Bryologist, 89, 91–98.

Culberson, C. F., Culberson, W. L., Gowan, S. and Johnson, A. 1987. New depsides from lichens: microchemical methodologies applied to the study of new natural products discovered in herbarium specimens. Amer. J. Bot., 74, 403–414.

Culberson, C. F., Culberson, W. L. and Johnson, A. 1988. Gene flow in lichens. Amer. J. Bot., 75, 1135–1139.

Culberson, W. L. 1986. Chemistry and sibling speciation in the lichen-forming fungi: ecological and biological considerations. Bryologist, 89, 123–131.

Czochanska, Z., Sheppard, C. M, Weston, R. J. and Woolhouse, A. D. 1987. A biological marker study of oils and sediments from the West Coast, South Island, New Zealand.

NZJ Geol. Geophys., 30, 1–17.

Dalziel, I. W. D. and Elliot, D. H. 1982. West Antarctica: problem child of Gondwanaland. Tectonics, 1, 3–19.

Dalziel, I. W. D. and Grunow, A. M. 1985. The Pacific margin of Antarctica: Terranes within terranes within terranes. In: Howell, D. G. (ed), Tectonostratigraphic Terranes of the Circum–Pacific Region, pp. 565–581. Circum–Pacific Council For Energy and Mineral Resources, Houston.

Darwin, C. 1845. Journal of researches into the natural history of the countries visited during the voyage of H.M.S. *Beagle* round the world. Murray, London.

Degelius, G. 1974. The lichen genus *Collema* with special reference to the extra-European species. Symb. Bot. Upsal., 20, 1–215.

Delise, D. F. 1825. Histoire des lichens: Genre *Sticta*. Mém. Soc. Linn. Calvados, 2, 1–167, 598–600.

Dodge, C. W. 1973. Lichen Flora of the Antarctic Continent and Adjacent Islands. Phoenix Publishing, Canaan.

Egan, R. S. 1986. Correlations and non-correlations of chemical variation patterns with lichen morphology and geography. Bryologist, 89, 99–110.

Elix, J. A. 1979. A taxonomic revision of the lichen genus *Hypogymnia* in Australasia. Brunonia, 2, 175–245.

Elix, J. A. 1982. Peculiarities of the australasian lichen flora: accessory metabolites, chemical and hybrid strains. J. Hattori Bot. Lab., 52, 407–415.

Elix, J. A. and Armstrong, P. M. 1983. Further new species of *Parmelia* subgen. *Xanthoparmelia* (lichens) from Australia and New Zealand. Aust. J. Bot., 31, 467–483.

Elix, J. A. and Child, 1987. A new species of *Chondropsis* (lichenized Ascomycotina) from Australia and New Zealand. Brunonia, 9, 113–115.

Elix, J. A. and Hale, M. E. 1987. *Canomaculina, Myelochroa, Parmelinella, Parmelinopsis* and *Parmotremopsis*, five new genera in the Parmeliaceae (lichenized Ascomycotina). Mycotaxon, 29, 233–244.

Elix, J. A. and Johnston, J. 1986. New species of *Relicina* (lichenized Ascomycotina) from Australasia. Mycotaxon, 27, 611–616.

Elix, J. A. and Johnston, J. 1987a. New species of *Parmelina* (lichenized Ascomycotina) from Australia and New Zealand. Brunonia, 9, 155–161. ["1986"].

Elix, J. A. and Johnston, J. 1987b. New species and new records of *Xanthoparmelia* (lichenized Ascomycotina) from Australia. Mycotaxon, 29, 359–372.

Elix, J. A. and Johnston, J. 1988. New species in the lichen family Parmeliaceae (Ascomycotina) from the Southern Hemisphere. Mycotaxon, 31, 491–510.

Elix, J. A., Johnston, J. and Armstrong, P. M. 1986. A revision of the lichen genus *Xanthoparmelia* in Australasia. Bull. Br. Mus. Nat. Hist. (Bot.), 15, 163–362.

Elix, J. A., Whitton, A. A. and Jones, A. J. 1982. Triterpenes from the lichen genus *Physcia*. Aust. J. Chem., 35, 641–647.

Elix, J. A., Whitton, A. A. and Sargent, M. V. 1984. Recent progress in the chemistry of lichen substances. Progr. Chem. Org. Nat. Prod., 45, 103–234.

Elliot, D. H. 1985. Physical geography-geological evolution. In: Bonner, W. N. and Walton, D. W. H. (eds), Key Environments Antarctica, pp. 39–61. Pergamon, Oxford.

Engelskjøn, T. and Jørgensen, P. M. 1986. Phytogeographical relations of the cryptogamic flora of Bouvetøya. Norsk Polarinst. Skr., 185, 71–79.

Esslinger, T. L. 1977. A chemosystematic revision of the brown Parmeliae. J. Hattori Bot. Lab., 42, 1–211.

Esslinger, T. L. 1978. A new status for the brown Parmeliae. Mycotaxon, 7, 45–54.

Esslinger, T. L. 1981. *Almbornia*, a new lichen genus from South Africa. Nord. J. Bot., 1, 125–127.

Esslinger, T. L. 1986a. Further reports on the brown Parmeliaceae of southern Africa. Nord. J. Bot., 6, 87–91.

Esslinger, T. L. 1986b. Notes on the chemistry and distribution of selected Parmeliaceae (Lichens) from the Southern Hemisphere. Bryologist, 89, 296–299.

Fahselt, D. 1985. Multiple enzyme forms in lichens. In: Brown, D. H. (ed), Lichen Physiology and Cell Biology, pp. 129–143. Plenum Press, New York.

Fahselt, D. A. and Jancey, R. C. 1977. Polyacrylamide gel electrophoresis of protein extracts from members of the *Parmelia perforata* complex. Bryologist, 80, 429–438.

Fife, A. J. 1986. The phytogeographic affinities of the alpine mosses of New Zealand. In: Barlow, B. A. (ed), Flora and Fauna of Alpine Australasia, pp. 337–355. CSIRO, Melbourne.

Filson, R. B. 1976. Australian lichenology: a brief history. Muelleria, 3, 183–190.

Filson, R. B. 1981a. Studies on Macquarie Island lichens 1: general. Muelleria, 4, 305–316.

Filson, R. B. 1981b. Studies on Macquarie Island lichens 2: the genera *Hypogymnia, Menegazzia, Parmelia* and *Pseudocyphellaria*. Muelleria, 4, 317–331.

Filson, R. B. 1981c. A revision of the lichen genus *Cladia* Nyl. J. Hattori. Bot. Lab., 49, 1–75.

Filson, R. B. 1982. A contribution on the genus *Parmelia* (lichens) in southern Australia. Aust. J. Bot., 30, 511–582.

Filson, R. B. 1986a. Index to type specimens of Australian lichens: 1800–1984. Bureau of Flora and Fauna, Canberra.

Filson, R. B. 1986b. Lichens of Macquarie Island 3: the genus *Sphaerophorus* Muelleria, 6, 169–172.

Filson, R. B. 1988a. The lichen genera *Heppia* and *Peltula* in Australia. Muelleria, 6, 495–517.

Filson, R. B. 1988b. Checklist of Australian lichens (3rd edn). National Herbarium of Victoria, Melbourne.

Filson, R. B. and Archer, A. W. 1986. Studies on Macquarie Island lichens 4: the genera *Cladia* and *Cladonia*. Muelleria, 6, 217–235.

Fineran, B. A. 1971. A catalogue of the bryophytes, lichens, and fungi collected on the Auckland Islands. J. R. Soc. NZ, 1, 215–229.

Fleming, C. A. 1987. Comments on Udvardy's biogeographical realm of Antarctica. J. R. Soc. NZ, 17, 195–200.

Francis, J. E. 1986. Growth rings in Cretaceous and Tertiary wood from Antarctica and their palaeoclimatic implications. Palaeontology, 29, 665–684.

French, D. D. and Smith, V. R. 1985. A comparison between Northern and Southern Hemisphere tundras and related ecosystems. Polar Biol., 5, 5–21.

Galloway, D. J. 1977. Additional notes on the lichen genus *Cladia* Nyl., in New Zealand. Nova Hedwigia, 28, 475–486.

Galloway, D. J. 1978. *Anzia* and *Pannoparmelia* (Lichenes) in New Zealand. NZJ Bot., 16, 261–270.

Galloway, D.J. 1979. Biogeographical elements in the New Zealand lichen flora. In: Bramwell, D. (ed.), Plants and Islands, pp. 201-224. Academic Press, London.

Galloway, D.J. 1980. The lichen genera *Argopsis* and *Stereocaulon* in New Zealand. Bot. Not., 133, 261–279.

Galloway, D.J. 1981a. The lichen genus *Lobaria* (Schreber) Hoffm., in New Zealand. Nova Hedwigia, 24, 317–331.

Galloway, D.J. 1981b. *Xanthoparmelia* and *Chondropsis* (Lichenes) in New Zealand. NZJ Bot., 18, 525–552 ["1980"].

Galloway, D.J. 1981c. Erik Acharius, Olof Swartz and the evolution of generic concepts in lichenology. In: Wheeler, A. and Price, J.H. (eds), History in the Service of Systematics. Soc. Bibl. Nat. Hist. Spec. Pub., 1, 119–127.

Galloway, D.J. 1985a. Flora of New Zealand Lichens. Government Printer, Wellington.

Galloway, D.J. 1985b. Lichenology in the South Pacific, 1790–1840. In: Wheeler, A. and Price, J.H. (eds), From Linnaeus to Darwin: commentaries on the history of biology and geology, pp. 205–214. Soc. Hist. Nat. Hist., London.

Galloway, D.J. 1986. Non-glabrous species of *Pseudocyphellaria* from southern South America. Lichenologist, 18, 105–168.

Galloway, D.J. 1987. Austral lichen genera: some biogeographical problems. Biblthca. Lichenol., 25, 385–399.

Galloway, D.J. 1988a. Plate tectonics and the distribution of cool temperate Southern Hemisphere macrolichens. Bot. J. Linn. Soc., 96, 45–55.

Galloway, D.J. 1988b. Studies in *Pseudocyphellaria* (lichens) I. The New Zealand species. Bull. Br. Mus. Nat. Hist. (Bot.) 17, 1-267.

Galloway, D.J. 1988c. La Billardière's Tasmanian lichens. Pap. Proc. R. Soc. Tasm., 122, 97–108.

Galloway, D.J. and Bartlett, J.K. 1983. The lichen genus *Thysanothecium* Mont. and Berk., in New Zealand. Nova Hedwigia, 36, 381–398 ["1982"].

Galloway, D.J. and Bartlett, J.K. 1986. *Arthrorhaphis* Th.Fr. (lichenised Ascomycotina) in New Zealand. NZJ Bot., 24, 393–402.

Galloway, D.J. and Elix, J.A. 1981. *Knightiella* Müll. Arg., a monotypic lichen genus from Australasia. NZJ Bot., 18, 481–486 ["1980"].

Galloway, D.J. and Elix, J.A. 1983. The lichen genera *Parmelia* Ach., and *Punctelia* Krog, in Australasia. NZJ Bot., 21, 397–420.

Galloway, D.J. and Elix, J.A. 1984. Additional notes on *Parmelia* and *Punctelia* (lichenised Ascomycotina) in Australasia. NZJ Bot., 22, 441–445.

Galloway, D.J. and Groves, E.W. 1987. Archibald Menzies MD, FLS (1754–1842), aspects of his life, travels and collections. Archs. Nat. Hist. 14, 3–43.

Galloway, D.J. and Guzman, G. 1988. A new species of *Phlyctis* from Chile. Lichenologist, 20, 393–397.

Galloway, D.J. and James, P.W. 1985. The lichen genus *Psoromidium* Stirton. Lichenologist, 17, 173–188.

Galloway, D.J. and James P.W. 1987. *Metus*, a new austral lichen genus and notes on an australasian species of *Pycnothelia*. Notes R. Bot. Gard. Edin., 44, 561–579.

Galloway, D.J., James, P.W. and Wilkins, A.L. 1983. Further nomenclatural and chemical notes on *Pseudocyphellaria* in New Zealand. Lichenologist, 15, 135–145.

Galloway, D.J., and Jørgensen, P.M. 1987. Studies in the lichen family Pannariaceae II. The genus *Leioderma* Nyl. Lichenologist, 19, 345–400.

Glasby, G.P. 1985. The future development of oceanography in New Zealand: scientific, potential uses and international comparisons. Tuatara, 28, 14–42.

Godley, E.J. 1960. The botany of southern Chile in relation to New Zealand and the subantarctic. Proc. R. Soc. B., 152, 457–475.

Goodwin, T.W. and Mercer, E.I. 1983. Introduction to Plant Biochemistry, (2nd edn)., Pergamon, Oxford.

Gradstein, S.R., Pócs, T. and Vána, J. 1983. Disjunct Hepaticae in tropical America and Africa. Acta Bot. Hung., 29, 127–171.

Hale, M.E. 1975. A monograph of the lichen genus *Relicina* (Parmeliaceae). Smiths. Contr. Bot., 26, 1–32.

Hale, M.E. 1976. A monograph of the lichen genus *Bulbothrix* Hale (Parmeliaceae). Smiths. Contr. Bot. 32, 1–29.

Hale, M.E. 1985. *Xanthomaculina* Hale, a new lichen genus in the Parmeliaceae (Ascomycotina). Lichenologist, 17, 255–265.

Hale, M.E. 1987. A monograph of the lichen genus *Parmelia* Acharius sensu stricto (Ascomycotina : Parmeliaceae). Smiths. Contr. Bot., 66, 1–55.

Hall, K. 1987. Periglacial landforms and processes of the subantarctic and antarctic islands. Palaeoecology of Africa and the Surrounding Islands, 18, 383–392.

Harborne, J.B. and Turner, B.L. 1984. Plant Chemosystematics. Academic Press, London.

Hawksworth, D.L. 1988. Coevolution of fungi with algae and cyanobacteria in lichen symbioses. In: Pirozynski, K.A. and Hawksworth, D.L. (eds), Coevolution of Fungi with Plants and Animals, pp. 125–148. Academic Press, London.

Henssen, A. 1984. *Placynthium arachnoideum*, a new lichen from Patagonia, and notes on other species of the genus in the Southern Hemisphere. Lichenologist, 16, 265–271.

Henssen, A. and Büdel, B. 1984. *Phylisciella*, a new genus of the Lichinaceae. Beih. Nova Hedwigia, 79, 381–398.

Henssen, A.M. and James, P.W. 1982. The lichen genus *Steinera*. Bull. Br. Mus. Nat. Hist. (Bot.), 10, 227–256.

Henssen, A. and Kantvilas, G. 1985. *Wawea fruticulosa*, a new genus and species from the Southern Hemisphere. Lichenologist, 17, 85–97.

Henssen, A., Vobis, G. and Renner, B. 1982. New species of *Roccellinastrum* with an emendation of the genus. Nord. J. Bot., 2, 587–599.

Hertel, H. 1984. Über saxicole, lecideoide Flechten der Subantarktis, Beih. Nova Hedwigia, 79, 399–499.

Hertel, H. 1985. New, or little-known New Zealand lecideoid lichens. Mitt. Bot. Münch., 21, 301–337.

Hertel, H. 1987a. Progress and problems in taxonomy of antarctic saxicolous lecideoid lichens. Biblthca. Lichenol., 25, 219–242.

Hertel, H. 1987b. Bemerkenswerte Funde sudhemisphärischer, saxicoler Arten der Sammelgattung *Lecidea*. Mitt. Bot. Münch., 23, 321–340.

Hervé, F. 1988. Late Paleozoic subduction and accretion in southern Chile. Episodes, 11, 183–188.

Hoffmann, G.F. 1801. Descriptio et adumbratio plantarum e

classe cryptogamica Linnaei quae lichenes dicuntur. III. Lipsiae.

Holloway, J. D. 1986. Origin of lepidopteran faunas in high mountains of the Indo-Australian tropics. In: Vuilleumier, F. and Monasterio, M. (eds). High-Altitude Tropical Biogeography, pp. 533–556. Oxford Univ. Press, Oxford.

Holtan-Hartwig, J. 1988. Two new species of *Peltigera*. Lichenologist, 20, 11–17.

Hooker, J. D. 1847. Lichenes. In: Hooker J. D. (ed.), Botany of the Antarctic Voyage. I. Flora Antarctica, Part II, pp. 519–542, 547. Reeve, London.

Hooker, J. D. 1853. Introductory essay to the Flora of New Zealand. Flora Novae Zelandiae, pp.i–xxxix. Reeve, London.

Hooker, J. D. and Taylor, T. 1844. Lichenes antarctici.... Hook. Lond. J. Bot., 3, 634–658.

Hooker, W. J. 1842. On *Cenomyce retipora*. Hook. Lond. J. Bot., 1, 292–294.

Howell, D. G., Jones, D. L. and Schermer, E. R. 1985. Tectonostratigraphic terranes of the circum-Pacific region. In: Howell, D. G. (ed.), Tectonostratigraphic Terranes of the Circum-Pacific Region, pp. 3–30. Circum-Pacific Council for Energy and Mineral Resources, Houston.

Humphries, C. J. and Parenti, L. R. 1986. Cladistic Biogeography. Clarendon Press, Oxford.

Humphries, C. J., Ladiges, P. Y., Roos, M. and Zandee, M. 1988. Cladistic biogeography. In: Myers, A. A. and Giller, P. S. (eds), Analytical Biogeography, pp. 371–404. Chapman and Hall, London.

Huneck, S., Sainsbury, M., Rickard, T. M. A. and Smith, R. I. Lewis 1984. Ecological and chemical investigations of lichens from South Georgia and the maritime antarctic. J. Hattori Bot. Lab., 56, 461–480.

Jørgensen, P. M. 1977. Foliose and fruticose lichens from Tristan da Cunha. Norske Vidensk. -Akad. I. Mat. Naturvitensk. Kl. Skr. II, 36, 1–40.

Jørgensen, P. M. The phytogeographical relationships of the lichen flora of Tristan da Cunha (excluding Gough Island) Can. J. Bot., 57, 2279–2282.

Jørgensen, P. M. 1983. Distribution patterns of lichens in the Pacific region. Aust. J. Bot. Suppl. 10, 43–66.

Jørgensen, P. M. 1986. Macrolichens of Bouvetøya. Norsk Polarinst. Skr., 185, 23–34.

Jørgensen, P. M. and Galloway, D. J. 1983. *Bryoria* (lichenised Ascomycotina) in New Zealand. NZJ Bot., 21, 335–340.

Jørgensen, P. M. and Galloway, D. J. 1984. A new subantarctic species of the lichen genus *Peltularia*. Lichenologist, 16, 189–196.

James, P. W. and White, F. J. 1987. Studies on the genus *Nephroma* I. The European and macaronesian species. Lichenologist, 19, 215–268.

Kantvilas, G. 1983. A brief history of lichenology in Tasmania. Pap. Proc. R. Soc. Tasm. 117, 41–51.

Kantvilas, G. 1987. *Siphula jamesii*, a new lichen from southwestern Tasmania. Nord. J. Bot., 7, 585–588.

Kantvilas, G. 1988. Tasmanian rainforest lichen communities: a preliminary classification. Phytocoenol, 16, 391–428.

Kantvilas, G. and Elix, J. A. 1987. A new species of *Cladia* (lichenized Ascomycotina) from Tasmania. Mycotaxon, 29, 199–205.

Kantvilas, G. and James, P. W. 1987. The macrolichens of Tasmanian rainforest: key and notes. Lichenologist, 19, 1–28.

Kantvilas, G., James, P. W. and Jarman, S. J. 1985. Macrolichens in Tasmanian rainforests. Lichenologist, 17, 67–83.

Kappen, L. 1988. Ecophysiological relationships in different climatic regions. In: Galun, M. (ed), CRC Handbook of Lichenology, Vol. III, pp. 37–100. Boca Raton.

Kappen, L. and Friedmann, E. I. 1983. Ecophysiology of lichens in the dry valleys of southern Victoria Land, Antarctica, II: CO2 gas exchange in cryptoendolithic lichens. Polar Biol., 1, 227–232.

Kärnefelt, I. 1979. The brown fruticose species of *Cetraria*. Opera Bot., 46, 1–150.

Kärnefelt, I. 1986. The genera *Bryocaulon, Coelocaulon* and *Cornicularia* and formerly associated taxa. Opera Bot., 86, 1–90.

Kärnefelt, I. 1987. *Cetraria* (Parmeliaceae) and some related genera on the African continent. Bothallia, 17, 45–49.

Kilias, H. 1987. Protein characters as a taxonomic tool in lichen systematics. Biblthca Lichenol., 25, 445–455.

Kilias, H. 1988. Isoenzyme patterns as a tool in taxonomy of chlorococcal algae. Electrophoresis, 9, 613–617.

Kirkpatrick, J. B. 1986. Tasmanian alpine biogeography and ecology and interpretations of the past. In: Barlow, B. A. (ed), Flora and Fauna of Alpine Australasia, pp. 229–242. CSIRO, Melbourne.

Knox, M. D. E. and Brusse, F. A. 1983. New *Xanthoparmeliae* (Lichenes) from southern and central Africa. J. S. Afr. Bot., 49, 143–159.

Korsch, R. J. and Wellman, H. W. 1988. The geological evolution of New Zealand and the New Zealand region. In: Nairn, A. E. M., Stehli, F. G. and Uyeda, S. (eds), The Ocean Basins and Margins, 7B: The Pacific Ocean, pp. 411–482. Plenum Press, New York.

Krog, H. 1976. *Lethariella* and *Protousnea*, two new lichen genera in Parmeliaceae. Norw. J. Bot., 23, 83–106.

Lamb, I. M. 1959. La vegetacion liquenica de los Parques Nacionales Patagonicos (Nahuel Huapi, Los Alerces, Lanin). An. Parque Na. B. Aires, 7, 1–188.

Lamb, I. M. 1964. Antarctic lichens I. The genera *Usnea, Ramalina, Himantormia, Alectoria, Cornicularia*. Scient. Rep. Br. Antarct. Surv., 38, 1–34.

Lamb, I. M. 1974. The lichen genus *Argopsis* Th.Fr. J. Hattori Bot. Lab., 38, 447–462.

Lamb, I. M. 1977. A conspectus of the lichen genus *Stereocaulon* (Schreb.) Hoffm. J. Hattori Bot. Lab., 43, 191–355.

Lawrey, J. D. 1984. Biology of Lichenized Fungi. Praeger Scientific, New York.

Lawver, L. A. and Scotese, C. R. 1987. A revised reconstruction of Gondwanaland. In: McKenzie, G. D. (ed), Gondwana Six: structure, tectonics and geophysics. Geophys. Monogr., 40, 17–23.

Lindsay, D. C. 1971. Vegetation of the South Shetland Islands. Br. Antarct. Surv. Bull., 25, 59–83.

Lindsay, D. C. 1975. The macrolichens of South Georgia. Br. Antarct. Surv. Sci. Rep., 89, 1–91.

Lindsay, D. C. 1977a. The lichens of Marion and Prince Edward Islands southern Indian Ocean. Nova Hedwigia, 28, 667–689.

Lindsay, D. C. 1977b. Lichens of cold deserts. In: Seaward, M. R. D. (ed), Lichen Ecology, pp. 183–209. Academic Press, London.

Longton R. E. 1985. Terrestrial habitats – vegetation. In: Bonner, W. N. and Walton, D. W. H. (eds), Key Environments Antarctica, pp. 73–105. Pergamon, Oxford.

Longton R. E. 1988. The biology of polar bryophytes and lichens. Cambridge Univ. Press, Cambridge.

Lumbsch, H. T. 1987. Eine neue Subspecies in der Flechtengattung *Diploschistes* aus der Sudhemisphäre. Herzogia, 7, 601–608.

Mattson, J. E. and Kärnefelt, I. 1986. Protein banding patterns in the *Ramalina siliquosa* group. Lichenologist, 18, 231–240.

Mayer, A. 1989. Plant–fungal interactions: a plant physiologist's viewpoint. Phytochemistry, 28, 311–317.

Mildenhall, D. C. 1980. New Zealand and late Cretaceous and Cenozoic plant biogeography: a contribution. Palaeogeogr. Palaeoclimatol. Palaeocol., 31, 197–232.

Mosbach, K. 1974. Biosynthesis of lichen substances. In: Ahmadjian, V. and Hale, M. E. (eds), The Lichens, pp. 523–546. Academic Press, New York.

Ollier, C. D. 1986. The origin of alpine landforms in Australasia. In: Barlow, B. A. (ed), Flora and Fauna of Alpine Australasia, pp. 3–26. CSIRO, Melbourne.

Ourisson, G., Rohmer, M. and Anton, R. 1979. From terpenes to sterols: macroevolution and microevolution. Rec. Adv. Phytochem., 13, 131–162.

Øvstedal, D. O. 1986. Crustose lichens of Bouvetøya. Norsk Polarinst. Skr., 185, 35–56.

Øvstedal, D. O. and Hawksworth, D. L. 1986. Lichenicolous ascomycetes from Bouvetøya. Norsk Polarinst. Skr., 185, 57–60.

Ramos, V. A. 1988. Late Proterozoic–early Paleozoic of South America – a collisional history. Episodes, 11, 168–174.

Ramsay, H. P., Streimann, H., Ratkowsky, A. V., Seppelt, R. and Fife, A. J. 1986. Australasian alpine bryophytes. In: Barlow, B. A. (ed), Flora and Fauna of Alpine Australasia, pp. 301–335. CSIRO, Melbourne.

Redon, J. 1972. Liquenes del Parque Nacional "Vicente Perez Rosales", provincia de Llanquihue, Chile. Anal. Mus. Hist. Nat. Valpso, 5, 117–126.

Redon, J. 1973 Beobachtungen zur Geographie und Ökologie der chilenischen Flechtenflora. J. Hattori Bot. Lab., 37, 153–167.

Redon, J. 1974. Observaciones sistematicas y ecologicas en liquenes del Parque Nacional "Vicente Perez Rosales". Anal. Mus. Hist. Nat. Valpso, 7, 169–225.

Redon, J. 1976. Fitogeographia de los liquenes chilenos. Anal. Mus. Hist. Nat. Valpso, 9, 7–22.

Redon J. and Lange, O. L. 1983. Epiphytische Flechten in Bereich einer chilenischen "Nebeloase" (Fray Jorge) I. Vegetationskundliche Gliederung und Standortsbedingungen. Flora, 174, 213–243.

Redon, J. and Quilhot, W. 1977. Los liquenes de las islas de Juan Fernandez 1: estudio preliminar. Anal. Mus. Hist. Nat. Valpso, 10, 15–26.

Redon, J. 1985. Liquenes antarcticos. INACH, Santiago.

Robin, G. de Q. 1988. The Antarctic ice sheet, its history and response to sea level and climatic changes over the past 100 million years. Palaeogeogr. Palaeoclimatol. Palaeoecol., 67, 31–50.

Rogers, R. W. 1986. The genus *Pyxine* (Physciaceae, lichenized Ascomycotina) in Australia. Aust. J. Bot., 34, 131–154.

Rogers, R. W. and Bartlett, J. K. 1986. The lichen genus *Haematomma* in New Zealand. Lichenologist, 18, 247–255.

Rogers, R. W. and Hafellner, J. 1987. *Sagenidiopsis*, a new genus of byssoid lichenized fungi. Lichenologist, 19, 401–408.

Rogers, R. W. and Stevens, G. N. 1981. Lichens. In: Keast, A. (ed), Ecological Biogeography of Australia, pp. 593–603. Junk, The Hague.

Rodman, J. E. 1987. Compound co-occurrence and biosynthetic inference. Biochem. Syst. Ecol., 15, 365–372.

Rosen, B. R. 1988. Biogeographic patterns: a perceptual overview. In: Myers, A. A. and Giller, P. S. (eds), Analytical Biogeography, pp. 23–55. Chapman and Hall, London.

Rundel, P. 1978. The ecological role of secondary lichen substances. Biochem. Syst. Ecol., 6, 157–170.

Seberg, O. 1986. A critique of the theory and methods of panbiogeography. Syst. Zool., 35, 369–380.

Seberg, O. 1988. Taxonomy, phylogeny, and biogeography of the genus *Oreobolus* R.Br. (Cyperaceae), with comments on the biogeography of the South Pacific continents. Bot. J. Linn. Soc., 96, 119–195.

Sipman, H. 1983. A monograph of the lichen family Megalosporaceae. Biblthca Lichenol., 18, 1–241.

Sipman, H. 1986. Additional notes on the lichen family Megalosporaceae. Willdenowia, 15, 557–564.

Smith, A. L. 1921. Lichens. Cambridge Univ. Press, Cambridge.

Smith, J. M. B. 1986a. Origins and history of the Malesian high mountain flora. In: Vuilleumier, F. and Monasterio M. (eds), High-Altitude Tropical Biogeography, pp. 469–477. Oxford Univ. Press, Oxford.

Smith, J. M. B. 1986b. Origins of Australasian tropicalpine and alpine floras. In: Barlow, B. A. (ed), Flora and Fauna of Alpine Australasia, pp. 109–128. CSIRO, Melbourne.

Smith, R. I. Lewis 1988. Classification and ordination of cryptogamic communities in Wilkes Land, continental Antarctica. Vegetatio, 76, 155–166.

Stevens, G. 1985. Lands in Collision. Discovering New Zealand's past geography. SIPC, Wellington.

Stevens, G. N. 1987. The lichen genus *Ramalina* in Australia. Bull. Br. Mus. Nat. Hist. (Bot.), 16, 107–223.

Swartz, O. P. 1781. Methodus muscorum illustrata. Edman, Uppsaliae.

Takhtajan, A. 1986. Floristic Regions of the World. Univ. of California Press, Berkeley.

Taylor, T. and Hooker, J. D. 1845. Lichenes. In: Hooker, J. D. (ed), The Botany of the Antarctic Voyage. Flora Antarctica, I, pp. 194–200. Reeve, London.

Tehler, A. 1983. The genera *Dirina* and *Roccellina* (Roccellaceae). Opera Bot. 70, 1–86.

Thorne, R. F. 1986. Antarctic elements in Australasian rainforests. Telopea, 2, 611–617.

Tibell, L. 1984. A reappraisal of the taxonomy of Caliciales. Beih. Nova Hedwigia, 79, 597–713.

Tibell, L. 1987. Australasian Caliciales. Symb. Bot. Upsal., 27, 1–279.

Tønsberg, T. and Holtan-Hartwig, J. 1983. Phycotype pairs in *Nephroma, Peltigera* and *Lobaria* in Norway. Nord. J. Bot., 3, 681–688.

Troll, C. 1973. The upper timberlines in different climatic zones. Arctic and Alpine Research, 5, A3–18.

Udvardy, M. D. F. 1987. The biogeographical realm Antarctica: a proposal. J. R. Soc. NZ, 17, 187–194.

Veevers, J. J. 1987. Earth history of the Southeast Indian Ocean and the conjugate margins of Australia and Antarctica. J. Proc. R. Soc. NSW, 120, 57–70.

Verdon, D. and Elix, J. A. 1986. *Myelorrhiza*, a new Australian lichen genus from North Queensland. Brunonia, 9, 193–214.

Vezda, A. and Kantvilas, G. 1988. *Sarrameana tasmanica*, a new Tasmanian lichen. Lichenologist, 20, 179–182.

Vitikainen, O. 1985. Three new species of *Peltigera* (lichenized Ascomycetes). Annls. Bot. Fenn., 22, 291–298.

Walker, F. J. 1985. The lichen genus *Usnea* subgenus *Neuropogon*. Bull. Br. Mus. Nat. Hist. (Bot.), 13, 1–130.

Wardle, P. 1978. Origin of the New Zealand mountain flora, with special reference to trans-Tasman relationships. NZJ Bot., 16, 535–550.

Waterhouse, J. B. and Sivell, W. J. 1987. Permian evidence for trans-Tasman relationships between East Australia, New Caledonia and New Zealand. Tectonophysics, 142, 227–240.

White, F. J. and James, P. W. 1988. Studies on the genus *Nephroma* II. The southern temperate species. Lichenologist, 20, 103–166.

Wilkins, A. L. and James, P. W. 1979. The chemistry of *Pseudocyphellaria impressa* s.lat., in New Zealand. Lichenologist, 11, 271–281.

Williamson, M. 1984. Sir Joseph Hooker's lecture on insular floras. Biol. J. Linn. Soc., 22, 55–77.

Yoshimura, Y. 1987. Taxonomy and speciation of *Anzia* and *Pannoparmelia*. Biblthca Lichenol., 25, 185–195.

Zahlbruckner, A. 1924. Die Flechten der Juan Fernandez–Inseln. In: Skottsberg, C. (ed.), The Natural History of Juan Fernandez and Easter Island, 2, pp. 315–408. Almqvist and Wiksells, Uppsala.

9. VICARIANCE AND CLINAL VARIATION IN SYNANTHROPIC VEGETATION

LADISLAV MUCINA

Department of Vegetation Ecology and Nature Conservancy, Institute of Plant Physiology, University of Vienna, Althanstrasse 14, A-1091 Wien, Austria

1. INTRODUCTION

Distribution of vegetation types is a subject of synchorology (Braun-Blanquet 1921, 1964; Westhoff and van der Maarel 1978; Géhu and Rivas-Martínez 1981), which is an interface between vegetation science and biogeography. Synanthropic vegetation covers extensive areas of disturbed habitats. It consists of weeds with prevailingly ruderal life-strategy (sensu Grime 1979). Several types of synanthropic vegetation can be recognized. In this paper, however, only two basic ones, the ruderal and agrestal (segetal) vegetation (Holzner 1978), will be paid particular attention. The synchorology of synanthropic vegetation offers special challenges for biogeographic discussions since weeds are considered as "botanical birds", mainly because of their great dispersal ability (Baker 1965) paralleled to the vagility of birds. The "weedy" character is thus expressed in their (sub)cosmopolitan distribution (Jehlík 1986). Accordingly, the weed communities should possess extensive distribution areas which are synecologically and historically difficult to interprete.

In order to tackle the basic question of the distribution patterns and variability within the synanthropic vegetation on a large (geographic), scale, the problems of geographic vicariance of vegetation types (termed also synvicariance) of the European synanthropic vegetation as well as clinal character of the intra-unit (within-unit) variation were reviewed.

In the sequel the classification and nomenclatural conventions of the Braun-Blanquet floristic-sociological approach (Braun-Blanquet 1964; Westhoff and van der Maarel 1978; Barkman et al. 1986) are followed as most of the reviewed material comes from countries dominated by that approach (West Germany, Czechoslovakia, France, Spain, Italy, etc.). The formal classification hierarchy of the approach recognizes association as the basic vegetation category, while alliance, order, class, and division (as well as association group, suballiance, suborder, subclass) are considered the high-ranked (superordered syntaxa). Subassociation and several other vegetation categories (variant, facies, "Ausbildungsform", "Ausprägung", race group, race, subrace, form) serve to describe the within-association variability. For definition of the latter units consult Gutte (1972), Moravec (1975) and Géhu and Rivas-Martínez (1981).

2. PHYTOGEOGRAPHIC AND SYNTAXONOMIC BACKGROUND

There are two main horizontal phytogeographic gradients in Europe. That one following the north–south direction is rather steep (in term of species turnover) and spans different floristic regions, such as the cold Boreal region, Eurosiberian region, and the Mediterranean region. The another gradient, running west–east, is less steep, thus the limits between the phytogeographic units it spans are of lesser biogeographic importance (hierarchically subordered to those between the floristic regions). This gradient includes the precipitation-rich Atlantic domain,

P. L. Nimis and T. J. Crovello (eds): Quantitative Approaches to Phytogeography, pp. 263–276.

the central European domain and finds its another pole in Pontic-Pannonian domain (for the extent of these phytogeographic units consult Peinado Lorca and Rivas-Martínez 1987) covering the (sub)continental territories of east and southeast Europe.

The limits of phytogeographic units are at the same time the distribution units of many species, and mark the transition zones between various high-ranked vegetation units and their complexes.

The Mediterranean is species rich and it is also more diversified in terms of number of vegetation units (Rivas-Martínez 1976; Pignatti and Pignatti 1984). The Eurosiberian region, although covering most of the European continent, is less rich and less diversified. The agrestal vegetation of this regions includes four orders of which the *Aperetalia* and *Chenopodietalia albi* found their distribution optima in that region. The *Secalietalia* and *Eragrostietalia* are better developed in the Mediterranean. The (sub)thermophilous vegetation of the *Artemisietea* s.str. includes the temperate *Artemisietalia vulgaris* and *Onopordetalia acanthii*. Analogous vegetation in the Mediterranean (*Scolymo-Onopordetalia nervosi*) should be classified in a separate class, the *Onopordetea nervosi* which is limited to Mediterranean distribution area. The hemicryptophyte ruderal vegetation of the *Galio-Urticetea* is almost exclusively limited to the Eurosiberian and Boreal regions, while the annual ruderal vegetation types (incl. *Brometalia rubenti-tectori, Chenopodietalia muralis*) are very abundant in the Mediterranean. The therophyte trampled vegetation of Europe belongs to one class and one order (*Polygono-Poetea, Poo-Polygonetalia*), but again, the highest diversity of types is found in the Mediterranean (Rivas-Martínez 1975).

3. GENERAL REMARKS ON VICARIANCE AND CLINAL VARIATION OF SYNTAXA

In phytogeography and idiotaxonomy the vicariant relation between taxa implies (e.g. Holub and Jirásek 1967): (a) that the taxa share a common ancestor; (b) that the taxa occupy exclusive biogeographically defined regions (thus geographic isolation and allopatry lead to the formation of the taxa).

Vicariance in vegetation science (e.g. Westhoff and van der Maarel 1978; also called synvicariance by Géhu and Rivas-Martínez 1981) is, to some extent, an analogy.

Two syntaxa are vicariants (or vicariads) provided that: (1) they are synsystematically related (very similar in such a manner that they share the common next super-ordered syntaxon); (2) their diagnostic taxa share similarities in growth and life form, and thus show common demographic and population-dynamic (life-history) traits; (3) the considered communities occupy habitats with a similar ecologic regime (characterized by similar set of ecologic factors); and (4) their distribution areas are exclusive, biogeographically (bioclimatically) defined regions.

When distribution areas of the considered syntaxa produce a transition zone in regions where they overlap, the syntaxa are considered as quasi-vicariants. Their centra of diversification (compare with genetic diversification centra, e.g. Zohary 1970) are clearly defined and separated.

The duality of a continuous versus discontinuous character of vegetation is reflected very well in the coupled concepts of vicariance and cline. Whereas the vicariance stands for discontinuous variation, the cline is the expression of continuity.

The term cline was coined in idiotaxonomy and plant geography and means a character gradient – in other words, a gradual change in character (trait) as you travel thought the geographic range of a species, rather than an abrupt change at a certain point where one form gives way to another (e.g. Stace 1980). This concept was taken over by vegetation scientists in a slightly altered form; it was introduced by Westhoff (1947) and became widely accepted as a merit of a review paper by Whittaker (1960). Coenocline is defined as a gradient of vegetation structure; in this particular case, studied along a wide geographic range with changing climate.

4. VICARIANCE AND CLINAL PHENOMENA ON THE ASSOCIATION LEVEL

Extensive distribution areas of several agrestal and ruderal communities motivated several authors to study their intra-unit variability. Synchorologic criteria were incorporated into syntaxonomic procedures especially by Passarge (1964, 1979, 1985). This author ranked as associations many of the vegetation units he recognized on a phytogeographic basis.

It is still rather problematic to decide if a unit detected by geographic analysis of a low-ranked syntaxon should be considered a regional association or a geographic race of broadly conceived "Hauptassoziation" (for an exhaustive discussion see Oberdorfer 1968; Schwickerath 1968; Moravec 1975; Westhoff and van der Maarel 1978). This is just because of lack of proper functional definitions of the respective concepts. The example of the *Sambucus ebulus* communities from Europe documents this problem. Brandes (1982) found three subunits well defined in synchorologic terms, which he also called "races" within the broadly conceived *Sambucetum ebuli*. The *Heracleum* race was distributed in western Europe, the *Carduus acanthoides* race in eastern Europe, and the *Ballota alba* race was found abundantly in Submediterranean regions (Fig. 1). Later the *Heracleum* race was upranked to association (Brandes 1985). Also, this race is sometimes considered an association, the *Urtico-Sambucetum*, by some authors (e.g. Braun-Blanquet 1967).

A geographical race is defined as a unit within an association which is segregated from others by a group of geographically interpretable differential taxa, and does not possess its own character species. A study of the *Lamio–Conietum (Arction lappae)* by Mucina (1991) documents another

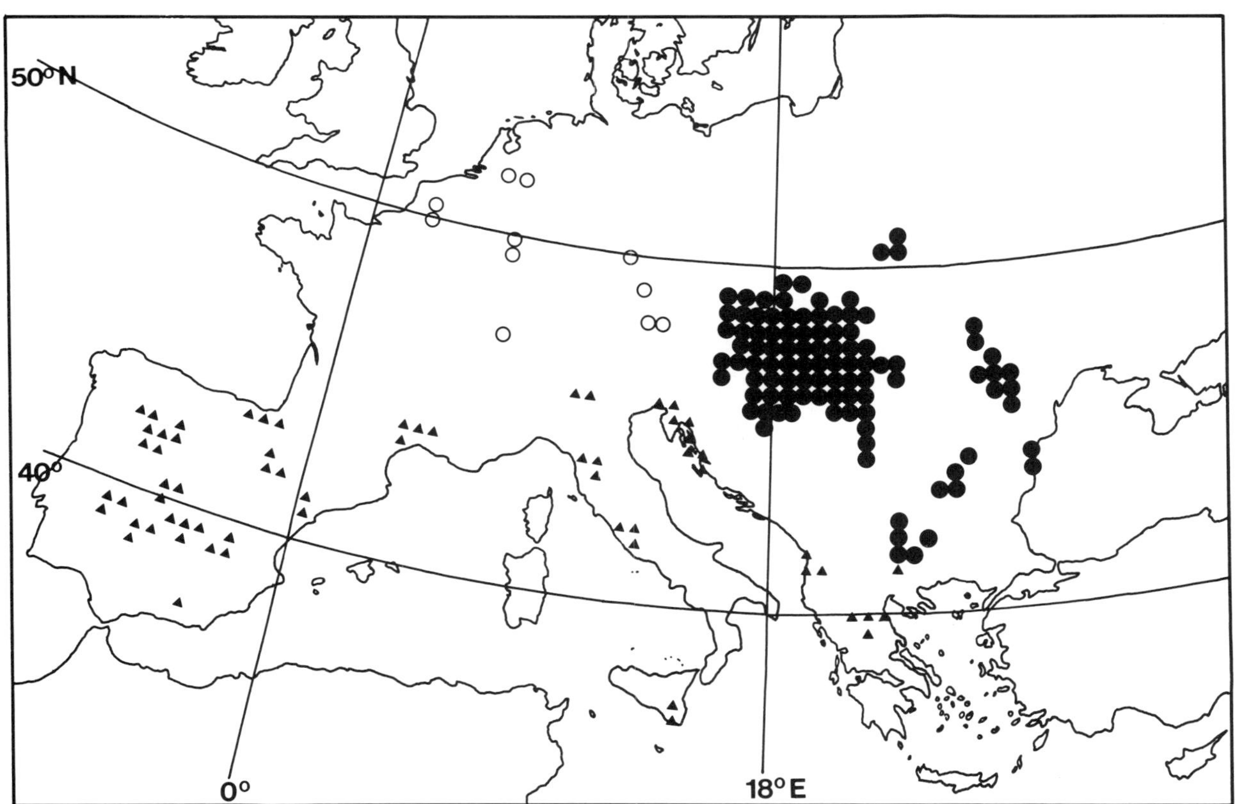

Fig. 1. Distribution of the *Sambucus ebulus* communities in Europe (compiled after various sources). Legend: filled circles – *Sambucetum ebuli* s.str. (*Carduus acanthoides* race in Brandes, 1982); empty circles – *Heracleo-Sambucetum* (*Heracleum sphondylium* race); triangles – *Urtico-Sambucetum* (*Ballota alba* race).

case of the geographic variability. The western race is mesic as it contains more *Glechometalia hederaceae* and *Galio–Urticetea* (nitrophilous fringes), *Molinio-Arrhenatheretea* (mesic grasslands) and *Bidentetea tripartitae* (therophyte water-edge vegetation) species while the eastern race, as in other cases (Mucina and Brandes 1985; Mucina 1989b), contains many xeric taxa – diagnostic of the *Onopordetalia* and *Sisymbrietalia*. The east European type is also classified in a separate association (Pop 1968).

A recent paper by Mucina (1989b) handled the problem of variability within the *Onopordum acanthium* communities. The dominant species takes part in forming a series of very different community types, separable into two groups such as a Mediterranean group (belonging to the *Scolymo–Onopordetalia nervosi*) and a Temperate group (within the *Onopordetalia acanthii*). The temperate relevés of the *Onopordetum acanthii sensu lato* were classified by a cluster analysis (Fig. 2), and the resulting clusters, interpreted in terms of geographic races, were mapped onto an ordination diagram (Mucina 1989b). The races of the *Carduo acanthoidis–Onopordetum* are well discernible in geographic terms. The limits between the respective races run approximately through Hungary and northern Yugoslavia (Mucina 1989b). Yet another geographic variability within the *Onopordetum acanthii* (classified

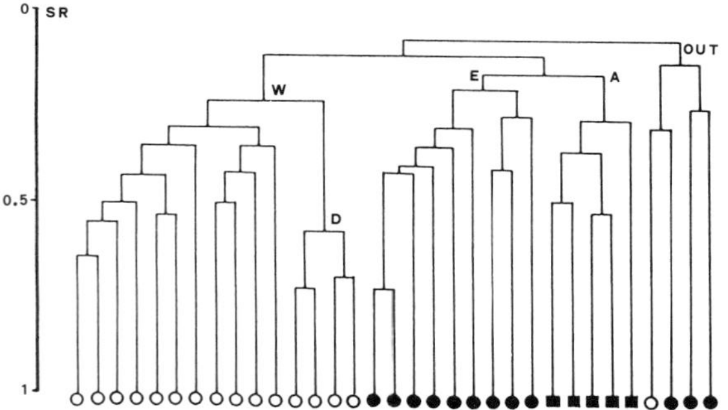

Fig. 2. Cluster analysis (Complete Linkage Clustering using the NCLAS program of Podani, 1988) of the *Onopordum acanthium* constancy tables from temperate Europe (after Mucina 1989b; courtesy of Kluwer Acad. Publ.). Legend: W (western) and E (eastern) races of the *Carduo acanthoidis-Onopordetum*; A: *Onopordetum acanthii* s.str., D: *Malva alcea* subrace within the W race (W. Germany).

Table 1. The differentiation of the *Onopordetum acanthii* (sensu lato) in SW Germany (after Seybold and Müller 1972). The scores in columns are in % per synoptic table.

		1	2	3	4
	No. of column	1	2	3	4
	No. of relevés	6	11	11	21
	Altitudinal span in m	190–240	95–470	170–580	450–760
A	*Onopordum acanthium*	100	64	100	81
	Hyoscyamus niger	67	45	27	52
	Anchusa officinalis	17	–	27	24
d	*Diplotaxis tenuifolia*	100	55	–	–
	Ballota alba	100	55	–	–
	Rhynchosinapis cheiranthos	50	–	–	–
d	*Carduus acanthoides*	–	55	91	90
	Ballota nigra	–	18	82	81
	Arctium tomentosum	–	–	36	81
D	*Cirsium eriophorum*	–	–	–	86
	Galeopsis tetrahit	–	–	–	33
V	*Carduus nutans*	67	45	73	100
	Cynoglossum officinale	83	64	73	86
	Reseda lutea	33	36	91	81
	Verbascum thapsiforme	83	64	45	43
	Artemisia absinthium	50	27	55	62
	Echinops sphaerocephalus	100	18	64	38
	Reseda luteola	100	27	45	24
	Malva alcea	17	9	18	24
	Datura stramonium	–	9	–	–

Legend: A – character species of association; V – alliance character species; d – differential species of geographic races; D – differential species of altitudinal form.

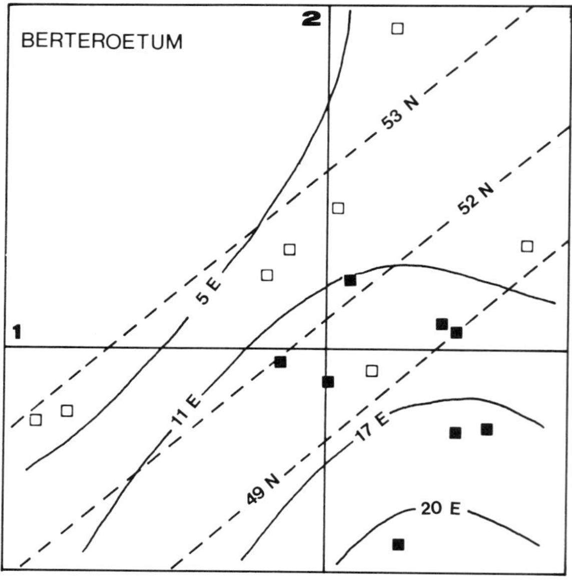

Fig. 3. Correspondence Analysis (the DECORANA program of Hill 1979) of the *Berteroetum incanae* local tables (modified after Mucina and Brandes 1985; courtesy of Kluwer Acad. Publ.). The centroids of the local tables are plotted. The solid and dashed lines represent geographical longitude and latitude resp. Filled squares: E race; empty squares: W race.

within the *Onopordion acanthii*) can be documented on a small-sized regional scale (Table 1). Seybold and Müller (1972) described two local races (and a transitional one) within that community and recognized a vicariant subunit (termed "form" in the latter paper) showing a floristic transition towards the *Cirsietum eriophori* (belonging also to the *Onopordion acanthii*). The latter association has also a vicariant counterpart described from northern Spain by Bellot Rodriguez (1966) and Castroviejo (1975) as *Cynoglosso–Cirsietum*.

In case of the *Berteroa incana* community group (Mucina and Brandes 1985), the local tables were plotted onto an ordination diagram on which the longitude and latitude were also superimposed (Fig. 3). This procedure allowed the position of a transition zone between the respective races to be identified.

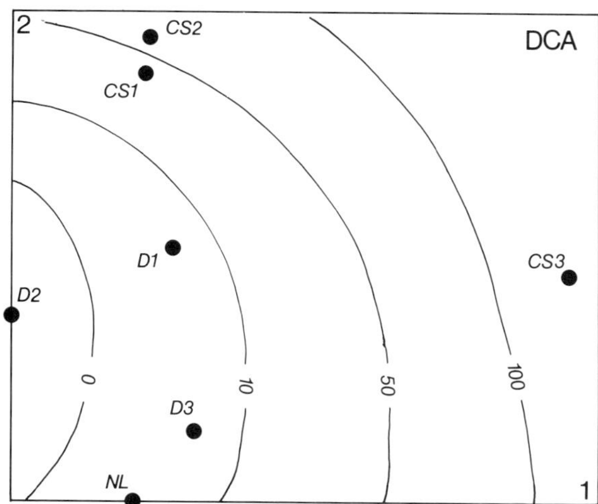

Fig. 4. Detrended Correspondence Analysis (Hill and Gauch 1980 using the CANOCO program of ter Braak 1987) of constancy tables of *Torilidetum japonicae* from Europe based on the most important coenologic groups. Average constancy within a coenologic group in a table was computed. Prior to ordination the variables were transformed to unit variance to remove the effect of differing size of coenologic groups.

The concentration of *Stellarietea mediae* on the ordination plane was plotted using isocoenes (after van der Maarel 1969). Origin of local tables: NL – Braakhekke and Braakhekke-Ilsink (1976), Holland; D1 – Dierschke (1974, Table 10), W. Germany, Leine-Werra-Bergland; D2 – Tüxen in Dierschke (1974, Table 14), NW Germany; D3 – Görs and Müller (1969), S. Germany; CS1 – Kopecký and Hejný (1973), E. Bohemia; CS2 – Mucina (1983), Slovakia; CS3 –

The west–east coenoclines within associations of ruderal vegetation appear to be the rule, as is also shown with the *Torilis japonica* communities (Fig. 4). There is a cline spanning the Dutch (suboceanic) and Slovak (subcontinental) records of the *Torilidetum japonicae*. The extreme poles of this cline differ in the presence of many *Festuco-Brometea, Sedo-Scleranthetea, Trifolio-Geranietea* and *Onopordetalia* species groups found on the xeric pole, and the *Galio-Urticetea* (particularly *Convolvuletalia*), *Molinio-Arrhenatheretea* and *Querco-Fagetea* species found prevailingly on the opposite pole of the diagram.

The communities of the *Aphanion* are an example of the quasi-vicariance of associations. The *Alchemillo-Matricarietum* (as amended by Passarge 1957), *Papaveretum argemones, Vicietum tetraspermae* and *Consolido-Brometum tectorum* (see Meisel 1967; Passarge 1957, 1964, 1985; Wójcik 1978, 1984) form a series of agrestal communities within the *Aperetalia* occurring on acid sandy soils of the north European lowlands. This series follows the trend of increasing thermic continentality (Degórski 1985). The distribution areas of the respective associations are not fully exclusive; they overlap, as shown for instance with the *Aphano-Matricarietum sensu lato* (i.e. *Papaveretum argemones sensu stricto*) and the *Vicietum tetraspermae* (Wójcik 1978; Matuszkiewicz 1980; Fig. 5a). Thus these syntaxa can be considered as quasi-vicariants. The *Consolido-Brometum* occurring in the boreal region of Poland (Fig. 5b) is another vicariant within the *Aphanion* communities of northern Europe.

5. VICARIANCE OF HIGH-RANKED SYNTAXA

Vicariance at the level of alliance and order is a common phenomenon. We can generalize that vicariance often occurs along steeper floristic gradients; along the less steep gradients clinal variation is more pronounced. While the clinal variation is common within associations along the west–east gradient (and the poles of the coenoclines can be interpreted as races), the

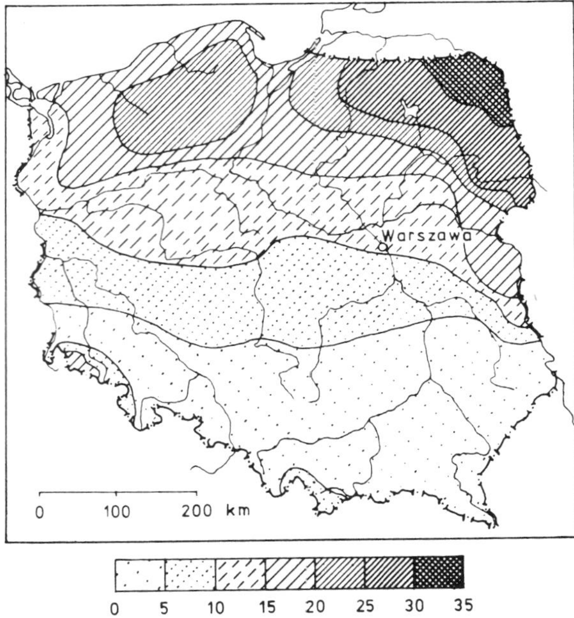

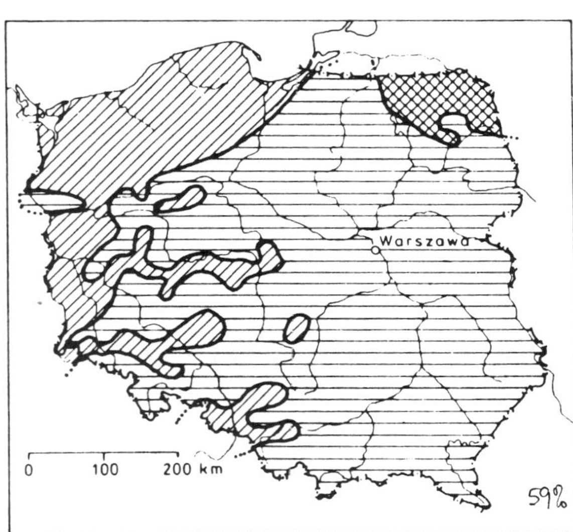

Aphanion

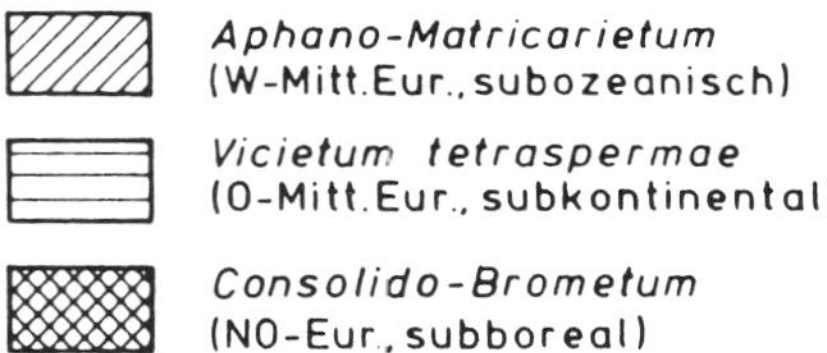

Fig. 5. (a) Distribution of the *Aphanion* communities in Poland (after Wójcik 1978); (b) the plant communities of boreal distribution in Poland (after Matuszkiewicz 1980).

vicariance of associations is more frequent along the north–south gradient.

As shown by an Emberger climate diagram (e.g. Daget 1977), where the pluviometric coefficient and mean temperature of the coldest month were plotted for several stations both in temperate and Mediterranean Europe, there is a clear climatic difference between distribution areas, with the *Sisymbrion officinalis* and *Hordeion leporini* therophyte ruderal communities growing in temperate Europe and in the Mediterranean, respectively. The distribution area of the *Sisymbrion officinalis* and the entire *Sisymbrietalia*, is characterized by a subhumid to perhumid (cold to temperate) climate. The *Hordeion leporini* is found under milder climatic conditions (Fig. 6). The vicariance of the respective alliances is also documented by their geographic distribution in Spain (Fig. 7).

Vicariance patterns among high-ranked syntaxa are very common when the temperate orders are compared with those of the Mediterranean

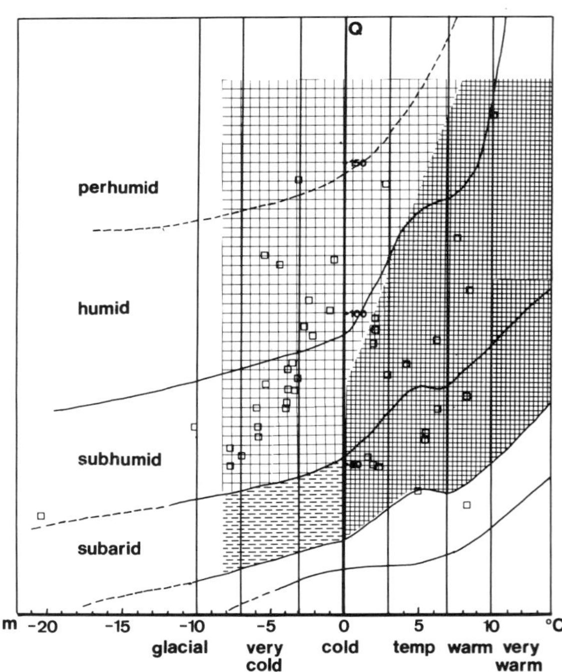

Fig. 6. A climate diagram of Emberger. The squares indicate position of various climate stations. Intensive mesh: climate regions of *Hordeion leporini*; loose mesh: climate regions of *Sisymbrion officinalis*; dashed area: no sufficient data available.

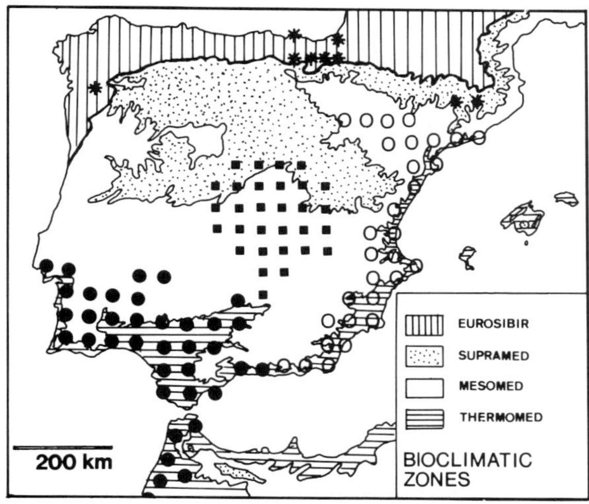

Fig. 7. Vicariance of several *Sisymbrion officinalis (Hordeetum murini)* and *Hordeion leporini* (*Bromo scoparii-Hordeetum leporini, Anacyclo-Hordeetum leporini* and *Asphodelo-Hordeetum leporini*) on the Iberian Peninsula. The phytogeographic division follows Rivas-Martínez (1981) and the phytosociologic data were compiled from Braun-Blanquet (1967), Bellot Rodriguez (1964), Castroviejo (1975) and Rivas-Martínez (1978). Stars: *Hordeetum murini*, filled circles: *Bromo scoparii-Hordeetum*, empty circles: *Anacyclo-Hordeetum leporini*, empty squares: *Asphodelo-Hordeetum leporini*.

Table 2. Pseudo-vicariance and true vicariance patterns among selected syntaxa on three syntaxonomic levels (association, alliance and order).

	Temperate Europe	Mediterranean Europe
Ass.	*Sclerochloo-Polygonetum*	*Sclerochloo-Coronopetum*
	Sagino-Bryetum	*Bryo-Saginetum apetalae*
	Hordeetum murini	*Anacyclo-Hordeetum leporini*
		Asphodelo-Hordeetum leporini
		Hordeetum leporini s.str
All.	*Polygonion avicularis*	*Polycarpion tetraphylli*
		Sclerochloo-Coronopion
	Malvion neglectae	*Chenopodion muralis*
		Malvion parviflorae
	Sisymbrion officinalis	*Hordeion leporini s.str*
	Onopordion acanthii	*Onopordion illyrici*
		Onopordion nervosi
		Carduo carpetani-Cirsion
	Galio-Alliarion	*Geranio-Anthriscion*
		Smyrnion olusatri
	Caucalidion lappulae	*Veronico-Scandicion graecae*
		Secalinion
Order	*Sisymbrietalia*	*Chenopodietalia muralis*
		Brometalia rubenti-tectori
	Onopordetalia acanthii	*Scolymo-Onopordetalia nervosi*
	Glechometalia hederaceae	*Geranio-Arabidetalia hirsutae*

(Table 2). Due to very diversified therophyte ruderal vegetation in the Mediterranean, numerous groups of vicariants (also quasi-vicariants) are encountered (Fig. 8) among the alliances, e.g. within the *Brometalia rubenti-tectori*.

A typical example of the east–west quasi-vicariance in high-ranked syntaxa is that between the *Onopordion acanthii* and *Dauco-Melilotion*. Both alliances belong to the *Onopordetalia* and are widely distributed in temperate Europe. The *Onopordion acanthii* has its distribution centre in southeast Europe. It gradually loses diagnostic species and occurs only in fragments in colder and wetter parts of northwest Europe. The distribution of the *Dauco-Melilotion* is more central European and the unit replaces the *Onopordion* under less favourable climatic conditions.

6. HYPSOMETRIC VICARIANCE AND ALTITUDINAL CLINES

Hyposometric (vertical) synvicariance is a less frequent phenomenon that can be ascribed to a generally lower diversity of weed species and communities as a consequence of less intensive human activity at higher altitudes as compared to agriculturally heavily-attacked lowlands and highlands. A classical example is that of perennial ruderal communities of temperate Europe within the *Artemisietalia*, namely between the *Arction*

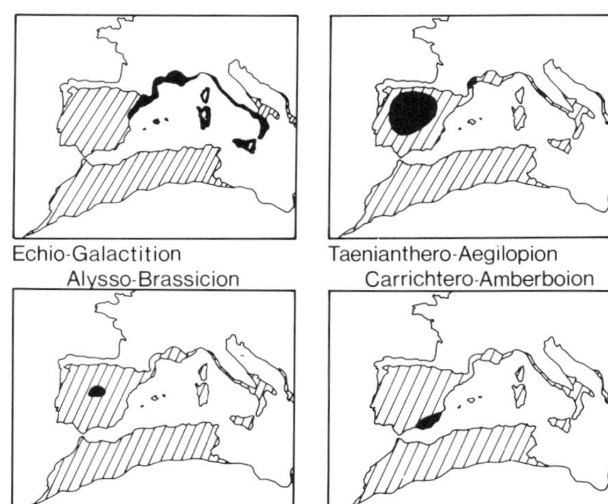

Fig. 8. Schematic distribution of the west-Mediterranean alliances of the *Brometalia rubenti-tectori* (compiled from Izco 1977; Rivas-Martínez and Izco 1977; Brullo 1983, Brullo and Marcenó 1985).

Table 3. Ecologic and geographic differentiation of the Lamio-Ballotettum albae and *Lamio-Ballotetum nigrae* (after Seybold and Müller 1972).

	No. of column	1	2	3	4	5	6	7	8	9	10	11	12	13	14	15	16
	No. of relevés	76	11	51	16	13	5	11	5	35	5	12	5	48	9	26	9
	Altitudinal span in m	150 to 470	150 to 430	150 to 430	150 to 430	340 to 720	370 to 720	340 to 590	340 to 590	180 to 430	190 to 330	180 to 330	190 to 320	350 to 775	410 to 707	350 to 570	410 to 570
A	*Ballota alba*	100	100	100	100	100	100	100	100	–	–	–	–	–	–	–	–
A	*Ballota nigra*	–	–	–	–	–	–	–	–	100	100	100	100	100	100	100	100
	Leonurus cardiaca	1	–	–	–	–	–	–	–	17	20	25	20	17	22	19	22
	Arctium tomentosum	–	–	–	–	–	–	–	–	51	60	58	60	71	44	65	67
	Galeopsis pubescens	–	–	–	–	–	–	–	–	20	20	25	20	35	44	46	56
	Artemisia absinthium	–	–	–	–	–	–	–	–	14	20	17	–	25	22	23	22
DA	*Verbena officinalis*	49	55	39	44	62	60	55	60	49	60	42	60	48	44	58	56
	Malva neglecta	45	36	25	44	46	60	36	40	49	40	50	40	44	33	42	44
D	*Galeopsis tetrahit*	–	–	–	–	77	80	82	80	–	–	–	–	58	67	69	78
	Chaerophyllum aureum	–	–	–	–	54	60	36	80	–	–	–	–	56	56	54	56
V	*Lamium album*	84	100	90	88	100	100	91	100	100	100	100	100	100	100	100	100
	Arctium minus	63	82	55	50	62	60	55	60	54	60	50	60	48	56	50	44
	Arctium lappa	9	27	25	31	23	20	36	60	11	20	25	20	17	22	31	33
dv	*Rumex obtusifolius*	–	–	88	94	–	–	82	100	–	–	83	100	–	–	92	100
	Lamium maculatum	–	–	65	94	–	–	64	100	–	–	75	100	–	–	69	67
	Polygonium persicaria	–	–	53	75	–	–	45	40	–	–	58	60	–	–	62	56
	Potentilla anserina	–	–	45	75	–	–	55	60	–	–	50	60	–	–	58	56
	Convolvulus sepium	–	–	43	50	–	–	36	60	–	–	42	60	–	–	46	67
	Glechoma hederacea	–	–	37	44	–	–	45	60	–	–	42	60	–	–	46	44
	Galium aparine	–	–	41	38	–	–	45	60	–	–	42	40	–	–	50	44
	Carduus crispus	–	–	22	38	–	–	36	40	–	–	8	20	–	–	27	33
	Polygonum lapathifolium	–	–	16	31	–	–	18	40	–	–	33	40	–	–	38	44
	Polygonum hydropiper	–	–	10	7	–	–	18	20	–	–	25	20	–	–	15	22
	Polygonum mite	–	–	20	31	–	–	9	–	–	–	8	20	–	–	12	–
	Armoracia rusticana	–	–	10	25	–	–	–	–	–	–	50	60	–	–	42	56
ds	*Chenopod. bonus-henricus*	–	100	–	100	–	100	–	100	–	100	–	100	–	100	–	100

Legend: A – character species of associations; V – alliance character species; DA – differential species of the association group; D – differential species of altitudinal form; ds – differential species of subassociations; dv – differential species of variants.

lappae of lowland habitats and the *Rumicion obtusifolii* in the submontane to montane zones (Gutte 1972). At higher altitudes these communities are substituted by the *Rumicion alpini* and *Carduo-Urticion dioicae* (Braun-Blanquet 1972; Hadač et al. 1969), which belong to another order. Thus the synchorologic relation between the orders should be called synvicariance while that between, for instance, the *Arction lappae* and *Rumicion alpini* should be designed as pseudo-vicariants. Both concepts are, however, clearly dependent on the syntaxonomic scheme adopted (for other options see Kopecký 1969; Hejný et al. 1979). With many anthropogenic communities, the local floristic deviations ascribed to altitudinal effects were described (Gutte 1972; Seybold and Müller 1972). These syntaxonomic units were either described as altitudinal races or altitudinal vicariant associations (Passarge 1979, 1985). In fact, the altitude-generated variability of vegetation types can be seen as the "third" dimension of the syntaxonomic variability (Passarge 1985). The so-called vertical-ecologic dimension of the syntaxonomic system underlies the variability generated by mainly pedologic deviations, and is reflected in syntaxonomic systems on the level of subassociations or variants. The other dimension is that of a historic-geographic (horizontal) variability (Passarge 1985) and generates geographic races or regional associations. A typical example of a multilevel description of variability within a vicariant group of associations is that by Seybold and Müller (1972). On the basis of two vicariant taxa, such as *Ballota alba* and *B. nigra*, two horizontally vicariant associations (the *Lamio-Ballotetum albae* and *Lamio-Ballotetum nigrae*; for syntaxonomy see also Mucina 1991) were set up. Further ecologic variability was described in

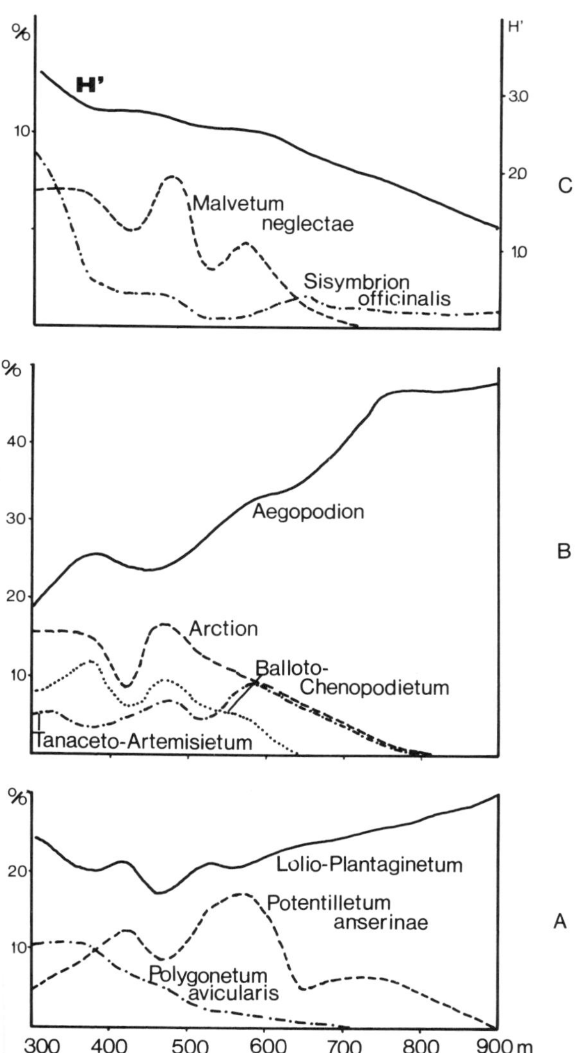

Fig. 9. Distribution pattern of selected ruderal communities from West Bohemia in relation to altitude (modified and compiled from Pyšek and Pyšek 1985, 1987): (a) communities of trampled habitats; (b) hemicryptophyte communities; (c) therophyte communities and the diversity (H') pattern.

subassociations and variants, whereas the hypsometric variability is reflected in the description of several "altitudinal forms" (Table 3). Another hypsometric series of syntaxa can be seen in the communities of river gravel banks. A set of apophytes of the *Dauco-Melilotion* (Mucina 1982) is supposed to have originated from natural gravel-bank vegetation types (Slavík 1978). In fact, there exist a coenocline from high altitudes towards lowlands and includes the *Epilobion fleicheri* (e.g. Oberdorfer 1977), the *Erysimo-Epilobion dodonaei* (with *Meliloto-Epilobietum dodonaei* and *Epilobio-Scrophularietum*) from middle altitudes, and the *Dauco-Melilotion* (with the *Echio-Verbascetum* and *Echio-Melilotetum* as typical communities). This synchorologic (quasi) vicariance pattern among the respective alliances also underlies the hypsometric vicariance of *Epilobium* and *Scrophularia* species (*E. fleischeri, E. dodonaei, Scrophularia canina, S. laciniata*; Slavík 1978, 1986). The hypsometric vicariance in trampled communities was studied by Passarge (1979) who pointed out the substitutive character of the distribution of the *Poa annua* versus *P. supina* dominated trampled swards (Tüxen 1970). Passarge (l.c.) introduced many new syntaxa (mainly as regional associations) and solved the problem of vicariance of the relevant high-ranked syntaxa on the suballiance level. He pointed out the richness of lower species in trampled communities of elevated habitats, which was documented by more exact data by Pyšek and Pyšek (1987) also for other ruderal syntaxa (Fig. 9).

7. GEOLOGIC VICARIANCE

Geologic (edaphic) synvicariance in synanthropic vegetation is a rather rare phenomenon encountered at high altitudes. The *Rumicion alpini* (on silicate substrata) and the *Carduo-Urticion dioicae* (on calcareous bedrocks) are considered as geologic vicariants (Hadač et al. 1969; Hejný et al. 1979). Both units comprise communities of disturbed montane to alpine habitats where the climatic and geologic factors play leading roles in structuring the coenoclines. In natural vegetation the geologic differences reflect syntaxonomically on the class level (*Caricetea currulae* and *Seslerietea variae*) or more often on the order level (e.g. *Thlaspietalia rotundifolli* and *Androsacetalia vandellii; Salicetalia herbaceae* and *Arabidetalia caeruleae*, etc.).

On the other hand, at low altitudes, the synanthropic vegetation units can be found both on silicate and calcareous substrata, as exemplified by the *Taenianthero-Aegilopion*, winter-annual grassland ruderal vegetation of central Spain (Izco 1977).

REFERENCES

Baker, H. G. 1965. Characteristics and modes of origin of weeds. In: Baker, H. G. and Stebbins, G. L. (eds), The Genetics of Colonizing Species, pp. 147–168. Academic Press, New York.

Barkman, J. J., Moravec, J. and Rauschert, S. 1986. Code of phytosociological nomenclature. Code der pflanzensoziologischen Nomenklatur. Code de nomenclatur phytosociologique. Vegetatio, 67, 145–195.

Bellot Rodriguez, F. 1966. La vegetación de Galicia. Anal. Inst. Bot. Cavanilles, Madrid, 24, 1–306.

Braakhekke, W. G. and Braakhekke-Ilsink, E. I. 1976. Nitrophile Saumgesellschaften im Südosten der Niederlande. Vegetatio, 32, 55–60.

Brandes, D. 1982. Das *Sambucetum ebuli* Felf. 1942 im südlichen Mitteleuropa und seine geographische Gliederung. Tuexenia, Göttingen, 2, 47–60.

Brandes, D. 1985. Das *Heracleo-Sambucetum ebuli* in West- und Mitteleuropa. Coll. Phytosociol., 12, 591–596.

Braun-Blanquet, J. 1921. Prinzipien einer Systematik der Pflanzengesellschaften auf floristischer Grundlage. Jahrb. St. Gallen. Naturwiss. Ges., St. Gallen, 57, 305–351.

Braun-Blanquet, J. 1964. Pflanzensoziologie. Grundzüge der Vegetationskunde (3. Aufl.). Springer, Wien, New York.

Braun-Blanquet, J. 1967. Vegetationskizzen aus dem Baskenland. II. Teil. Vegetatio, 14, 1–126.

Braun-Blanquet, J. 1972. Die Gänsefussweiden der Alpen (*Chenopodion subalpinum*). Saussurea, Genève, 3, 141–156.

Brullo, S. 1983. Le associazioni subnitrofile dell'*Echio-Galactition tomentosae* in Sicilia. Boll. Acc. Gioenia Sci. Nat., Catania, 15, 405–452.

Brullo, S. and Marcenó, C. 1985. Contributo alla conoscenza della vegetazione nitrofila della Sicilia. Coll. Phytosociol. 12, 23–148.

Castroviejo, S. 1975. Algunos datos sobre las comunidades nitrófilas vivaces (*Artemisietea vulgaris*) de Galicia. Anal. Inst. Bot. Cavanilles, Madrid, 32, 489–502.

Daget, P. 1977. Le bioclimat mediterranéen: analyse des formes climatiques par le systéme d'Emberger. Vegetatio, 34, 87–103.

Degórski, M. L. 1985. An investigation into the spatial variability of continentality in west and central Europe by the Ellenberg method. Doc. Phytosoc., Camerino, 9, 337–349.

Dierschke, H. 1974. Saumgesellschaften in Vegetations- und Standortsgefälle an Waldrändern. Scripta Geobot., Göttingen, 6, 1–246.

Géhu, J.-M. and Rivas-Martínez, S. 1981. Notions fondamentales de phytosociologie. In: Dierschke, H. (ed.), Syntaxonomie, pp. 5–33. Cramer, Vaduz.

Görs, S. and Müller, T. 1969. Beitrag zur Kenntnis der nitrophilen Saumgesellschaften Südwestdeutschlands. Mitt. Florist.-Soziol. Arbeitsgem. NF, 14, 153–168.

Grime, J. P. 1979. Plant Strategies and Vegetation Processes. Edwards, Chichester.

Gutte, P. 1972. Ruderalgesellschaften West- und Mittelsachsens. Feddes Repert., Berlin, 83, 11–122.

Hadač, E. et al. 1969. Die Pflanzengesellschaften des Tales "Dolina Siedmich prameňov" in der Belaer Tatra. Vegetácia ČSSR, Ser. B, Bratislava, 2, 1–343.

Hejný, S., Kopecký, K., Jehlík, V. and Krippelová, T. 1979. Survey of ruderal plant communities of Czechoslovakia. Rozpr. Čs. Akad. Věd., R. Mat. Přír., Praha, 89 (2), 1–100 (in Czech).

Hill, M. O. 1979. DECORANA – A Fortran program for detrended correspondence analysis and reciprocal averaging. Ecology and Systematics. Cornell Univ., Ithaca.

Hill, M. O. and Gauch. H. G., Jr 1980. Detrended correspondence analysis: An improved ordination technique. Vegetatio, 42, 47–58.

Holub, J. and Jirásek, V. 1967. Zur Vereinheitlichung der Terminologie in der Phytogeographie. Fol. Geobot. Phytotax., Praha, 2, 69–113.

Holzner, W. 1978. Weed species and weed communities. Vegetatio, 38, 13–20.

Izco, J. 1977. Revisión sintética de los pastizales del suborden *Bromenalia rubenti-tectori*. Coll. Phytosociol. 6, 37–54.

Jehlík, V. 1986. Konstruktion chorologischer Spektren von synanthropen Pflanzengesellschaften nach der Synanthropie ihrer Komponenten. Tuexenia, Göttingen, 6, 99–103.

Kopecký, K. 1969. Zur Syntaxonomie der natürlichen nitrophilen Saumgesellschaften in der Tschechoslowakei und zur Gliederung der Klasse *Galio-Urticetea*. Fol. Geobot. Phytotax., Praha, 4, 235–259.

Kopecký, K. and Hejný, S. 1973. Neue syntaxonomische Auffassung der Gesellschaften ein- bis zweijähriger Pflanzen der *Galio-Urticetea* in Böhmen. Fol. Geobot. Phytotax., Praha, 8, 49–66.

Matuszkiewicz, W. 1980. Synopsis und geographische Analyse der Pflanzengesellschaften von Polen. Mitt. Florist.-Soziol. Arbeitsgem. NF, 22, 19–50.

Meisel, K. 1967. Über die Artenverbindung des *Aphanion arvensis* J. et R.Tx. 1960 im west- und nordwestdeutschen Flachland. Schriftenrh. Vegetkde., Bad Godesberg, 2, 123–133.

Moravec, J. 1975. Die Untereinheiten der Assoziation. Beitr. Naturkd. Forsch. NW-Deutsch., Karlsruhe, 34, 225–232.

Mucina, L. 1982. Die Ruderalvegetation des nördlichen Teils der Donau-Tiefebene. 3. Gesellschaften des Verbandes

Dauco-Melilotion auf natürlichen Standorten. Fol. Geobot. Phytotax., Praha, 17, 21–47.

Mucina, L. 1983. *Torilidetum japonicae* in western Slovakia. Biológia, Bratislava, 38, 889–895 (in Slovak).

Mucina, L. 1989a. Engangered ruderal plant communities in Slovakia and their preservation. Phytocoenologia, Stuttgart, 17, 271–286.

Mucina, L. 1989b. Syntaxonomy of the *Onopordum acanthium* communities in temperate and continental Europe. Vegetatio, 81, 107–115.

Mucina, L. 1991. Ruderalvegetation der westlichen Slowakei. (submitted)

Mucina, L. and Brandes, D. 1985. Communities of *Berteroa incana* in Europe and their geographical differentiation. Vegetatio, 59, 125–136.

Oberdorfer, E. 1968. Assoziation, Gebietsassoziation, geographische Rasse. In: Tüxen, R. (ed.), Pflanzensoziologische Systematik, pp. 124–141. Junk, The Hague.

Oberdorfer, E. (ed.). 1977. Süddeutsche Pflanzengesellschaften. Teil I (2. Aufl.). Pflanzensoziologie, Jena, 10, 1–311.

Passarge, H. 1957. Zur geographischen Gliederung der *Agrostion spica-venti*-Gesellschaften im nordostdeutschen Flachland. Phyton 7, 22–31.

Passarge, H. 1964. Pflanzengesellschaften des nordostdeutschen Flachlandes. I. Pflanzensoziologie, Jena, 13, 1–324.

Passarge, H. 1979. Über mitteleuropäisch-montane Trittpflanzengesellschaften. Vegetatio, 39, 7–84.

Passarge, H. 1985. Syntaxonomische Wertung chorologischer Phänomene. Vegetatio, 59, 137–144.

Peinado Lorca, M. and Rivas-Martínez, S. (eds). 1987. La vegetación de España. Universidad de Alcala de Henares.

Pignatti, E. and Pignatti, S. 1984. Sekundäre Vegetation und floristische Vielfalt im Mittelmeerraum. Phytocoenologia, 12, 351–358.

Podani, J. 1988. SYN-TAX III. User's manual. Abstr. Bot., Budapest, 12 (Suppl 1), 1–183.

Pop, I. 1968. Flora şi vegetaţia Cîmpiei Crişulilor. Edit. Acad. Sci. RPR, Bucureşti.

Pyšek, A. and Pyšek, P. 1987. Die Methode der Einheitsflächen beim Studium der Ruderalvegetation. Tuexenia, Göttingen, 7, 479–485.

Pyšek, P. and Pyšek, A. 1985. Die Ausnutzung der Ruderalvegetation zur quantitativen Indikation von Standortsverhältnissen mit Hilfe von Einheitsflächen. Fol. Mus. Rerum Nat. Bohem. Occident., Ser. Bot., Plzeň, 22, 1–35.

Rivas-Martínez, S. 1975. Sobre la nueva clase *Polygono-Poetea annuae*. Phytocoenologia, 2, 123–140.

Rivas-Martínez, S. 1976. Phytosociological and chorological aspects of the Mediterranean region (I). Doc. Phytosociol., 15–18, 137–145.

Rivas-Martínez, S. 1978. La vegetación del *Hordeion leporini* en España. Doc. Phytosociol., 2, 377–392.

Rivas-Martínez, S. 1981. Les étages bioclimatiques de la végétation de la Peninsule Ibérique. Ann. Jard. Bot. Madrid, 37, 251–268.

Rivas-Martínez, S. and Izco, J. 1977. Sobre la vegetación terofítica subnitrófila mediterránea (*Brometalia rubenti-tectori*). Anal. Inst. Bot. Cavanilles, Madrid, 34, 355–381.

Schwickerath, M. 1968. Begriff und Bedeutung der geographischen Differentialarten. In: Tüxen, R. (ed.), Pflanzensoziologische Systematik, pp. 78–84. Junk, The Hague.

Seybold, S. and Müller, T. 1972. Beitrag zur Kenntnis der Schwarznessel (*Ballota nigra* agg.). Veröff. Natursch. Landschaftspfl. Bad.-Württ., Ludwigsburg, 40, 51–126.

Slavík, B. 1978. *Epilobio dodonaei-Melilotetum albi*, eine neue Pflanzenassoziation. Fol. Geobot. Phytotax., Praha, 13, 381–395.

Slavík, B. 1986. *Epilobium dodonaei* Vill. in der Tschechoslowakei. Preslia, Praha, 59, 307–338.

Stace, C. A. 1980. Plant Taxonomy and Biosystematics. Arnold, London.

ter Braak, C. J. F. 1987. CANOCO – a FOTRAN program for canonical community ordination by (patrial) (detrended) (canonical) correspondence analysis, principal components analysis and redundancy analysis (version 2.1). TNO Institute of Applied Computer Science, Wageningen.

Tüxen, R. 1970. Zur Syntaxonomie des europäischen Wirtschafts-Grünlandes (Wiesen, Weiden, Tritt- und Flutrasen). Ber. Naturhist. Ges. Hannover, 114, 77–85.

Tüxen, R. 1979. Die Pflanzengesellschaften Nordwestdeutschlands, (2. Aufl.). Cramer, Vaduz.

van der Maarel, E. 1969. On the use of ordination methods in phytosociology. Vegetatio, 19, 21–46.

Westhoff, V. 1947. The vegetation of dunes and salt marshes on the Dutch islands of Terschelling, Vlieland and Texel. Thesis, Univ. of Utrecht.

Westhoff, V. and van der Maarel, E., 1978. The Braun-Blanquet approach. In: Whittaker, R. H. (ed.), Classification of Plant Communities. Handb. Veget. Sci. 5, 287–399. Junk, The Hague.

Whittaker, R. H. 1960. Vegetation of the Siskiyou Mountains, Oregon and California. Ecol. Monogr. 30, 279–338.

Wójcik, Z. 1978. Plant communities of Poland cereal field. Preliminary results of comparative studies. Acta Bot. Slov. Acad. Sci. Slov., Ser. A, Bratislava, 3, 229–238.

Wójcik, Z. 1984. *Consolido-Brometum* in northeastern Poland. Acta Bot. Slov. Acad. Sci. Slov., Ser. A, Bratislava, Suppl. 1, 327–339.

Zohary, D. 1970. Centers of diversity and centers of origin. In: Frankel, O. H. and Bennett, E. (eds), Genetic Resources in Plants – their Exploration and Conservation. IBP Handbook, 11, 33–42. Blackwell, Oxford.

APPENDIX

Table 4. System of high-ranked syntaxa of synanthropic (nitropilous semiterrestrial, aquatic weed, trampled, ruderal and agrestal) vegetation in Europe.

---

1. Oryzetea sativae Miyawaki 1960
1.1. Cypero–Echinochloetalia oryzoidis de Bolós et Masclans 1955
1.1.1 Oryzo–Echinochloion oryzoidis de Bolós et Masclans 1955

2. Bidentetea tripartitae R.Tx., Lohm. et Preising in R.Tx. 1950
2.1. Bidentetalia tripartitae Br.–Bl. et R.Tx. 1943
2.1.1. Bidention tripartitae Nordhagen 1940 em. R.Tx. in Poli et J.Tx. 1960
2.1.2. Chenopodion rubri (R.Tx. in Poli et J.Tx. 1960) Soó 1960

3. Polygono–Poetea annuae Rivas–Martínez 1975
3.1. Poo–Polygonetalia R.Tx. et Ohba in Géhu et al. 1972
3.1.1. Polygonion avicularis Br.–Bl. ex Aichinger 1933
3.1.2. Sclerochloo–Coronopion Rivas–Martínez 1975
3.1.3. Polycarpion tetraphylli Rivas–Martínez 1975
3.1.4. Saginion procumbentis R.Tx. et Ohba in Géhu et al. 1972

4. Molinio–Arrhenatheretea R.Tx. 1937
4.1. Plantagini–Lolietalia Mucina 1991
4.1.1 Lolio–Plantaginion Sissingh 1969
4.1.2. Alchemillo–Ranunculion repentis Passarge 1979
4.2. Potentillo–Polygonetalia R.Tx. 1947
4.2.1. Lolio–Potentillion anserinae R.Tx. 1947
4.2.2. Loto–Trifolion (Westhoff et van Leeuwen ex Vicherek 1973) Passarge 1978
4.2.3. Juncion inflexi (Knapp 1971) Mucina 1991
4.2.4. Juncion effusi Westhoff et van Leeuwen ex Hejny et al. 1979
4.3. Paspalo–Heleochloetalia Br.–Bl. et al. 1952
4.3.1. Paspalo–Polypogonion semiverticillati Br.–Bl. et al. 1952
4.3.2. Trifolio–Cynodontion Br.–Bl. et de Bolós 1957

5. Stellarietea mediae R.Tx., Lohm. et Preising in R.Tx. 1950
5.1. Secalietalia Br.–Bl. et al. 1936
5.1.1. Caucalidion lappulae R.Tx. 1950
5.1.2. Secalinion Br.–Bl. et al. 1936
5.1.3. Veronico chaubardii–Scandicion graecae Ferro et Scammacca 1985
5.2. Aperetalia spica–venti J. et R.Tx. in Malato–Beliz et al. 1960
5.2.1. Arnoseridion minimae Malato–Beliz et al. 1960
5.2.2. Agrostion spica–venti R.Tx. 1950
5.2.3. Lolio remotae–Linion J. Tx. 1966
5.3. Chenopodietalia albi R.Tx. (1937) 1950
5.3.1. Eu–Polygono–Chenopodion Koch 1926
5.3.2. Fumario–Euphorbion Müller in Görs 1966
5.3.3. Fumario wirtgenii–agrariae Brullo in Brullo et Marcenó 1985
5.4. Eragrostietalia J.Tx. ex Lohm. et al. 1962
5.4.1. Amarantho–Chenopodion albi Morariu 1943
5.4.2. Diplotaxidion erucoidis Br.–Bl.et al. 1936
5.4.3. Heliotropion Oberd. 1953/1954
5.4.4. Salsolion ruthenicae Philippi 1971

5.4.5. Eragrostio–Amaranthion crispi Mucina 1991
5.4.6. Amaranthion prostratae Rivas–Martínez 1976
5.5. Sisymbrietalia J.Tx. 1966
5.5.1. Malvion neglectae (Gutte 1966) Hejny 1978
5.5.2. Atriplici–Sisymbrion Hejny 1978
5.5.3. Chenopodio–Atriplicion tataricae (Mucina in Krippelová et Mucina 1988) Mucina 1991
5.5.4. Sisymbrion officinalis R.Tx. et al. in R.Tx. 1950
5.6. Chenopodietalia muralis (Br.–Bl. et al. 1936) de Bolós 1962
5.6.1. Chenopodion muralis (Br.–Bl. et al. 1936) de Bolós 1967
5.6.2. Silybo–Urticion Sissingh 1950
5.7. Brometalia rubenti–tectori (Rivas Goday et Rivas–Martínez 1963) Rivas–Martínez et Izco 1977
5.7.1. Hordeion leporini Br.–Bl. 1947
5.7.2. Laguro–Bromion rigidi J.–M. Géhu et J. Géhu 1983
5.7.3. Taenianthero–Aegilopion geniculatae Rivas–Martínez et Izco 1977
5.7.4. Linario viscosae–Vulpion alopecuroidis Rivas–Martínez et al. 1980
5.7.5. Echio–Galactition de Bolós et Molinier 1969
5.7.6. Alysso–Brassicion barrelieri Rivas–Martínez et Izco 1977
5.7.7. Carrichtero–Amberboion lippii Rivas Goday et Rivas–Martínez ex Esteve Chueca 1973
5.7.8. Cerintho–Mandragorion Rivas Goday et Rivas–Martínez ex Izco 1977
5.8. Geranio–Cardaminetalia hirsutae Brullo in Brullo et Marcenó 1985
5.8.1. Geranio–Anthriscion caucalidis Rivas–Martínez 1978
5.8.2. Valantio–Galion muralis Brullo in Brullo et Marcenó 1985
5.9. Urtico–Scrophularietalia peregrinae Brullo in Brullo et Marcenó 1985
5.9.1. Allion triquetri de Bolós 1967
5.9.2. Veronico–Urticion urentis Brullo in Brullo et Marcenó 1985

6. Artemisietea vulgaris Lohm., Preising et R.Tx. in R.Tx. 1950
6.1. Onopordetalia acanthii Br.–Bl. et R.Tx. 1943
6.1.1. Onopordion acanthii Br.–Bl. et al. 1936
6.1.2. Dauco–Melilotion Görs ex Oberd. et al. 1967
6.1.3. Erysimo wittmanni–Hackelion Bernátová 1987
6.1.4. Cirsion richterano–chodati (Rivas–Martínez in Rivas–Martínez et al. 1984) Mucina 1991
6.2. Agropyretalia repentis Oberd., Müller et Görs in Oberd. et al. 1967
6.2.1. Convolvulo–Agropyrion Görs 1966
6.2.2. Agropyro–Kochion Soó 1959
6.3. Artemisietalia vulgaris Lohm. in R.Tx. 1947
6.3.1. Arction lappae R.Tx. 1937 em. Gutte 1971
6.3.2. Rumicion obtusifolii Gutte 1972
6.3.3. Balloto–Conion maculati Brullo in Brullo et Marcenó 1985

7. Galio–Urticetea Passarge ex Kopecky 1969
7.1. Glechometalia hederaceae R.Tx. in Brun–Hool et R.Tx. 1975
7.1.1. Alliarion (Oberd. 1962) Hejny in Holub et al. 1967
7.1.2. Anthriscion nemorosae Brullo in Brullo et Marcenó 1985
7.1.3. Aegopodion podagrariae R.Tx. 1967
7.1.4. Ranunculo–Impatiention noli–tangere Passarge 1967
7.2. Convolvuletalia sepium R.Tx. 1950
7.2.1. Convolvulion sepium R.Tx. 1947
7.2.2. Angelicion littoralis R.Tx. 1950
7.2.3. Cynancho–Calystegion sepium Rivas Goday et Rivas–Martínez ex Rivas–Martínez 1977
7.2.4. Bromo ramosi–Eupatorion cannabini de Bolós et Masalles in de Bolós 1983

8. Betulo–Adenostyletea Br.–Bl. et R. Tx. 1943
8.1. Rumicetalia alpini Mucina 1991
8.1.1. Rumicion alpini Rübel ex Klika in Klika et Hadac 1944

8.1.2. Carduo-Urticion dioicae Hadac et al. 1969

9. Onopordetea nervosi Rivas-Martínez 1977 em. Mucina 1990
9.1. Scolymo hispanici-Onopordetalia nervosi Rivas-Martínez in Ladero et al. 1983
9.1.1. Onopordion arabici Br.-Bl. et de Bolós 1957
9.1.2. Onopordion illyrici Oberd. 1953/1954
9.1.3. Scolymo-Carthamion lanati (Rivas Goday 1961) Ladero et al. 1981
9.1.4. Carduo carpetani-Cirsion odontolepis Rivas-Martínez et al. 1986

10. Pegano-Salsoletea Br.-Bl. et de Bolós 1957
10.1. Salsolo-Peganetalia Br.-Bl et de Bolós 1957
10.1.1. Salsolo-Peganion Br.-Bl. et de Bolós 1957
10.2. Helichryso-Santolinetalia Peinado et Martínez-Parras 1984
10.2.1. Artemisio-Santolion Costa 1975
10.2.2. Santolion pectinato-canescentis Peinado et Martínez-Parras 1984

INDEX